MATHEMATICAL PROGRAMMING

MATHEMATICAL PROGRAMMING:
Structures and Algorithms

JEREMY F. SHAPIRO

Alfred P. Sloan School of Management
Massachusetts Institute of Technology

A Wiley-Interscience Publication
JOHN WILEY & SONS
New York · Chichester · Brisbane · Toronto

Library of Congress Cataloging in Publication Data

Shapiro, Jeremy F 1939–
Mathematical programming.

"A Wiley-Interscience publication."
Includes bibliographical references.
1. Programming (Mathematics) I. Title.

QA402.5.S5 519.7 79-4478
ISBN 0-471-77886-9

Printed in the United States of America

10 9 8 7 6 5 4 3 2 1

To my parents

PREFACE

The field of mathematical programming has reached a mature, although by no means moribund, state. The 1950s saw the development of many of the basic results and techniques, such as the simplex method and the Kuhn-Tucker theorem, and the identification of many important research problems. The 1960s saw the specialization of the field into subfields including integer programming, nonlinear programming, dynamic programming, and so on, with researchers working exclusively within one subfield on the development of special-purpose algorithms.

The 1970s have seen the development of conceptual results facilitating the reintegration of the various subfields. Of particular importance in this regard is Lagrangean duality and the related topic of convex analysis, which has played a central role in our understanding of all types of mathematical programming problems. By contrast, we have probably reached a point of diminishing returns in the development of new algorithms, although integer and nonlinear programming problems remain difficult to solve.

Applications of mathematical programming models have proliferated in the 1970s, in large part because of the existence of effective computer codes for solving them. These applications often involve a mixture of mathematical programming problems; for example, a plant location model for a manufacturing firm can include a network optimization problem describing the distribution of its products, nonlinear functions describing demand for the products in various markets as functions of their selling prices, and a mixed integer programming model describing the timing, sizing, and locational options available for new plants. Thus practical applications as well as theoretical developments have provided an impetus toward the integration of mathematical programming.

The mature state of mathematical programming suggests that the time is appropriate to try to present in one volume most of the basic results about the structure of mathematical programming problems and descriptions of many of the algorithms for solving them. In addition, such a volume should contain a development of Lagrangean duality and convex analysis that relates these subjects to the spectrum of mathematical programming problems, thereby exposing their differences and similarities and indicating how complex model integration can be achieved. These are the goals of this book. The reader should be cautioned, however, that the mathematical development is uniformly terse and will sometimes require a careful reading of new material.

The wide coverage of topics should permit this book to be used as the basic text in a variety of intermediate and advanced courses in mathematical programming. The following are suggested one-quarter or one-semester courses based on various combinations of the chapters:

Linear Programming	Chapters 1 and 2
Network Optimization	Chapters 3 and 4
Combinatorial Optimization	Chapter 3, Chapter 5 (Sections 5.1 to 5.3), and Chapter 8
Large Scale Optimization	Chapters 5 and 6
Nonlinear Programming	Chapters 5 and 7
Convex Analysis and the Theory of Optimization	Chapters 5 and 6, Appendix A

The book can provide the basic results for a course on a particular subject. These could be supplemented by the instructor's own notes, specific papers from the literature, or another book, according to taste.

The content and organization of the book also serve as an aide to students and researchers in mathematical programming as the field continues to grow in the years ahead. There are currently two major directions of growth. One is the development of fields of pure and applied mathematics that use mathematical programming constructs as basic building blocks; for example, the field of combinatorics that has grown out of combinatorial optimization, or abstract theories of optimization and convex analysis involving more complex vector spaces than finite dimensional spaces. This book provides students and researchers in these new fields with a broad survey of the basic results from which the mathematical generalizations are taken.

A second area of growth results from the infiltration of mathematical programming into other scientific disciplines such as economics, statistics, and engineering design, to name only a few. Many new theoretical and applied research problems have been discovered as the result of the use of mathematical programming in these new fields; for example, studies resulting from the integration of forecasting models with mathematical programming models of the efficient production and distribution of the commodities being forecasted. Again, this book provides much of the background necessary to study the use of mathematical programming in new fields.

The ideas and research accomplishments of many people influenced the development and content of this book. George Dantzig's prodigious contribution to mathematical programming must be singled out as having profoundly influenced us all. Ralph Gomory's work had a particularly strong influence on me, especially his research on the use of group theory to analyze integer programming problems. My initiation into research is due in large part to Harvey Wagner, who served as my dissertation advisor and encouraged me to pursue an academic career. My personal viewpoint of mathematical programming was shaped most by direct interaction with and reading the papers of David Bell, Tony Fiacco, Marshall Fisher, Tony Gorry, Ellis Johnson, Art Geoffrion, Richard Grinold, Jerry Gould, Mike Held, Dick Karp, Tom Magnanti, George Nemhauser, Bill Northup, Herb Scarf, and Larry Wolsey.

Contracts from the Army Research Office and the National Science Foundation to the Operations Research Center at M.I.T. supported research that subsequently motivated me to write this book.

Tom Magnanti showed devotion beyond the call of duty by reading the final draft of this book and by making many insightful comments as well as catching numerous errors. I also profited by comments received from Gabriel Bitran, Dorothy Elliott, Bruce Faaland, Eduardo Modiano and Paulo Villela. The remaining errors are my own responsibility. The title of the book was suggested by an anonymous reviewer who I also wish to thank. Finally, a debt of gratitude is owed to Kathy Sumera for having done most of the typing of the final manuscript. Joan Kargel and Betsy Sherman did the typing of earlier drafts.

JEREMY F. SHAPIRO

Cambridge, Massachusetts
April 1979

MATHEMATICAL CONVENTIONS

Whenever possible, we will not define explicitly a vector as a row or column vector and use it in both senses in matrix multiplication. The dimensions of the matrix and the nature of the matrix multiplication, pre- or postmultiplication, will usually determine unambiguously the sense of the vector. Thus, we will indicate vector transposes only where it is necessary for clarity. We will use the notation G^c to denote the complement of a set G. The notation $[X]^c$, where X is a subset of R^n, denotes the convex hull of X. The empty set will be denoted by $\varnothing$. Occasionally, we will also use $\varnothing$ to denote a mathematical function. The context will make the meaning clear. Other conventions will be explained as they arise.

CONTENTS

MATHEMATICAL PROGRAMMING

1
LINEAR PROGRAMMING

1.1 INTRODUCTION

Linear programming is fundamental to the study of mathematical programming because many important applications are linear programming problems, linear programming approximations are used in the solution of nearly all mathematical programming problems, and concepts and insights derived from linear programming are the basis for much of the general theory of mathematical programming. In this chapter we consider the well-known simplex algorithm for solving linear programming problems. The algorithm has been a great practical success in solving real-life problems as well as providing a mathematical tool that is of paramount importance to all areas of mathematical programming.

1.2 THE SIMPLEX METHOD

The linear programming problem in the form we will study it in this chapter is

$$\begin{aligned} \min z &= \mathbf{cx} \\ \text{s.t. } \mathbf{Ax} &= \mathbf{b} \\ \mathbf{x} &\geq 0 \end{aligned} \tag{1.1}$$

where $\mathbf{A}$ is a $m \times n$ matrix of rank m, $\mathbf{c}$ is a $1 \times n$ vector, and $\mathbf{b}$ is an $m \times 1$ vector. The notation "s.t." is an abbreviation for "subject to." This linear programming problem is said to have m rows and n columns. The matrix $\mathbf{A}$ is called the coefficient matrix, $\mathbf{c}$ the cost vector, $\mathbf{b}$ the right-hand-side vector. The linear function $\mathbf{cx}$ is called the objective function. The $m \times 1$ vectors $\mathbf{a}_j$, which are the columns of $\mathbf{A}$ will sometimes be called activities. An $n \times 1$ vector $\mathbf{x}$ satisfying $\mathbf{Ax} = \mathbf{b}$, and $\mathbf{x} \geq \mathbf{0}$, is called a feasible solution, and a minimal cost feasible solution is called an optimal solution. The set $\{\mathbf{x} | \mathbf{Ax} = \mathbf{b}, \mathbf{x} \geq \mathbf{0}\}$ is called the *feasible region* and it is a convex set; properties of convex sets and functions are reviewed in Appendix A. If (1.1) does not have a feasible solution, we take the minimum to be $+\infty$. If there exists a sequence of feasible solutions to (1.1) with objective function cost approaching $-\infty$, we take the minimum to be $-\infty$.

The constraints of a linear programming problem can be presented originally in the inequality form and converted to the equality form by the addition of slack variables or surplus variables. Specifically, an inequality $\Sigma_{j=1}^{n} a_{ij} x_j \leq b_i$ is converted to the equality $\Sigma_{j=1}^{n} a_{ij} x_j + s_i = b_i$ by the addition of the *slack* variable s_i that is constrained to be nonnegative. The reverse inequality $\Sigma_{j=1}^{n} a_{ij} x_{ij} \geq b_i$ is converted to $\Sigma_{j=1}^{n} a_{ij} x_j - s_i = b_i$ by the addition of the nonnegative *surplus* variable s_i. Similarly, a maximization problem can be converted to a minimization problem by the rule $\max cx = -\min - cx$; that is, we minimize $-cx$ and the optimal objective function value of our maximization problem is the negative of this quantity. Finally, if a variable x_j is unconstrained in sign, then we make the substitution $x_j = x_j^+ - x_j^-$ where x_j^+ and x_j^- are constrained to be nonnegative.

One of the major strengths of linear programming as a model and the simplex method for solving it is the fact that so many decision problems can be formulated and solved in exactly the form (1.1). This is in contrast to the nonlinear and integer programming problems studied in later chapters that can also be stated in general form but their effective optimization often requires the use of specific algorithms that exploit special structures of particular problems.

The simplex method for solving the linear programming problem (1.1) is derived from classical theory for characterizing the solutions $\mathbf{x}$ satisfying $\mathbf{Ax} = \mathbf{b}$ by transforming this system to a more convenient equivalent form. A system $\mathbf{A'x} = \mathbf{b'}$ is said to be *equivalent* to $\mathbf{Ax} = \mathbf{b}$ if their solution sets are equal. The simplex method adapts the classical theory to take into account the nonnegativity constraints $\mathbf{x} \geq \mathbf{0}$ and to select a minimal cost solution from among the feasible solutions.

Let $\mathbf{B}$ denote an $m \times m$ nonsingular submatrix of $\mathbf{A}$; without loss of generality, assume we have reordered the columns of $\mathbf{A}$ so that $\mathbf{A} = (\mathbf{B}, \mathbf{N})$. The matrix $\mathbf{B}$ is called a *basis*. Let $\mathbf{x}$ be similarly partitioned as $(\mathbf{x_B}, \mathbf{x_N})$; the variables $\mathbf{x_B}$ are called *basic* variables and the $\mathbf{x_N}$ are called *nonbasic* variables. A characterization of the solutions to the system $\mathbf{Ax} = \mathbf{b}$ or equivalently $\mathbf{Bx_B} + \mathbf{Nx_N} = \mathbf{b}$ is given by the vectors $(\mathbf{x_B}, \mathbf{x_N})$ in R^n where $\mathbf{x_N}$ is any vector in R^{n-m} and

$$\mathbf{x_B} = \mathbf{B}^{-1}\mathbf{b} - \mathbf{B}^{-1}\mathbf{N}\mathbf{x_N} \tag{1.2}$$

The solution $(\mathbf{x_B}, \mathbf{x_N}) = (\mathbf{B}^{-1}\mathbf{b}, 0)$ is called a *basic solution*. It should be clear that a basic solution is unique since $\mathbf{B}$ is a nonsingular matrix. A basis $\mathbf{B}$ is called a *feasible basis* if $\mathbf{B}^{-1}\mathbf{b} \geq 0$. We say a basic feasible solution is *degenerate* if one or more components $\bar{b}_i$ of the vector $\bar{\mathbf{b}} = \mathbf{B}^{-1}\mathbf{b}$ is zero. If all components $\bar{b}_i$ are positive, then the basic feasible solution is called *nondegenerate*.

It is necessary to study in detail the process whereby $\mathbf{Ax}=\mathbf{b}$ can be transformed with respect to a basis inverse to the equivalent form (1.2). Specifically, we want to study the elementary operations whereby the original system

$$\sum_{j=1}^{n} a_{ij}x_j = b_i \qquad i=1,\ldots,m \tag{1.3}$$

is transformed to the basic system

$$x_i + \sum_{j=m+1}^{n} \bar{a}_{ij}x_j = \bar{b}_i \qquad i=1,\ldots,m \tag{1.4}$$

The $\bar{a}_{ij}$ are the elements of the $m\times(n-m)$ matrix $\mathbf{B}^{-1}\mathbf{N}$ and the $\bar{b}_i$ are the elements of the m-vector $\mathbf{B}^{-1}\mathbf{b}$.

The transformation from (1.3) to (1.4) can be accomplished by a series of *pivot operations*, which we now define. We leave it to the reader to verify that a pivot operation on a linear system produces an equivalent system.

Definition 1.1. A pivot operation on a linear system consists of m elementary operations, which transform the system to an equivalent system in which a specified variable has a coefficient of unity in one equation and zero elsewhere. The specific operations are:

1. Select a term a_{rs} in the system (1.3) such that $a_{rs}\neq 0$. This term is called the *pivot* term.
2. Replace equation r by the equation r multiplied by $(1/a_{rs})$.
3. For $i=1,2,\ldots,m$, except $i=r$, replace equation i by the sum of equation i and the replaced equation r multiplied by $(-a_{is})$.

The transformation of (1.3) to the basic system (1.4) is accomplished by a sequence of m pivots. The first pivot term can be any a_{rs} such that $a_{rs}\neq 0$. After the first pivot operations has been completed, the second pivot term is selected using a nonzero element from any equation except r, say equation r^1. After the second pivot operation has been completed the third pivot term is selected in the resulting system from any equation except r and r^1. The general pivot operation is identical with the pivot term chosen from equations that do not correspond to equations previously selected. For simplicity, we have assumed that the form (1.4) can be achieved without interchanging rows or columns.

A fundamental property of the simplex method is the transformation from one basic system to another, which results when a basic variable is replaced by a nonbasic variable. This is accomplished by selecting any element $\bar{a}_{rs} \neq 0$ in (1.4) and pivoting on it. The result is

$$x_i + \frac{(-\bar{a}_{is})}{\bar{a}_{rs}} x_r + \sum_{\substack{j=m+1 \\ j \neq s}}^{n} \left(\bar{a}_{ij} - \frac{\bar{a}_{rj}}{\bar{a}_{rs}} \cdot \bar{a}_{is} \right) x_j = \bar{b}_i - \frac{\bar{b}_r}{\bar{a}_{rs}} \cdot \bar{a}_{is}$$

$$i = 1, \ldots, m, \ i \neq r \qquad (1.5)$$

$$\frac{1}{\bar{a}_{rs}} x_r + \sum_{\substack{j=m+1 \\ j \neq s}}^{n} \frac{\bar{a}_{rj}}{\bar{a}_{rs}} x_j + x_s = \frac{\bar{b}_r}{\bar{a}_{rs}}$$

The indicated basic solution obtained by setting the nonbasic variables to zero in (1.5) is

$$x_i = \bar{b}_i - \frac{\bar{b}_r}{\bar{a}_{rs}} \cdot \bar{a}_{is} \qquad i = 1, \ldots, m, \ i \neq r$$

$$x_s = \frac{\bar{b}_r}{\bar{a}_{rs}} \qquad (1.6)$$

To see that this pivoting operation is equivalent to substituting the column $\mathbf{a}_s$ for the column $\mathbf{a}_r$ in the basis, let

$$\mathbf{B}_0 = (\mathbf{a}_1, \ldots, \mathbf{a}_{r-1}, \mathbf{a}_r, \mathbf{a}_{r+1}, \ldots, \mathbf{a}_m)$$

and

$$\mathbf{B}_1 = (\mathbf{a}_1, \ldots, \mathbf{a}_{r-1}, \mathbf{a}_s, \mathbf{a}_{r+1}, \ldots, \mathbf{a}_m).$$

We can readily verify that

$$\mathbf{B}_1^{-1} = \mathbf{E}\mathbf{B}_0^{-1} \qquad (1.7)$$

where

$$\mathbf{E} = \begin{pmatrix} 1 & & & & \frac{-\bar{a}_{1s}}{\bar{a}_{rs}} & & & \\ & 1 & & & & & & \\ & & \ddots & & \vdots & & & \\ & & & 1 & \frac{-\bar{a}_{r-1,s}}{\bar{a}_{rs}} & & & \\ & & & & \frac{1}{\bar{a}_{rs}} & & & \\ & & & & \frac{-\bar{a}_{r+1,s}}{\bar{a}_{rs}} & 1 & & \\ & & & & \vdots & & \ddots & \\ & & & & \frac{-\bar{a}_{ms}}{\bar{a}_{rs}} & & & 1 \end{pmatrix}$$

The non-unit column of $\mathbf{E}$ is column r. In particular, (1.7) follows by verifying $\mathbf{EB}_0^{-1}\mathbf{a}_i=\mathbf{e}_i$ for $i=1,\ldots,m$, $i\neq r$, and $\mathbf{EB}_0^{-1}\mathbf{a}_s=\mathbf{e}_r$ where $\mathbf{e}_i$ is the ith unit vector in R^m.

To complete the argument, note that (1.4) can be expressed as

$$\mathbf{B}_0^{-1}(\mathbf{Ax})=\mathbf{B}_0^{-1}\mathbf{b}$$

and (1.5) as

$$\mathbf{EB}_0^{-1}(\mathbf{Ax})=\mathbf{EB}_0^{-1}\mathbf{b}$$

where the latter equation can be verified by direct calculation. Thus, since $\mathbf{B}_1^{-1}=\mathbf{EB}_0^{-1}$, pivoting on $\bar{a}_{rs}$ in (1.4) to obtain (1.5) is equivalent to changing the basis representation of $\mathbf{Ax}=\mathbf{b}$ with respect to $\mathbf{B}_0$ to one with respect to $\mathbf{B}_1$.

With this background, we show how the simplex method proceeds from a basic feasible solution to an optimal basic feasible solution. We will show later in this section how an initial basic feasible solution is obtained. The first result is a test for optimality of a basic feasible solution. Let the vector $\mathbf{c}$ be partitioned as $(\mathbf{c_B},\mathbf{c_N})$ conformally as $(\mathbf{x_B},\mathbf{x_N})$.

LEMMA 1.1. A basic feasible solution $(\mathbf{x_B},\mathbf{x_N})=(\mathbf{B}^{-1}\mathbf{b},\mathbf{0})$ to the linear programming problem (1.1) is a minimal cost solution if

$$\bar{\mathbf{c}}_\mathbf{N}=\mathbf{c_N}-\mathbf{c_B}\mathbf{B}^{-1}\mathbf{N}\geq\mathbf{0} \tag{1.8}$$

PROOF. We use (1.2) to substitute for the dependent basic variables $\mathbf{x_B}$ in the objective function

$$z=\mathbf{c_B}\mathbf{x_B}+\mathbf{c_N}\mathbf{x_N}=\mathbf{c_B}\mathbf{B}^{-1}\mathbf{b}+(\mathbf{c_N}-\mathbf{c_B}\mathbf{B}^{-1}\mathbf{N})\mathbf{x_N}=\mathbf{c_B}\mathbf{B}^{-1}\mathbf{b}+\bar{\mathbf{c}}_\mathbf{N}\mathbf{x_N}$$

The cost of the basic feasible solution with $\mathbf{x_N}=\mathbf{0}$ is $\mathbf{c_B}\mathbf{B}^{-1}\mathbf{b}$. If $\mathbf{c_N}-\mathbf{c_B}\mathbf{B}^{-1}\mathbf{N}\geq\mathbf{0}$, then clearly $\mathbf{cx}\geq\mathbf{c_B}\mathbf{B}^{-1}\mathbf{b}$ for any feasible solution $(\mathbf{x_B},\mathbf{x_N})$. ■

Lemma 1.1 gives a sufficient condition for optimality of a basic feasible solution. The coefficients

$$\bar{c}_j=c_j-\mathbf{c_B}\mathbf{B}^{-1}\mathbf{a}_j,\qquad j=1,\ldots,n$$

are called *reduced cost coefficients*; note that the reduced cost coefficient of a basic variable is 0 since $\mathbf{B}^{-1}\mathbf{a}_j$ is simply a unit vector that selects the coefficient c_j from the vector $\mathbf{c_B}$. If we define the m-vector $\mathbf{u}=\mathbf{c_B}\mathbf{B}^{-1}$, then each reduced cost coefficient $\bar{c}_j$ is derived from the original cost coefficient c_j by subtracting the quantity $\sum_{i=1}^m u_i a_{ij}$. In the next chapter, we give an economic interpretation of these u_i, which are called *shadow prices*. For future reference, we note from the proof of Lemma 1.1 that the objective function cost as a function of the reduced cost coefficients is

$$z=\bar{z}_0+\sum_{j=1}^{n}\bar{c}_jx_j \tag{1.9}$$

where $\bar{z}_0=\mathbf{c_B}\mathbf{B}^{-1}\mathbf{b}$ (recall the $\bar{c}_j$ are zero for the basic x_j).

If the condition (1.8) does not hold, then a feasible solution with lower cost is indicated. Specifically, suppose the basic feasible solution $(\mathbf{x_B}, \mathbf{x_N}) = (\mathbf{B}^{-1}\mathbf{b}, \mathbf{0})$ has the property that $\bar{c}_s < 0$ for some nonbasic activity $\mathbf{a}_s$.

LEMMA 1.2. If, for a basic feasible system, there is a nonbasic variable x_s with the properties $\bar{c}_s < 0$ and $\bar{a}_{is} \le 0$ for $i = 1, \ldots, m$, then the objective function of the linear programming problem (1.1) can be driven to $-\infty$.

PROOF. From (1.4) we have the following relation between the dependent basic variables x_i and the independent nonbasic variable x_s

$$x_i = \bar{b}_i - \bar{a}_{is} x_s \qquad i = 1, \ldots, m$$

If $\bar{a}_{is} \le 0$ for all i, then clearly x_s can be increased from zero without bound and the x_i will remain nonnegative. Since $\bar{c}_s < 0$, the objective function $z = \mathbf{c_B}\mathbf{B}^{-1}\mathbf{b} + \bar{c}_s x_s$ goes to $-\infty$ as x_s goes to $+\infty$. ■

LEMMA 1.3. If, for a nondegenerate basic feasible system, there is a nonbasic variable x_s with $\bar{c}_s < 0$ and $\bar{a}_{is} > 0$ for some i, then a new basic feasible solution can be constructed with strictly lower objective function cost.

PROOF. The new solution with lower value is constructed by increasing x_s as much as possible while maintaining feasibility; that is, while maintaining $x_i = \bar{b}_i - \bar{a}_{is} x_s \ge 0$ for all i. This is accomplished by letting $x_s = \theta_s$, where

$$\theta_s = \min_{\bar{a}_{is} > 0} \frac{\bar{b}_i}{\bar{a}_{is}} = \frac{\bar{b}_r}{\bar{a}_{rs}} > 0 \tag{1.10}$$

The value $\theta_s > 0$, because $\bar{b}_i > 0$ for all i by our nondegeneracy assumption. Since $\theta_s > 0$, the objective function has strictly decreased in the amount $(-\bar{c}_s)\theta_s > 0$.

We must still demonstrate that the new solution we have obtained is a basic solution. This is true because, in effect, the nonbasic variable x_s has been substituted in the basis for the basic variable x_r by the choice of θ_s. Specifically, the new solution is

$$\bar{x}_i = \bar{b}_i - \bar{a}_{is} \frac{\bar{b}_r}{\bar{a}_{rs}} \qquad i = 1, \ldots, m, \quad i \ne r$$

$$\bar{x}_r = 0$$

$$\bar{x}_s = \frac{\bar{b}_r}{\bar{a}_{rs}} = \theta_s$$

$$\bar{x}_j = 0 \qquad j = m+1, \ldots, n, \quad j \ne s$$

This solution is precisely the unique basic feasible solution (1.6) which would result if we pivoted on $\bar{a}_{rs}$ in the basic system (1.4). ∎

The simplex method indicated by Lemmas 1.1 to 1.3 is summarized in the following steps. One pass through Steps 1 to 3 is called an iteration. The rule (1.11) for selecting an entering nonbasic activity is a heuristic that has worked well in practice (see Exercise 1.4).

Simplex Algorithm

Step 1. Determine if there is a nonbasic activity a_j in a basic feasible system with reduced cost coefficient $\bar{c}_j < 0$. If all $\bar{c}_j \geq 0$, stop because the current basic feasible solution is optimal (minimal cost) by Lemma 1.1. If the basic feasible solution does not satisfy the optimality conditions, select a nonbasic activity a_s such that $\bar{c}_s < 0$, for example,

$$\bar{c}_s = \min \bar{c}_j < 0 \tag{1.11}$$

The activity a_s wants to enter the basis.

Step 2. If $\bar{a}_{is} \leq 0$, $i = 1, \ldots, m$, then the objective function cost is $-\infty$ by Lemma 1.2, and the method terminates. Otherwise, determine which basic activity a_r is to be replaced by activity a_s by the rule

$$\theta_s = \frac{\bar{b}_r}{\bar{a}_{rs}} = \min_{\bar{a}_{is} > 0} \frac{\bar{b}_i}{\bar{a}_{is}}$$

Step 3. Pivot on the term $\bar{a}_{rs}$ creating a new basic feasible solution, which, barring degeneracy, will have a lower objective function cost according to Lemma 1.3. Return to Step 1.

THEOREM 1.1. Starting from a nondegenerate basic feasible solution and assuming nondegeneracy at each iteration, the simplex algorithm will in a finite number of iterations either (1) discover that the objective function cost can be driven to $-\infty$, or (2) terminate with an optimal basic feasible solution.

PROOF. If the simplex algorithm did not terminate in a finite number of iterations, it would be necessary to repeat some basic feasible solution because the number of distinct bases is finite. An upper bound on this number is $\binom{n}{m}$. Since basic solutions are unique, repeating one would imply that the objective function is at the same cost at the start of two different iterations. But this is impossible by Lemma 1.3 and our nondegeneracy assumption. ∎

We want to consider briefly the so-called phase one procedure of linear programming for finding an initial basic feasible solution to (1.1). Without loss of generality, we can assume that the right-hand side vector **b** in (1.1) is nonnegative. The phase one procedure consists of the introduction of an m-vector **w** of artificial vectors and the solution of the linear programming problem

$$\min \phi = \sum_{i=1}^{m} w_i$$

$$\text{s.t. } \mathbf{Ax} + \mathbf{Iw} = \mathbf{b} \tag{1.12}$$

$$\mathbf{x} \geq \mathbf{0}, \qquad \mathbf{w} \geq \mathbf{0}$$

Clearly, the original linear programming problem (1.1) has a feasible solution if and only if the minimal objective function value of the problem (1.12) is zero.

The starting basic feasible solution for (1.12) is $\mathbf{w}=\mathbf{b}$, $\mathbf{x}=\mathbf{0}$ and the simplex method proceeds as described above. If the minimal phase one objective function value equals zero, then we have a basic feasible solution $\mathbf{x} \geq \mathbf{0}$, $\mathbf{w} \geq \mathbf{0}$ to $\mathbf{Ax}+\mathbf{Iw}=\mathbf{b}$ with $\mathbf{w}=\mathbf{0}$; that is, a feasible solution $\mathbf{x} \geq \mathbf{0}$ to $\mathbf{Ax}=\mathbf{b}$. If none of the artificial variables remain in the basis at this point, then we have the desired starting basic feasible solution. In this case the phase two procedure for finding an optimal solution to the original problem (1.1) is initiated simply by deleting the w_i and reinstating the objective function **cx**. If one or more of the w_i are in the basis at phase one optimality with $\phi=0$, then it is necessary to ensure that ϕ remains identically equal to zero during phase two, when the objective function **cx** is reinstated. The correct step to take becomes evident if we consider the phase one objective function at optimality as a function of the variables

$$\phi = \sum_{j=1}^{n} \bar{d}_j x_j + \sum_{i=1}^{m} \bar{f}_i w_i$$

where $\bar{d}_j$ and $\bar{f}_i$ are the phase one reduced cost coefficients and are equal to zero for the variables x_j and w_i that are basic. Since phase one optimality was obtained, all the $\bar{d}_j$ and $\bar{f}_i$ are nonnegative. Thus, if we delete from the phase two problem all variables x_j with $\bar{d}_j > 0$ and all nonbasic variables w_i, then the phase one objective function remains identically equal to zero throughout phase two.

The simplex algorithm just described is the first and most important of a number of algorithms we will be studying for solving mathematical programming problems. The algorithm has proved to be quite efficient in practice. Computational experience has indicated that the total number of phase one and phase two iterations it requires is a linear function of the number of rows: approximately $2m$ iterations for a typical problem regardless of the magnitude of m and n. In general, mathematical programming algorithms can converge in a finite or infinite number of iterations to an optimal solution where infinite convergence means that any limit point of a sequence of solutions is an optimal solution. Some of the algorithms in Chapter 7 exhibit infinite convergence. Finite convergence does not necessarily imply efficiency, however, because the number of iterations required can be excessively large even for the most powerful computers. The simplex algorithm is an enigma in this sense because theoretically it could require in the worst case $\binom{n}{m}$ iterations to find an optimal basic feasible solution, although as we have said it always requires far fewer in practice.

Recently, a new area of applied mathematics has emerged, called the theory of *computational complexity*, devoted to a mathematical study of algorithms. Although a rigorous development of computational complexity is beyond the intended scope of this book, we occasionally relate in a qualitative way our study of mathematical programming to it. A major concern of the theory of computational complexity is whether or not the number of arithmetic operations required by an algorithm can always be bounded by a polynomial in the parameters of the problem. If so, then the algorithm is said to be "good" theoretically, and we can also expect it to perform well empirically. This is in contrast to those algorithms that are not polynomially bounded and can theoretically require an exponentially increasing number of arithmetic operations as problem size increases. The simplex algorithm is of great interest to researchers in computational complexity because of its good empirical and its poor theoretical behavior.

EXAMPLE 1.1. Consider the following problem faced by the manager of an oil refinery. He has 8 million barrels of crude oil A and 5 million barrels of crude oil B. He can use these resources to make either gasoline, with a profit of \$4/barrel, or heating oil, with a profit of \$3/barrel. He has at his disposal three production processes with the following characteristics.

	Process 1	Process 2	Process 3
Input crude A	3	1	5
Input crude B	5	1	3
Output gasoline	4	1	3
Output heating oil	3	1	4

The manager wishes to choose the production levels of these processes to maximize profit. This problem can be formulated as the linear programming problem

$$\max z = 25x_1 + 7x_2 + 24x_3$$

$$\text{s.t.} \qquad 3x_1 + 1x_2 + 5x_3 \leq 8 \times 10^6$$

$$5x_1 + 1x_2 + 3x_3 \leq 5 \times 10^6$$

$$x_1 \geq 0,\ x_2 \geq 0,\ x_3 \geq 0$$

where the variables x_1, x_2, x_3 are the production levels for the three processes.

For pedagogical purposes, the simplex calculations for this example will be performed using the format defined in Table 1.1 where the first row corresponds to the objective function $-z + \Sigma_{j=1}^{n} \bar{c}_j x_j = -\bar{z}_0$ from (1.9) and the remaining m rows correspond to the basic system (1.4). The basic variables are shown on the left side of the Table 1.1 and their values in the basic solution are given in the rightmost column. If the basic feasible solution does not prove to be optimal by the test of Lemma 1.1 and a pivot is made on the pivot element $\bar{a}_{rs}$, the tableau must be changed by the change of basis formulae (1.5). Similarly, the objective function row of the tableau changes to

$$-z + \frac{\bar{c}_s}{\bar{a}_{rs}} x_r + \sum_{\substack{j=m+1 \\ j \neq s}}^{n} \left(\bar{c}_j - \frac{\bar{a}_{rj}}{\bar{a}_{rs}} \bar{c}_s \right) x_j = -\bar{z}_0 - \frac{\bar{b}_r}{\bar{a}_{rs}} \bar{c}_s$$

Let x_4 and x_5 be the slack variables in the numerical problem; the slack basic solution $x_4 = 8 \times 10^6$, $x_5 = 5 \times 10^6$ is feasible and is the starting point for our illustration of the simplex method. The initial tableau is shown in Table 1.2. Since we are maximizing instead of minimizing, the optimality condition for this problem is $\bar{c}_j \leq 0$ for all nonbasic activities. This is clearly not the case in Table 1.2 and we choose x_1 to enter the basis by the usual

TABLE 1.1 TABLEAU REPRESENTATION OF A LINEAR PROGRAMMING PROBLEM

Basic Variables	Current Basic Variables $-z$	$x_1 \ldots$	x_m	Current Nonbasic Variables $x_{m+1} \ldots$	x_n	Basic Variable Values
$-z$	1			$\bar{c}_{m+1}$	$\bar{c}_n$	$-\bar{z}_0$
x_1		1		$\bar{a}_{1,m+1}$	$\bar{a}_{1,n}$	$\bar{b}_1$
x_m			1	$\bar{a}_{m,m+1}$	$\bar{a}_{m,n}$	$\bar{b}_m$

TABLE 1.2 OIL REFINERY PROBLEM—ITERATION ONE

Basic Variables	$-z$	x_1	x_2	x_3	x_4	x_5	Basic Variable Values
$-z$	1	25	7	24			0
x_4		3	1	5	1		8×10^6
x_5		(5)	1	3		1	5×10^6

TABLE 1.3 OIL REFINERY PROBLEM—ITERATION TWO

Basic Variables	$-z$	x_1	x_2	x_3	x_4	x_5	Basic Variable Values
$-z$	1		2	9		-5	-25×10^6
x_4			2/5	(16/5)	1	$-3/5$	5×10^6
x_1		1	1/5	3/5		1/5	1×10^6

TABLE 1.4 OIL REFINERY PROBLEM—ITERATION THREE

Basic Variables	$-z$	x_1	x_2	x_3	x_4	x_5	Basic Variable Values
$-z$	1		14/16		$-45/16$	$-53/16$	$-\frac{625}{16}\times 10^6$
x_3			2/16	1	5/16	$-3/16$	$\frac{25}{16}\times 10^6$
x_1		1	(2/16)		$-3/16$	5/16	$\frac{1}{16}\times 10^6$

TABLE 1.5 OIL REFINERY PROBLEM—ITERATION FOUR

Basic Variables	$-z$	x_1	x_2	x_3	x_4	x_5	Basic Variable Values
$-z$	1	$-14/2$			$-3/2$	$-11/2$	$-79/2\times 10^6$
x_3		$-2/2$		1	1/2	$-1/2$	$3/2\times 10^6$
x_2		16/2	1		$-3/2$	5/2	$1/2\times 10^6$

simplex rule of selecting the nonbasic with maximal $\bar{c}_j$. The pivot element as computed in Step 2 of the simplex algorithm is the number 5 circled on the third row indicating that x_5 leaves the basis. The resulting new tableau is shown in Table 1.3. Tables 1.4 and 1.5 result by successive pivots on the indicated elements. The solution $x_2=\frac{3}{2}\times 10^6$, $x_3=\frac{1}{2}\times 10^6$ with objective function value $z=(79/2)\times 10^6$ shown in Table 1.5 is optimal. ▲

1.3 CONVERGENCE OF THE SIMPLEX METHOD: DEGENERATE CASE

The previous section contained the essence of the simplex method, and convergence of the simplex algorithm was proved assuming nondegeneracy at each iteration. If degeneracy occurs it is possible in principle for the simplex method to cycle forever through a repeating sequence of basic solutions with the same nonoptimal objective function value. Reasonable conditions do not exist for ensuring nondegeneracy and degeneracy occurs often. From a practical point of view, the simplex algorithm almost always converges anyway, although large-scale (that is extensive) degeneracy can slow convergence. Nevertheless, it is important to establish without the nondegeneracy assumption that the simplex algorithm can be made to converge finitely if the pivot selection rule for changing bases is modified appropriately. This permits the algorithm to be used as a tool in establishing other mathematical programming results including convergence of mathematical programming methods which use the algorithm as a subroutine.

Degeneracy occurs at the start of phase one if one of the right-hand-side elements in (1.1) is zero. It also occurs when more than one basic variable is made zero by pivoting a selected nonbasic variable into the basis [see (1.5) and (1.6)]. Once it occurs, degeneracy persists when there is a basic variable x_i with basic solution value 0 and $\bar{a}_{is} > 0$ for the entering nonbasic variable x_s. The idea in resolving degeneracy is to develop secondary basis change rules that exclude the possibility that a basic solution will ever be repeated. The secondary rules use the following concept.

Definition 1.2. A vector $\mathbf{a} \in R^m$ is *lexico-positive*, denoted $\mathbf{a} \succ \mathbf{0}$, if $\mathbf{a} \neq \mathbf{0}$ and the first nonzero component is positive. A vector $\mathbf{a}$ is *lexico-greater* than $\mathbf{b}$, denoted by $\mathbf{a} \succ \mathbf{b}$, if $\mathbf{a} - \mathbf{b} \succ \mathbf{0}$.

It is easy to establish that this lexicographic ordering of vectors is complete and transitive; that is, $\mathbf{a} \succ \mathbf{b}$ or $\mathbf{b} \succ \mathbf{a}$ for any vectors $\mathbf{a}$ and $\mathbf{b}$ and $\mathbf{a} \succ \mathbf{b}$ and $\mathbf{b} \succ \mathbf{c}$ imply $\mathbf{a} \succ \mathbf{c}$. Thus we can meaningfully speak of the lexico-min of a finite set of distinct vectors.

The procedure for resolving degeneracy is to replace the right-hand-side vector $\mathbf{b}$ in (1.1) by the $m \times (m+1)$ matrix $(\mathbf{b}, \mathbf{I})$. The basic feasible system (1.2) thereby becomes

$$\mathbf{x_B} = (\mathbf{B}^{-1}\mathbf{b}, \mathbf{B}^{-1}) - (\mathbf{B}^{-1}\mathbf{N}\mathbf{x_N}, \mathbf{0}_m)$$

where $\mathbf{0}_m$ is an $m \times m$ matrix of zeros, and we can think of the basic variables as taking on vector values, which we call the *basic solution vectors*

$$x_i = \bar{\mathbf{v}}_i = (\bar{\mathbf{b}}_i, \beta_{i1}, \ldots, \beta_{im}) \qquad i = 1, \ldots, m \tag{1.13}$$

where $\mathbf{B}^{-1}=(\beta_{ij})$. As with the scalar case, the basic solution vectors for a given basis are unique. The objective function also takes on a vector value, called the *objective function vector* [cf. (1.9)]

$$z=\bar{\mathbf{w}}=\mathbf{c_B}\mathbf{B}^{-1}(\mathbf{b},\mathbf{I})=(\bar{z}_0,\Pi_1,\ldots,\Pi_m) \tag{1.14}$$

Suppose some nonbasic variable x_s with $\bar{c}_s<0$ is selected to enter the basis; for example, x_s selected according to the usual rule (1.11). The leaving basic variable is selected by the generalized rule [cf. (1.10)]

$$\frac{1}{\bar{a}_{rs}}\bar{\mathbf{v}}_r=\underset{\bar{a}_{is}>0}{\text{lexico-min}}\left\{\frac{1}{\bar{a}_{i,s}}\bar{\mathbf{v}}_i\right\} \tag{1.15}$$

LEMMA 1.4. The basic variable selected to leave the basis by the lexico-min rule (1.15) is unique.

PROOF. The lexico-min rule implies the following elimination process. Begin by eliminating as the leaving variable all those basic variables x_i with $\bar{a}_{is}>0$ such that $\bar{b}_i/\bar{a}_{is}$ is not minimal. If the minimum of these scalar terms is attained by more than one basic variable, eliminate those basic variables x_i with $\bar{a}_{is}>0$ such that $\bar{b}_i/\bar{a}_{is}$ is minimal but $\beta_{i1}/\bar{a}_{is}$ is not. Continue until a unique basic variable x_r remains. A tie cannot occur in all components of two vectors $(1/\bar{a}_{i_1,s})\bar{\mathbf{v}}_{i_1}$ and $(1/\bar{a}_{i_2,s})\bar{\mathbf{v}}_{i_2}$ because then

$$\frac{1}{\bar{a}_{i_1,s}}(\beta_{i_1,1},\ldots,\beta_{i_1,m})=\frac{1}{\bar{a}_{i_2,s}}(\beta_{i_2,1},\ldots,\beta_{i_2,m})$$

contradicting the nonsingularity of $\mathbf{B}^{-1}$. ■

After the leaving basic variable is selected uniquely by (1.15), the next step is to make the usual pivot on the element $\bar{a}_{rs}$. This has the result of updating the basic solution vectors [cf. (1.6)]

$$\bar{\mathbf{v}}'_r=\frac{1}{\bar{a}_{rs}}\bar{\mathbf{v}}_r \tag{1.16}$$

$$\bar{\mathbf{v}}'_i=\bar{\mathbf{v}}_i-\frac{\bar{a}_{is}}{\bar{a}_{rs}}\bar{\mathbf{v}}_r \qquad i=1,\ldots,m,\ i\neq r$$

and the objective function vector

$$\bar{\mathbf{w}}'=\bar{\mathbf{w}}+\frac{\bar{c}_s}{\bar{a}_{rs}}\bar{\mathbf{v}}_r \tag{1.17}$$

LEMMA 1.5. Starting with the initial phase one basic solution, the basic solution vectors are lexico-positive at every iteration when the leaving basic variable is selected by the lexico-min rule (1.15).

PROOF. The proof is by induction. The first basic solution vectors at the start of phase one are $(b_i, \mathbf{e}_i), i=1,\dots,m$, where $b_i \geq 0$ and $\mathbf{e}_i$ is the ith unit vector. These vectors are obviously lexico-positive.

To complete the induction proof, we demonstrate that if the basic solution vectors $\bar{\mathbf{v}}_i$ are lexico-positive at a given iteration, then the $\bar{\mathbf{v}}_i'$ given by (1.16) are lexico-positive at the next iteration. The vector $\bar{\mathbf{v}}_r' \succ 0$ since $\bar{\mathbf{v}}_r \succ 0$ by the induction assumption and $\bar{a}_{rs} > 0$. For i such that $\bar{a}_{is} > 0$, we have by the rule (1.15)

$$\frac{1}{\bar{a}_{is}}\bar{\mathbf{v}}_i - \frac{1}{\bar{a}_{rs}}\bar{\mathbf{v}}_r \succ 0$$

and this implies $\bar{\mathbf{v}}_i' \succ 0$. For i such that $\bar{a}_{is} \leq 0$, we have from (1.16) that

$$\bar{\mathbf{v}}_i' = \bar{\mathbf{v}}_i + (-\bar{a}_{is})\bar{\mathbf{v}}_r'$$

implying $\bar{\mathbf{v}}_i' \succ 0$ since $\bar{\mathbf{v}}_i \succ 0$ and $\bar{\mathbf{v}}_r' \succ 0$. ■

The following lemma is established for the phase two objective function vector, but it is clearly applicable to the phase one problem as well.

LEMMA 1.6. The objective function vector decreases lexicographically at each iteration if the nonbasic variable selected to enter the basis has a negative reduced cost coefficient.

PROOF. By Lemma 1.5, at every iteration the basic solution vector $\bar{\mathbf{v}}_r \succ 0$ corresponding to the leaving basic variable. Since the entering variable x_s has the property that $\bar{c}_s < 0$, and the pivot element $\bar{a}_{rs} > 0$, we have $(\bar{c}_s / \bar{a}_{rs})\bar{\mathbf{v}}_r \prec 0$. This implies from (1.17) for updating the objective function vector that $\bar{\mathbf{w}}' \prec \bar{\mathbf{w}}$. ■

THEOREM 1.2. If the lexico-min rule (1.15) is used at each iteration to select the leaving basic variable, the simplex algorithm will in a finite number of phase one and phase two iterations, either (1) discover that the objective function cost can be drive to $-\infty$, or (2) terminate with an optimal basic feasible solution.

PROOF. If termination in phase one or phase two were not finite, then some basis and corresponding objective function vector would have to be repeated. But this is not possible since by Lemma 1.6 the objective function vector decreases lexicographically at each iteration. ■

1.4 GEOMETRY OF LINEAR PROGRAMMING

In this section, we discuss some geometrical insights into how the simplex method searches the convex polytope

$$X = \{\mathbf{x} | \mathbf{A}\mathbf{x} = \mathbf{b}, \mathbf{x} \geq \mathbf{0}\}$$

for an optimal linear programming solution. We will be using some definitions and concepts from Appendix A about convex sets. Additional geometrical insights about linear programming will be given in the next chapter after we discuss linear programming duality theory.

LEMMA 1.7. Let $\tilde{\mathbf{x}}$ be a basic feasible solution to the linear system $\mathbf{Ax}=\mathbf{b}$. Then $\tilde{\mathbf{x}}$ is an extreme point of the convex polytope X.

PROOF. Consider the basic system (1.2)

$$\mathbf{x_B}=\mathbf{B}^{-1}\mathbf{b}-\mathbf{B}^{-1}\mathbf{N}\mathbf{x_N}$$

and the associated basic solution $\tilde{\mathbf{x}}=(\tilde{\mathbf{x}}_\mathbf{B},\tilde{\mathbf{x}}_\mathbf{N})=(\mathbf{B}^{-1}\mathbf{b},\mathbf{0})$ where feasibility of $\tilde{\mathbf{x}}$ means $\mathbf{B}^{-1}\mathbf{b}\geq\mathbf{0}$. Suppose $\tilde{\mathbf{x}}$ is not an extreme point. This implies the existence of distinct solutions $\mathbf{x}^1$ and $\mathbf{x}^2$ satisfying $\mathbf{Ax}=\mathbf{b},\mathbf{x}\geq\mathbf{0}$, and a scalar, $0<\lambda<1$, such that

$$\tilde{\mathbf{x}}_\mathbf{B}=\lambda\mathbf{x}_\mathbf{B}^1+(1-\lambda)\mathbf{x}_\mathbf{B}^2$$

$$\tilde{\mathbf{x}}_\mathbf{N}=\lambda\mathbf{x}_\mathbf{N}^1+(1-\lambda)\mathbf{x}_\mathbf{N}^2$$

Since $\mathbf{x}_\mathbf{N}^1\geq 0$ and $\mathbf{x}_\mathbf{N}^2\geq 0$ and $0<\lambda<1$, we must have $\mathbf{x}_\mathbf{N}^1=\mathbf{x}_\mathbf{N}^2=0$ in order that $\tilde{\mathbf{x}}_\mathbf{N}=0$. Thus, $\mathbf{B}\mathbf{x}_\mathbf{B}^1=\mathbf{b}$ and $\mathbf{B}\mathbf{x}_\mathbf{B}^2=\mathbf{b}$ implying $\mathbf{x}_\mathbf{B}^1=\mathbf{x}_\mathbf{B}^2=\tilde{\mathbf{x}}_\mathbf{B}=\mathbf{B}^{-1}\mathbf{b}$ and therefore $\mathbf{x}^1=\mathbf{x}^2=\tilde{\mathbf{x}}$. This is a contradiction to the distinctness of $\mathbf{x}^1$ and $\mathbf{x}^2$. ■

LEMMA 1.8. Let $\tilde{\mathbf{x}}$ be an extreme point of the convex polytope X. Then the columns of $\mathbf{A}$ corresponding to the positive components of $\tilde{\mathbf{x}}$ are linearly independent.

PROOF. Without loss of generality, assume $\tilde{x}_j>0$ for $j=1,\ldots,k$, and $\tilde{x}_j=0$ for $j=k+1,\ldots,n$; therefore, $\sum_{j=1}^{k}\mathbf{a}_j\tilde{x}_j=\mathbf{b}$. Suppose the $\mathbf{a}_j$ were linearly dependent; that is, suppose there exist λ_j for $j=1,\ldots,k$, not all zero such that

$$\sum_{j=1}^{k}\mathbf{a}_j\lambda_j=0 \tag{1.18}$$

Define

$$\delta=\min\left\{\frac{\tilde{x}_j}{|\lambda_j|}\;\Big|\;\lambda_j\neq 0 \text{ and } j=1,\ldots,k\right\}$$

Clearly $\delta>0$ and if we select any ε satisfying $0<\varepsilon<\delta$, then

$$\tilde{x}_j+\varepsilon\lambda_j>0 \qquad \text{and} \qquad \tilde{x}_j-\varepsilon\lambda_j>0, \qquad j=1,\ldots,k \tag{1.19}$$

Define a vector $\boldsymbol{\lambda}\in R^n$ by setting $\lambda_j=0$ for $j=k+1,\ldots,n$, and the other λ_j as above. By construction, we have two solutions $\mathbf{x}^1=\tilde{\mathbf{x}}+\varepsilon\boldsymbol{\lambda},\mathbf{x}^2=\tilde{\mathbf{x}}-\varepsilon\boldsymbol{\lambda}$ which satisfy $\mathbf{Ax}=\mathbf{b}$ from $\mathbf{A}\tilde{\mathbf{x}}=\mathbf{b}$ and (1.18), and $\mathbf{x}\geq\mathbf{0}$ from (1.19). Moreover, $\tilde{\mathbf{x}}=\frac{1}{2}\mathbf{x}^1+\frac{1}{2}\mathbf{x}^2$ which is impossible if $\tilde{\mathbf{x}}$ is an extreme point. ■

If every extreme point of X has m positive components, then there is a one-to-one correspondence between basic feasible solutions and extreme points. If an extreme point $\tilde{\mathbf{x}}$ has $k<m$ positive components, we say it is degenerate. A basis can be constructed with the property that such an $\tilde{\mathbf{x}}$ is the corresponding basic solution by adding $m-k$ linearly independent columns of $\mathbf{A}$ to the k linearly independent columns corresponding to the positive components of $\tilde{\mathbf{x}}$. Of course, the choice of the $m-k$ columns of $\mathbf{A}$ to complete a basis is generally not unique and therefore there will be many bases corresponding to a degenerate extreme point. This can cause computational difficulties because the simplex method may iterate through a number of different basic representations of a degenerate extreme point before finding a different extreme point with strictly lower cost.

We want to characterize in detail the set that is generated as the entering nonbasic variable is increased from 0 to its new value in the basis. This set is

$$E=\left\{\mathbf{x}|x_i=\bar{b}_i-\bar{a}_{is}x_s, i=1,\ldots,m, 0\le x_s\le\theta,\right.$$
$$\left.x_j=0, j=m+1,\ldots,n, j\neq s\right\}$$

where

$$\theta=\min_{\bar{a}_{is}>0}\frac{\bar{b}_i}{\bar{a}_{is}}$$

The set E is contained in the supporting hyperplane of X given by

$$\left\{\mathbf{x}\Big|\sum_{\substack{j=m+1\\ j\neq s}}^{n} x_j=0\right\}$$

If $\theta=0$, then E consists of a single point, which is a degenerate extreme point of X. If $\theta>0$ and finite, then E contains a bounded set of points on the supporting hyperplane, and it is an edge of X. If $\theta=+\infty$, then E is an extreme ray of the convex polytope X.

We can depict these concepts graphically if we consider problem (1.1) with $n=m+2$. A basic system for this problem contains two nonbasic variables, which we assume are x_{m+1} and x_{m+2}

$$\begin{aligned}
&\min z=\bar{z}_0+\bar{c}_{m+1}x_{m+1}+\bar{c}_{m+2}x_{m+2}\\
&\text{s.t.}\quad x_i=\bar{b}_i-\bar{a}_{i,m+1}x_{m+1}-\bar{a}_{i,m+2}x_{m+2}\qquad i=1,\ldots,m\\
&\qquad\quad x_j\ge 0,\qquad j=1,\ldots,m+2.
\end{aligned}$$

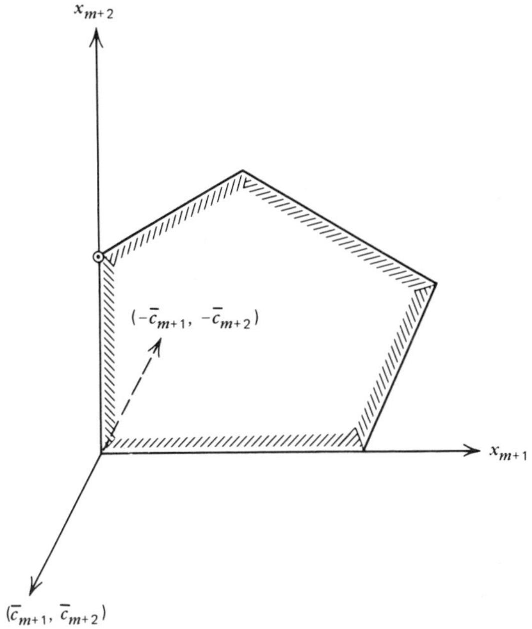

Figure 1.1. Simplex method as descent along an edge, nondegenerate case.

The feasible region for this problem consists of $x_{m+1} \geq 0, x_{m+2} \geq 0$ satisfying

$$\bar{a}_{i,m+1}x_{m+1} + \bar{a}_{i,m+2}x_{m+2} \leq \bar{b}_i \qquad i=1,\ldots,m$$

This region is the shaded area in Figure 1.1 and the basic solution $x_i = \bar{b}_i$, $i=1,\ldots,m, x_{m+1}=x_{m+2}=0$ is the origin. The vector $(-\bar{c}_{m+1}, -\bar{c}_{m+2})$ is also shown in the figure, and it represents the direction of greatest decrease in the objective function per unit movement from the origin. The usual simplex rule of choosing an entering variable to be the one for which $\min(\bar{c}_{m+1}, \bar{c}_{m+2})$ is attained can be seen as selecting the edge of steepest descent per unit. In Figure 1.1, the entering basic variable is x_{m+2} and the next extreme point basic solution is the circled point. Figure 1.2 illustrates the situation when there is degeneracy and the choice x_{m+2} for entering variable results in a degenerate pivot.

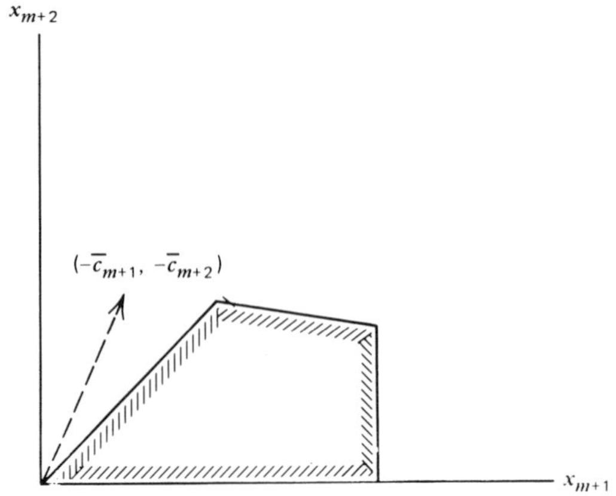

Figure 1.2. Simplex method as descent along an edge, degenerate case.

1.5 ADDITIONAL FEATURES AND USES OF THE SIMPLEX ALGORITHM

There are many important extensions of the simplex algorithm as it has been described in this chapter. In later chapters we discuss important variants of the algorithm, its specialization to network optimization, and its use in approximating convex and even nonconvex programming problems. Our purpose here is to discuss briefly the issue of the matrix inversion of bases and also to show how certain structures of linear and separable nonlinear programming problems can be handled implicitly by the simplex method.

The calculation of a basic feasible solution requires knowledge of the basis inverse. The explicit calculation of this inverse at each iteration would be very time consuming on a computer for problems of any size. Fortunately, a complete inversion at each iteration is unnecessary because the basis changes in only one column from one iteration to the next. The formula (1.7) gives the relation between successive basis inverses, and, since the initial basis at the start of phase one is the identity matrix, we have at iteration k

$$\mathbf{B}_k^{-1} = \mathbf{E}_k \mathbf{E}_{k-1} \ldots \mathbf{E}_1 \mathbf{I}$$

where the $\mathbf{E}_l$ can be compactly stored because they resemble identity matrices except for one column. Thus, the new basic solution at iteration k can be calculated by premultiplying the old basic solution by $\mathbf{E}_k$. The matrix $\mathbf{E}_k$ can also be used to compute the new shadow prices and the new reduced cost coefficients of the nonbasic variables. This method of updating linear programming basic solutions is called the *revised simplex method*.

Periodic reinversions of the basis inverse are still required, however, because of the build-up of numerical error and also because of the accumulation of the columns characterizing the $\mathbf{E}_l$ matrices. Highly sophisticated numerical analytic procedures have been devised for accomplishing these inversions, but they are beyond the intended scope of this book.

In most applications, there are upper bound as well as nonnegativity constraints on some of the variables. It is not necessary to add upper bounds explicitly to the constraint set because they can be handled implicitly in the same way that nonnegativity constraints are handled. We sketch the procedure by considering the linear programming problem

$$\begin{aligned}&\min \mathbf{cx}\\ &\text{s.t. } \mathbf{Ax}=\mathbf{b}\\ &\quad 0\le x_j\le 1 \qquad j=1,\ldots,n\end{aligned}$$

where $\mathbf{A}$ is $m\times n$. The upper bound of 1 on the variables is not restricting because any positive upper bound can be converted to 1 by appropriate scaling.

The upper bound constraint $x_j\le 1$ is rewritten as

$$x_j+x_j'=1 \qquad j=1,\ldots,n$$

At each iteration, the idea is to consider making one of the *upper bound substitutions*

$$x_j=1-x_j'$$
$$x_j'=1-x_j$$

as well as the usual replacement rule in Step 2 of the simplex algorithm. As before, suppose we wish to increase from zero the nonbasic variable x_s with $\bar{c}_s<0$. It is also possible that the selected nonbasic variable is x_j', in which case increasing it from zero is the same as decreasing x_j from 1. The reader can verify the validity of the following rule: Compute

$$\theta_1=\min_{\bar{a}_{is}>0}\frac{\bar{b}_i}{\bar{a}_{is}}$$

and

$$\theta_2=\min_{\bar{a}_{is}<0}\frac{1-\bar{b}_i}{(-\bar{a}_{is})}$$

and let

$$\theta=\min(\theta_1,\theta_2)$$

Suppose $\theta \leq 1$, say $\theta = \theta_1 = \bar{b}_r / \bar{a}_{rs}$; then the procedure is to pivot on $\bar{a}_{rs}$ exactly as before and the variable x_s replaces x_r in the basis. If $\theta = \theta_2 \leq 1$, then the procedure is to pivot on $\bar{a}_{rs}$ and afterward make an upper bound substitution on the variable x_r (or x_r'), which has just left the basis. Finally, if $\theta > 1$, then we simply make an upper bound substitution on x_s.

An important application of the simplex method is to the *separable programming problem*

$$\begin{aligned} \min \sum_{j=1}^{n} f_j(x_j) & \\ \text{s.t. } \mathbf{Ax} = \mathbf{b} & \\ \mathbf{x} \geq 0 & \end{aligned} \tag{1.20}$$

where the f_j are convex functions. For ease of exposition, we suppress the subscript j and demonstrate the procedure for a generic function $f(x)$ of the scalar argument x. The function is shown in Figure 1.3, where we approximate it by a piecewise linear function with the indicated slopes Δ_i. The piecewise linear function is used to approximate the separable func-

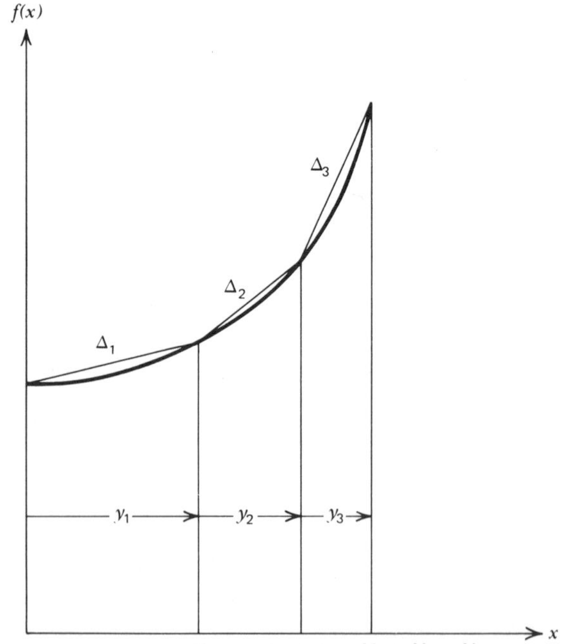

Figure 1.3. Piecewise linear approximation to separable convex objective function.

tion by making the following types of substitutions in (1.20)

$$x = y_1 + y_2 + y_3$$
$$0 \le y_1 \le M_1 \qquad 0 \le y_2 \le M_2 \qquad 0 \le y_3 \le M_3 \tag{1.21}$$
$$f(x) \approx \Delta_1 y_1 + \Delta_2 y_2 + \Delta_3 y_3$$

Since f is convex, we have $\Delta_1 < \Delta_2 < \Delta_3$ implying the variable y_1 is preferred to y_2 and the variable y_2 is preferred to y_3. Thus, at optimality we can be assured that $y_2 > 0$ implies $y_1 = M_1$ and $y_3 > 0$ implies $y_2 = M_2$. Moreover, this can be guaranteed only if f is convex.

There are a number of important applications of mathematical programming where the functions f_j in (1.20) are increasing but concave, that is, where there are economies of scale in the use of the activities a_j. The separable programming scheme just described does not work in this case. For example, if Δ_2 were less than Δ_1 in Figure 1.3, then in trying to minimize cost, the simplex algorithm would prefer y_2 to y_1 and we could have $y_1 = 0$, $y_2 > 0$ at optimality. Such a difficulty can be resolved by the introduction of additional variables constrained to be either zero or one. The method for doing this is described in Chapter 8.

1.6 EXERCISES

1.1 Construct an example of a feasible linear programming problem with an activity that must be at a zero level in any feasible solution.

1.2 (J.D.C. Little) Suppose you have been given a large square matrix $\mathbf{D}$ and asked to find its inverse or establish that it has none. The only computer code available is a linear programming system that uses the simplex algorithm on a given linear programming problem to produce (1) an optimal solution, (2) a statement that the solution is unbounded, or (3) a statement that the problem is infeasible. In case (1), the final tableau (Table 1.1) of coefficients and right-hand-side constants is available as output. However, no intermediate output can be obtained. How would you use this linear programming system to compute $\mathbf{D}^{-1}$ or prove $\mathbf{D}$ is singular?

1.3 Let x_r be the variable that leaves the basis in Step 2 of the simplex algorithm. Prove that x_r will not be a candidate by the rule (1.11) to enter the basis at the next iteration.

1.4 Show by appropriate scaling that any nonbasic variable $\bar{c}_j < 0$ can be the one selected to enter the basis in Step 1 of the simplex algorithm by the rule (1.11). Illustrate the idea with a figure similar to Figure 1.1.

1.5 Consider the numerical example of Section 1.2. Suppose a new production process is available with the following characteristics:

input crude A	2
input crude B	5
output gasoline	2
output heating oil	9

Using the optimal solution without this process, determine whether or not the process can be used to increase the profits of the oil refinery. If so, resolve the problem starting with the previously optimal solution.

1.6 A *fractional programming problem* is one of the form

$$\max z = \frac{\mathbf{cx} + c_0}{\mathbf{dx} + d_0}$$

$$\text{s.t.} \quad \mathbf{Ax} = \mathbf{b}$$

$$\mathbf{x} \geq \mathbf{0}.$$

where **c** and **d** are $1 \times n$, **A** is $m \times n$, **b** is $m \times 1$, c_0 and d_0 are scalars, and it is assumed that $\mathbf{dx} + d_0 > 0$ for all feasible **x**.

(1) We can convert this problem to an equivalent linear programming problem as follows: Let $v(\mathbf{x}) = \mathbf{dx} + d_0 > 0$ and write $y = \mathbf{x}/v(\mathbf{x})$, $y_0 = 1/v(\mathbf{x}) > 0$. Show that the constraints $\mathbf{Ax} = \mathbf{b}$ become

$$\mathbf{Ay} - \mathbf{b}y_0 = 0$$

$$\mathbf{dy} + d_0 y_0 = 1,$$

and therefore that the original problem becomes the linear programming problem

$$\max z = \mathbf{cy} + c_0 y_0$$

$$\text{s.t.} \quad \mathbf{Ay} - \mathbf{b}y_0 = 0$$

$$\mathbf{dy} + d_0 y_0 = 1$$

$$\mathbf{y} \geq 0 \qquad y_0 \geq 0$$

(2) Let $\mathbf{y}^*$, y_0^* be an optimal solution to this linear programming problem and suppose $y_0^* \neq 0$. Show that we can use this solution to obtain an optimal solution to the original problem.

1.7 (Srinivasan and Shocker, 1973) Mathematical programming models have been used in a variety of ways to "optimally" estimate utility functions of decisionmakers. This exercise addresses one method that has been proposed and implemented. Suppose there are t quantifiable attributes relevant to the decisionmakers' preferences. We present the decisionmaker with n t-vectors of attributes $\mathbf{a}_j$ and for each pair $\mathbf{a}_j$, $\mathbf{a}_k$ ask him to state which one he prefers. The result is the collection of $n(n-1)/2$ ordered pairs where $(j,k)\in\Omega$ indicates that $\mathbf{a}_j$ is preferred to $\mathbf{a}_k$. We posit the existence of an ideal point $\mathbf{x}\in R^t$ and a vector of nonnegative weights $\mathbf{w}\in R^t$ such that the disutility of an arbitrary vector $\mathbf{a}\in R^t$ is given by $\sum_{p=1}^t (a_p - x_p)^2 w_p$. The following mathematical programming problem has been proposed for estimating $\mathbf{x}$ and $\mathbf{w}$ from the a_j, $j=1,\ldots,n$, and the ordering Ω.

$$v=\min \sum_{(j,k)\in\Omega} z_{jk}$$

$$\text{s.t. } s_j = \sum_{p=1}^{t} \{(a_{jp}-x_p)^2 w_p\} \qquad j=1,\ldots,n$$

$$z_{jk} \geq s_j - s_k \qquad \text{for all } (j,k)\in\Omega$$

$$\sum_{(j,k)\in\Omega} (s_k - s_j) = 1$$

$$z_{jk} \geq 0 \qquad \text{for all } j,k\in\Omega$$

$$w_p \geq 0 \qquad \text{for } p=1,2,\ldots,t$$

$$x_p \text{ unconstrained in sign} \qquad \text{for } p=1,2,\ldots,t$$

(1) Give an interpretation of this problem. What would happen if the constraint $\sum_{(i,j)\in\Omega} (s_k - s_j) = 1$ were omitted?

(2) Show how the problem can be transformed to a linear programming problem by appropriate changes of variable.

(3) Suppose we changed the problem by setting $\sum_{(j,k)\in\Omega} (s_k - s_j) = h > 0$. Show that this has no effect upon the relative magnitudes of the w_p or the magnitudes of the x_p.

(4) Suppose $v=0$ is the optimal objective function value to this problem; that is $z_{jk}=0$ for all $(j,k)\in\Omega$, is feasible for appropriate values of the w_p and the x_p. What secondary optimality criterion would you propose for selecting the w_p and x_p while maintaining $z_{jk}=0$ for all $(j,k)\in\Omega$?

(5) Apply this model to estimate an individual's utility for automobiles. The utility is measured with respect to four attributes: cost, style, economy, and safety. The individual was presented with the following attributes of five different automobiles

	Auto 1	Auto 2	Auto 3	Auto 4	Auto 5
Cost (×$500)	8	9	12	14	21
Style	5	6	8	9	10
Economy (miles per gal.)	24	28	23	28	19
Safety	7	5	9	10	10

The style and safety measures were determined by experts in these fields. The individual responded to these vectors of attributes by the following piecewise preferences [recall (j,k) indicates auto j is preferred to auto k]:

(2, 1)	(3, 2)	(4, 3)	(4, 5)
(3, 1)	(2, 4)	(3, 5)	
(4, 1)	(2, 5)		
(1, 5)			

1.8 Suppose you are given the problem

$$\min \mathbf{cx}$$

$$\text{s.t. } \mathbf{Ax} \le \mathbf{b}$$

$$\sum_{j=1}^{n} g_j(x_j) \le q$$

$$\mathbf{x} \ge \mathbf{0}$$

where A is $m \times n$ and q is a scalar. What property of the functions $g_j(\cdot)$ is required for you to be able to use the simplex algorithm to solve the approximation to this problem which results when the $g_j(\cdot)$ are replaced by piecewise linear approximations?

1.7 NOTES

SECTION 1.2. Dantzig (1963) devotes a chapter to the origins and history of the simplex method. It is not possible to reference, much less discuss, the significant books and articles pertaining to linear programming. A sampling of books that influenced the presentation here are Bradley et al. (1977), Dantzig (1963), Hadley (1962), Karlin (1959), Lasdon (1970), Simonnard (1966), and Wagner (1975). Wagner (1975) contains a number of numerical exercises.

SECTION 1.2. The lexicographic method for resolving degeneracy was discovered by Dantzig et al. (1955). An alternative method due to Charnes

(1952) is to perturb the right-hand-side vector when degeneracy occurs; details are given in Chapter 10 of Dantzig (1963). A recent combinatorial method for resolving degeneracy is given by Bland (1977). Lawler (1976) discusses computational complexity and the simplex algorithm.

SECTION 1.4. Hadley (1961) contains material on the geometry of convex sets arising in linear programming. The simplex method viewed as the steepest descent along edges is presented in Section 7-2 of Dantzig (1963).

SECTION 1.5. Numerical analysis and the simplex method is a diffuse subject treated in a variety of books and articles. Two examples of work in this area are contained in Bartels et al. (1970) and Magnanti (1976). An important numerical topic not covered is generalized upper bounding, which is a way to treat implicitly in the simplex method constraints of the form $\sum_{j \in J} x_j = 1$; see Dantzig and Van Slyke (1967). Chapter six of Lasdon (1970) also contains a discussion of this technique along with other compact inverse methods. Separable programming is another practical extension of the simplex method; see also chapter four in Hadley (1964). Orchard-Hays (1968) treats a number of linear programming implementation issues.

2
LINEAR PROGRAMMING DUALITY AND SENSITIVITY ANALYSIS

2.1 INTRODUCTION

Linear programming duality theory is concerned with the symmetric relationship that exists between a given linear programming problem, called the primal problem, and a related linear programming problem, called the dual problem. Duality theory permits an economic interpretation of the simplex method that not only adds insight but also indicates how important variants and generalizations of the method can be constructed. Moreover, linear programming duality theory can be extended to a duality theory for any mathematical programming problem. The validity of linear programming approximations to more general mathematical programming problems is intimately related to duality.

2.2 DUAL THEOREMS

Consider the following linear programming problems, called the primal problem and the dual problem, respectively,

$$\begin{aligned} \text{Primal: } v = \min\ & \mathbf{cx} \\ \text{s.t. } & \mathbf{Ax} \geq \mathbf{b} \\ & \mathbf{x} \geq \mathbf{0} \end{aligned} \tag{P}$$

$$\begin{aligned} \text{Dual: } d = \max\ & \mathbf{ub} \\ \text{s.t. } & \mathbf{uA} \leq \mathbf{c} \\ & \mathbf{u} \geq \mathbf{0} \end{aligned} \tag{D}$$

The matrix $\mathbf{A}$ is $m \times n$ with columns $\mathbf{a}_j$ for $j = 1, \ldots, n$, the vector $\mathbf{c}$ is $1 \times n$, and the matrix $\mathbf{b}$ is $m\times 1$. Define the sets

$$X=\{\mathbf{x}|\mathbf{Ax}\geq\mathbf{b},\ \mathbf{x}\geq\mathbf{0}\}$$
$$U=\{\mathbf{u}|\mathbf{uA}\leq\mathbf{c},\ \mathbf{u}\geq\mathbf{0}\} \tag{2.1}$$

LEMMA 2.1. (Linear Programming Weak Duality). Suppose $\tilde{\mathbf{x}}\in X$ and $\tilde{\mathbf{u}}\in U$. Then

$$\mathbf{c}\tilde{\mathbf{x}}\geq\tilde{\mathbf{u}}\mathbf{b}$$

PROOF. Since $\tilde{\mathbf{x}}\in X$, $\tilde{\mathbf{u}}\in U$, we have $\mathbf{A}\tilde{\mathbf{x}}\geq\mathbf{b}$ and $\tilde{\mathbf{u}}\geq\mathbf{0}$ implying $\tilde{\mathbf{u}}\mathbf{A}\tilde{\mathbf{x}}\geq\tilde{\mathbf{u}}\mathbf{b}$. Similarly, we have $\tilde{\mathbf{u}}\mathbf{A}\leq\mathbf{c}$ and $\tilde{\mathbf{x}}\geq\mathbf{0}$ implying $\tilde{\mathbf{u}}\mathbf{A}\tilde{\mathbf{x}}\leq\mathbf{c}\tilde{\mathbf{x}}$. Thus, $\tilde{\mathbf{u}}\mathbf{b}\leq\tilde{\mathbf{u}}\mathbf{A}\tilde{\mathbf{x}}\leq\mathbf{c}\tilde{\mathbf{x}}$ and the lemma is proven. ■

COROLLARY 2.1. If $\mathbf{x}^*\in X$ and $\mathbf{u}^*\in U$ satisfy $\mathbf{cx}^*=\mathbf{u}^*\mathbf{b}$, then $\mathbf{x}^*$ is optimal in (P) and $\mathbf{u}^*$ is optimal in (D).

PROOF. The result is obvious since

$$\mathbf{cx}^*=\mathbf{u}^*\mathbf{b}\leq\mathbf{cx}\qquad\text{for all }\mathbf{x}\in X$$

and

$$\mathbf{u}^*\mathbf{b}=\mathbf{cx}^*\geq\mathbf{ub}\qquad\text{for all }\mathbf{u}\in U.\ ■$$

The following theorem says that when feasible solutions exist for both the primal and the dual, then the conditions of Corollary 2.1 obtain; namely, the optimal objective function values are equal. The proof of the theorem is a constructive one based on the simplex algorithm. The proof is valid because the simplex algorithm converges. An alternate proof of this fundamental theorem that does not rely on the simplex algorithm follows the first proof. We give the alternate proof to illustrate how classical results of linear algebra and analysis are related to the more modern theory of linear programming.

THEOREM 2.1. (Linear Programming Strong Duality). Suppose $X\neq\phi$ and $U\neq\phi$. Then there exists an $\mathbf{x}^*$ optimal in (P) and a $\mathbf{u}^*$ optimal in (D) and $\mathbf{cx}^*=\mathbf{u}^*\mathbf{b}$.

PROOF. Because of Corollary 2.1, it suffices to show the existence of $\mathbf{x}^*\in X$ and $\mathbf{u}^*\in U$ such that $\mathbf{cx}^*=\mathbf{u}^*\mathbf{b}$. We know (P) has a (bounded) optimal solution by Lemma 2.1 since $X\neq\phi$ and $U\neq\phi$. By Theorem 1.2, the simplex method converges to an optimal solution to (P), which we rewrite as

$$\begin{aligned}&\min\ \mathbf{cx}\\ &\text{s.t. }\mathbf{Ax}-\mathbf{Is}=\mathbf{b}\\ &\quad\mathbf{x}\geq\mathbf{0},\qquad\mathbf{s}\geq\mathbf{0}\end{aligned}$$

or

$$\begin{aligned} &\min \mathbf{c}'\mathbf{w} \\ &\text{s.t. } \mathbf{A}'\mathbf{w} = \mathbf{b} \\ &\qquad \mathbf{w} \geq \mathbf{0} \end{aligned}$$

Let $\mathbf{B}'$ denote the optimal basis found for this problem, let $\mathbf{x}^*$ denote the optimal values of the $\mathbf{x}$ variables, and let $\mathbf{u}^* = (\mathbf{c}_{\mathbf{B}'})(\mathbf{B}')^{-1}$. Since $\mathbf{B}'$ is an optimal basis, we have $c_j - \mathbf{u}^*\mathbf{a}_j \geq 0$ for $j = 1,\ldots,n$, or $\mathbf{u}^*\mathbf{A} \leq \mathbf{c}$. Among the columns of $\mathbf{A}'$ are all the vectors $-\mathbf{e}_i$ for $i = 1,\ldots,m$, where $\mathbf{e}_i$ is the ith unit vector in R^m. Again since $\mathbf{B}'$ is optimal, we have $0 - \mathbf{u}^*(-\mathbf{e}_i) \geq 0$, or $u_i^* \geq 0$ for $i = 1,\ldots,m$. Thus $\mathbf{u}^* \in U$. Finally,

$$\mathbf{u}^*\mathbf{b} = (\mathbf{c}_{\mathbf{B}'})(\mathbf{B}')^{-1}\mathbf{b} = \mathbf{c}\mathbf{x}^*$$

and the theorem is proven. ■

The alternate proof of Theorem 2.1 relies on a fundamental result characterizing dual systems of linear inequalities.

LEMMA 2.2. (Farkas). Let $\mathbf{Q}$ be a $k \times l$ matrix and $\mathbf{q} \in R^l$. The following statements are equivalent.

1. For all $\mathbf{x} \in R^l$, $\mathbf{Q}\mathbf{x} \geq \mathbf{0}$ implies $\mathbf{q}\mathbf{x} \geq 0$.
2. There exists $\mathbf{u} \geq \mathbf{0}$, $\mathbf{u} \in R^k$ such that $\mathbf{u}\mathbf{Q} = \mathbf{q}$.

PROOF. Statement 2⇒1: If there exists a $\mathbf{u} \geq \mathbf{0}$ such that $\mathbf{u}\mathbf{Q} = \mathbf{q}$, then $\mathbf{u}\mathbf{Q}\mathbf{x} = \mathbf{q}\mathbf{x}$ for any $\mathbf{x}$. If $\mathbf{Q}\mathbf{x} \geq \mathbf{0}$, then $\mathbf{u}\mathbf{Q}\mathbf{x} \geq \mathbf{0}$ since $\mathbf{u} \geq \mathbf{0}$.

Statement 1⇒2: We prove the contrapositive; namely, if there does not exist a $\mathbf{u} \geq \mathbf{0}$ such that $\mathbf{u}\mathbf{Q} = \mathbf{q}$, then there exists an $\mathbf{x}$ such that $\mathbf{Q}\mathbf{x} \geq \mathbf{0}$, but $\mathbf{q}\mathbf{x} < 0$. To this end, consider the set

$$Z = \{\mathbf{z} | \mathbf{z} = \mathbf{u}\mathbf{Q}, \mathbf{u} \geq \mathbf{0}\}.$$

The set Z is a closed convex set, and we have $\mathbf{q} \notin Z$ by assumption. The fact that Z is closed is difficult to demonstrate, but we leave it to the reader to do so (see Exercise 2.3). Thus, by the separating hyperplane theorem (see Appendix A) there exists an $\mathbf{x} \in R^l$ and a $\beta \in R$ such that

$$\mathbf{q}\mathbf{x} < \beta$$

and

$$\mathbf{z}\mathbf{x} \geq \beta \qquad \text{for all } \mathbf{z} \in Z$$

Since $\mathbf{0} \in Z$, we can conclude $0 \geq \beta$ and $\mathbf{q}\mathbf{x} < \beta \leq 0$. On the other hand,

$$\mathbf{u}\mathbf{Q}\mathbf{x} \geq \beta \qquad \text{for all } \mathbf{u} \geq \mathbf{0}$$

from which we can conclude $\mathbf{Q}\mathbf{x} \geq \mathbf{0}$. For suppose some component of $\mathbf{Q}\mathbf{x}$ were negative; then we could select a sequence $\mathbf{u}^k \geq 0$ for $k = 1,2,\ldots,$ such that $\mathbf{u}^k\mathbf{Q}\mathbf{x}$ converges to $-\infty$ contradicting $\mathbf{u}^k\mathbf{Q}\mathbf{x} \geq \beta$. ■

ALTERNATE PROOF OF THEOREM 2.1. Again, we want to show the existence of an $\mathbf{x}^* \in X$ and a $\mathbf{u}^* \in U$ such that $\mathbf{c}\mathbf{x}^* = \mathbf{u}^*\mathbf{b}$, or equivalently

$\mathbf{u}^*\mathbf{b}-\mathbf{c}\mathbf{x}^* \geq 0$. We introduce surplus variables $\mathbf{s} \in R^m$ satisfying $\mathbf{A}\mathbf{x}-\mathbf{I}_m\mathbf{s}=\mathbf{b}$ for the constraints of (P), and slack variables $\mathbf{t} \in R^n$ satisfying $\mathbf{u}\mathbf{A}+\mathbf{t}\mathbf{I}_n=\mathbf{c}$ for (D). We also introduce the surplus variable $r \in R$ satisfying $\mathbf{u}\mathbf{b}-\mathbf{c}\mathbf{x}-r=0$. We can rewrite the conditions we seek to establish as a solution $\mathbf{x} \geq \mathbf{0}$, $\mathbf{s} \geq \mathbf{0}$, $\mathbf{u} \geq \mathbf{0}$, $\mathbf{t} \geq \mathbf{0}$, $r \geq 0$ to

$$\begin{bmatrix} \mathbf{A} & -\mathbf{I}_m & & & \\ & & \mathbf{A}^T & \mathbf{I}_n & \\ -\mathbf{c} & & \mathbf{b}^T & & -1 \end{bmatrix} \begin{bmatrix} \mathbf{x} \\ \mathbf{s} \\ \mathbf{u}^T \\ \mathbf{t}^T \\ r \end{bmatrix} = \begin{bmatrix} \mathbf{b} \\ \mathbf{c}^T \\ 0 \end{bmatrix}$$

By the Farkas lemma, this is possible only if

$$\xi\mathbf{b}+\gamma\mathbf{c}^T \geq 0 \tag{2.2}$$

for all $\xi \in R^m$, $\gamma \in R^n$, $\theta \in R$ satisfying

$$\begin{aligned} \xi\mathbf{A}-\theta\mathbf{c} &\geq 0 \qquad \xi \leq \mathbf{0} \\ \gamma\mathbf{A}^T+\theta\mathbf{b}^T &\geq 0 \qquad \gamma \geq \mathbf{0} \\ \theta &\leq 0 \end{aligned} \tag{2.3}$$

Case 1. Suppose (ξ,γ,θ) satisfies (2.3) and $\theta<0$. Then $(\xi/\theta) \in U$ and $(\gamma/-\theta) \in X$ so that by Lemma 2.1 $(\xi/\theta)\mathbf{b} \leq (\gamma/-\theta)\mathbf{c}^T$ which is equivalent to (2.2) since $\theta<0$.

Case 2. Suppose (ξ,γ,θ) satisfies (2.3) and $\theta=0$, implying $\xi\mathbf{A} \geq \mathbf{0}$, $\xi \leq \mathbf{0}$, $\gamma\mathbf{A}^T \geq \mathbf{0}$, $\gamma \geq \mathbf{0}$. By the hypothesis of Theorem 2.1, there exists an $\tilde{\mathbf{x}} \in X$, $\tilde{\mathbf{u}} \in U$. Thus, $\mathbf{A}\tilde{\mathbf{x}} \geq \mathbf{b}$, $\tilde{\mathbf{x}} \geq \mathbf{0}$, and since $-\xi \geq \mathbf{0}$ and $(-\xi)\mathbf{A} \leq \mathbf{0}$, we have $0 \geq -\xi\mathbf{A}\mathbf{x} \geq (-\xi)\mathbf{b}$. Similarly, $\mathbf{A}^T\tilde{\mathbf{u}}^T \leq \mathbf{c}^T$, $\tilde{\mathbf{u}}^T \geq \mathbf{0}$, and since $\gamma \geq \mathbf{0}$ and $\gamma\mathbf{A}^T \geq \mathbf{0}$, we have $0 \leq \gamma\mathbf{A}^T\tilde{\mathbf{u}}^T \leq \gamma\mathbf{c}^T$. Thus, we can conclude $\gamma\mathbf{c}^T \geq 0 \geq (-\xi)\mathbf{b}$ and (2.2) holds. ■

There are a few points about linear programming duality as we have studied it thus far that need amplification. First, the primal problem (P) was stated in a way such that the dual problem (D) could be symmetrically stated. Other forms of (P) are permissible, of course, and the results of Lemma 2.1, Corollary 2.1, and Theorem 2.1 hold for an appropriately modified dual problem. A summary of the relationships required is

Primal (Minimization)	Dual (Maximization)
ith relation is an inequality $(\geq)$	$u_i \geq 0$
ith relation is an inequality $(\leq)$	$u_i \leq 0$
ith relation is an equality	u_i unconstrained in sign
$x_j \geq 0$	jth relation is an inequality $(\leq)$
$x_j \leq 0$	jth relation is an inequality $(\geq)$
x_j unconstrained in sign	jth relation is an equality

The validity of these relationships can be established by transforming an arbitrary primal problem into the standard form (P). For example, if the ith relation in (P) is the equality $\Sigma_{j=1}^{n} a_{ij}x_j = b_i$, we replace it by two inequalities $\Sigma_{j=1}^{n} a_{ij}x_j \geq b_i$ and $\Sigma_{j=1}^{n}(-a_{ij})x_j \geq -b_i$. According to the duality theory already developed, there are two variables, say u_i^+ and u_i^-, corresponding to these two inequalities. The variables are constrained to be nonnegative and appear in the dual constraints in the term $a_{ij}(u_i^+ - u_i^-)$ for all j. Thus, we can replace u_i^+ and u_i^- by the single variable $u_i = u_i^+ - u_i^-$, which is unconstrained in sign.

Lemma 2.1 and Theorem 2.1 have not explicitly addressed the cases when X and/or U are empty. For example, if $X \neq \phi$ and $U = \phi$, then Theorem 2.1 tells us that (P) has no optimal solution because otherwise U would be nonempty. By our convention on unbounded and infeasible linear programming problems, we still have equality of primal and dual objective function values, namely, $d = v = -\infty$. The same argument applies when $U \neq \phi$ and $X = \phi$, in which case we must have that (D) has no optimal solution and $v = d = +\infty$. Finally, it is possible to construct a linear programming problem (P) such that $X = \phi$ and its dual (D) is such that $U = \phi$. For example, this is the case for the problem

$$\begin{aligned} \min \quad & x_1 - 2x_2 \\ \text{s.t.} \quad & x_1 - x_2 \geq 2 \\ & -x_1 + x_2 \geq -1 \\ & x_1 \geq 0, \qquad x_2 \geq 0 \end{aligned}$$

The consequences of duality theory to the construction of simplex and nonsimplex algorithms for linear programming is best summarized by the following corollary to Lemma 2.1 and Theorem 2.1.

COROLLARY 2.2. The solutions $\tilde{\mathbf{x}} \in X$ and $\tilde{\mathbf{u}} \in U$ are optimal in (P) and (D), respectively, if and only if

$$\tilde{\mathbf{u}}(\mathbf{A}\tilde{\mathbf{x}} - \mathbf{b}) = 0$$

and (2.4)

$$(\mathbf{c} - \tilde{\mathbf{u}}\mathbf{A})\tilde{\mathbf{x}} = 0$$

PROOF. From the proof of Lemma 2.1, $\tilde{\mathbf{x}} \in X$, $\tilde{\mathbf{u}} \in U$ implies

$$\tilde{\mathbf{u}}\mathbf{b} \leq \tilde{\mathbf{u}}\mathbf{A}\tilde{\mathbf{x}} \leq \mathbf{c}\tilde{\mathbf{x}}$$

If $\tilde{\mathbf{u}}(\mathbf{A}\tilde{\mathbf{x}} - \mathbf{b}) = 0$ and $(\mathbf{c} - \tilde{\mathbf{u}}\mathbf{A})\tilde{\mathbf{x}} = 0$, we have $\tilde{\mathbf{u}}\mathbf{b} = \mathbf{c}\tilde{\mathbf{x}}$ and $\tilde{\mathbf{x}}$ and $\tilde{\mathbf{u}}$ are optimal in (P) and (D) by Corollary 2.1. If one of the conditions (2.4) does not hold, then we must have $\tilde{\mathbf{u}}\mathbf{b} < \mathbf{c}\tilde{\mathbf{x}}$ and at least one of the solutions $\tilde{\mathbf{x}}, \tilde{\mathbf{u}}$ is not optimal in its respective problem. ∎

Conditions (2.4) are called *complementary slackness conditions* and they state that a primal (dual) variable can be positive only if the slack (surplus) variable in the corresponding dual (primal) constraint is zero. Thus, there are three sets of conditions that must be met in order to conclude that a solution $\mathbf{x}$ is optimal in a given (primal) linear programming problem. These are primal feasibility ($\mathbf{x} \in X$), dual feasibility ($\mathbf{u} \in U$) and complementary slackness [$\mathbf{u}(\mathbf{Ax}-\mathbf{b})=0$, $(\mathbf{c}-\mathbf{uA})\mathbf{x}=0$]. The simplex algorithm as it was developed in Chapter 1 takes implicitly at each iteration the m-vector of shadow prices as trial dual variables to complement the primal basic feasible solution. This choice of a primal-dual pair ensures that the algorithm maintains at each iteration primal feasibility and complementary slackness (see Exercise 2.7). The algorithm terminates with optimal solutions to the primal and the dual when dual feasibility of the shadow prices is also attained; dual feasibility is precisely the sufficient condition for optimality given in Lemma 1.1.

Variants of the simplex method are based on maintaining some of these conditions and performing simplex iterations until all of them are satisfied. For example, the dual simplex algorithm maintains dual feasibility and complementary slackness until primal feasibility is achieved. The primal-dual simplex algorithm maintains dual feasibility until primal feasibility and complementary slackness are simultaneously obtained. Finally, in Chapter 6 we give a method for solving (P) and (D) by establishing, at least approximately, the conditions of Corollary 2.2 without resorting to the simplex method in any of its forms.

Another approach to the construction of the dual linear programming problem (D) from the primal problem (P) is to weight the constraints $\mathbf{Ax} \geq \mathbf{b}$ by the vector of dual variables $\mathbf{u} \geq \mathbf{0}$ and place them in the objective function. The result is the *Lagrangean function*

$$L(\mathbf{u}) = \min_{\mathbf{x} \geq \mathbf{0}} \{\mathbf{cx} + \mathbf{u}(\mathbf{b} - \mathbf{Ax})\} = \mathbf{ub} + \min_{\mathbf{x} \geq \mathbf{0}} (\mathbf{c} - \mathbf{uA})\mathbf{x} \tag{2.5}$$

It is apparent that $L(\mathbf{u})=\mathbf{ub}$ if $\mathbf{c}-\mathbf{uA} \geq \mathbf{0}$ because then $\mathbf{x}=\mathbf{0}$ is optimal. Conversely, we have $L(\mathbf{u})=-\infty$ if $\mathbf{c}-\mathbf{uA} \ngeq \mathbf{0}$, say $c_k - \mathbf{ua}_k < 0$, because then we can drive $L(\mathbf{u})$ to $-\infty$ by letting x_k go to $+\infty$ while holding $x_j=0$ for $j \neq k$. By the same argument used in the weak duality Lemma 2.1, we can conclude in the former case that $L(\mathbf{u})$ is a lower bound on the minimal primal objective function cost, and it is trivially a lower bound in the latter case. The dual problem (D) can be derived by considering the problem of finding the best lower bound

$$\begin{aligned} &\max L(\mathbf{u}) \\ &\text{s.t. } \mathbf{u} \geq \mathbf{0} \end{aligned} \tag{2.6}$$

According to the analysis above, the only nontrivial lower bounds in (2.6)

$[L(\mathbf{u}) > -\infty]$ occur when $\mathbf{c} - \mathbf{uA} \geq 0$. Thus, we can write (2.6) as

$$\begin{aligned} &\max \mathbf{ub} \\ &\text{s.t. } \mathbf{uA} \leq \mathbf{c} \\ &\quad \mathbf{u} \geq \mathbf{0} \end{aligned}$$

which is precisely the dual problem (D).

This dual analysis in terms of the Lagrangean function produced an explicit linear programming dual problem from the given problem (P). In Chapter 5 and later chapters, we study dual problems of the form (2.6) derived from more general mathematical programming problems for which explicit representations are more difficult to obtain. In other words, we consider the maximization over the nonnegative orthant of Lagrangean functions L that are evaluated at each point by solving minimization problems of which (2.5) is a simple example. Implicit optimization in this way might appear difficult to do, but it turns out that dual problems like (2.6) are well behaved and convergent algorithms can be constructed for their solution.

2.3 INVESTMENT EXAMPLE OF DUALITY

An investor is faced with consumption versus investment alternatives over a finite planning horizon consisting of T periods. In period t he can consume a nonnegative quantity x_t of his total capital with present value $\beta_t x_t$ where $0 < \beta_t < 1$; or he can invest a nonnegative quantity y_t of his total capital and thereby add $a_t y_t$ ($a_t > 0$) to his available capital in periods $t+1, \ldots, T$. From past investments, he has exogenous cash flows $M_t > 0$ for $t = 1, \ldots, T$. The investor's objective is to maximize the total present value of his consumption over the T periods.

We can write this decision problem as the linear programming problem

$$\begin{aligned} &\max \sum_{t=1}^{T} \beta_t x_t \\ &\text{s.t. } x_t + y_t - \sum_{s=1}^{t-1} a_s y_s \leq M_t \qquad t = 1, \ldots, T \qquad (2.7) \\ &\quad x_t \geq 0 \qquad y_t \geq 0 \qquad t = 1, \ldots, T \end{aligned}$$

The constraints state that the sum of consumption x_t and investment y_t in period t cannot exceed the capital available which totals $M_t + \Sigma_{s=1}^{t-1} a_s y_s$.

The dual to problem (2.7) is

$$\min \sum_{t=1}^{T} u_t M_t$$

$$\text{s.t. } \left.\begin{aligned} u_t &\geq \beta_t \\ u_t &\geq a_t \sum_{s=t+1}^{T} u_s \end{aligned}\right\} \quad t=1,\ldots,T \tag{2.8}$$

$$u_t \geq 0 \qquad t=1,\ldots,T$$

The dual problem can be interpreted as the decision problem faced by the market that wishes to minimize the cost of payments to the investor who has exogenous flows M_t, the given investment alternatives, and the given discount factor β_t. Intuitively, problem (2.8) says the market can minimize its cost by selecting the u_t to be as small as possible subject to the constraints imposed by the investor's activities. This goal can be accomplished by choosing the u_t to satisfy

$$u_t^* = \max\left\{\beta_t, a_t \sum_{s=t+1}^{T} u_s^*\right\} \qquad t=T, T-1,\ldots,1 \tag{2.9}$$

where the backward indexing indicates that the u_t^* are to be found recursively starting with u_T^*. Each u_t^* has the interpretation as an effective discount factor that takes into account not only the manifest discount factor β_t but also the investment alternatives in period t and later periods, and the discount factors in later periods when the returns from current and future investments are realized. For example, we clearly have $u_T^* = \beta_T$, and suppose $u_{T-1}^* = \beta_{T-1}$; according to (2.9), u_{T-2}^* is the maximum of β_{T-2} and $a_{T-2}(\beta_{T-1}+\beta_T)$. Thus, if the second term is sufficiently great, $u_{T-2}^* = a_{T-2}(\beta_{T-1}+\beta_T)$ and consumption in period $T-2$ should be deferred in favor of investment.

We can demonstrate that the u_t^* given by (2.9) are optimal by selecting a corresponding primal feasible solution x_t^*, y_t^* satisfying the complementary slackness conditions for $t=1,\ldots,T$,

$$u_t^*\left(M_t + \sum_{s=1}^{t-1} a_s y_s^* - x_t^* - y_t^*\right) = 0 \tag{2.10a}$$

$$(u_t^* - \beta_t)x_t^* = 0 \tag{2.10b}$$

$$\left(u_t^* - a_t \sum_{s=t+1}^{T} u_s^*\right) y_t^* = 0 \tag{2.10c}$$

Since the u_t^* from (2.9) are positive, the condition (2.10a) implies the obvious requirement that $x_t^* + y_t^* = M_t + \Sigma_{s=1}^{t-1} a_s y_s^*$; that is, that the investor uses all of his money. Conditions (2.10b) and (2.10c) state that the investor chooses x_t^* or y_t^* to be zero depending on whether $u_t^* = a_t \Sigma_{s=t+1}^{T} u_s$ or β_t, respectively, with indifference when $a_t \Sigma_{s=t+1}^{T} u_s^* = \beta_t$. The specific values of x_t^* and y_t^* can be calculated using this rule by forward recursion.

Note that the calculation of optimal dual variables and the selection of one of the variables x_t and y_t to be positive in each period does not depend on the exogenous cash flows M_t. Problem (2.7) is an example of a general class of linear programming problems called Leontief substitution models with the property that optimal solutions to the dual do not depend on the right-hand vector in the primal (see Exercise 2.21). It should also be noted that the linear objective function of problem (2.7) is the reason for the optimal consumption/investment strategy with $x_t^* y_t^* = 0$. If the objective function were nonlinear, say $\Sigma_{t=1}^{T} \beta_t U(x_t)$ with U concave, then this all-or-nothing strategy would not obtain.

2.4 ECONOMIC INTERPRETATION OF DUAL VARIABLES

The economic interpretation of dual variables is closely related to parametric analysis of the primal problem (P) as the right-hand-side vector $\mathbf{b}$ varies over R^m. The constructive methods we will develop and use in this section to study right-hand-side parametrics will be used again in the development of decomposition methods for large-scale linear programming, and in a more general form, in the development of a duality theory for arbitrary mathematical programming problems in Chapter 5.

Let $v(\mathbf{b})$ denote the minimal cost of the primal problem (P) as a function of its right-hand-side vector $\mathbf{b} \in R^m$; that is,

$$\begin{aligned} v(\mathbf{b}) = \min\ & \mathbf{cx} \\ \text{s.t. } & \mathbf{Ax} \geq \mathbf{b} \\ & \mathbf{x} \geq \mathbf{0} \end{aligned}$$

LEMMA 2.3. The function $v(\mathbf{b})$ is convex.

PROOF. Let $\mathbf{b}^1, \mathbf{b}^2$ be any points in R^m. If $v(\mathbf{b}^1)$ or $v(\mathbf{b}^2) = +\infty$, then it follows trivially that $v[\alpha \mathbf{b}^1 + (1-\alpha)\mathbf{b}^2] \leq \alpha v(\mathbf{b}^1) + (1-\alpha) v(\mathbf{b}^2)$ for $0 \leq \alpha \leq 1$. Thus, we assume $v(\mathbf{b}^1) < +\infty$ and $v(\mathbf{b}^2) < +\infty$. If either $v(\mathbf{b}^1)$ or $v(\mathbf{b}^2) = -\infty$, then the dual problem does not have a feasible solution which implies $v(\mathbf{b}) = -\infty$ for all $\mathbf{b} \in R^m$ and again there is nothing to prove.

Consider the case when $v(\mathbf{b}^1)$ and $v(\mathbf{b}^2)$ are both finite, say, $v(\mathbf{b}^1) = \mathbf{cx}^1$ and $v(\mathbf{b}^2) = \mathbf{cx}^2$ for $\mathbf{x}^1, \mathbf{x}^2$ satisfying $\mathbf{Ax} \geq \mathbf{b}$, $\mathbf{x} \geq \mathbf{0}$. For $0 \leq \alpha \leq 1$, $\mathbf{A}[\alpha \mathbf{x}^1 + (1-$

$\alpha)\mathbf{x}^2] \ge \alpha\mathbf{b}^1 + (1-\alpha)\mathbf{b}^2$, $\alpha\mathbf{x}^1 + (1-\alpha)\mathbf{x}^2 \ge \mathbf{0}$ implying $v(\alpha\mathbf{b}^1 + (1-\alpha)\mathbf{b}^2) \le \alpha\mathbf{c}\mathbf{x}^1 + (1-\alpha)\mathbf{c}\mathbf{x}^2 = \alpha v(\mathbf{b}^1) + (1-\alpha)v(\mathbf{b}^2)$. ■

The convex function $v(\mathbf{b})$ is nondifferentiable at points where there are alternative optima in the dual; for example, the point $\mathbf{b}^2$ in Figure 2.1. However, directional derivatives exist everywhere and, as we show in this section, are derived from the set of optimal dual solutions. Our analysis of these directional derivatives is a specific derivation of a general property of convex, nondifferentiable functions discussed in Appendix A.

The analysis is interesting only if the dual feasible region

$$U = \{\mathbf{u} | \mathbf{u}\mathbf{A} \le \mathbf{c}, \mathbf{u} \ge \mathbf{0}\}$$

is assumed to be nonempty. By Theorem 2.1, we know

$$v(\mathbf{b}) = \max_{\mathbf{u} \in U} \mathbf{u}\mathbf{b} \tag{2.11}$$

where we admit the added possibility that $v(\mathbf{b}) = +\infty$ [we must have $v(\mathbf{b}) > -\infty$ because U is assumed nonempty]. The convex polytope U can be written as

$$\begin{aligned} U = \Big\{ \mathbf{u} \in R^m | \mathbf{u} = \sum_{t=1}^{T} \lambda_t \mathbf{u}^t + \sum_{s=1}^{S} \omega_s \mathbf{u}^s \\ \sum_{t=1}^{T} \lambda_t = 1 \\ \lambda_t \ge 0 \qquad t = 1, \ldots, T \\ \omega_s \ge 0 \qquad s = 1, \ldots, S \Big\} \end{aligned} \tag{2.12}$$

where the vectors $\mathbf{u}^t$ are the extreme points of U and the vectors $\mathbf{u}^s$ are its extreme rays. The extreme points satisfy $\mathbf{u}^t\mathbf{A} \le \mathbf{c}$, $\mathbf{u}^t \ge \mathbf{0}$ whereas the extreme rays satisfy $\mathbf{u}^s\mathbf{A} \le \mathbf{0}$, $\mathbf{u}^s \ge \mathbf{0}$. This characterization of convex polytopes is discussed in more detail in Appendix A.

Combining (2.11) and (2.12), we have

$$\begin{aligned} v(\mathbf{b}) = \max \sum_{t=1}^{T} \lambda_t \mathbf{u}^t \mathbf{b} + \sum_{s=1}^{S} \omega_s \mathbf{u}^s \mathbf{b} \\ \text{s.t.} \sum_{t=1}^{T} \lambda_t = 1 \\ \lambda_t \ge 0 \qquad t = 1, \ldots, T \\ \omega_s \ge 0 \qquad s = 1, \ldots, S \end{aligned} \tag{2.13}$$

It is clear from problem (2.13) that $v(\mathbf{b})$ is finite if and only if $\mathbf{u}^s\mathbf{b} \le 0$ for $s = 1, \ldots, S$. For if $\mathbf{u}^s\mathbf{b} \le 0$, then it is optimal to take $\omega_s = 0$ for $s = 1, \ldots, S$, and $v(\mathbf{b})=\mathbf{u}^t\mathbf{b}$ for any extreme point $\mathbf{u}^t$ such that $\mathbf{u}^t\mathbf{b}$ is maximal. On the other hand, if $\mathbf{u}^s\mathbf{b}>0$ for some s, then $v(\mathbf{b})\geq(\mathbf{u}^t+\omega_s\mathbf{u}^s)\mathbf{b}$ for any $\omega_s\geq 0$ and any extreme point $\mathbf{u}^t$ implying $v(\mathbf{b})=+\infty$.

Consider a point $\mathbf{b}\in R^m$ such that $v(\mathbf{b})$ is finite. A nonzero direction $\mathbf{d}\in R^m$ is called a *feasible direction* if $v(\mathbf{b}+\theta\mathbf{d})$ is finite for all $0\leq\theta\leq\theta^*$ where $\theta^*>0$. Since $v(\mathbf{b})$ is convex, a feasible direction will always exist at a point if there is more than one $\mathbf{b}\in R^m$ such that $v(\mathbf{b})$ is finite.

THEOREM 2.2. Suppose $v(\mathbf{b})$ is finite and $\mathbf{d}\neq 0$ is a feasible direction of v at $\mathbf{b}$. Then the directional derivative of v at $\mathbf{b}$ in the direction $\mathbf{d}$ exists and is given by

$$\nabla v(\mathbf{b};\mathbf{d})=\max\{\mathbf{ud}|\mathbf{u}\text{ optimal in (D)}\} \tag{2.14}$$

PROOF. It can be shown for problem (2.13) with $\mathbf{b}$ replaced by $\mathbf{b}+\theta\mathbf{d}$ that $v(\mathbf{b}+\theta\mathbf{d})$ is finite for θ sufficiently small, and therefore that $\mathbf{d}$ is a feasible direction, if and only if $\mathbf{u}^s\mathbf{d}\leq 0$ for all extreme rays such that $\mathbf{u}^s\mathbf{b}=0$ (see Exercise 2.17). This implies that the maximum in (2.14) is attained. The optimal points in (D) are convex combinations of optimal extreme points plus nonnegative combinations of rays satisfying $\mathbf{u}^s\mathbf{b}=0$.

The definition of $\nabla v(\mathbf{b};\mathbf{d})$, if it exists, is

$$\nabla v(\mathbf{b};\mathbf{d})=\lim_{\theta\to 0^+}\frac{v(\mathbf{b}+\theta\mathbf{d})-v(\mathbf{b})}{\theta}$$

For any $\hat{\mathbf{u}}$ optimal in (D) and any $\theta>0$,

$$v(\mathbf{b}+\theta\mathbf{d})\geq\hat{\mathbf{u}}(\mathbf{b}+\theta\mathbf{d}),$$

which implies

$$\frac{v(\mathbf{b}+\theta\mathbf{d})-v(\mathbf{b})}{\theta}\geq\hat{\mathbf{u}}\mathbf{d}$$

Thus

$$\liminf_{\theta\to 0^+}\frac{v(\mathbf{b}+\theta\mathbf{d})-v(\mathbf{b})}{\theta}\geq\max\{\mathbf{ud}|\mathbf{u}\text{ optimal in (D)}\} \tag{2.15}$$

since we could take $\hat{\mathbf{u}}$ to be optimal in the right-hand expression.

To complete the proof, let $\{\theta_k\}_{k=1}^{\infty}$ be a sequence of positive scalars converging to zero such that

$$\lim_{k=1,2,\ldots}\frac{v(\mathbf{b}+\theta_k\mathbf{d})-v(\mathbf{b})}{\theta_k}=\limsup_{\theta\to 0^+}\frac{v(b+\theta\mathbf{d})-v(\mathbf{b})}{\theta}$$

Since $\mathbf{d}$ is a feasible direction, $v(\mathbf{b}+\theta_k\mathbf{d})$ is finite for θ_k sufficiently small. It also equals $\mathbf{u}^k(\mathbf{b}+\theta_k\mathbf{d})$ for some extreme point, say u^*, repeated infinitely often because there are only a finite number of extreme points. Thus, there is a subsequence $\{\theta_k\}_{k\in K}$ such that $v(\mathbf{b}+\theta_k\mathbf{d})=\mathbf{u}^*(\mathbf{b}+\theta_k\mathbf{d})$.

Now it must be that $v(\mathbf{b})=\mathbf{u}^*\mathbf{b}$. For suppose the contrary, that is, $v(\mathbf{b})>\mathbf{u}^*\mathbf{b}$, say $v(\mathbf{b})=\mathbf{u}^{**}\mathbf{b}>\mathbf{u}^*\mathbf{b}$. For all $k\in K$, we have

$$v(\mathbf{b}+\theta_k\mathbf{d})=\mathbf{u}^*(\mathbf{b}+\theta_k\mathbf{d})\geq \mathbf{u}^{**}(\mathbf{b}+\theta_k\mathbf{d})$$

because $\mathbf{u}^{**}$ may not be optimal at $\mathbf{b}+\theta_k\mathbf{d}$. Rearranging terms we have

$$\theta_k(\mathbf{u}^*-\mathbf{u}^{**})\mathbf{d}\geq(\mathbf{u}^*-\mathbf{u}^{**})\mathbf{b}>0$$

But this is impossible since θ_k goes to zero and we can conclude $v(\mathbf{b})=\mathbf{u}^*\mathbf{b}$.

Thus, for $k\in K$,

$$\frac{v(\mathbf{b}+\theta_k\mathbf{d})-v(\mathbf{b})}{\theta_k}=\mathbf{u}^*\mathbf{d}$$

and

$$\lim_{k\in K}\frac{v(\mathbf{b}+\theta_k\mathbf{d})-v(\mathbf{b})}{\theta_k}=\mathbf{u}^*\mathbf{d}$$

This implies

$$\lim_{\theta\to 0^+}\sup\frac{v(\mathbf{b}+\theta\mathbf{d})-v(\mathbf{b})}{\theta}=\mathbf{u}^*\mathbf{d}$$

$$\leq\max\{\mathbf{ud}|\mathbf{u}\text{ optimal in (D)}\} \tag{2.16}$$

Comparing (2.15) and (2.16), we can conclude $\nabla v(\mathbf{b};\mathbf{d})$ exists and satisfies (2.14). ■

An implication of Theorem 2.2 is that when the set of optimal solutions to (D) is the unique m-vector $\mathbf{u}^*$, then $\nabla v(\mathbf{b};\mathbf{d})=\mathbf{u}^*\mathbf{d}$ for all feasible directions $\mathbf{d}$. If, in addition, the set of feasible directions is all of R^m, except $\mathbf{0}$, then v is differentiable and $\partial v(\mathbf{b})/\partial b_i=u_i^*$. In words, the u_i^* in this case measure the rate of increase at $\mathbf{b}$ of the minimal cost in (P) per unit increase in the ith component of $\mathbf{b}$. This is the reason that optimal dual variables are called *shadow prices*.

Unfortunately, the optimal dual solution is often not unique and the optimal dual solution found by the simplex method may give misleading information about the rates of change of v in various directions. Moreover, decomposition methods studied in Chapter 6 that use the pricing mechanism of duality theory tend to go to points where the optimal dual solution is not unique. When this happens, the characterization of the directional derivatives given in Theorem 2.2 must be used.

Figure 2.1 depicts the numerical example of Section 1.2 in the space of the right-hand-side vector $\mathbf{b}$. The right-hand sides have been scaled by a factor of 10^{-6} for convenience; this has the effect of scaling the optimal solutions by the same factor and does not invalidate the geometry of the figure. The vector $\mathbf{b}^1$ corresponds to the right-hand side in the example and it lies in the cone C^1 defined by all nonnegative combinations of the

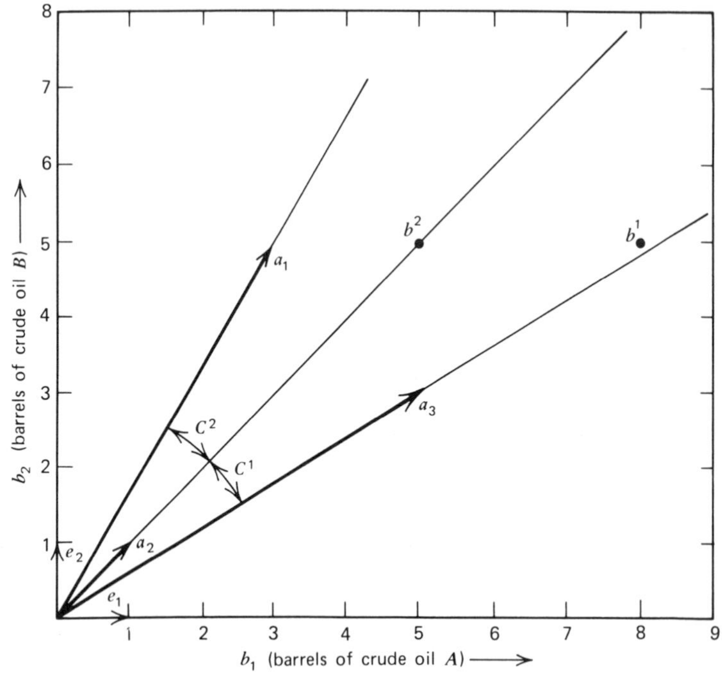

Figure 2.1. Resource space of linear programming example.

optimal basis activities $\mathbf{a}_2$ and $\mathbf{a}_3$. For all $\mathbf{b} \in C^1$, $\mathbf{a}_2$ and $\mathbf{a}_3$ constitute an optimal basis implying $v(\mathbf{b}) = 3/2b_1 + 11/2b_2$ where $\mathbf{u}^1 = (3/2, 11/2)$ is the optimal dual solution given by $\mathbf{c_B B}^{-1}$ for this basis.

Nonuniqueness in the optimal dual solution occurs, for example, at the point $b^2 = \binom{5}{5}$ where the basis consisting of the activities $\mathbf{a}_1$ and $\mathbf{a}_2$, as well as the basis of $\mathbf{a}_2$ and $\mathbf{a}_3$, is optimal. Thus, $\mathbf{u}^2 = (5, 2)$ corresponding to the cone C^2 is an alternative optimal dual solution at $\mathbf{b}^2$, and so is any convex combination of $\mathbf{u}^1$ and $\mathbf{u}^2$. Note that the rate of change at that point in the component b_2 in the up direction into C^2 is 2 per unit, and the rate of change in the down direction into C^1 is $-11/2$ per unit.

2.5 SENSITIVITY ANALYSIS

Linear programming and the simplex method have been used successfully in large part because it is possible to test the sensitivity of an optimal solution to parametric variation in the right-hand-side and cost vectors, and to a lesser extent, the coefficient matrix **A**. It will be convenient in this section and the remainder of the chapter to consider the given (primal) linear programming problem in the equality form of Chapter 1; namely,

$$\begin{aligned} &\min \mathbf{cx} \\ &\text{s.t. } \mathbf{Ax}=\mathbf{b} \\ &\qquad \mathbf{x}\geq \mathbf{0} \end{aligned} \tag{2.17}$$

The dual problem is

$$\begin{aligned} &\max \mathbf{ub} \\ &\text{s.t. } \mathbf{uA}\leq \mathbf{c} \end{aligned} \tag{2.18}$$

Corollary 2.2 becomes the following: The primal and dual feasible solutions $\tilde{\mathbf{x}}$ and $\tilde{\mathbf{u}}$ are optimal if and only if $(\mathbf{c}-\tilde{\mathbf{u}}\mathbf{A})\tilde{\mathbf{x}}=0$.

Sensitivity analysis is performed after (2.17) has been solved by the simplex method. Following the notation of Section 1.2, the simplex method transforms (2.17) to

$$\begin{aligned} &\min z=\mathbf{c_B B^{-1} b}+(\mathbf{c_N}-\mathbf{c_B B^{-1} N})\mathbf{x_N} \\ &\text{s.t. } \mathbf{x_B}+\mathbf{B^{-1}N x_N}=\mathbf{B^{-1}b} \\ &\qquad \mathbf{x_B}\geq \mathbf{0} \qquad \mathbf{x_N}\geq \mathbf{0} \end{aligned} \tag{2.19}$$

where $\mathbf{A}=(\mathbf{B},\mathbf{N})$, $\mathbf{x}=(\mathbf{x_B},\mathbf{x_N})$ and $\mathbf{c}=(\mathbf{c_B},\mathbf{c_N})$. By Lemma 1.1, the solution $(\tilde{\mathbf{x}}_\mathbf{B},\tilde{\mathbf{x}}_\mathbf{N})=(\mathbf{B}^{-1}\mathbf{b},\mathbf{0})$ and the basis $\mathbf{B}$ are optimal if $\bar{\mathbf{b}}=\mathbf{B}^{-1}\mathbf{b}\geq\mathbf{0}$ and $\bar{\mathbf{c}}_\mathbf{N}=\mathbf{c_N}-\mathbf{c_B B^{-1}N}\geq\mathbf{0}$. In duality terms, the primal feasible solution $(\tilde{\mathbf{x}}_\mathbf{B},\tilde{\mathbf{x}}_\mathbf{N})$ is optimal because (1) the solution $\mathbf{u_B}=\mathbf{c_B B}^{-1}$ is dual feasible since $c_j-\mathbf{u_B a}_j\geq 0$ for $j=1,\ldots,n$ and (2) the primal and dual feasible solutions obey complementary slackness since $\tilde{x}_j>0$ implies the variable is basic and therefore $c_j-\mathbf{u_B a}_j=c_j-c_j=0$. Sensitivity analysis is concerned with bounds on coefficient changes such that primal and dual feasibility and complementary slackness are maintained, and also on systematic methods for reestablishing these conditions when they are violated by larger coefficient changes.

For example, suppose the right-hand-side vector $\mathbf{b}$ is changed to $\mathbf{b}+\Delta\mathbf{e}_k$. The basic solution $(\tilde{\mathbf{x}}_\mathbf{B},\tilde{\mathbf{x}}_\mathbf{N})=[\mathbf{B}^{-1}(\mathbf{b}+\Delta\mathbf{e}_k),\mathbf{0}]$ remains optimal as long as $\mathbf{B}^{-1}(\mathbf{b}+\Delta\mathbf{e}_k)\geq\mathbf{0}$ because dual feasibility and complementary slackness are unaffected. Upper and lower bounds on Δ such that primal feasibility is maintained are

$$\max_{\beta_{ik}>0}\frac{(-\bar{b}_i)}{\beta_{ik}}\leq\Delta\leq\min_{\beta_{ik}<0}\frac{\bar{b}_i}{(-\beta_{ik})} \tag{2.20}$$

where the m-vector $\boldsymbol{\beta}_k$ is the kth column of $\mathbf{B}^{-1}$ and $\bar{\mathbf{b}}=\mathbf{B}^{-1}\mathbf{b}$. If Δ is set to some value outside of this range, say $\Delta=\tilde{\Delta}$, then the dual simplex method discussed in the next section can be used directly on the basic system (2.19) with $\mathbf{B}^{-1}\mathbf{b}$ replaced by $\mathbf{B}^{-1}(\mathbf{b}+\tilde{\Delta}\mathbf{e}_k)$.

Next, suppose the cost coefficient c_k of a nonbasic variable is changed to $c_k+\Delta$. The current solution clearly remains optimal as long as $\Delta \geq -\bar{c}_k$. If the cost coefficient of a basis variable is changed to $c_k+\Delta$, then the dual vector $\mathbf{u_B}=(\mathbf{c_B}+\Delta\mathbf{e}_k)\mathbf{B}^{-1}$, and the current solution remains optimal as long as $\mathbf{c_N}-\mathbf{c_B}\mathbf{B}^{-1}\mathbf{N}-\Delta\mathbf{e}_k\mathbf{B}^{-1}\mathbf{N}\geq\mathbf{0}$ or

$$\max_{\rho_k a_j<0} \frac{\bar{c}_j}{\rho_k a_j} \leq \Delta \leq \min_{\rho_k a_j>0} \frac{\bar{c}_j}{\rho_k a_j} \qquad (2.21)$$

where ρ_k is the kth row of $\mathbf{B}^{-1}$. If Δ is set to a value outside of these ranges, the problem can be reoptimized using the primal simplex method.

If a coefficient a_{ik} of a nonbasic activity $\mathbf{a}_k$ changes to $a_{ik}+\Delta$, the current solution remains optimal as long as $c_k-\mathbf{u_B}(\mathbf{a}_k+\Delta\mathbf{e}_i)=\bar{c}_k-u_i\Delta\geq 0$, or as long as $\Delta\leq\bar{c}_k/u_i$, unless $u_i=0$, in which case the value of Δ does not influence the optimal solution. If a coefficient a_{ik} of a basic activity $\mathbf{a}_k$ changes, the analysis is much more complicated and we omit details.

Finally, suppose a constraint

$$\sum_{j=1}^{n} a_{m+1,j}x_j \geq b_{m+1}$$

is added to the primal problem (2.17) after it has been solved. The current solution clearly remains optimal as long as it satisfies the new constraint. If it does not then we add a new row

$$s_{m+1}+\sum_{j=1}^{n}(-a_{m+1,j})x_j=-b_{m+1}$$

to the basic system and reoptimize using the dual simplex algorithm described in Section 2.7.

2.6 MULTIOBJECTIVE LINEAR PROGRAMMING

A generalization of the notion of sensitivity analysis of objective function coefficients is to admit the possibility that there are multiple objective functions to be minimized. This can arise when a single decisionmaker has multiple decision criteria or when there are several parties interested in the outcome of a linear programming problem, each party with his own objective function but the same feasible region. Viewed another way, multiobjective linear programming might arise from a desire to calculate a variety of good, but not necessarily optimal, solutions to a given linear programming problem with a single objective function rather than a single optimal solution. For example, if a linear programming model is concerned with expansion of electric power operating capacity to meet given demand at minimum cost, we might also be interested in examining a number of low-cost feasible solutions with respect to reliability of the generating system at peak loads. Some linear measure of reliability could be used for this purpose as a second objective function using the methods about to be described.

Suppose we have K objective functions $\mathbf{c}^k$ for $k=1,\ldots,K$, for the linear programming problems

$$\begin{aligned} &\min \mathbf{c}^k\mathbf{x} \\ &\text{s.t. } \mathbf{Ax}=\mathbf{b} \\ &\mathbf{x}\geq \mathbf{0} \end{aligned} \tag{2.22}$$

where $\mathbf{A}$ is $m\times n$. A feasible solution $\tilde{\mathbf{x}}$ is called *minimal* if $\mathbf{c}^k\tilde{\mathbf{x}}\leq \mathbf{c}^k\mathbf{x}$ for all feasible $\mathbf{x}$ and all k; in other words, $\tilde{\mathbf{x}}$ is minimal if it is simultaneously optimal in the K linear programming problems (2.22). A minimal solution for the K problems will usually not exist and we must compromise in our analysis of these problems. A feasible solution $\tilde{\mathbf{x}}$ is called *efficient* if there does not exist a feasible solution $\mathbf{y}$ such that $\mathbf{c}^k\mathbf{y}\leq \mathbf{c}^k\tilde{\mathbf{x}}$ for $k=1,\ldots,K$, with strict inequality for some k. In other words, $\tilde{\mathbf{x}}$ is efficient if it is not possible to lower some objective function cost without increasing some other. Efficient solutions are also known as *Pareto optima*.

The set of all efficient solutions can be generated, at least in theory, by systematic procedures for combining the K objectives into one. Let $\mathbf{C}$ denote the $K\times n$ matrix with row $\mathbf{c}^k$ and for any K-vector $\boldsymbol{\lambda}\geq \mathbf{0}$, define

$$\begin{aligned} v(\boldsymbol{\lambda}) = &\min \boldsymbol{\lambda}\mathbf{Cx} \\ &\text{s.t. } \mathbf{Ax}=\mathbf{b} \\ &\mathbf{x}\geq \mathbf{0} \end{aligned} \tag{2.23}$$

THEOREM 2.3. If $\tilde{\mathbf{x}}$ is optimal in (2.23) for $\boldsymbol{\lambda}$ with all positive components, then $\tilde{\mathbf{x}}$ is efficient in (2.22). Conversely, any efficient $\tilde{\mathbf{x}}$ in (2.22) is optimal in (2.23) for some $\boldsymbol{\lambda}\geq \mathbf{0}$.

PROOF. To prove that $\tilde{\mathbf{x}}$ is optimal in (2.23) for $\boldsymbol{\lambda}$ with all positive components, we suppose the contrary; namely, there exists a feasible $\mathbf{y}$ such that $\mathbf{c}^k\mathbf{y}\leq \mathbf{c}^k\tilde{\mathbf{x}}$ and for some l, $\mathbf{c}^l\mathbf{y}<\mathbf{c}^l\tilde{\mathbf{x}}$. Then it must be that $\boldsymbol{\lambda}\mathbf{Cy}<\boldsymbol{\lambda}\mathbf{C}\tilde{\mathbf{x}}$ since $\lambda_l>0$ contradicting the optimality of $\tilde{\mathbf{x}}$.

Suppose $\tilde{\mathbf{x}}$ is efficient in (2.22) and define the set in R^K

$$W=\{\mathbf{w}=\mathbf{Cx}-\mathbf{C}\tilde{\mathbf{x}} \mid \mathbf{Ax}=\mathbf{b}, \mathbf{x}\geq \mathbf{0}\}$$

The set W is convex and intersects the negative orthant in R^K only in the point $\mathbf{0}$ since $\tilde{x}$ is efficient. Thus, by the separating hyperplane theorem, we can find in R^K a hyperplane $\boldsymbol{\lambda}\mathbf{w}=\lambda_0$ with $\boldsymbol{\lambda}\neq \mathbf{0}$ which separates W from the interior of the negative orthant; that is,

$$\begin{aligned} \boldsymbol{\lambda}\mathbf{w}\leq \lambda_0 \quad &\text{for all } \mathbf{w} \text{ in the interior of} \\ &\text{the negative orthant of } R^K \\ \boldsymbol{\lambda}\mathbf{w}\geq \lambda_0 \quad &\text{for all } w\in W \end{aligned}$$

The first inequality implies $\boldsymbol{\lambda} \geq \mathbf{0}$ because otherwise, say $\lambda_l < 0$, we could let w_l go to $-\infty$ and violate the inequality. Moreover, $\lambda_0 \geq 0$ since we can take $\mathbf{w}$ in the interior of the negative orthant to be arbitrarily close to the zero vector implying $\boldsymbol{\lambda}\mathbf{w} \leq \mathbf{0}$ is arbitrarily close to zero. But the second inequality implies $\lambda_0 \leq \mathbf{0}$ since $\mathbf{0} \in W$. In conclusion, we have found in R^K a vector $\boldsymbol{\lambda}$ satisfying $\boldsymbol{\lambda} \geq \mathbf{0}$ and $\boldsymbol{\lambda}\mathbf{w} \geq 0$ for all $\mathbf{w} \in W$, or $\boldsymbol{\lambda}\mathbf{C}\mathbf{x} \geq \boldsymbol{\lambda}\mathbf{C}\tilde{\mathbf{x}}$ for all feasible $\mathbf{x}$ implying $\tilde{\mathbf{x}}$ is optimal in (2.23) with this $\boldsymbol{\lambda}$. ■

The $\boldsymbol{\lambda} \geq \mathbf{0}$ in problem (2.23) can be limited to satisfy $\Sigma_{k=1}^{K} \lambda_k = 1$ since any $\boldsymbol{\lambda} \neq \mathbf{0}$ can be scaled to satisfy this constraint. Systematic procedures for searching over this bounded set to generate efficient solutions can be devised using the sensitivity analysis results of the previous section. The number of efficient solutions can be prohibitively large, however, when the number of objective functions is three or greater.

2.7 DUAL SIMPLEX ALGORITHM AND THE CUTTING PLANE METHOD FOR INTEGER PROGRAMMING

As was mentioned in Section 2.5, situations can arise in linear programming where there is a basic linear system and corresponding basic solution satisfying dual feasibility and complementary slackness, but for which primal feasibility does not hold. For such problems, the dual simplex algorithm can be used, starting with the given basic system, and without need for a phase one procedure. Exercise 2.15 at the end of the chapter shows how a dual feasible basic system can be derived from any basic system if an upper bound on $\Sigma_{j=1}^{n} x_j$ is known. Thus, the dual simplex can be used as an alternative to the phase one and phase two primal simplex algorithm. At the end of this section, we show how the dual simplex algorithm can be used to solve integer programming problems.

At each iteration, the dual simplex algorithm makes a pivot to a new dual feasible basic system with a strictly higher cost (barring degeneracy). Convergence to an optimal solution, if one exists, is thereby ensured by the same arguments as those used in Chapter 1 for the primal simplex method.

For completeness, we write out the dual feasible basic system

$$\min z = \bar{z}_0 + \sum_{j=m+1}^{n} \bar{c}_j x_j$$

$$\text{s.t. } x_i = \bar{b}_i - \sum_{j=m+1}^{n} \bar{a}_{ij} x_j \qquad i = 1, \ldots, m \tag{2.24}$$

where $\bar{c}_j \geq 0, j = m+1, \ldots, n$, but $\bar{b}_i < 0$ for at least one i. The following steps describe the dual simplex algorithm; we prove that the algorithm maintains dual feasibility in the theorem which follows. Convergence of the dual simplex algorithm is established by the same argument used in Theorems 1.1 and 1.2 to prove convergence of the primal simplex algorithm.

Dual Simplex Algorithm

Step 1. Determine if there is a basic variable in a dual feasible basic system with value $\bar{b}_i < 0$. If all $\bar{b}_i \geq 0$, stop because the current basic solution is optimal. Otherwise, select a basic variable x_r such that $\bar{b}_r < 0$; for example,

$$\bar{b}_r = \min \bar{b}_i < 0$$

The variable x_r is to be taken out of the basis.

Step 2. If $\bar{a}_{rj} \geq 0$ for $j = m+1, \ldots, n$, then there is no primal feasible solution and the method terminates. Otherwise, select the nonbasic variable x_s to enter the basis and replace the variable x_r by the rule

$$\frac{\bar{c}_r}{(-\bar{a}_{rs})} = \min_{\bar{a}_{rj} < 0} \frac{\bar{c}_j}{(-\bar{a}_{rj})}$$

Step 3. Pivot on the term $\bar{a}_{rs}$ creating a new dual feasible basic solution which, barring degeneracy, will have a higher objective function cost. Return to step 1.

THEOREM 2.4. Consider the dual feasible basic system (2.24) with $\bar{b}_r < 0$ for some r. If a pivot is made on the term $\bar{a}_{rs}$ satisfying

$$\frac{\bar{c}_r}{(-\bar{a}_{rs})} = \min_{\bar{a}_{rj} < 0} \frac{\bar{c}_j}{(-\bar{a}_{rj})} \tag{2.25}$$

the resulting basic system is dual feasible.

PROOF. The dual feasible solution before the pivot is made is $\mathbf{u}_{\mathbf{B}_0} = \mathbf{c}_{\mathbf{B}_0}\mathbf{B}_0^{-1}$ where $\mathbf{c}_{\mathbf{B}_0} = (c_1, \ldots, c_m)$ and $\mathbf{B}_0 = (\mathbf{a}_1, \ldots, \mathbf{a}_m)$, that is, $c_j - \mathbf{u}_{\mathbf{B}_0}\mathbf{a}_j \geq 0$ for $j = 1, \ldots, n$. The new dual solution which we want to show is dual feasible is $\mathbf{u}_{\mathbf{B}_1} = \mathbf{c}_{\mathbf{B}_1}\mathbf{B}_1^{-1}$ where $\mathbf{c}_{\mathbf{B}_1} = (c_1, \ldots, c_{r-1}, c_s, c_{r+1}, \ldots, c_m)$ and $\mathbf{B}_1 = (\mathbf{a}_1, \ldots, \mathbf{a}_{r-1}, \mathbf{a}_s, \mathbf{a}_{r+1}, \ldots, \mathbf{a}_m)$. We established in Chapter 1 that $\mathbf{B}_1^{-1} = \mathbf{E}\mathbf{B}_0^{-1}$ where the elementary transformation matrix $\mathbf{E}$ is defined in (1.7). Thus, $\mathbf{u}_{\mathbf{B}_1} = \mathbf{c}_{\mathbf{B}_1}\mathbf{E}\mathbf{B}_0^{-1}$, which gives us by direct calculation

$$\mathbf{u}_{\mathbf{B}_1} = \left\{ c_1, \ldots, c_{r-1}, \frac{1}{\bar{a}_{rs}} \left\{ c_s - \sum_{\substack{i=1 \\ i \neq r}}^{m} \bar{a}_{is} c_i \right\}, c_{r+1}, \ldots, c_m \right\} \mathbf{B}_0^{-1}$$

By definition of the reduced cost coefficient

$$\bar{c}_s = c_s - \sum_{i=1}^{m} \bar{a}_{is} c_i$$

we have

$$\mathbf{u}_{\mathbf{B}_1} = \left(c_1, \ldots, c_{r-1}, c_r + \frac{\bar{c}_s}{\bar{a}_{rs}}, c_{r+1}, \ldots, c_m \right) B_0^{-1}$$

Thus, for the nonbasic variables x_j for $j = m+1, \ldots, n$, $j \neq s$, the new reduced cost coefficient

$$\bar{c}_j' = c_j - \mathbf{u}_{\mathbf{B}_1}\mathbf{a}_j = c_j - \mathbf{u}_{\mathbf{B}_0}\mathbf{a}_j - \frac{\bar{c}_s}{\bar{a}_{rs}} \mathbf{e}_r \mathbf{B}_0^{-1} \mathbf{a}_j = \bar{c}_j + \frac{\bar{c}_s}{(-\bar{a}_{rs})} \bar{a}_{rj}$$

For the nonbasics with $\bar{a}_{rj} \geq 0$, the $\bar{c}_j' \geq 0$ since $\bar{c}_j \geq 0$ because (2.24) is dual feasible, and $\bar{c}_s/(-\bar{a}_{rs}) \geq 0$ because $\bar{a}_{rs} < 0$ in the pivot selection rule. For nonbasics with $\bar{a}_{rj} < 0$, the pivot selection rule (2.25) produces

$$\frac{\bar{c}_s}{(-\bar{a}_{rs})} \leq \frac{\bar{c}_j}{(-\bar{a}_{rj})}$$

which again implies $\bar{c}_j' \geq 0$. Finally,

$$\bar{c}_r' = c_r - \mathbf{u}_{\mathbf{B}_1}\mathbf{a}_r = c_r - \mathbf{u}_{\mathbf{B}_0}\mathbf{a}_r - \frac{\bar{c}_s}{\bar{a}_{rs}} \mathbf{e}_r \mathbf{B}_0^{-1} \mathbf{a}_r = -\frac{\bar{c}_s}{\bar{a}_{rs}} \cdot 1 \geq 0$$

Thus, the reduced cost coefficients are nonnegative for all the nonbasic variables after the pivot on $\bar{a}_{rs}$. ■

We illustrate the dual simplex algorithm as it can be applied in conjunction with the *cutting plane method* to the integer programming problem

$$\begin{aligned} &\min \mathbf{cx} \\ &\text{s.t. } \mathbf{Ax} = \mathbf{b} \\ &\qquad \mathbf{x} \geq \mathbf{0} \text{ and integer} \end{aligned} \tag{2.26}$$

The idea of the cutting plane method is as follows. We solve (2.26) ignoring the integrality restriction. If the optimal linear programming solution found by the simplex algorithm is integer, then it is optimal *a fortiori* in (2.26). If the optimal linear programming solution is fractional, then we derive a new constraint, called a *cut*, to be added to (2.26), which is not satisfied by this solution but is satisfied by any feasible integer solution. Problem (2.26) with the cut added is then resolved as a linear programming problem and the procedure is repeated. Convergence of the cutting plane method to an optimal integer solution to (2.26) can be established under the assumption that cx is bounded over the feasible region.

Suppose we have an optimal basic system of the form (2.24) for (2.26) with the integrality restriction ignored, and suppose further that $\bar{b}_k$ is fractional. A cut can be derived from the equation

$$x_k = \bar{b}_k - \sum_{j=m+1}^{n} \bar{a}_{kj} x_j \tag{2.27}$$

Let $\bar{b}_k$ be decomposed into two parts,

$$\bar{b}_k = \left[\bar{b}_k\right] + f_k$$

where [ ] denotes "integer part of"; that is, $[t]$=largest integer $\leq t$. By definition, $f_k < 1$ and also $f_k > 0$ since we assumed $\bar{b}_k$ is fractional. Similarly, let $\bar{a}_{kj} = [\bar{a}_{kj}] + f_{kj}$. Substituting in (2.27), we have

$$x_k = \left[\bar{b}_k\right] + f_k - \sum_{j=m+1}^{n} \left[\bar{a}_{kj}\right] x_j - \sum_{j=m+1}^{n} f_{kj} x_j$$

Rearranging terms, we have

$$x_k - \left[\bar{b}_k\right] + \sum_{j=m+1}^{n} \left[\bar{a}_{kj}\right] x_j = f_k - \sum_{j=m+1}^{n} f_{kj} x_j \tag{2.28}$$

For any feasible integer solution, we must have

$$\sum_{j=m+1}^{n} f_{kj} x_j \geq 0$$

since $f_{kj} \geq 0$ and $x_j \geq 0$. Thus

$$f_k - \sum_{j=m+1}^{n} f_{kj} x_j \leq f_k < 1$$

Moreover, $f_k - \sum_{j=m+1}^{n} f_{kj} x_j$ must be integer for any feasible integer solution because the left-hand side of (2.28) is integer for x_j integer. These two facts permit us to conclude that the constraint

$$f_k - \sum_{j=m+1}^{n} f_{kj} x_j \leq 0$$

does not eliminate any feasible integer solution. Finally, the constraint

$$\sum_{j=m+1}^{n} f_{kj} x_j \geq f_k \tag{2.29}$$

is not satisfied by the optimal linear programming solution from (2.24) with $x_j = 0$ for $j = m+1, \ldots, n$ because $f_k > 0$. Thus, (2.29) is a valid cut.

Although convergence of the cutting plane method for integer programming can be established, it has proven to be a relatively ineffective method. This is due in part to the ambiguity about which cut or cuts from the many cuts possible to add to the linear programming approximation. For example, a cut can be written on any row k such that $\bar{b}_k$ is fractional. Moreover, different cuts can be derived from a given row. For example, a different cut is obtained from row k if we multiply the row by 2 before deriving it. In Chapter 8 we discuss this point again, and show that the procedure for defining and selecting strong cuts makes the need for them largely superfluous.

EXAMPLE 2.1. Consider the integer programming problem

$$\begin{aligned} \min\ & 6x_1+2x_2+11x_3+3x_4 \\ \text{s.t.}\ & 2x_1-1x_2+5x_3+1x_4=30 \\ & 2x_1+5x_2+8x_3+3x_4=60 \\ & x_1,x_2,x_3,x_4 \text{ nonnegative integer} \end{aligned} \tag{2.30}$$

Table 2.1 gives the optimal linear programming solution for this problem with the integrality restriction ignored.

TABLE 2.1 EXAMPLE OF THE CUTTING PLANE METHOD

Basic Variable Values	$-z$	x_1	x_2	x_3	x_4	Basic Variable Values
$-z$	1	16/7	3/7			$-510/7$
x_3		4/7	$-8/7$	1		30/7
x_4		$-6/7$	33/7		1	60/7

The cut derived directly from row one is $(\frac{4}{7})x_1+(\frac{6}{7})x_2\geq\frac{2}{7}$ while the cut from row two is $(\frac{1}{7})x_1+(\frac{5}{7})x_2\geq\frac{4}{7}$. When added to the linear programming problem of Table 2.1, neither of these cuts produces an optimal integer solution. We consider instead the cut that results if we multiply row 2 in Table 2.1 by 3 before generating a cut. The result is $3x_4=\frac{180}{7}+(\frac{18}{7})x_1-(\frac{99}{7})x_2$, and the same reasoning used to derive (2.29) applied to this equation give us the cut

$$\tfrac{3}{7}x_1+\tfrac{1}{7}x_2\geq\tfrac{5}{7}$$

Table 2.2 gives the initial dual feasible tableau derived from Table 2.1 by adding this cut.

TABLE 2.2 EXAMPLE OF THE CUTTING PLANE METHOD

Basic Variables	$-z$	x_1	x_2	x_3	x_4	x_5	Basic Variable Values
$-z$	1	16/7	3/7				−510/7
x_3		4/7	−8/7	1			30/7
x_4		−6/7	33/7		1		60/7
x_5		−3/7	(−1/7)			1	−5/7

The variable x_5 is the surplus variable on the cut. According to the dual simplex criterion (2.25), x_2 is pivoted into the basis for x_5. After another iteration, the optimal linear programming solution is obtained as shown in Table 2.3. This solution is integer implying that it is optimal in the integer programming problem (2.30) as well. ▲

TABLE 2.3 EXAMPLE OF THE CUTTING PLANE METHOD

Basic Variables	$-z$	x_1	x_2	x_3	x_4	x_5	Basic Variable Values
$-z$	1				1/15	26/5	−76
x_3				1	4/15	4/5	6
x_1		1			−1/15	−33/15	1
x_2			1		1/5	33/5	2

2.8 PRIMAL-DUAL SIMPLEX ALGORITHM

The primal-dual algorithm has not been used extensively on general linear programming problems. It has been used mainly to solve some of the linear programming network problems discussed in the next chapter. It will also serve as the constructive basis for studying mathematical programming duality in Chapter 6.

Consider the dual linear programs (2.17) and (2.18). Suppose we have a dual feasible solution $\tilde{\mathbf{u}}$; i.e., $\tilde{\mathbf{u}}\mathbf{A} \leq \mathbf{c}$. We try to establish optimality of $\tilde{\mathbf{u}}$ in (2.18) by finding an $\tilde{\mathbf{x}}$ that is feasible in the primal problem (2.17) and that also satisfies the complementary slackness condition $(\mathbf{c}-\tilde{\mathbf{u}}\mathbf{A})\tilde{\mathbf{x}}=0$. To this end, we let

$$J(\tilde{\mathbf{u}}) = \{ j | \tilde{\mathbf{u}}\mathbf{a}_j = c_j \} \tag{2.31}$$

and construct the following restricted phase one primal problem

$$\begin{aligned} \phi(\tilde{\mathbf{u}}) = \min \sum_{i=1}^{m} s_i^+ + \sum_{i=1}^{m} s_i^- \\ \text{s.t.} \sum_{j \in J(\tilde{\mathbf{u}})} a_{ij} x_j + s_i^+ - s_i^- = b_i \qquad i = 1, \ldots, m \\ x_j \geq 0 \qquad j \in J(\tilde{\mathbf{u}}) \\ s_i^+ \geq 0 \qquad s_i^- \geq 0 \qquad i = 1, \ldots, m \end{aligned} \tag{2.32}$$

Let $\tilde{x}_j$, $j \in J(\tilde{\mathbf{u}})$ denote the optimal values in (2.32) found by the primal simplex algorithm, and define $\tilde{x}_j = 0$, $j \notin J(\tilde{\mathbf{u}})$. If $\phi(\tilde{\mathbf{u}}) = 0$, then $\tilde{\mathbf{x}}$ is primal feasible, that is, $\mathbf{A}\tilde{\mathbf{x}} = \mathbf{b}$, $\tilde{\mathbf{x}} \geq \mathbf{0}$, and in addition $(\mathbf{c} - \tilde{\mathbf{u}}\mathbf{A})\tilde{\mathbf{x}} = 0$ by construction. Thus, $\tilde{\mathbf{x}}$ and $\tilde{\mathbf{u}}$ are optimal in the primal and dual because complementary slackness holds.

If $\phi(\tilde{\mathbf{u}}) > 0$, the primal-dual algorithm proceeds by moving to a new dual feasible solution $\tilde{\mathbf{u}}'$ such that $\phi(\tilde{\mathbf{u}}') < \phi(\tilde{\mathbf{u}})$ (barring degeneracy). In discussing how this is done, we will make use of the dual to (2.32)

$$\begin{aligned} \phi(\tilde{\mathbf{u}}) = \max \mathbf{v}\mathbf{b} \\ \text{s.t. } \mathbf{v}\mathbf{a}_j \leq 0 \qquad j \in J(\tilde{u}) \\ -1 \leq v_i \leq 1 \qquad i = 1, \ldots, m \end{aligned} \tag{2.33}$$

We will show that, if one exists, problem (2.33) finds a direction of increase of the dual objective function in (2.18) from the dual feasible point $\tilde{\mathbf{u}}$.

Primal-Dual Simplex Algorithm

Step 0. Select a dual feasible solution $\tilde{\mathbf{u}}$.

Step 1. Solve the restricted phase one primal problem (2.32) defined at the point $\tilde{\mathbf{u}}$ using the primal simplex algorithm. If this problem has minimal objective function value $\phi(\tilde{\mathbf{u}}) = 0$, stop because the indicated primal feasible solution is optimal.

Step 2. Let $\tilde{\mathbf{v}}$ be an optimal solution to (2.33). If $\tilde{\mathbf{v}}\mathbf{a}_j \leq 0$ for all $j \notin J(\tilde{\mathbf{u}})$, stop because the primal problem has no feasible solution. Otherwise, compute

$$\tilde{\theta} = \min \left\{ \frac{c_j - \tilde{\mathbf{u}}\mathbf{a}_j}{\tilde{\mathbf{v}}\mathbf{a}_j} | \tilde{\mathbf{v}}\mathbf{a}_j > 0 \text{ and } j \notin J(\tilde{\mathbf{u}}) \right\}$$

move to $\tilde{\mathbf{u}}' = \tilde{\mathbf{u}} + \tilde{\theta}\tilde{\mathbf{v}}$ and return to Step 1 with this new dual feasible solution.

There are several points about the primal-dual algorithm that need elaboration. First, we need to substantiate that the primal problem (2.17) has no feasible solution if $\phi(\tilde{\mathbf{u}})>0$ and $\tilde{\mathbf{v}}\mathbf{a}_j \leq 0, j \notin J(\tilde{\mathbf{u}})$. The second condition implies $(\tilde{\mathbf{u}}+\theta\tilde{\mathbf{v}})\mathbf{A} \leq \mathbf{c}$ for all $\theta \geq 0$; that is, $\tilde{\mathbf{u}}+\theta\tilde{\mathbf{v}}$ is dual feasible in (2.18) for all $\theta \geq 0$. The dual objective function value for $\tilde{\mathbf{u}}+\theta\tilde{\mathbf{v}}$ is $(\tilde{\mathbf{u}}+\theta\tilde{\mathbf{v}})\mathbf{b}$ and since $\phi(\tilde{\mathbf{u}})=\tilde{\mathbf{v}}\mathbf{b}>0$, the maximal dual objective function value is $+\infty$ implying that the primal has no feasible solution by Lemma 2.1.

Suppose then, that there is some $j \notin J(\tilde{\mathbf{u}})$ such that $\tilde{\mathbf{v}}\mathbf{a}_j>0$, say

$$\theta = \frac{c_s - \tilde{\mathbf{u}}\mathbf{a}_s}{\tilde{\mathbf{v}}\mathbf{a}_s} > 0$$

We want to demonstrate that changing $\tilde{\mathbf{u}}$ to $\tilde{\mathbf{u}}+\tilde{\theta}\tilde{\mathbf{v}}$ in defining (2.32) has the effect of allowing $\phi(\tilde{\mathbf{u}})$ to be strictly decreased, barring degeneracy. Finite convergence of the primal-dual simplex algorithm is thereby ensured because there are only a finite number of basic solutions to (2.32). For $j \in J(\tilde{\mathbf{u}})$ and x_j basic in the optimal solution to (2.32), we have by complementary slackness that $\tilde{\mathbf{v}}\mathbf{a}_j=0$ and therefore $(\tilde{\mathbf{u}}+\tilde{\theta}\tilde{\mathbf{v}})\mathbf{a}_j=\tilde{\mathbf{u}}\mathbf{a}_j=c_j$ implying $j \in J(\tilde{\mathbf{u}}+\tilde{\theta}\tilde{\mathbf{v}})$. Moreover, $(\tilde{\mathbf{u}}+\tilde{\theta}\tilde{\mathbf{v}})\mathbf{a}_s=c_s$ by the definition of $\tilde{\theta}$ implying $s \in J(\tilde{\mathbf{u}}+\tilde{\theta}\mathbf{v})$. Thus, the solution of (2.32) at $\tilde{\mathbf{u}}+\tilde{\theta}\tilde{\mathbf{v}}$ can begin with the optimal basic variables at $\tilde{\mathbf{u}}$ retained at their basic values, plus the new nonbasic variable x_s. Finally, the new nonbasic has the reduced cost coefficient $-\tilde{\mathbf{v}}\mathbf{a}_s<0$, and it can be pivoted into the previously optimal basis thereby lowering the objective function value in (2.32). In addition, the dual objective function in the dual problem (2.18) has increased from $\tilde{\mathbf{u}}\mathbf{b}$ at $\tilde{\mathbf{u}}$ to $(\tilde{\mathbf{u}}+\tilde{\theta}\tilde{\mathbf{v}})\mathbf{b}$ at $\tilde{\mathbf{u}}+\tilde{\theta}\tilde{\mathbf{v}}$ since $\phi(\tilde{\mathbf{u}})=\mathbf{v}\tilde{\mathbf{b}}>0$ and $\tilde{\theta}>0$.

2.9 EXERCISES

2.1 Use the strong duality theorem 2.1 to prove Farkas' lemma 2.1.

2.2 Consider the primal problem

$$\begin{aligned} &\min \mathbf{cx} \\ &\text{s.t. } \mathbf{Ax}=\mathbf{b} \end{aligned} \qquad (\mathrm{P}')$$

Show that the dual to this problem is

$$\begin{aligned} &\max \mathbf{ub} \\ &\text{s.t. } \mathbf{uA}=\mathbf{c} \end{aligned} \qquad (\mathrm{D}')$$

in the sense that Lemma 2.1 and Theorem 2.1 hold for (P′) and (D′).

2.3 Prove that the set $Z=\{\mathbf{z}|\mathbf{z}=\mathbf{uQ},\ \mathbf{u}\geq\mathbf{0}\}$ is closed.

2.4 A random variable Y assumes one of the values $1,2,\ldots,m$. There are two possible probability distributions of Y, viz., $\{a_k\}$ and $\{b_k\}$ where

$$a_k > 0 \qquad b_k > 0$$

$$\sum_{k=1}^{m} a_k = 1 \qquad \sum_{k=1}^{m} b_k = 1$$

Assume that the random variable Y is defined so that

$$\frac{a_1}{b_1} < \frac{a_2}{b_2} < \cdots < \frac{a_m}{b_m}$$

Suppose now that on the basis of one observed value of Y it is desired to test the hypothesis: H_a: $\{a_k\}$ is the distribution of Y versus the alternative hypothesis H_b: $\{b_k\}$ is the distribution of Y. A decision rule for this purpose is a set of numbers $x_1,\ldots,x_m$, $0 \le x_j \le 1$, such that if Y is observed equal to k, then H_a is accepted with probability x_k and H_b is accepted with probability $1-x_k$. The decision rule must be chosen so that the probability of accepting H_a when it is false (so H_b is true) does not exceed a fixed number β, $0<\beta<1$. Among all such rules, a rule that maximizes the probability of accepting H_a when it is true will be termed optimal.

(1) Formulate the problem of finding an optimal decision rule as a linear programming problem.

(2) Show by using the principle of complementary slackness that an optimal rule $\{x_k^*\}$, say, must have the property that $x_j^* = 1$ for $j > k$ whenever $x_k^* > 0$. (This implies that for some integer K, $x_j^* = 0$ for $j < K$ and $x_j^* = 1$ for $j > K$.)

2.5 Assume that we are given the primal problem

$$\begin{aligned} \max \quad & z = \mathbf{cx} \\ \text{s.t.} \quad & \mathbf{Ax} = \mathbf{b} \\ & \mathbf{x} \ge \mathbf{0} \end{aligned}$$

and an optimal solution $\mathbf{u}^*$ to the dual problem

$$\begin{aligned} & \min \mathbf{ub} \\ & \text{s.t. } \mathbf{uA} \ge \mathbf{c} \end{aligned}$$

Suppose that we form a new primal problem by adding ($\lambda \ne 0$) times constraint k to constraint r. What is an optimal solution to the dual of this new problem?

2.6 Consider the linear programming problem

$$\begin{aligned} & \min \mathbf{cx} \\ & \text{s.t. } \mathbf{Ax} \ge \mathbf{b} \\ & \qquad \mathbf{x} \ge \mathbf{0} \end{aligned}$$

Show that this problem can be converted to the following existence problem: given the p vector $\mathbf{q}$ and the $p \times p$ matrix $\mathbf{M}$, find $\mathbf{w}$, $\mathbf{z}$ satisfying

$$\mathbf{w} = \mathbf{q} + \mathbf{M}\mathbf{z}$$
$$\mathbf{w} \geq \mathbf{0} \qquad \mathbf{z} \geq \mathbf{0} \qquad \text{and} \qquad \mathbf{z}^T\mathbf{w} = 0$$

2.7 Suppose the simplex algorithm as stated in Chapter 1 is applied to the linear programming problem (P). For each basic feasible solution, show that the corresponding vector of shadow prices as dual variables and the basic feasible solution satisfy the complementary slackness conditions (2.4). Recall that the vector of shadow prices induced by a basis equals the vector of cost coefficients of the basic activities postmultiplied by the basis inverse.

2.8 Use the strong duality theorem to prove *the theorem of the alternative*: let $\mathbf{A}$ be a $m \times n$ matrix and $\mathbf{c}$ a $1 \times n$ vector. Then either there is a vector $\mathbf{z}$ such that $\mathbf{zA} \geq \mathbf{c}$ or there is a vector $\mathbf{y}$ satisfying $\mathbf{Ay} = \mathbf{0}$, $\mathbf{y} \geq \mathbf{0}$ for which $\mathbf{cy} > 0$, but not both.

2.9 Let $\mathbf{A}$ be a $m \times n$ matrix and consider the set $F = \{\mathbf{x} | \mathbf{Ax} \geq \mathbf{b},\ \mathbf{x} \geq \mathbf{0}\}$. Assume F is nonempty. Prove that F is a bounded subset of R^n if and only if there is a m-vector $\mathbf{u}$ satisfying $\mathbf{u} \geq \mathbf{0}$ and $\mathbf{ua}_j < 0$ for every column $\mathbf{a}_j$ of $\mathbf{A}$.

2.10 Let A be a $m \times n$ matrix and consider the parametric family of linear programming problems

$$\begin{aligned} v(\lambda) = \min\ & \mathbf{cx} \\ \text{s.t. } & \mathbf{Ax} = \mathbf{b} + \lambda\Delta\mathbf{b} \\ & \mathbf{x} \geq \mathbf{0} \end{aligned}$$

Let

$$\Omega = \{\lambda | \mathbf{Ax} = \mathbf{b} + \lambda\Delta\mathbf{b},\ \mathbf{x} \geq \mathbf{0} \text{ has a solution}\}.$$

Prove that the set Ω is convex and $v(\lambda)$ is convex on Ω.

2.11 For the primal problem

$$\begin{aligned} \max\ & \mathbf{cx} \\ \text{s.t. } & \mathbf{Ax} = \mathbf{b} \\ & \mathbf{x} \geq \mathbf{0} \end{aligned}$$

assume that the optimal solution is unique and degenerate. Will the dual problem have alternative optima? Be sure to support your answer.

2.12 There are n types of parts labeled $1, 2, \ldots, n$. An assembly consists of n parts, exactly one of each type. Each part can be manufactured on each of m machines labeled $1, 2, \ldots, m$. Denote by a_{ij} the given hourly rate of production of part i on machine j (when producing part i exclusively).

Assume $a_{ij}>0$ for all i,j. Denote by x_{ij} the fraction of each hour that machine j produces part i. Let y be the total number of assemblies produced per hour. Let z_i denote the total number of parts of type i produced per hour in excess of those used in the assemblies. Let w_j denote the fraction of each hour during which machine j is idle. The problem is to produce in such a way so as to maximize the total production of assemblies. Formally, we seek x_{ij}, y, z_i, w_j, that

$$\text{maximize } y$$

$$\text{s.t.} \sum_{j=1}^{m} a_{ij}x_{ij} - z_i - y = 0 \qquad i=1,\ldots,n$$

$$\sum_{i=1}^{n} x_{ij} + w_j = 1 \qquad j=1,\ldots,m$$

$$x_{ij} \geq 0 \qquad z_i \geq 0 \qquad w_j \geq 0 \qquad y \geq 0 \qquad \text{all } i,j$$

(1) Prove that an optimal solution to the above problem exists.

(2) Use duality theory including the principle of complementary slackness to prove that in every optimal solution found by the simplex algorithm
   (i) $w_j=0 \quad j=1,\ldots,m$
   (ii) $z_i=0 \quad i=1,\ldots,n$

2.13 Consider problem (P) and define the function

$$L(\mathbf{x},\mathbf{u}) = \mathbf{ub} + (\mathbf{c} - \mathbf{uA})\mathbf{x}$$

Assume the sets X and U defined in (2.1) are non-empty.

(1) Prove

$$\max_{\mathbf{u}\geq\mathbf{0}} \min_{\mathbf{x}\geq\mathbf{0}} L(\mathbf{x},\mathbf{u}) = \min_{\mathbf{x}\geq\mathbf{0}} \max_{\mathbf{u}\geq\mathbf{0}} L(\mathbf{x},\mathbf{u})$$

(2) Use part (1) to show that there exists $\bar{\mathbf{x}}, \bar{\mathbf{u}}$ such that

$$L(\bar{\mathbf{x}},\bar{\mathbf{u}}) \leq L(\mathbf{x},\bar{\mathbf{u}}) \qquad \text{for all } \mathbf{x}\geq\mathbf{0}$$

$$L(\bar{\mathbf{x}},\bar{\mathbf{u}}) \geq L(\bar{\mathbf{x}},\mathbf{u}) \qquad \text{for all } \mathbf{u}\geq\mathbf{0}$$

This result is called the *saddlepoint theorem* for linear programming.

2.14 The function $v(\mathbf{b})$ defined in Section 2.4 is convex on R^m, but not everywhere differentiable. It is still reasonably well behaved at points where it is finite but not differentiable. A vector $\bar{\gamma} \in R^m$ is a *subgradient* of v at $\bar{\mathbf{b}}$ if

$$v(\mathbf{b}) \geq v(\bar{\mathbf{b}}) + \bar{\gamma}(\mathbf{b}-\bar{\mathbf{b}}) \qquad \text{for all } \mathbf{b}$$

Prove that a vector $\bar{\gamma}$ is a subgradient of v at $\bar{\mathbf{b}}$ if and only if $\bar{\gamma}$ is an optimal solution in (D) with $\mathbf{b}=\bar{\mathbf{b}}$.

2.15 Consider a basic system for a linear programming problem:

$$\min z = z_0 + \sum_{j=m+1}^{n} c_j x_j$$

$$x_i = b_i - \sum_{j=m+1}^{n} a_{ij} x_j \qquad i = 1, \ldots, m$$

where the c_j and the b_j may each be positive or negative. Suppose an upper bound q is known on the sum of the nonbasic x_j; that is, $\Sigma_{j=m+1}^{n} x_j \leq q$. Show how a dual feasible basic solution can be obtained by adding $\Sigma_{j=m+1}^{n} x_j \leq q$ as a constraint on row $m+1$ and then making a pivot on row $m+1$.

2.16 Show that the function $v(\boldsymbol{\lambda})$ in the problem (2.23) is concave.

2.17 Prove that a nonzero vector $\mathbf{d} \in R^m$ is a feasible direction at $\mathbf{b}$ of the function $v(\mathbf{b})$ defined in (2.11) (equivalently (2.13)) if and only if $\mathbf{u}^s\mathbf{d} \leq 0$ for all extreme rays $\mathbf{u}^s$ satisfying $\mathbf{u}^s\mathbf{b} = 0$.

2.18 Consider the problem

$$\begin{aligned} &\min \mathbf{cx} + c_0 y \\ &\text{s.t. } \mathbf{Ax} + \mathbf{a}^0 y \geq \mathbf{b} \\ &\qquad \mathbf{x} \geq \mathbf{0} \qquad y \text{ nonnegative integer} \end{aligned}$$

where $\mathbf{A}$ is $m \times n$ and $\mathbf{a}^0 \in R^m$. Let $\mathbf{x}^*, y^*$ denote an optimal solution to this problem with the integrality restriction on y omitted. Show that the optimal solution to the problem can be found by setting y equal to either $[y^*]$ or $[y^*]+1$ and solving for x. $[t]$ denotes the largest integer less than or equal to t, for example, $[3.4]=3$; $[-3.8]=-4$.

2.19 Consider problem (2.26) with the additional restriction that the variables are upper bounded by one. Suppose we have an optimal linear programming solution with the integrality restriction ignored. Consider a row in the optimal linear programming tableau

$$x_i = \bar{b}_i - \sum_{j=1}^{n} \bar{a}_{ij} x_j$$

with $\bar{b}_i$ fractional and thus, $x_i = \bar{b}_i$ is fractional in the optimal linear programming solution. Prove that the minimal objective function cost of the integer programming problem is at least as great as the linear programming minimal cost plus δ where $\delta = \min(\delta_1, \delta_2)$ and

$$\delta_1 = \min\left\{ \frac{\bar{b}_i \bar{c}_j}{\bar{a}_{ij}} \;\middle|\; \bar{a}_{ij} > 0 \right\}$$

$$\delta_2 = \min\left\{ (\bar{b}_i - 1)\frac{\bar{c}_j}{\bar{a}_{ij}} \;\middle|\; \bar{a}_{ij} < 0 \right\}$$

The term $\bar{c}_j$ is the reduced cost of variable x_j relative to the optimal linear programming dual variables. The quantities δ_1 and δ_2 are called integer programming *penalties*.

2.20 Consider the following game, called a *two-person, zero-sum game*, between player 1 and player 2. Player 1 has m strategies and player 2 has n strategies. If player 1 chooses strategy i and player 2 chooses strategy j, then the payoff to player 1 is a_{ij} and the payoff to player 2 is $-a_{ij}$. The game is called a zero-sum game because the sum of payoffs is zero. Each player uses a randomized policy where he chooses each of his strategies with a fixed probability. In other words, player 1 has a policy characterized by a vector $\mathbf{u}=(u_i,\ldots,u_m)$ satisfying $\sum_{i=1}^{m}u_i=1$, $u_i\geq 0$, where u_i is player 1's probability of choosing strategy i on a given play of the game. Similarly, player 2 selects a randomized policy characterized by the vector $\mathbf{x}=(x_i,\ldots,x_n)$ satisfying $\sum_{j=1}^{n}x_j=1$, $x\geq 0$, where x_j is player 2's probability of selecting strategy j on a given play of the game. Thus, for fixed $\mathbf{x}$ and $\mathbf{u}$, the expected payoff to player 1 is $\sum_{i=1}^{m}\sum_{j=1}^{n}u_ia_{ij}x_j$ and the expected payoff to player 2 is $-\sum_{i=1}^{m}u_ia_{ij}x_j$. Player 1 attempts to find a randomized policy $\mathbf{u}$ such that in the worst case when player 2 makes his best choice of $\mathbf{x}$, he will maximize his expected payoff. Conversely, player 2 attempts to find a randomized policy $\mathbf{x}$ such that when player 1 makes his best choice of $\mathbf{u}$, his expected payoff is maximized, or equivalently, the payoff to player 1 is minimized. Use the strong duality theorem 2.1 to prove

$$\max_{\mathbf{u}}\min_{\mathbf{x}}\sum_{i=1}^{m}\sum_{j=1}^{n}u_ia_{ij}x_j=\min_{\mathbf{x}}\max_{\mathbf{u}}\sum_{i=1}^{m}\sum_{j=1}^{n}u_ia_{ij}x_j$$

where $\mathbf{x}$ and $\mathbf{u}$ are constrained to be randomized policies. This result is called the *minimax theorem*.

2.21 A *Leontief substitution model* is a linear programming problem of the form

$$\begin{aligned}&\min \mathbf{cx}\\ &\text{s.t. } \mathbf{Ax}=\mathbf{b}\\ &\qquad \mathbf{x}\geq\mathbf{0}\end{aligned}$$

where $\mathbf{b}$ is an m vector with positive components and $\mathbf{A}$ is a $m\times n$ matrix ($n\geq m$) with columns $\mathbf{a}_j$ having exactly one positive component equal to 1 and all other components nonpositive. If $\mathbf{A}$ is square ($n=m$) and the positive component in each column is on a different row, then it is called an *input-output matrix*. This is because each column or activity $\mathbf{a}_j$ has one output corresponding to the row with component equal to 1, say $a_{kj}=1$, and uses as input the quantities $a_{ij}\leq 0$, $i\neq k$, from other activities. You can assume the problem has an optimal solution.

(1) Prove that if the nonnegative matrix $\mathbf{Q}$ has the property that $\mathbf{Q}^n$ converges to zero, then $(\mathbf{I}-\mathbf{Q})^{-1}$ exists and moreover,

$$(\mathbf{I}-\mathbf{Q})^{-1}=\mathbf{I}+\mathbf{Q}+\mathbf{Q}^2\ldots=\sum_{k=0}^{\infty}\mathbf{Q}^k$$

(2) Suppose we have an optimal basis $\mathbf{B}$ to the Leontief model with substitution; say, $\mathbf{A}=(\mathbf{B},\mathbf{N})$, $\mathbf{x}=(\mathbf{x_B},\mathbf{x_N})$, and $\mathbf{c}=(\mathbf{c_B},\mathbf{c_N})$. Use part (1) to show that $(\mathbf{x_B},\mathbf{x_N})=(\mathbf{B}^{-1}\bar{\mathbf{b}},\mathbf{0})$ is an optimal solution to the problem for all $\bar{\mathbf{b}}\geq 0$.

(3) Prove that the optimal dual solution $\mathbf{u}=\mathbf{c_B}\mathbf{B}^{-1}$ from part (2) is the maximal vector of the set $\{\mathbf{u}|\mathbf{uA}\leq\mathbf{c}\}$.

2.10 NOTES

SECTION 2.1. Nearly all books treating linear programming include material on linear programming duality. Karlin (1959) gives the classical mathematical and game theoretic results pertaining to the subject. Gale (1960) discusses some of the economic consequences of duality. Wagner (1975) gives a more lengthy and simplified discussion.

SECTION 2.2. Dantzig (1963) contains a brief discussion on the history of the development of linear programming duality theory. The alternative proof of the strong duality theorem 2.1 follows the derivation given by Varaiya (1972).

SECTION 2.3. The investment example is due to A. S. Manne who extended the model and the analysis to the nonlinear case (see Manne, 1970).

SECTION 2.4. This development is a special case of more general duality results to be discussed in chapters five and six.

SECTION 2.5. Sensitivity analysis is another subject treated in nearly all books on linear programming. Simonnard (1966) has a lengthier discussion on the subject treated in a manner similar to the way it is treated here.

SECTION 2.6. There is no single approach that is universally accepted for dealing with multiobjectives in mathematical programming. The volume edited by Starr and Zeleny (1977) contains a broad collection of papers on multiobjective programming.

SECTION 2.7. The cutting plane method for integer programming was devised by Gomory (1958) who provided a convergence proof. Since that time, a large number of papers have been written on the subject; Garfinkel and Nemhauser (1972) have a more recent treatment of the method. Further insights into the method will be given in Chapter 8 when we study integer programming in detail.

SECTION 2.8. The primal-dual algorithm is also discussed in Simonnard (1966).

SECTION 2.9. Dantzig (1963) has a chapter on two-person zero-sum games and linear programming. The mathematical structure of Leontief and related matrices is studied in detail by Veinott (1968).

3 NETWORK OPTIMIZATION PROBLEMS

3.1 INTRODUCTION

An important large class of linear programming problems have coefficient matrices with special structure that admits a network interpretation and analysis. Specifically, these are the problems for which all activities other than slacks and surpluses can be represented as arcs connecting pairs of nodes in a network. The values of the variables associated with these activities can be interpreted as flows in the network and the constraints as flow balance equations. Formally, a network optimization problem is a linear programming problem with a coefficient matrix consisting of a node-arc incidence matrix plus some slack and surplus variables. The reader is referred to Appendix B for the definition of the node-arc incidence matrix and other definitions and constructs from graph theory that we use in this chapter.

Network optimization problems arise in a wide variety of operations research applications, particularly in models of distribution systems, but also in the analysis of communications networks, machine scheduling, gas and oil pipelines, to name a few. Our purpose here is to study the mathematical properties of these problems rather than the applications, but we will consider some representative models, both in this chapter and later ones.

Linear programming bases for network optimization problems correspond to sets of trees in the network, which in turn implies they are triangular matrices. As a result, a linear programming basic solution can be computed without matrix inversion because the triangular invertible system can be solved by direct substitution. The simplex method can be specialized to exploit this property.

Network optimization problems have the additional property that all nonzero coefficients are $+1$ or -1. This property coupled with the triangularity property implies that all bases are unimodular (that is, all basis determinants are $+1$ or -1). Thus, any basic solution, including the optimal basic solution, is integer if the right-hand-side vector is integer. In other words, every extreme point of the linear programming polytope is integer. For some applications such as the assignment problem, integer solutions are required. Thus, the fact that the simplex method finds an optimal extreme point solution ensures *a fortiori* that an optimal integer solution is obtained when a network optimization problem is solved as a linear programming problem.

Some optimization problems have a natural description in terms of networks but cannot be formulated precisely as network optimization problems in the sense that they have been defined above. An example is the fixed cost transportation problem where there is a fixed cost associated with positive flow in each arc. Such a problem requires for solution the integer programming methods discussed in Chapter 8. Complex optimization problems with network optimization subproblems can often be usefully analyzed by decomposition methods that permit the special substructures to be exploited. The network optimization subproblems are solved separately, but must be solved many times, and the optimal solutions to the subproblems are synthesized in some global decomposition scheme such as the ones discussed in Chapter 6. The nonconvex nature of these problems sometimes requires the use of branch and bound searches to ensure global optimization; branch and bound is discussed in Chapter 8.

The class of network optimization problems contains problems with still further special structure; included are the maximal flow, circulation, transportation, and assignment problems. Transformations from one special network structure to another are often possible and sometimes useful when the problem is to be solved by a specific algorithm.

Finally, we will study two optimization problems with structure closely related to the network optimization problem. One is the flow with gains problem where a unit of flow leaving a node along an arc becomes more or less than a unit of flow by the time it reaches the node at the other end of the arc. The other related problem is the graph optimization problem where the coefficient matrix is the node-edge incidence matrix of a graph plus some slacks and surplus columns.

3.2 EXAMPLES OF NETWORK OPTIMIZATION PROBLEMS

We begin with a formal definition of network optimization problems. Let G be a network with node set $\mathcal{N}=\{1,\ldots,m\}$ and arc set $\mathcal{A}$ with arcs (i,j). There is an implied direction from node i to node j associated with an arc (i,j). The notation $\langle i,j\rangle$ will be used to indicate an edge or undirected connection between nodes. It will be convenient for the analysis in this section to have an enumeration of the arcs, say $\mathscr{Q}=\{r_1,\ldots,r_n\}$. In later chapters, natural labelings of the nodes different from the first m integers will arise in specific applications; for example, node labels i,t denoting i units of inventory at the start of period t in a dynamic programming problem in Chapter 4, or node labels corresponding to elements of a finite abelian group in a group optimization problem arising in integer programming discussed in Chapter 8.

The generic network optimization problem that we consider is the linear programming problem

$$\begin{aligned} \min z &= \mathbf{cx}+\mathbf{ds} \\ \text{s.t.} \quad \mathbf{Ax}+\mathbf{Us} &= \mathbf{b} \\ \mathbf{0}\le\mathbf{x}\le\mathbf{q} \qquad & \mathbf{0}\le\mathbf{s}\le\mathbf{p} \end{aligned} \tag{3.1}$$

where $\mathbf{A}$ is the $m\times n$ node-arc incidence matrix of G, and $\mathbf{U}$ is an $m\times\bar{n}$ matrix with the property that each column has exactly one nonzero entry equal to $+1$ or -1. If a column of $\mathbf{U}$ has a $+1$ on row i, the corresponding slack variable is denoted by s_i^+ with cost coefficient d_i^+; if a column of U has a -1 on row i, the corresponding surplus variable is denoted by s_i^- with cost coefficient d_i^-. These variables are called logicals. There can be rows in (3.1) with logicals of both signs.

Column k of $\mathbf{A}$ corresponds to the arc r_k of the network and the variable x_k can be viewed as the flow in this arc. A column of $\mathbf{U}$ with nonzero entry on row i corresponds to node i; the variable s_i^+ can be viewed as an exogenous flow out of the node, and s_i^- can be viewed as an exogenous flow into the node.

The components q_k of the n-vector q of upper bounds are called *arc capacities*; the components p_i^+ and/or p_i^- of the $\bar{n}$-vector p of upper bounds are called *node capacities*. Upper bounds on the variables play an important role in many applications of the network optimization model; for example, the maximal flow problem studied in Section 3.4. Thus, the bounds are included explicitly in our analysis.

EXAMPLE 3.1. We consider a typical single commodity distribution problem as our first example of a network optimization problem. Suppose warehouses 1 and 2 have supplies of 9 and 8 units, respectively, of a commodity. Storage capacity is 5 units at warehouse 1 with a cost of \$3 per unit per week, and storage capacity is 6 units at warehouse 2 with a cost of \$4 per unit per week. Suppose factories 5 and 6 will need 9 and 6 units, respectively, in the coming week. The factories have no excess storage space and must meet these demands by shipments from warehouses 1 and 2. The nodes 3 and 4 are transshipment nodes. The objective is to minimize

TABLE 3.1 TRANSSHIPMENT PROBLEM

Row Description	x_1	x_2	x_3	x_4	x_5	x_6	x_7	x_8	x_9	x_{10}	s_1^+	s_2^+	Right-hand Sides
Node 1	1	1									1		9
Node 2			1	1								1	8
Node 3	−1				1	1	−1						0
Node 4			−1		−1		1	1					0
Node 5		−1				−1			1	−1			−9
Node 6				−1				−1	−1	1			−6
Costs	4	5	3	6	1	2	7	5	8	1	3	4	
Capacities	5	5	4	5	2	6	2	4	2	2	5	6	

the cost of distribution and storage in the coming week. Table 3.1 gives the algebraic statement of the problem. Figure 3.1 is a network representation of the problem illustrating the routes connecting warehouses to factories. The slacks s_1^+ and s_2^+ are indicated by arrows out of nodes 1 and 2. ▲

The classical *transportation problem* is a special case of the network optimization problem (3.1). Specifically, the transportation problem node set $\mathfrak{N}=\{1,\ldots,m\}$ is partitioned into a set of origins $S=\{1,\ldots,k\}$ and destinations $T=\{k+1,\ldots,m\}$ with arcs (i,j) restricted to $i\in S$ and $j\in T$ for some or all pairs i and j in $\mathfrak{N}$ (i.e., the network is bipartite). The nonnegative quantity b_i is the supply of an item at origin i . The nonnegative quantity $-b_j$ is the demand for the item at destination j; b_j is nonpositive because of our sign conventions. A cost per unit transported c_{ij} is associated with each arc (i,j). The type of logical variables included depends on the given application. For example, if items are available in

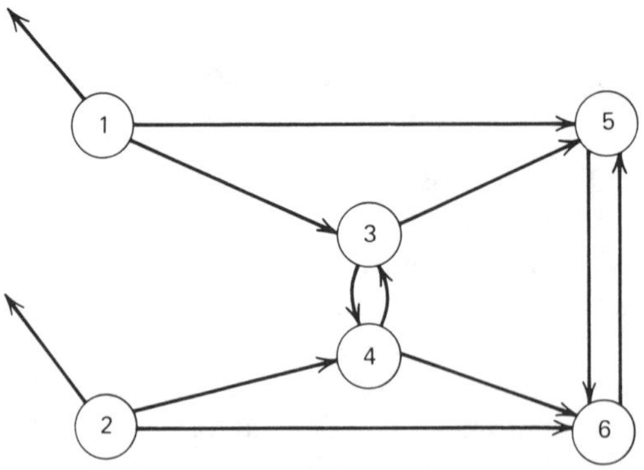

Figure 3.1. Transshipment problem.

unlimited supply at destination j from an outside vendor, then there is an s_j^- variable on row j, cost per unit supplied d_j^- and upper bound $p_j^- = +\infty$. The *transshipment problem* has a transportation substructure. In addition to origin and destination nodes, there are transshipment nodes without supply or demand that lie along routes between origins and destinations.

The *assignment problem* is a transportation problem with m even, the number of origins equal to the number of destinations, and $b_i = 1$ for $i = 1,\ldots,m/2$, $b_j = -1$ for $j = j/2+1,\ldots,m$. The name arises from the application of this model to the problem of assigning $m/2$ jobs to $m/2$ people where c_{ij} is the cost of assigning job i to person j. In practice, the assignment problem arises as a substructure in models such as the traveling salesman problem discussed in Chapter 8.

We will be studying a number of other specific network optimization problems later in this chapter and in Chapter 4. These include the maximal flow problem discussed in Section 3.5, the flow with gains problem in Section 3.7, the shortest route and multicommodity flow problems in Chapter 4.

3.3 GRAPH STRUCTURE OF COEFFICIENT MATRICES

Our study of network optimization problems does not treat explicitly the transportation, transshipment, and assignment problems discussed in the previous section, although a number of specific algorithms have been proposed for them. Instead, we concern ourselves with the structure of the network optimization problem in the general form (3.1). To this end, let $\mathbf{A}^0$ denote the coefficient matrix $(\mathbf{A},\mathbf{U})$ of problem (3.1). As we shall see, linear programming bases derived from $\mathbf{A}^0$ have a graph structure that can be exploited to specialize the simplex method.

Our analysis proceeds by considering arbitrary square submatrices of $\mathbf{A}^0$ and deriving conditions characterizing when they are nonsingular. Let $\mathbf{T}$ be a $t\times t$ submatrix of $\mathbf{A}^0$ and consider the following graph $F_\mathbf{T}$ defined by $\mathbf{T}$. The graph consists of the k nodes $i_1,\ldots,i_k$ corresponding to rows of $\mathbf{A}^0$ in $\mathbf{T}$ for which there is at least one nonzero entry. The graph has edges $\langle i,j\rangle$ for each arc (i,j) corresponding to a column of $\mathbf{T}$; $\mathbf{T}$ must contain both row i and row j. If arcs (i,j) and (j,i) are both present, we include only one edge $\langle i,j\rangle$. We ignore directions of arcs in constructing $F_\mathbf{T}$ because the properties of linear independence we wish to study in this section do not depend on the direction of connections between pairs of nodes, but rather on whether or not connections exist. The directions become important again in the following section when we use these properties to solve the network optimization problem (3.1).

Any node i of F_T corresponding to a column of T with exactly one nonzero entry on row i is called a *root* of F_T. We say the graph F_T is a *spanning* graph if it consists of t nodes; that is, if no row of T contains all zeroes.

LEMMA 3.1. If the $t \times t$ submatrix T of A^0 has rank t, then the graph F_T is a spanning forest.

PROOF. Clearly, F_T must be spanning because otherwise T has a row of zeros, and it cannot have rank t. To show F_T is a forest on the nodes $i_1, \ldots, i_t$, we show that it contains no cycles. For suppose F_T has a cycle $\langle i_1, i_2, \ldots, i_k, i_1 \rangle$ where each arc r_j is either (i_j, i_{j+1}) or (i_{j+1}, i_j). We show a contradiction by showing that the t-dimensional columns $a_1, \ldots, a_k$ of T corresponding to $r_1, \ldots, r_k$ are linearly dependent.

To this end define the weights

$$\lambda_j = \begin{cases} 1 & \text{if } r_j = (i_j, i_{j+1}) \\ -1 & \text{if } r_j = (i_{j+1}, i_j) \end{cases}$$

Then the matrix $(\lambda_1 a_1, \ldots, \lambda_k a_k)$ has the form

	$\lambda_1 a_1$	$\lambda_2 a_2 \ldots$	$\lambda_{k-1} a_{k-1}$	$\lambda_k a_k$
node i_1	1			-1
node i_2	-1	1		
node i_3		-1		
$\vdots$				
node i_{k-1}			1	
node i_k			-1	1
node i_{k+1}	0	0	0	0
$\vdots$	$\vdots$	$\vdots$	$\vdots$	$\vdots$
node i_t	0	0	0	0

It is evident from this schema that $\sum_{j=1}^{k} \lambda_j a_j = 0$ contradicting the assumption that T has rank t. ∎

Figure 3.2 denotes a spanning forest for the problem of Example 3.1. The 6×6 matrix T in this case consists of columns in A corresponding to the arcs $(1,5)$, $(2,6)$, $(4,3)$, $(4,6)$, and $(6,5)$ and the root on node 1. Notice

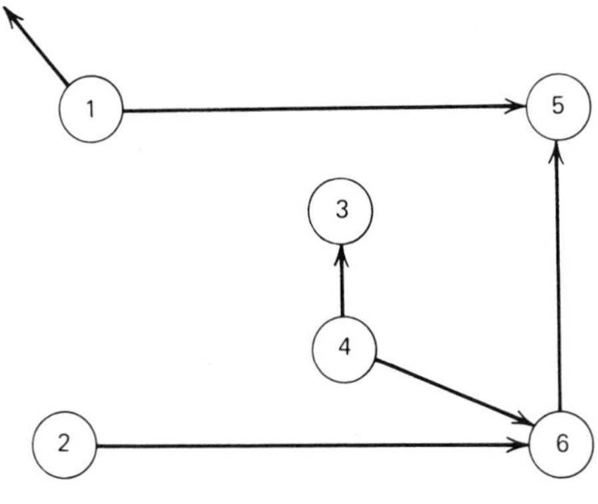

Figure 3.2. Spanning forest example.

that this spanning forest consists of one spanning tree of $m-1$ arcs. The presence of the slack, corresponding to the root, increased the rank of the matrix to 6. The importance of the slack activities to the rank of bases is the next point to be studied.

LEMMA 3.2. If the $t \times t$ submatrix $\mathbf{T}$ of $\mathbf{A}^0$ has rank t, then each tree of the spanning forest $F_{\mathbf{T}}$ has at most one root.

PROOF. Suppose some tree had two roots i_1 and i_k. Since a tree is connected, there is a simple path from i_1 to i_k in the tree, say $\langle i_1, i_2, \ldots, i_k \rangle$. Thus, there are $k+1$ t-dimensional columns of $\mathbf{T}$, say $\mathbf{a}_1, \ldots, \mathbf{a}_{k+1}$, corresponding to the arcs $r_1, r_2, \ldots, r_{k-1}$ and the roots i_1, i_k, respectively, where $r_j = (i_j, i_{j+1})$ or (i_{j+1}, i_j) for $j = 1, \ldots, k-1$. We define the weights

$$\lambda_j = \begin{cases} +1 \text{ if } r_j = (i_j, i_{j+1}) \\ -1 \text{ if } r_j = (i_{j+1}, i_j) \end{cases} \quad \text{for } j = 1, \ldots, k-1$$

$$\lambda_k = \begin{cases} +1 \text{ if } \mathbf{a}_k \text{ has a } -1 \text{ nonzero component} \\ -1 \text{ if } \mathbf{a}_k \text{ has a } +1 \text{ nonzero component} \end{cases}$$

$$\lambda_{k+1} = \begin{cases} +1 \text{ if } \mathbf{a}_{k+1} \text{ has a } +1 \text{ nonzero component} \\ -1 \text{ if } \mathbf{a}_{k+1} \text{ has a } -1 \text{ nonzero component} \end{cases}$$

As in Lemma 3.1, the matrix $(\lambda_1\mathbf{a}_1,\ldots,\lambda_{k+1}\mathbf{a}_{k+1})$ can be expressed schematically as

$$
\begin{array}{l} \\ \text{node } i_1 \\ \text{node } i_2 \\ \text{node } i_3 \\ \vdots \\ \text{node } i_{k-1} \\ \text{node } i_k \\ \text{node } i_{k+1} \\ \vdots \\ \text{node } i_t \end{array}
\begin{array}{c} \begin{array}{ccccc} \lambda_1\mathbf{a}_1 & \lambda_2\mathbf{a}_2\ldots & \lambda_{k-1}\mathbf{a}_{k-1} & \lambda_k\mathbf{a}_k & \lambda_{k+1}\mathbf{a}_{k+1} \end{array} \\
\left[\begin{array}{ccccc}
1 & & & -1 & \\
-1 & 1 & & & \\
 & -1 & & & \\
 & & & & \\
 & & 1 & & \\
 & & -1 & & 1 \\
0 & 0 & 0 & 0 & 0 \\
\vdots & \vdots & \vdots & \vdots & \vdots \\
0 & 0 & 0 & 0 & 0
\end{array}\right] \end{array}
$$

Clearly, $\Sigma_{j=1}^{k+1}\lambda_j\mathbf{a}_j=0$ contradicting the assumption that $\mathbf{T}$ has rank t. ■

With these results, we can give the desired characterization of square invertible submatrices of $\mathbf{A}^0$. On the one hand, the graph of the submatrix must be a *rooted spanning forest*, that is, a spanning forest with the additional property that each tree of the forest has exactly one root. On the other hand, the submatrix must possess the special property of triangularity. A *triangular* matrix is a square matrix with nonzeroes on the main diagonal and all zeroes above the main diagonal, or a square matrix that can be put into such a form by interchanging rows and interchanging columns. An equivalent inductive characterization to be used in the proofs below is as follows: For $t \geq 2$, a $(t \times t)$ matrix $\mathbf{T}$ is triangular if and only if it is possible to define a $(t-1)\times(t-1)$ triangular matrix $\mathbf{S}$ derived from $\mathbf{T}$ by finding a row in R^k with only one nonzero entry and deleting that row and column. The definition is complete if a 1×1 matrix is defined to be triangular if and only if it is nonzero.

THEOREM 3.1. Suppose $\mathbf{T}$ is a $t\times t$ submatrix of $\mathbf{A}^0$. Then the following three properties are equivalent:

1. $\mathbf{T}$ has rank t
2. $\mathbf{T}$ is triangular
3. $F_{\mathbf{T}}$ is a rooted spanning forest

PROOF. Property 2 ($\mathbf{T}$ is triangular)$\Rightarrow$Property 1 ($\mathbf{T}$ has rank t). Let $\mathbf{a}_1,\ldots,\mathbf{a}_t$ denote the columns of $\mathbf{T}$ in explicit triangular form, that is, with nonzeroes on the main diagonal and zeroes above it. If $\Sigma_{j=1}^{t}\lambda_j\mathbf{a}_j=0$, then clearly $\lambda_1=0$ since $a_{11}\neq 0$ but $a_{1j}=0$ for $j=2,\ldots,t$. This implies $\Sigma_{j=2}^{t}\lambda_j\mathbf{a}_j=0$ which in turn implies $\lambda_2=0$ since $a_{22}\neq 0$ but $a_{2j}=0$ for $j=3,\ldots,t$. This line of reasoning permits us to conclude that $\lambda_1=\lambda_2=\ldots=\lambda_t=0$ implying linear independence of the columns of t. Note that an alternate proof is to observe that the determinant of $\mathbf{T}$ is nonzero because it equals the product of the nonzero numbers on the diagonal of T.

Property 3 ($F_{\mathbf{T}}$ is a rooted spanning forest)⇒Property 2 ($\mathbf{T}$ is triangular). The proof that the square matrix $\mathbf{T}$ is triangular is by induction on t, the number of rows of $\mathbf{T}$. For $t=1$, the graph $F_{\mathbf{T}}$ is a rooted spanning forest consisting of one node and one root. Thus, $\mathbf{T}$ is a 1×1 nonzero matrix which is triangular by definition.

Assume $\mathbf{T}$ is triangular for any submatrix of $\mathbf{A}^0$ having k rows for $k=1,\ldots,t-1$ for which $F_{\mathbf{T}}$ is a rooted spanning forest. Consider a submatrix $\mathbf{T}$ of $\mathbf{A}^0$ of dimension $t\times t$ and suppose $F_{\mathbf{T}}$ is a rooted spanning forest. If $F_{\mathbf{T}}$ consists of t trees, each having a root, then $\mathbf{T}$ is a matrix with t nonzero elements, each on a different row and column, and it is triangular. If $F_{\mathbf{T}}$ has fewer than t trees, then there is a tree with at least one edge and spanning at least two nodes. Such a nontrivial tree must have at least two nodes that are ends (see Appendix B), one of which is not a root, say i. Recall that an end of a tree is a node with only one edge drawn to it implying row i of $\mathbf{T}$ has only one nonzero entry. Let $\overline{\mathbf{T}}$ denote the $(t-1)\times(t-1)$ matrix derived from $\mathbf{T}$ by deleting row i and the column corresponding to the single edge $\langle j,i\rangle$ drawn to i. Let $F_{\mathbf{T}}$ denote the graph of $\overline{\mathbf{T}}$ derived from $F_{\mathbf{T}}$ by deleting the node i and the edge $\langle j,i\rangle$. By construction, $F_{\overline{\mathbf{T}}}$ is a rooted spanning forest and by the induction hypothesis, $\overline{\mathbf{T}}$ is triangular. Since $\mathbf{T}$ had only one nonzero entry on row i which was deleted to form $\overline{\mathbf{T}}$, $\mathbf{T}$ is clearly triangular by our inductive definition.

Property 1 ($\mathbf{T}$ has rank t)⇒Property 3 ($F_{\mathbf{T}}$ is a rooted spanning forest). By Lemma 3.1, $F_{\mathbf{T}}$ is a spanning forest and by Lemma 3.2, every tree of the spanning forest has at most one root. Thus, each tree of $F_{\mathbf{T}}$ spanning k nodes consists of $k-1$ edges and corresponds to a $k\times(k-1)$ submatrix of $\mathbf{T}$ if it has no root, and a $k\times k$ submatrix of $\mathbf{T}$ if it has a root. It is clear that $\mathbf{T}$ cannot be square unless every tree has a root. Since $\mathbf{T}$ is assumed to be a square matrix, the desired result obtains. ■

COROLLARY 3.1. A sufficient condition that the rank of $\mathbf{A}^0=(\mathbf{A},\mathbf{U})$ is m is that there is a slack column in $\mathbf{U}$ for at least one node of each connected component of the network G defining $\mathbf{A}$.

PROOF. Select a tree from each connected component and a root for the tree. Consider the $m\times m$ matrix corresponding to this graph; the graph of the matrix is a rooted spanning tree implying the matrix has rank m by Theorem 3.1. ■

COROLLARY 3.2. The matrix $\mathbf{A}^0$ is *totally unimodular*; that is, every square submatrix of $\mathbf{A}^0$ has determinant equal to 0, +1, or −1.

PROOF. If a square submatrix of $\mathbf{A}^0$ has nonzero determinant, then by Theorem 3.1, it is triangular. The submatrix can be put into explicit triangular form by interchanging rows and columns, which does not change the magnitude of the determinant. Since all elements along the diagonal of the explicit triangular matrix are +1 or −1, the determinant of the triangular matrix must be +1 or −1. ■

COROLLARY 3.3. Any basic feasible solution to the network optimization problem (3.1) is integer if the right-hand-side vector $\mathbf{b}$ is integer.

PROOF. The basic solution $\mathbf{x_B}=\mathbf{B}^{-1}\mathbf{b}$, $\mathbf{x_N}=\mathbf{0}$ is integer because $|\det \mathbf{B}|=1$ and $\mathbf{B}$ is integer which implies $\mathbf{B}^{-1}$ is a matrix of integers. ■

The triangular property of bases naturally implies the result of Corollary 3.3. The property also makes it possible to compute directly the values of the basic variables without a matrix inversion. For, suppose the triangular system $\mathbf{Ty}=\mathbf{q}$ is written as

$$\begin{aligned}
&a_{11}y_1 &&= q_1\\
&a_{21}y_1 + a_{22}y_2 &&= q_2\\
&\vdots\\
&a_{t1}y_1 + a_{t2}y_2 + \ldots + a_{tt}y_t &&= q_t
\end{aligned}$$

We can solve directly for $y_1, y_2, \ldots, y_t$ in that order by the formulae

$$y_1 = \frac{1}{a_{11}} q_1$$

$$y_k = \frac{1}{a_{kk}}\left[q_k - \sum_{j=1}^{k-1} a_{kj}y_j\right], \qquad \text{for } k=2,\ldots,t$$

Since $|a_{ii}|=1$ for $i=1,\ldots,t$, the y_k computed in this way will be integer when q_k is integer for $k=1,\ldots,t$.

3.4 PRIMAL SIMPLEX METHOD FOR NETWORK OPTIMIZATION PROBLEMS

In this section, we show how the primal simplex method discussed in Section 1.2 can be modified to exploit the graph structure of linear programming bases for the network optimization problem (3.1). For convenience, we assume that the coefficient matrix $\mathbf{A}^0=(\mathbf{A}, \mathbf{U})$ has rank m; a sufficient condition for this to be the case is given in Corollary 3.1. Theorem 3.1 tells us that each $m \times m$ linear programming basis derived from $\mathbf{A}^0$ corresponds to a rooted spanning forest and is triangular, which means that the basic feasible solution and dual variables can be found without explicit calculation of a basis inverse.

Since we use the graph representation of $\mathbf{A}^0$ and no longer need to refer to columns of $\mathbf{A}$ and $\mathbf{U}$ in studying (3.1), our notation can be simplified. The node set $\mathcal{N} = \{1, \ldots, m\}$ and arc set $\mathcal{A}$ with arcs (i,j) remain unchanged. However, we let the variable x_{ij} denote the flow in (i,j), c_{ij} denote the cost per unit of this flow, and q_{ij} denote the capacity of the arc.

Let $\langle i_1, i_2, \ldots, i_n \rangle$ denote a simple path ignoring arc directions in the network G. If the kth arc in this path is (i_k, i_{k+1}), it is called a *forward arc*, and if the kth arc is (i_{k+1}, i_k), it is called a *reverse arc*. In a rooted spanning forest, each node is spanned by a unique tree and there is a unique simple path connecting the root of this tree to the node. An edge will either be a forward arc or a reverse arc in all such paths. A forward arc is called an *up arc* with respect to the tree, and a reverse arc is called a *down arc* with respect to the tree. Thus, each arc in a rooted spanning forest can be designated as an up arc or a down arc. For the rooted tree of Figure 3.2, the arcs $(1,5)$ and $(4,3)$ are up arcs, and the arcs $(6,5)$, $(2,6)$, and $(4,6)$ are down arcs.

Given below are the three steps of the primal simplex algorithm from Chapter 1 adapted for the network optimization problem (3.1). An iteration begins with some nonbasic variables at their upper bounds as well as some nonbasic variables at zero. The upper bounded variables are netted out of (3.1) and it is assumed that we have an $m \times m$ feasible basis $\mathbf{B}$ for the residual problem with right-hand-side $\mathbf{b}'$. The iteration begins with knowledge of the variable values for this basic solution and the corresponding dual solution. The algorithm works with the rooted, spanning forest $F_{\mathbf{B}}$ either to prove the corresponding basic feasible solution is optimal or to find a new basic feasible solution with lower cost. The phase one problem of finding an initial basic feasible solution can be formulated as a maximal flow problem (see Exercise 3.8).

Network Primal Simplex Algorithm

Step 1. (Price Out Nonbasic Variables). For the network optimization problem (3.1), a column corresponding to the arc (i,j) has a 1 in row i, a -1 in row j, and zero elsewhere. The logical variables s_i^+ and s_i^- correspond to columns with 1 and -1, respectively, in row i, and zero elsewhere. Thus, a search for a nonbasic variable to enter the basis because it is at zero with a negative reduced cost or it is at its upper bound with a positive reduced cost is equivalent to a search for an arc (i,j) or a node i

such that one of the following holds:

1. $x_{ij}=0$ and $u_i-u_j>c_{ij}$
2. $x_{ij}=q_{ij}$ and $u_i-u_j<c_{ij}$
3. $s_i^+=0$ and $u_i>d_i^+$
4. $s_i^-=p_i^-$ and $u_i>-d_i^-$
5. $s_i^-=0$ and $u_i<-d_i^-$
6. $s_i^+=p_i^+$ and $u_i<d_i^+$

If any one of the conditions 1–6 obtains, then the corresponding variable has a negative reduced cost relative to the dual variables and should be increased from its lower bound or decreased from its upper bound. If none of conditions 1–6 applies, then the basic feasible solution corresponding to **B**, along with the variables at their upper bounds, is optimal and we exit from the algorithm.

Step 2. (Find a Blocking Variable and Compute New Basic Feasible Solution). One of the cases in Step 1 obtains in the selection of a variable to consider entering into the basis. As with the primal simplex algorithm of Chapter 1, this variable is changed by a nonnegative quantity θ in such a way that the feasibility conditions $0 \leq x \leq q$, and $0 \leq s \leq p$ are maintained. However, since each basis matrix corresponds to a rooted spanning forest, the introduction into the current basis of an arc variable corresponds to the addition of an edge to a tree, and the introduction of a logical variable corresponds to the addition of a root to a tree. The new rooted spanning forest that results by making θ as large as possible while maintaining feasibility is constructed by applying one of several operations to the current rooted spanning forest. To *reroot* a tree means to designate another node as its root and drop the old root. To *cut off the top* of a rooted tree at an edge means to delete the edge from the tree. In this case, part of the tree becomes a separate tree, which has no root and is called the *top* of the tree. Any one of its nodes can be designated as its root, or it can be *grafted* onto a rooted tree by adjoining an edge from it to a rooted tree.

There are five ways the rooted spanning forest can change as a result of these operations. For each of these cases, it is possible to compute explicitly the unique values of the basic variables determined by setting the entering nonbasic at a specified value. This is accomplished by looking at the graph structure of the basis and applying the principle of conservation of flow.

i. An arc enters with both nodes in the same tree. Then another arc in the tree must be dropped and the tree is unchanged except for the addition of the entering arc and the deletion of the blocking arc.

ii. An arc enters with nodes in two different trees, and an arc drops. Dropping the arc cuts off the top of one of the two trees. The top is grafted onto the other tree by the entering arc.

iii. An arc enters with nodes in two different trees and a root drops. The tree from which the root dropped is grafted onto the other tree by the entering arc.

iv. A root enters and a root drops. In this case, the tree is rerooted.

v. A root enters and an arc drops. Dropping the arc cuts off the top of the tree and the entering root becomes the root at this top.

We demonstrate these changes in the rooted spanning forest by considering cases 1–6 in Step 1 under the condition that the entering variable causes a basic variable to be dropped.

For case 1 of Step 1, suppose i and j are in the same tree of $F_{\mathbf{B}}$. Let $P=\langle j,j+1,\ldots,i\rangle$ be the path in $F_{\mathbf{B}}$ from j to i. The situation is depicted in Figure 3.3 where the addition of the arc has created a cycle in the tree. The cycle is broken by the elimination of some other arc in the cycle. This is change i in the rooted spanning forest. The entering variable is shown as well as the corresponding changes in the values of the other basic variables in the cycle. To balance this flow around the cycle, the flow x_{kl} in (k,l) is increased by θ if (k,l) is a forward arc in the path P, and x_{kl} is decreased by θ if (k,l) is a backward arc in P.

Consider case 1 of Step 1 again, but suppose the nodes i and j are in different trees of $F_{\mathbf{B}}$. This situation is depicted in Figure 3.4 where the arc wishing to enter the rooted spanning forest is (i,j). The flow in (i,j) is increased to θ. The flows in the arcs along the path from node 1 to node i in one tree, and in the arcs along the paths from node j to node r in the other tree, are increased by θ if they are forward arcs and decreased by θ if they are backward arcs. The change in the variables s_1^+ and s_r^- is also $-\theta$ because of the sense of these roots. If s_1^- were basic instead of s_1^+, then the direction of the arc at node 1 would be toward node 1 and the change in value of s_1^- would be $+\theta$; Similarly, if s_r^+ were basic instead of s_r^-, the change would be $+\theta$.

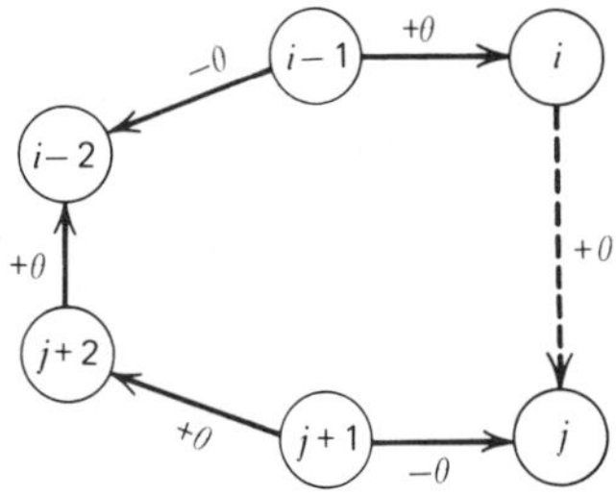

Figure 3.3. Changing a basis—increasing flow around a cycle.

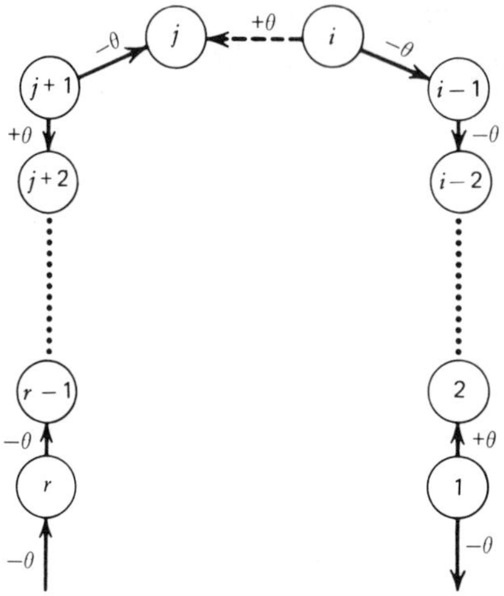

Figure 3.4. Changing a basis—flow between two trees.

If either s_1^+ or s_r^- is the blocking variable, then change iii in the rooted spanning forest occurs. On the other hand, if one of the x_{kl} is the blocking variable, then change ii in the rooted spanning forest occurs.

Suppose case 2 in Step 1 occurs. If the nodes i and j are in the same tree, the situation is depicted in Figure 3.5, which is similar to Figure 3.3. If the nodes i and j are in different trees, the situation is similar to the one depicted in Figure 3.4 except that the flow change is $-\theta$.

The network diagram for case 3 of Step 1 is depicted in Figure 3.6 where node 1 is the root of the tree spanning node i to which there is a root that wants to enter the basis. The adjustment in flow along the unique path from node 1 to node i is $+\theta$ along forward arcs and $-\theta$ along backward arcs, whereas the change in s_1^+ is $-\theta$. If s_1^- corresponded to the root at node 1, then it would be increased by θ. If the variable serving as the root at node 1 is the blocking variable, then change iv in the rooted spanning forest occurs. If one of the arcs along the path from node 1 to node i

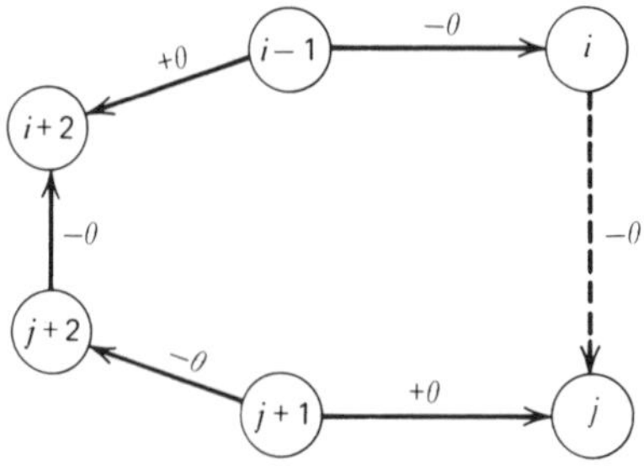

Figure 3.5. Changing a basis—decreasing flow around a cycle.

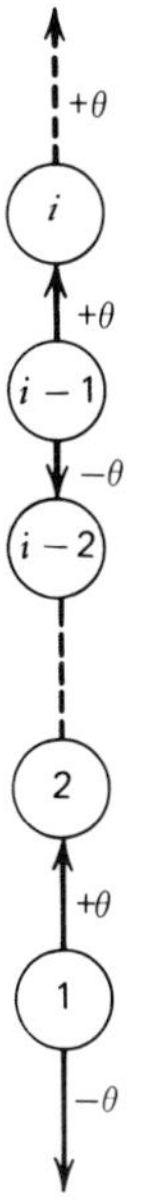

Figure 3.6. Changing a basis—introducing a new root.

corresponds to the blocking variable, then change v in the rooted spanning forest occurs.

Cases 4–6 are entirely analogous to case 3. Further details are omitted.

Step 3. (Compute Dual Variables). The dual variables are calculated by starting at the root of each tree of the rooted spanning forest and working upwards. Specifically, the dual variable u_i at a root attached to the node i is given by

$$u_i = \begin{cases} d_i^+ & \text{if } s_i^+ \text{ is basic} \\ -d_i^- & \text{if } s_i^- \text{ is basic} \end{cases}$$

Once the u_i is determined for the root of a tree in the spanning forest, the dual variables for the other nodes spanned by the forest can be calculated as follows. Starting at the root node i for which u_i is known, for an up arc (i,j), u_j is calculated by $u_j = -c_{ij} + u_i$, and for a down arc (j, i), u_j is calculated by $u_j = c_{ji} + u_i$. Since the rooted tree is connected and has no cycles, the u_i are thereby uniquely determined. In terms of the basis $\mathbf{B}$, the vector of dual variables $\mathbf{u} = c_{\mathbf{B}}\mathbf{B}^{-1}$ where the triangular property of $\mathbf{B}$ has been used to calculate the components of $\mathbf{u}$ without explicit inversion of $\mathbf{B}$. Return to Step 1.

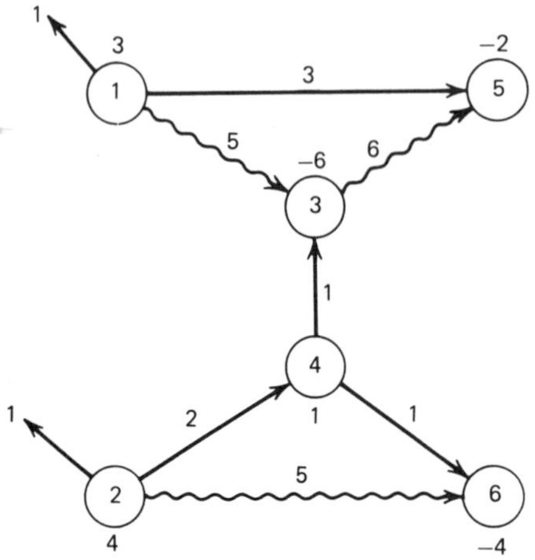

Figure 3.7. Transshipment example—starting basic feasible solution.

EXAMPLE 3.2. We illustrate the above steps by solving Example 3.1 starting with the basic feasible solution shown in Figure 3.7. The numbers next to the arcs are the values of the x_{ij} and s_i^+ variables. The numbers next to the nodes are the values of the dual variables. The variables at upper bounds are denoted by the serpentine arcs.

Relative to the dual variables shown in Figure 3.7, the reduced cost of x_{35} is 6 indicating it should be reduced from its upper bound which is also 6 by coincidence. The situation is depicted in Figure 3.8 where we see the entering arc (3, 5) has nodes in two trees. The blocking variable is x_{43} when $\theta = 1$. (The variable s_1^+ is an alternate choice for blocking variable.) This is

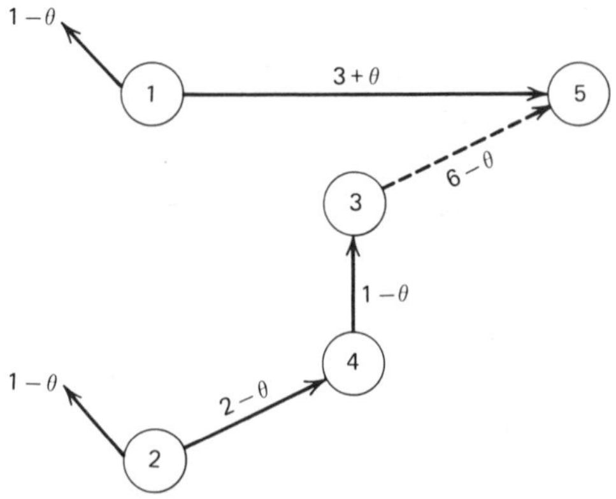

Figure 3.8. Transshipment example—changing the starting basis.

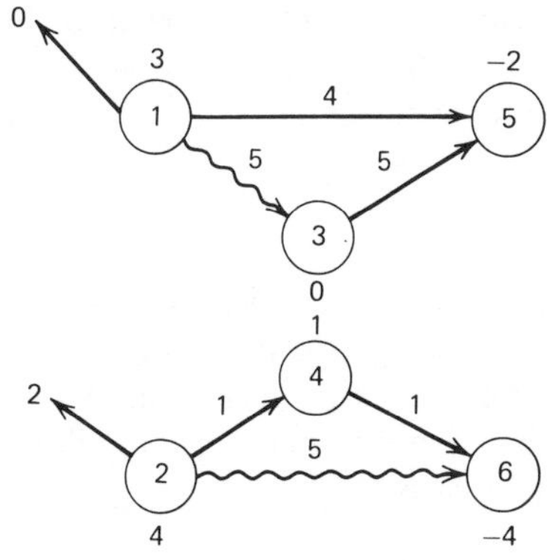

Figure 3.9. Transshipment example—intermediate basic feasible solution.

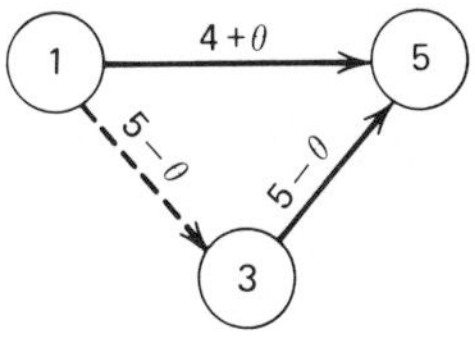

Figure 3.10. Transshipment example—changing the intermediate basic feasible solution.

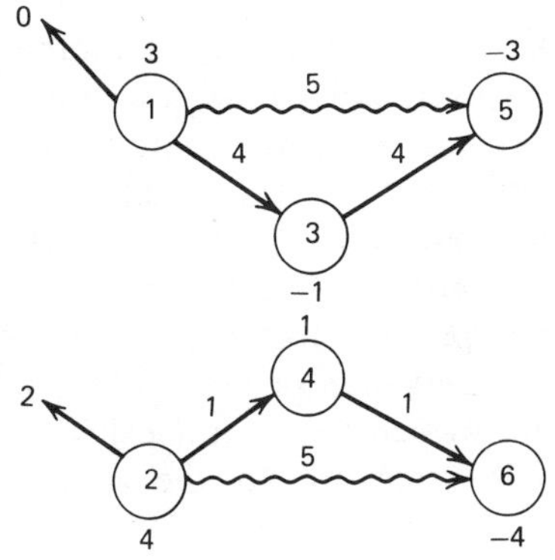

Figure 3.11. Transshipment example—optimal basic feasible solution.

case ii of Step 2; namely, the entering arc cuts off the top of one of the trees (node 3) and the top is grafted onto the other tree.

For the feasible basis of Figure 3.9, the variable x_{13} has positive reduced cost indicating it should be reduced from its upper bound. The calculation of how far it can be reduced consistent with feasibility is given in Figure 3.10. When $\theta = 1$, the variable x_{15} reaches its upper bound, so x_{13} enters the basis at a level of 4, and x_{15} leaves the basis at its upper bound.

The basic feasible solution depicted in Figure 3.11 is optimal because all nonbasic variables at upper bounds have nonpositive reduced cost, and all nonbasic variables at zero have nonnegative reduced cost. ▲

3.5 MAXIMAL FLOW PROBLEM

The maximal flow problem is a special case of the network optimization problem (3.1). For this problem, the node 1 is designated the *source node*, node m is designated the *sink node*, and every arc (i,j) has a finite arc capacity q_{ij} on the flow x_{ij} in the arc. The objective is to maximize the flow from source to sink. The mathematical statement is

$$\begin{aligned} &\max F \\ \text{st.} \quad &\sum_{(i,j)\in\mathcal{A}} x_{1j} - F \geq 0 \\ &\sum_{(i,j)\in\mathcal{A}} x_{ij} - \sum_{(k,i)\in\mathcal{A}} x_{ki} = 0 \qquad \text{for } i=2,\ldots,m-1 \\ &\sum_{(k,m)\in\mathcal{A}} -x_{km} + F = 0 \\ &0 \leq x_{ij} \leq q_{ij} \qquad \text{for all } (i,j)\in\mathcal{A} \end{aligned} \tag{3.2}$$

We assume that every node in the network $[\mathcal{N},\mathcal{A}]$ is connected to both the source and the sink. Note that the variable F corresponds to the flow in the return arc $(m,1)$. The first relation in (3.2) has been written for convenience as an inequality to ensure that the rank of the linear system is m. This follows from Corollary 3.1 because the connectivity of the network and the slack on row 1 implies that we can construct a rooted spanning tree of $[\mathcal{N},\mathcal{A}]$ with node 1 as the root. The surplus variable on row 1 will be zero in any feasible solution because of conservation of flow; in other words, the net flow F out of node 1 cannot be greater than F at the sink as determined by the last equation because there is nowhere for the surplus flow to go.

Maximal flow problems can have multiple sources and sinks, and there can be arcs directed toward sources and directed away from sinks. Such problems can be transformed to the form (3.2) by the addition of a fictitious new source with arcs directed toward the original sources, and a fictitious new sink with arcs from the original sinks directed to it.

Figure 3.12 depicts a maximal flow problem that serves as a numerical example in this section. For this problem, node 1 is the source, node 6 is the sink, and the arc $(6,1)$ corresponds to the variable F. The numbers drawn next to the arcs are the capacities of these arcs; there is no capacity constraint on arc $(6,1)$.

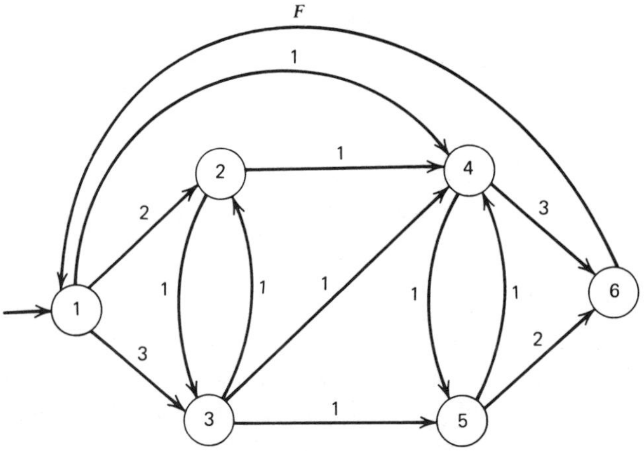

Figure 3.12. Maximal flow example.

Since the maximal flow problem (3.2) has rank m but only one root, Theorem 3.1 implies that every basic solution will correspond to a rooted spanning tree of the node set $\mathcal{N}$. Moreover, the variable F corresponding to $(m, 1)$ can be chosen to be basic in every basic feasible solution since it is unconstrained in sign. In general, F will be positive in any basic feasible solution. In addition, the variable s_1^- will always be basic at a value of zero.

Suppose for the moment that we remove the arc $(m, 1)$ from any rooted spanning trees. It is easy to see that two nonoverlapping subtrees are created that span the node set $\mathcal{N}$, one subtree connected to the source node 1 and the other subtree connected to the sink node m. Thus, for every basic feasible solution with the arc $(m, 1)$ removed, the node set $\mathcal{N}$ is partitioned into two subsets, say $\mathcal{N}_0$ and $\mathcal{N}_1$, where $\mathcal{N}_0$ contains the source and the nodes connected to it, and $\mathcal{N}_1$ contains the sink and the nodes connected to it. As we shall see, $\mathcal{N}_0$ also refers to the nodes with dual variable value 0, and $\mathcal{N}_1$ refers to the nodes with dual variable value 1, where the dual variables are computed in the usual way from a feasible basis, or equivalently, a rooted spanning tree. Let $(\mathcal{N}_0, \mathcal{N}_1)$ denote all the arcs $(i,j) \in \mathcal{A}$ with $i \in \mathcal{N}_0, j \in \mathcal{N}_1$; this set is called a *cut* separating source and sink. More generally, a cut can be defined with respect to any pair P, Q of node sets that partition $\mathcal{N}$ with the additional property that the source node is in P and the sink node is in Q. The relationship of cuts to the maximal flow problem will be developed below after the maximal flow simplex algorithm is presented.

An iteration of the maximal flow simplex algorithm begins with some variables at their upper bounds, and a basic feasible solution for the residual maximal flow problem that results when these variables are netted out of (3.2). The algorithm is simply a specialization of the network primal simplex algorithm given in the previous section.

Maximal Flow Simplex Algorithm

Step 1. (Price Out Nonbasic Arcs). Search for an arc (i,j) such that either

1. $x_{ij}=0$ with $i\in\mathcal{N}_0, j\in\mathcal{N}_1$; or
2. $x_{ij}=q_{ij}$ with $i\in\mathcal{N}_1, j\in\mathcal{N}_0$

The set $\mathcal{N}_0$ contains all the nodes with dual variables equal to zero, and the set $\mathcal{N}_1$ contains all the nodes with dual variables equal to one. If neither of these conditions obtains, the current basic feasible solution is optimal, and we exit from the algorithm.

Step 2. (Find a Blocking Variable and Compute New Basic Feasible Solution). Suppose case 1 in Step 1 obtained. Find the unique path $\langle 1,\dots,i_t\rangle$, where $i_t=i$, connecting the source to node i in the given rooted spanning tree. Similarly, find the unique path $\langle i_{t+1},\dots,m\rangle$, where $i_{t+1}=j$, connecting j to the sink. The path $P=\langle 1,\dots,i_t,i_{t+1},\dots,m\rangle$ connects the source to the sink, and it is called a *flow augmenting path*. The variable $x_{ij}=x_{i_t,i_{t+1}}$ is increased from 0 to the maximal value such that the feasibility condition $0\le x_{kl}\le q_{kl}$ is maintained for variables corresponding to arcs in the path P. If x_{ij} equals θ, the adjusted value of the variable corresponding to the edge $\langle k,l\rangle$ in the path P is $x_{kl}+\theta$ if $\langle k,l\rangle$ corresponds to a forward arc (k,l) and $x_{kl}-\theta$ if it corresponds to a reverse arc (l,k). The maximal value of θ consistent with feasibility is attained when a variable is driven to 0 or to its upper bound. This causes the cycle $\langle 1,i_2,\dots,m,1\rangle$ to be broken up as the edge $\langle k,l\rangle$ is not included as a basic variable in the new rooted spanning tree. Note that the edge $\langle k,l\rangle$ can be $\langle i,j\rangle$ corresponding to the entering arc (i,j). If case 2 in Step 1 obtained, the analysis is very similar. The main differences are that the edge $\langle i_t,i_{t+1}\rangle=\langle j,i\rangle$ in the path P corresponds to the backward arc (i,j) and the variable x_{ij} is decreased from its upper bound q_{ij} by a nonnegative quantity θ. For a variable corresponding to a forward arc (k,l) in the path, the value is $x_{kl}+\theta$, whereas a variable corresponding to a reverse arc (l,k) is $x_{kl}-\theta$.

Step 3. (Compute the Dual Variables). The rule for computing dual variables in Step 3 of the network primal simplex algorithm of the previous section has a special interpretation here. First, the source node has a dual variable value equal to 0 since the root at the source node is basic with objective function cost 0. In addition, the sink node has a dual variable value equal to 1 since it is connected to the source by the arc $(m, 1)$ with objective function cost 1. Since the objective function costs of all the other arcs in the tree are zero, it follows that the dual variable values for nodes $2,\ldots,m-1$ are either zero or one depending on whether the node is connected to the source or the sink when $(m, 1)$ is removed. Let $\mathfrak{N}_0$ denote the nodes connected to the source node and $\mathfrak{N}_1$ denote the nodes connected to the sink node. Return to Step 1.

EXAMPLE 3.3. The maximal flow simplex algorithm is illustrated by solving the example of Figure 3.12. The starting rooted spanning tree is given in Figure 3.13a where the numbers next to the arcs are the flows in the arcs, and the numbers next to the nodes are the dual variables. The flow augmenting path selected for this tree is shown in Figure 3.13b. The maximal value θ of flow increase along this path is 1 driving the flow in $(2,4)$ to its upper bound.

Figures 3.14a and 3.14b show the next iteration of the algorithm; as before, the serpentine arc denotes flow in the arc at its upper bound. Note that the flow augmenting path in Figure 3.14b is degenerate; that is, the reverse arc $(3,2)$ limits the increase in flow to zero and the arc leaves the basis. The labeling algorithm given below overcomes this difficulty and

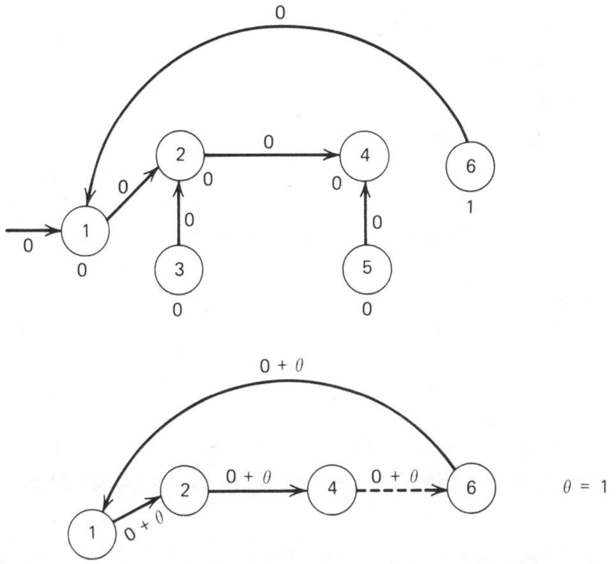

Figure 3.13. Maximal flow example. (a) Starting basic feasible solution. (b) Changing the starting basis.

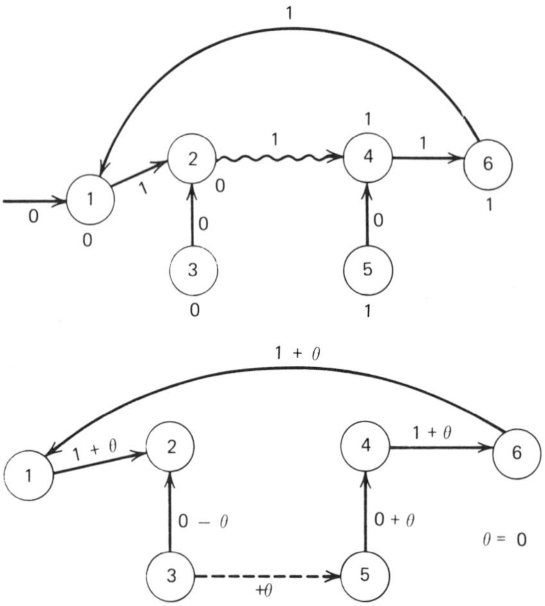

Figure 3.14. Maximal flow example. (a) Intermediate basic feasible solution. (b) Changing the intermediate basic feasible solution.

finds a nondegenerate flow augmenting path (or proves optimality) at each iteration by considering simultaneously more than one path from source to sink. The optimal solution is depicted in Figure 3.15. ▲

The maximal flow problem has been studied as a special combinatorial optimization problem distinct from the network optimization problem (3.1). We discuss briefly how some of the special results obtained for it can be derived from the simplex method analysis above. In addition, we give the labeling algorithm for the maximal flow problem and discuss how it differs from the maximal flow simplex algorithm.

Given any cut (P,R) of the arc set $\mathcal{Q}$, we define the *cut capacity*

$$q(P,R) = \sum_{(i,j)\in(P,R)} q_{ij}$$

In the numerical example of Figure 3.12, the cut (P,R) with $P=\{1,2\}$ and $R=\{3,4,5,6\}$ consists of the arcs $(1,3)$, $(1,4)$, $(2,3)$, $(2,4)$ with cut capacity 6.

LEMMA 3.3. Let (P,R) be any cut of the arc set $\mathcal{Q}$. The maximal flow in the network $[\mathfrak{N},\mathcal{Q}]$ is less than or equal to the cut capacity of (P,R).

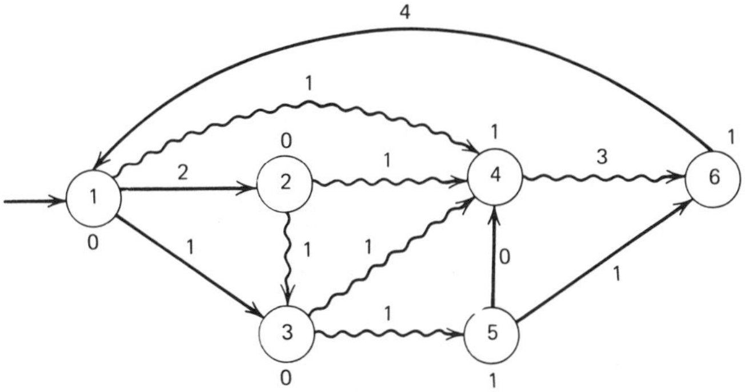

Figure 3.15. Maximal flow example—optimal basic feasible solution.

PROOF. Let x_{ij} for all $(i,j) \in \mathcal{A}$, and F be any feasible solution to (3.2). We have from (3.2)

$$\sum_{(1,j) \in \mathcal{A}} x_{1j} = F$$

and

$$\sum_{(i,j) \in \mathcal{A}} x_{ij} - \sum_{(k,i) \in \mathcal{A}} x_{ki} = 0 \qquad \text{for } i \in P,\ i \neq 1$$

Recall that the inequality constraint in (3.2) must be equality by conservation of flow. The flow variable x_{ij} for $i \in P$ and $j \in P$ appears in two equations, once as x_{ij} in equation i and once as $-x_{ij}$ in equation j. Thus, if we sum the above equations, these flows cancel out with the result

$$\begin{aligned} F &= \sum_{(i,j) \in (P,R)} x_{ij} - \sum_{(k,i) \in (R,P)} x_{ki} \\ &\leq \sum_{(i,j) \in (P,R)} x_{ij} \\ &\leq \sum_{(i,j) \in (P,R)} q_{ij} \end{aligned} \tag{3.3}$$

where the first inequality follows from the nonnegativity of the x_{ki} and the second inequality follows because $x_{ij} \leq q_{ij}$. ■

The result of Lemma 3.3 is the analogue of the weak linear programming duality Lemma 2.1 for the maximal flow problem. The following theorem is the analogue of the strong linear programming duality theorem 2.1.

THEOREM 3.2. (Max-Flow Min-Cut Theorem). The maximal flow in the network $[\mathfrak{N}, \mathcal{A}]$ equals the minimal cut capacity of all cuts separating source and sink.

PROOF. Because of Lemma 3.3, the theorem is established if we can exhibit a feasible flow in (3.2) equal to the cut capacity of some cut. A maximal flow exists because we have assumed all arc capacities to be finite. Specifically, consider the maximal flow F and flow variables x_{ij} produced by the maximal flow simplex algorithm. By Step 1 of the algorithm, both cases 1 and 2 fail to hold for the cut set $(\mathfrak{N}_0, \mathfrak{N}_1)$ corresponding to this solution. Consider any arc (i,j) with $i \in \mathfrak{N}_0, j \in \mathfrak{N}_1$. The variable x_{ij} is nonbasic by definition of $\mathfrak{N}_0$ and $\mathfrak{N}_1$ and must be at a value of 0 or at a value of q_{ij}. But, by assumption, case 1 does not hold implying $x_{ij} = q_{ij}$.

Similarly, the variable x_{ki} with $k \in \mathfrak{N}_1$, $i \in \mathfrak{N}_0$, is nonbasic and the failure of case 2 to hold implies that variable $x_{ki} = 0$. The proof is immediate using these properties and the construction (3.3) of Lemma 3.3 with $P = \mathfrak{N}_0$, $R = \mathfrak{N}_1$, namely,

$$\begin{aligned} F &= \sum_{(i,j) \in (\mathfrak{N}_0, \mathfrak{N}_1)} x_{ij} - \sum_{(k,i) \in (\mathfrak{N}_1, \mathfrak{N}_0)} x_{ki} \\ &= \sum_{(i,j) \in (\mathfrak{N}_0, \mathfrak{N}_1)} q_{ij} \\ &= q(\mathfrak{N}_0, \mathfrak{N}_1) \quad \blacksquare \end{aligned}$$

Convergence of the simplex algorithm is required in the above proof of the max-flow min-cut theorem. In Section 1.3 convergence was guaranteed even when there is degeneracy, and we saw in Figure 3.14b that degenerate flow augmenting paths causing degenerate basis changes can in fact occur. It is possible, however, to give a related algorithm, called the maximal flow labeling algorithm, which generates a nondegenerate flow augmenting path at each iteration or establishes optimality. Although it is similar to the maximal flow simplex algorithm given above, the maximal flow labeling algorithm is qualitatively different in the following sense. If properly implemented, the labeling algorithm requires a number of iterations and arithmetic operations that is bounded by a polynomial in the parameters m and n of the problem; that is, it is a theoretically "good" algorithm. This is in contrast to the simplex algorithm for the maximal flow problem, which as we have said in Section 1.2, is not a theoretically "good" problem.

The labeling algorithm stated below is due to Ford and Fulkerson, and we present it in a form similar to the one they give in their book *Flows in Networks*. Finite convergence of this algorithm for the maximal flow problems requires integrality of the arc capacities q_{ij}. If this assumption is made, then the final labeling of the maximal flow network when the algorithm terminates could be used to prove the max-flow min-cut theorem. The cut set for which the cut capacity equals the maximal flow is derived from partitioning the nodes into two sets, labeled and unlabeled. The return arc $(m, 1)$ is not used in the labeling algorithm.

Maximal Flow Labeling Algorithm

Step 1. (Label Nodes). The source 1 is given the label $[0, \Delta(1) = \infty]$. We say the source is labeled and unscanned; all other nodes are unlabeled.

Select any labeled, unscanned node j with label $[i^{\pm}, \Delta(j)]$. For all unlabeled nodes k such that $x_{jk} < q_{jk}$, assign the label $[j^{+}, \Delta(k)]$, where

$$\Delta(k) = \min\left[\Delta(j), q_{jk} - x_{jk}\right]$$

Such nodes are now labeled and unscanned. For all unlabeled nodes i such that $x_{ij} > 0$, assign the label $[j^{-}, \Delta(i)]$, where

$$\Delta(i) = \min\left[\Delta(j), x_{ij}\right]$$

Such nodes are now also labeled and unscanned. The node j is labeled and scanned.

Repeat the labeling step either until the sink m is labeled and unscanned, or until no more labels can be assigned and the sink is unlabeled. In the former case, go to Step 2. In the latter case, exit with the maximal flow solution (optimality can be established by appeal to the max-flow min-cut theorem).

Step 2. (Augment Flow). The sink m has label $[k^{\pm}, \Delta(m)]$. If the label is $[k^{+}, \Delta(m)]$, replace x_{km} by $x_{km} + \Delta(m)$. If the label is $[k^{-}, \Delta(m)]$, replace x_{km} by $x_{km} - \Delta(m)$. Then consider the label $[j^{\pm}, \Delta(k)]$ on node k. As before, either replace x_{jk} by $x_{jk} + \Delta(m)$ if the label is $[j^{+}, \Delta(k)]$, or replace x_{jk} by $x_{jk} - \Delta(m)$ if the label is $[j^{-}, \Delta(k)]$. Continue these changes until the source node 1 is reached. Return to Step 1.

EXAMPLE 3.4. We illustrate in Figure 3.16 the labeling algorithm by considering the numerical example of Figure 3.12 with the flows given in Figure 3.14a which led to the selection of a degenerate flow augmenting path in Figure 3.14b by the maximal flow simplex algorithm. The algorithm finds the flow augmenting path $\langle 1, 4, 6 \rangle$. Figure 3.17 illustrates the

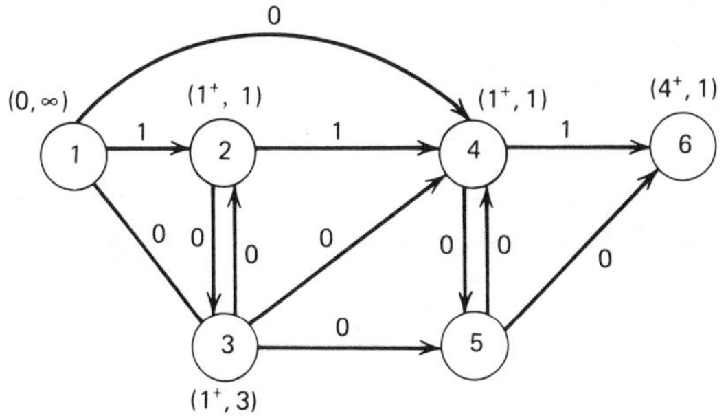

Figure 3.16. Maximal flow example—finding a flow augmenting path by the labeling algorithm.

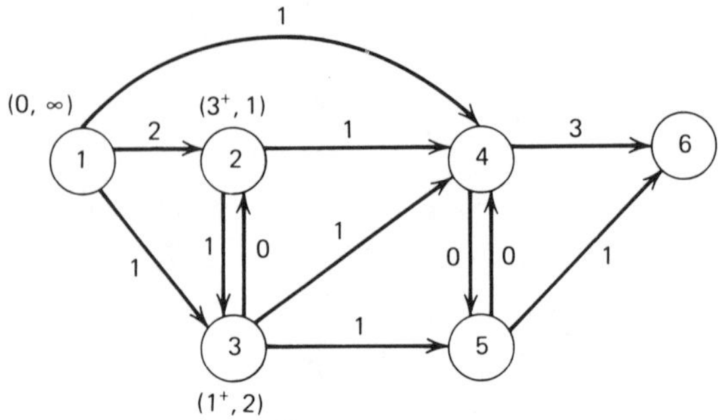

Figure 3.17. Maximal flow example—termination of the labeling algorithm with a maximal flow.

labeling when the flow in the given network is maximal. Note that the indicated cut (P,R) in Figure 3.17 is $P=\{1,2,3\}$ and $R=\{4,5,6\}$ with cut capacity 4 equal to the maximal flow. ▲

EXAMPLE 3.5. (Chain Decomposition by Maximal Flow: An Application to Airline Scheduling). An airline wishes to minimize the number of aircraft of a given type to meet a schedule of m flights, which repeats itself daily. The problem has a network description with nodes $1,\ldots,m$, one for each flight, and arcs (i,j) for each pair of flights i and j such that the origin city of flight j is the destination city of flight i and flight j departs after the arrival of flight i. This network is acyclic because time does not repeat itself. Figure 3.18 is an 8 node example of such a network.

Any acyclic network can be decomposed into chains that partition the node set into subsets, each of which include the nodes visited by the given chain; degenerate chains consist of single nodes, and no arcs. It is clear that the network can always be decomposed into m single node chains. The scheduling problem is solved by finding the chain decomposition consisting of a minimal number of chains because each chain corresponds to a sequence of flights covered by a single aircraft.

In order to find a minimal chain decomposition of the acyclic network, we solve a related maximal flow problem implied by the following theorem. ▲

THEOREM 3.3. (Dilworth). Let G be a chain decomposed acyclic network consisting of m nodes. Then

$$c+p=m$$

where c denotes the number of chains in the decomposition and p denotes the number of arcs that are components of chains.

PROOF. Let the chains in the decomposition be indexed k for $k=1,\ldots,c$. The number of nodes belonging to chain k is denoted by m_k. Since

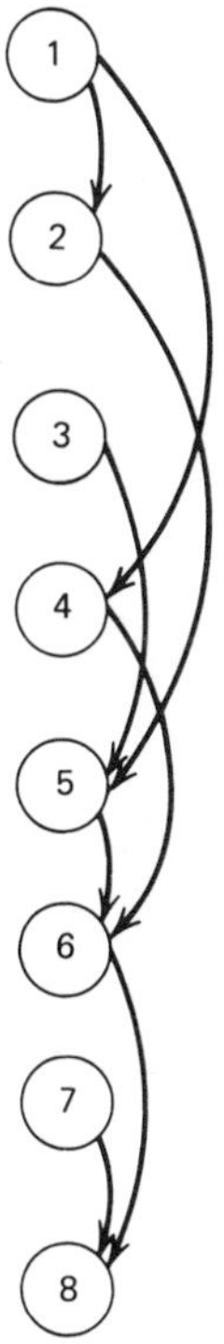

Figure 3.18. Airline scheduling problem.

each node belongs to exactly one chain, we have $\Sigma_{k=1}^{c} m_k = m$. Each chain contains one less arc than nodes implying

$$m = \Sigma_{k=1}^{c} m_k = \Sigma_{k=1}^{c} (m_k - 1) + c = p + c \quad \blacksquare$$

The direct implication of Theorem 3.3 is that we can find a minimal chain decomposition of the network by maximizing the number of arcs used in a chain decomposition. This is accomplished by constructing a network G^* derived from the acyclic network G and solving a maximal flow problem over G^*. The node set of G^* is $1, \ldots, 2m$ and the arc set consists of all arcs $(i, m+j)$ for (i,j) in the arc set of G. The capacity of each of these arcs is one. In addition, there is a source node 0 and a sink node $2m+1$ and arcs $(0,i)$ for $i = 1, \ldots, m$, and $(j, 2m+1)$ for $j = m+1, \ldots, 2m$ with capacities also equal to one.

It is easy to see that the maximal flow from source to sink in the network G^* yields a minimal chain decomposition. Each arc $(i, m+j)$ in G^* with a flow of one is an arc (i,j) in the chain decomposition of G. If nodes i and $m+i$ in G^* have no flow into or out of them, then node i in G is a single node chain. The fact that the flow in G^* is maximal implies that the corresponding chain decomposition is minimal by Theorem 3.3. Figure

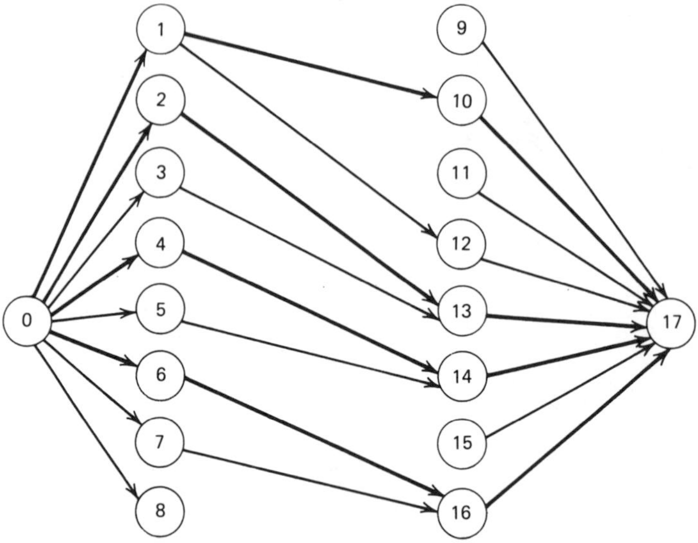

Figure 3.19. Maximal flow representation of airline scheduling problem.

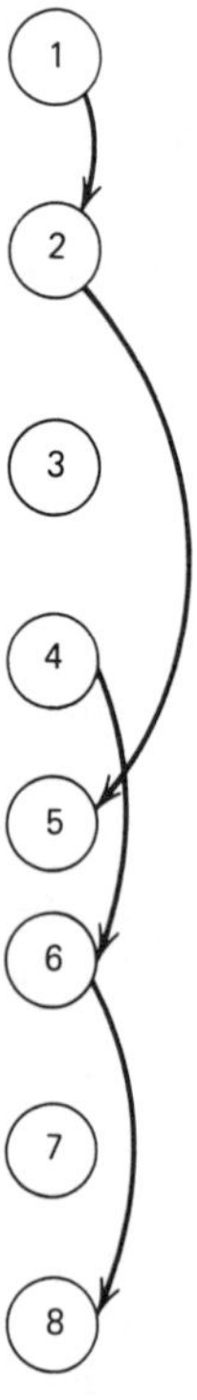

Figure 3.20. Minimal chain decomposition of airline scheduling problem.

3.19 gives the network G^* derived from G with the maximal flow of 4 indicated by the heavy lines. Figure 3.20 is the corresponding minimal chain decomposition of G consisting of 4 chains.

3.6 PRIMAL-DUAL AND OUT-OF-KILTER NETWORK OPTIMIZATION ALGORITHMS

There are alternative algorithmic approaches to the primal simplex algorithm for the network optimization problem. These approaches are based on a different use of the necessary and sufficient primal-dual optimality conditions of linear programming as discussed in Section 2.2. Complete detail about the variety of algorithmic approaches is beyond the intended scope of this book and we simply sketch the basic ideas.

Both the primal-dual algorithm and the out-of-kilter algorithm rely on iterative use of the maximal flow labeling algorithm given in the previous section. As noted, the labeling algorithm is a good algorithm, both theoretically and empirically, and thus there is reason to believe that the algorithms given in this section will perform better, at least on some problems, than the primal simplex algorithm given in Section 3.3. Empirically, however, this belief has been neither substantiated nor shown to be incorrect and the selection of a specific algorithm for a specific network optimization problem remains a question for which there is not yet a completely satisfactory answer.

For expositional convenience, we will work with a special class of network problems defined on the network G called *circulation* problems. This problem is

$$\min \sum_{(i,j)\in\mathcal{A}} c_{ij}x_{ij} \tag{3.4a}$$

$$\text{s.t.} \sum_{(i,j)\in\mathcal{A}} x_{ij} - \sum_{(k,i)\in\mathcal{A}} x_{ki} = 0 \qquad \text{for } i=1,\ldots,m \tag{3.4b}$$

$$l_{ij} \le x_{ij} \le q_{ij} \qquad \text{for all } (i,j)\in\mathcal{A} \tag{3.4c}$$

where $l_{ij} \ge 0$. A wide variety of network optimization problems can be converted to this form. For example, Figure 3.21 is the network representation as a circulation problem of the problem of Table 3.1 and Figure 3.1. The network in Figure 3.21 differs from the network in Figure 3.1 by the addition of the source and sink nodes 0 and 7, and the arcs (0,1), (0,2), (5,7), (6,7), and (7,0). The costs and capacities of the other arcs in the network are the same as in Table 3.1. Table 3.2 contains this information for the new arcs. The flow x_{01} in the arc (0,1) is constrained to lie between 4 and 9 because the capacity of the slack variable s_1^+ is 5 in Table 3.1, and

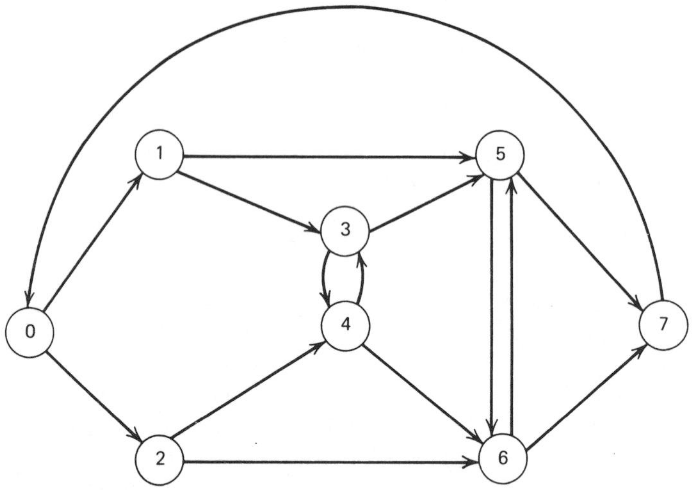

Figure 3.21. Reformulation of transshipment problem as a circulation problem.

the flow out of node 1 is constrained to be 9. The cost per unit flow in (0, 1) is -3 because this is the per unit reduction in cost due to less slack at node 1. Since there is no slack permitted in Table 3.1 at nodes 5 and 6, we have $l_{ij} = q_{ij}$ for the arcs (5, 7), (6, 7). The arc (7, 0) is a return arc with zero cost and upper capacity greater than the flow of 15 it carries for any feasible solution. A general construction along these lines is possible for many network optimization problems.

Listed below are the primal-dual complementary slackness conditions (of Corollary 2.2) for the circulation problem (3.4). Let $\mathbf{u} \in R^m$ be any dual solution; there are no constraints on the dual variables. The dual solution is optimal and the following primal solution x_{ij} for $(i,j) \in \mathcal{A}$, is also optimal

TABLE 3.2

(i,j)	c_{ij}	l_{ij}	q_{ij}
(0, 1)	−3	4	9
(0, 2)	−4	2	8
(5, 7)	0	9	9
(6, 7)	0	6	6
(7, 0)	0	0	16

if and only if

$$\sum_{(i,j)\in\mathcal{Q}} x_{ij} - \sum_{(k,i)\in\mathcal{Q}} x_{ki} = 0 \quad \text{for } i=1,\ldots,m \tag{3.5a}$$

$$c_{ij} - u_i + u_j < 0 \Rightarrow x_{ij} = q_{ij} \tag{3.5b}$$

$$c_{ij} - u_i + u_j = 0 \Rightarrow l_{ij} \le x_{ij} \le q_{ij} \tag{3.5c}$$

$$c_{ij} - u_i + u_j > 0 \Rightarrow x_{ij} = l_{ij} \tag{3.5d}$$

The primal-dual algorithm maintains (3.5b), (3.5c), (3.5d) at each iteration while trying to attain the feasibility condition (3.5a). The out-of-kilter algorithm takes the opposite approach and maintains only (3.5a) and seeks to attain the others.

The primal-dual algorithm for the circulation problem is a direct adaptation of the primal-dual simplex algorithm given in Section 2.7 for ordinary linear programming problems. The algorithm starts with a dual solution $\mathbf{u}\in R^m$. The arc set $\mathcal{Q}$ is partitioned into three sets relative to $\mathbf{u}$

$$\mathcal{Q}^-(\mathbf{u}) = \{(i,j)\in\mathcal{Q} \mid c_{ij} - u_i + u_j < 0\}$$

$$\mathcal{Q}^0(\mathbf{u}) = \{(i,j)\in\mathcal{Q} \mid c_{ij} - u_i + u_j = 0\}$$

$$\mathcal{Q}^+(\mathbf{u}) = \{(i,j)\in\mathcal{Q} \mid c_{ij} - u_i + u_j > 0\}$$

The variables x_{ij} for $(i,j)\in\mathcal{Q}^-(\mathbf{u})$ are set at their upper bounds q_{ij} and the variables x_{ij} for $(i,j)\in\mathcal{Q}^+(\mathbf{u})$ are set at their lower bounds l_{ij}. The variables x_{ij}, $(i,j)\in\mathcal{Q}^0(\mathbf{u})$ are free to vary between upper and lower bounds and the primal-dual algorithm tries to select them so that the flow equations (3.5a) are satisfied in which case an optimal solution to (3.4) has been found. To this end, define

$$b_i(\mathbf{u}) = \sum_{(k,i)\in\mathcal{Q}^-(\mathbf{u})} q_{ki} + \sum_{(k,i)\in\mathcal{Q}^+(\mathbf{u})} l_{ki} - \sum_{(i,j)\in\mathcal{Q}^-(\mathbf{u})} q_{ij} - \sum_{(i,j)\in\mathcal{Q}^+(\mathbf{u})} l_{ij} \quad \text{for } i=1,\ldots,m$$

The selection of the x_{ij}, $(i,j)\in\mathcal{Q}^0(\mathbf{u})$ is according to the phase one network optimization problem

$$\min \sum_{i=1}^{m} y_i^+ + y_i^-$$

$$\text{s.t.} \quad \sum_{(i,j)\in\mathcal{Q}^0(\mathbf{u})} x_{ij} - \sum_{(k,i)\in\mathcal{Q}^0(\mathbf{u})} x_{ki} + y_i^+ - y_i^- = b_i(\mathbf{u}) \quad \text{for } i=1,\ldots,m \tag{3.6}$$

$$l_{ij} \le x_{ij} \le q_{ij} \quad \text{for } (i,j)\in\mathcal{Q}^0(\mathbf{u})$$

$$y_i^+ \ge 0, \quad y_i^- \ge 0 \quad \text{for } i=1,\ldots,m$$

Problems of this type can be solved by the maximal flow labeling algorithm (see Exercise 3.8). As a result of the lower and upper bound substitution for $(i,j)\in\mathscr{A}^-(\mathbf{u})\cup\mathscr{A}^+(\mathbf{u})$ and the maximal flow calculation, we have a primal solution x_{ij} and dual solution u_i to the circulation problem which satisfies all of the conditions (3.5) except probably (3.5a). If this solution satisfies the circulation equations (3.5a), or equivalently, if the minimal objective function value in (3.6) is zero, then it is optimal.

If the minimal objective function value in (3.6) is greater than zero, then by necessity we have an optimal solution $\mathbf{w}\neq 0$ to the dual of (3.6) which satisfies the following conditions

$$x_{ij}=q_{ij} \quad \text{for } (i,j)\in\mathscr{A}^0(\mathbf{u}) \quad \text{and} \quad w_j-w_i<0 \qquad (3.7a)$$

$$x_{ij}=l_{ij} \quad \text{for } (i,j)\in\mathscr{A}^0(\mathbf{u}) \quad \text{and} \quad w_j-w_i>0 \qquad (3.7b)$$

$$l_{ij}\le x_{ij}\le q_{ij} \quad \text{for } (i,j)\in\mathscr{A}^0(\mathbf{u}) \quad \text{and} \quad w_j-w_i=0 \qquad (3.7c)$$

The m-vector $\mathbf{w}$ serves as a direction of change of the dual solution $\mathbf{u}$. Specifically, $\mathbf{u}$ is replaced by $\mathbf{u}+\theta^*\mathbf{w}$ where θ^* is the largest positive value of θ consistent with

$$c_{ij}-(u_i+\theta w_i)+(u_j+\theta w_j)\begin{cases}\ge 0 & \text{if } (i,j)\in\mathscr{A}^+(\mathbf{u})\\ \le 0 & \text{if } (i,j)\in\mathscr{A}^-(\mathbf{u})\end{cases}$$

If $\theta^*=+\infty$, the circulation problem has no feasible solution because the dual objective function can be driven to $+\infty$.

Suppose $\theta^*<+\infty$ and consider the above analysis at the new dual solution $\mathbf{u}+\theta^*\mathbf{w}$. By construction, the solution x_{ij}, $(i,j)\in\mathscr{A}$, given by (3.7) plus $x_{ij}=q_{ij}$ for $(i,j)\in\mathscr{A}^-(\mathbf{u})$, $x_{ij}=l_{ij}$ for $(i,j)\in\mathscr{A}^+(\mathbf{u})$, satisfies the conditions (3.5), except (3.5a), at $\mathbf{u}+\theta^*\mathbf{w}$. To see why this is so, consider x_{ij} defined in (3.7a). Since $c_{ij}-u_i+u_j=0$, we have $c_{ij}-(u_i+\theta w_i)+(u_j+\theta w_j)=\theta(w_j-w_i)<0$ for all $\theta>0$ implying $(i,j)\in\mathscr{A}^-(\mathbf{u}+\theta^*\mathbf{w})$ and $x_{ij}=q_{ij}$ is the correct setting at $\mathbf{u}+\theta^*\mathbf{w}$. The same argument shows $(i,j)\in\mathscr{A}^+(\mathbf{u}+\theta^*\mathbf{w})$ for x_{ij} defined in (3.7b) and $(i,j)\in\mathscr{A}^0(\mathbf{u}+\theta^*\mathbf{w})$ for x_{ij} defined in (3.7c). Similarly, $\theta^*>0$ is chosen small enough that, except for one arc, $(i,j)\in\mathscr{A}^-(\mathbf{u})$ implies $(i,j)\in\mathscr{A}^-(\mathbf{u}+\theta^*\mathbf{w})$ and $(i,j)\in\mathscr{A}^+(\mathbf{u})$ implies $(i,j)\in\mathscr{A}^+(\mathbf{u}+\theta^*\mathbf{w})$. The exceptional arc, say $(s,t)\in\mathscr{A}^-(\mathbf{u})$ is chosen such that $c_{st}-(u_s+\theta^*w_s)+(u_t+\theta^*w_t)=0$. Thus $x_{st}=q_{st}$ is a correct setting at $\mathbf{u}+\theta^*\mathbf{w}$ since the variable prices out zero.

The analysis at $\mathbf{u}+\theta^*\mathbf{w}$ proceeds by exploiting the variable x_{st} since now it is free to vary in the range $l_{st}\le x_{st}\le q_{st}$. In terms of the phase one arc network optimization problem (3.6), the variable x_{st} is added, it prices out negatively, and the minimization of the artificial variables continues. It can be shown that the maximal flow labeling algorithm can continue from its previous termination by the addition of the arc (s,t). The entire process described above is repeated. The primal-dual algorithm converges because there is a monotone decrease in the phase one objective function value. Degeneracy can be resolved by the lexicographic methods of Section 1.3.

We consider briefly the out-of-kilter algorithm. A set of flows x_{ij} for all $(i,j)\in\mathscr{Q}$ satisfying the flow equations (3.4b) is called a *circulation*. The algorithm begins with any circulation and any dual solution $\mathbf{u}\in R^m$. Thus, the optimality conditions (3.5b), (3.5c), (3.5d) may not hold for a given arc. At each iteration, an arc (i,j) is in exactly one of the states

$$
\begin{array}{lll}
(\alpha) & c_{ij}-u_i+u_j>0 & \text{for } x_{ij}=l_{ij} \\
(\beta) & c_{ij}-u_i+u_j=0 & \text{for } l_{ij}\le x_{ij}\le q_{ij} \\
(\gamma) & c_{ij}-u_i+u_j<0 & \text{for } x_{ij}=q_{ij} \\
(\alpha_1) & c_{ij}-u_i+u_j>0 & \text{for } x_{ij}<l_{ij} \\
(\beta_1) & c_{ij}-u_i+u_j=0 & \text{for } x_{ij}<l_{ij} \\
(\gamma_1) & c_{ij}-u_i+u_j<0 & \text{for } x_{ij}<q_{ij} \\
(\alpha_2) & c_{ij}-u_i+u_j>0 & \text{for } x_{ij}>l_{ij} \\
(\beta_2) & c_{ij}-u_i+u_j=0 & \text{for } x_{ij}>q_{ij} \\
(\gamma_2) & c_{ij}-u_i+u_j<0 & \text{for } x_{ij}>q_{ij}
\end{array}
$$

An arc (i,j) is said to be *in-kilter* if it is in any one of the states α,β,γ; otherwise, the arc is said to be *out-of-kilter*. As above, the circulation x_{ij} and the dual variables u_i are optimal in the circulation problem (3.4), and its linear programming dual, if all arcs are in-kilter. Otherwise, the out-of-kilter algorithm systematically reduces the out-of-kilter aspect of a primal-dual pair of solutions until feasibility and optimality are obtained.

To this end, we associate a nonnegative number, called the kilter number, with each arc (i,j) at each solution. If the arc (i,j) is in-kilter, the kilter number is 0. If the arc is out-of-kilter, the positive kilter numbers are as follows

$$
\begin{array}{rl}
(\alpha_1) \text{ or } (\beta_1): & l_{ij}-x_{ij} \\
(\gamma_1): & (c_{ij}-u_i+u_j)(x_{ij}-q_{ij}) \\
(\alpha_2): & (c_{ij}-u_i+u_j)(x_{ij}-l_{ij}) \\
(\beta_2) \text{ or } (\gamma_2): & x_{ij}-q_{ij}
\end{array}
$$

For states $\alpha_1,\beta_1,\beta_2,\gamma_2$, the kilter numbers measure the infeasibility of the circulation in the arc (i,j). For states γ_1,α_2, the kilter numbers measure the lack of optimality of the flow in (i,j).

At each iteration of the out-of-kilter algorithm, an out-of-kilter arc is selected and an attempt is made to put it into kilter. This is done using the maximal flow algorithm in such a way that each in-kilter arc remains in-kilter, whereas the kilter number of an out-of-kilter arc either remains unchanged or is decreased. The circulation problem is found to be infeasible when an arc in one of the states (α_1), (β_1), (β_2), or (γ_2) is shown to have a positive out-of-kilter number that cannot be reduced. Otherwise, convergence of the out-of-kilter algorithm is ensured by the monotonic decrease in the sum of the out-of-kilter numbers until zero is reached. Resolution of degeneracy is handled in a special way that exploits the property of the algorithm that each maximal flow calculation begins where the previous one left off.

3.7 NETWORK FLOWS WITH GAINS

Thus far, we have considered only network optimization problems for which a unit of flow leaving a node i along the arc (i,j) reaches node j intact. There are applications, however, where the flow reaching node j is greater than that leaving node i; for example, if the flow is invested money and a unit of time is associated with the transition from i to j, then a unit of flow leaving i becomes $1+k$ units upon reaching j where $k>0$ is the rate of interest or dividend. Similarly, the flow reaching node j along (i,j) can be less per unit than that which left i; for example, if the flow represents water in a dam or a chemical in inventory which can evaporate.

There are several interesting possibilities that can occur in flows with gains that do not occur in the network problems discussed in previous sections. First, flow around a circuit can be destroyed if the net gain around the circuit is less than one; see Figure 3.22. Similarly, flow around a circuit can be created as depicted in Figure 3.23. Finally, flow around a circuit can be cancelled as depicted in Figure 3.24.

An example of a management science model using flows with gains is shown in Figure 3.25. The problem is one of purchasing, storing, and selling a single commodity so as to maximize cash output at the end of a T period planning horizon. Flows with gains are used to advantage in a number of ways in this network. First, one dollar leaving a cash node c_t along the purchase arc (c_t, i_t) becomes $w(c_t, i_t)$ units of the commodity upon reaching the inventory node i_t. One unit of commodity leaving the inventory node c_t along the inventory arc (i_t, i_{t+1}) becomes $w(i_t, i_{t+1})$ upon

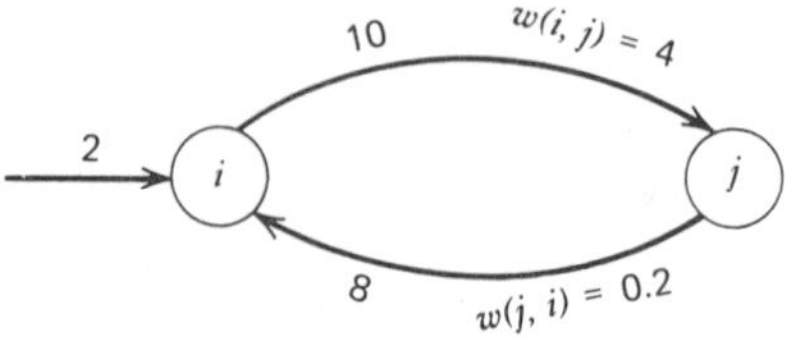

Figure 3.22. Flow with gains—flow destroyed around a circuit.

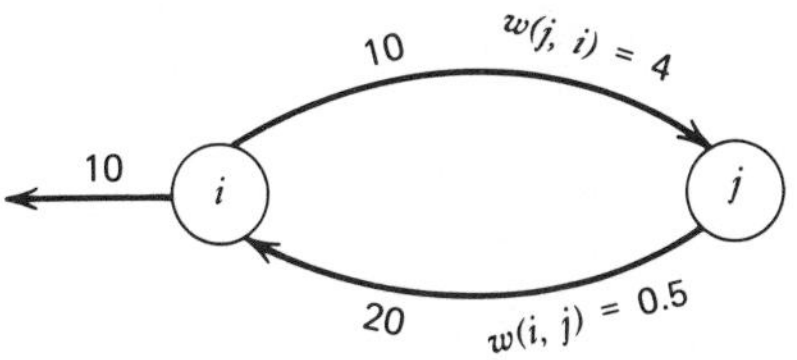

Figure 3.23. Flow with gains—flow created around a circuit.

reaching the inventory node i_{t+1}. The quantity $w(i_t, i_{t+1})$ can be less than one in the case of evaporation or loss, and greater than one in the case of growth when the commodity is agricultural. The gain $w(i_t, c_{t+1})$ along a sell arc depends upon the selling price in period t. Finally, there are the financial gains $w(c_t, c_{t+1}) > 1$ along the cash balance arcs due to bank interest, $w(c_t, c_T) > 1$ and $w(c_T, c_t) < 1$ along lending and borrowing arcs due to loan interest.

Formally, we treat flows with gains by associating a nonzero quantity $w(r_l)$ with the flow in each arc $r_l = (i,j)$ in the network. If x_l is the flow out of node i, then $w(r_l)x_l$ is the flow into node j. Notice that $w(r_l) = 1$ corresponds to the network optimization problem (3.1) studied in previous sections. A problem analogous to (3.1) is constructed by defining the matrix H whose nonzero components in column l are

$$h_{il} = 1$$

$$h_{jl} = -w(r_l)$$

The network optimization problem with gains is

$$\begin{aligned} &\min \mathbf{cx} + \mathbf{ds} \\ &\text{s.t. } \mathbf{Hx} + \mathbf{Us} = \mathbf{b} \\ &\mathbf{0} \le \mathbf{x} \le \mathbf{q} \qquad \mathbf{0} \le \mathbf{s} \le \mathbf{p} \end{aligned} \tag{3.8}$$

An important special case of (3.8) arises when $w(r_l) = -1$ for all arcs r_l. Then the matrix $\mathbf{H}$ is the node-edge incidence matrix of the graph derived from the network G by ignoring directions. These problems are treated in greater detail in the following section.

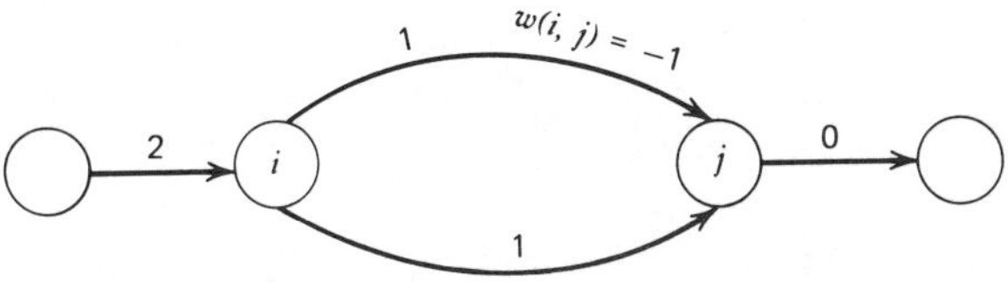

Figure 3.24. Flow with gains—flow cancellation.

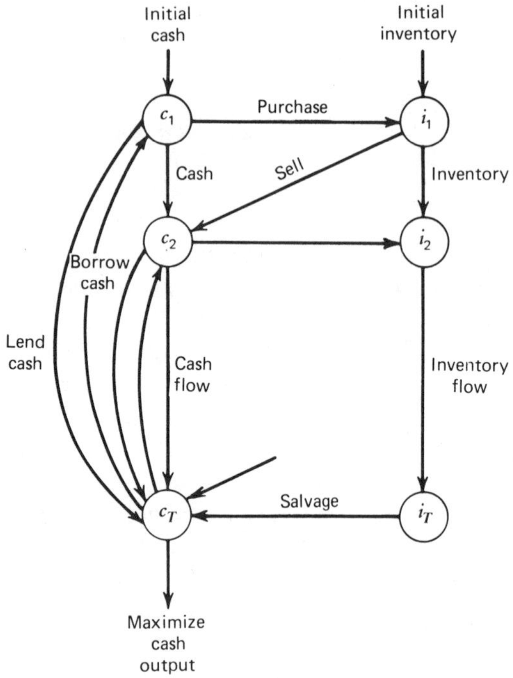

Figure 3.25. Cash management by network optimization.

As in Section 3.2, we let $\mathbf{H}^0 = [\mathbf{H}, \mathbf{U}]$ and consider the properties of $t \times t$ submatrices $\mathbf{T}$ of $\mathbf{H}^0$ of rank t. The graph $F_\mathbf{T}$ relative to $\mathbf{T}$ is the same one defined in Section 3.3.

THEOREM 3.4. Suppose $\mathbf{T}$ is a $t \times t$ submatrix of $\mathbf{H}^0$ with rank t. Then each connected component of the graph $F_\mathbf{T}$ is either a rooted tree or a subgraph of $F_\mathbf{T}$ with the same number of nodes and edges and having no roots.

PROOF. The proof of Lemma 3.2 can be extended to show that every connected component of $F_\mathbf{T}$ can have at most one root. The idea is to adjust the weights λ in that proof to take into account the gains. Further details are omitted. The proof of Lemma 3.1 that every connected component is a tree cannot be extended to the case of flows with gains, and this is the reason that cycles can occur.

Consider a connected component of $F_\mathbf{T}$ spanning k nodes. If this component corresponds to $k+1$ or more columns of $\mathbf{T}$, then those columns cannot be linearly independent because they have only k nonzero rows.

This contradicts the assumption that the square matrix $\mathbf{T}$ is invertible. Thus, if a connected component of $F_{\mathbf{T}}$ spanning k nodes has a root, then it cannot have more than $k-1$ edges. Since it is connected, it cannot have less than $k-1$ edges. Thus, it must have exactly $k-1$ edges and the connected component is a rooted tree. On the other hand, if the connected component has no root and k edges, the connected component must contain a cycle. ∎

The network primal simplex algorithm can be adapted to take into account the gains and the possibility of cycles in the basis structure. We omit further details here.

3.8 GRAPH OPTIMIZATION PROBLEMS: MAXIMAL MATCHING, MINIMUM COVERING AND MINIMUM SPANNING TREE PROBLEMS

In this section, we consider optimization problems analogous to the network optimization problem (3.1) but where the connections between nodes are (undirected) edges. The coefficient matrix $\mathbf{A}$ in (3.1) is now the node-edge incidence matrix of the graph $G=(\mathfrak{N},\mathcal{E})$ where $\mathfrak{N}$ is the same node set as before, $\mathfrak{N}=\{i,\ldots,m\}$, and $\mathcal{E}$ is the edge set containing edges $\langle i,j\rangle$ for some pairs $i,j\in\mathfrak{N}$. We assume there is an edge incident to each node. As in Section 3.2, we will work with an enumeration of the edges $e_1,\ldots,e_n$, and the columns $\mathbf{a}_1,\ldots,\mathbf{a}_n$ of $\mathbf{A}$ have two nonzero entries equal to $+1$ on the rows corresponding to the incident nodes of the edges. The graph optimization problem is a special case of the flows with gains problem studied in the previous section.

The lack of direction on the edges is appropriate when optimizing, for example, communications networks for which costs and capacities depend only on the traffic in a link and not the direction of the traffic. If the graph corresponds to a map and cost is a function of geographical distance between nodes, then once again direction does not matter. Some graph optimization problems can be readily recast and solved as network optimization problems. For example, maximizing the flow from a source to a sink over a graph can be converted to a problem of the form (3.2) by replacing each edge $\langle i,j\rangle$ with capacity q_{ij} by two arcs (i,j) and (j,i), each with capacity q_{ij}. Other graph optimization problems, however, cannot be converted to network optimization problems and are not solvable, at least directly, by linear programming methods. Nevertheless, the structure of graph optimization problems and solution methods are closely related to the network optimization problem structure and methods above, and it is appropriate to discuss them here.

As before, let $\mathbf{A}^0=(\mathbf{A},\mathbf{U})$ and for any square submatrix $\mathbf{T}$ of $\mathbf{A}^0$ define the graph $F_\mathbf{T}$ with node set corresponding to rows of $\mathbf{T}$ that are not zero, and edges corresponding to columns of $\mathbf{T}$ that are edges with both incident nodes in the node set. The roots of $F_\mathbf{T}$ are the nodes corresponding to columns of $\mathbf{T}$ with exactly one nonzero entry. Theorem 3.3 tells us that if $\mathbf{T}$ is nonsingular, then each connected component of $F_\mathbf{T}$ spanning nodes is either a rooted tree or consists of k edges and contains a cycle. This result can be specialized for the graph optimization problem.

LEMMA 3.4. Suppose $\mathbf{T}$ is a nonsingular submatrix of $\mathbf{A}^0$ for the graph optimization problem. Then the cycles in $F_\mathbf{T}$ are of odd length.

PROOF. Let $\langle i_1, i_2, \ldots, i_{2k}, i_1\rangle$ be a cycle of even length. The submatrix of $\mathbf{A}$ corresponding to this cycle is

$$\begin{array}{r} \text{row } i_1 \\ \\ \\ \\ \\ \\ \text{row } i_{2k} \end{array} \begin{bmatrix} 1 & & & & & 1 \\ 1 & 1 & & & & \\ & 1 & 1 & & & \\ & & 1 & & & \\ & & & \ddots & & \\ & & & & 1 & \\ & & & & 1 & 1 \end{bmatrix} \tag{3.9}$$

Clearly, the weights $\lambda_j=1$ for j even, $\lambda_j=-1$ for j odd, produce a linear combination of these columns equal to zero. ∎

If the graph G has no odd cycles, then $F_\mathbf{B}$ for every linear programming basis $\mathbf{B}$ for the graph optimization problem has no cycles implying $F_\mathbf{B}$ is a rooted forest. Moreover, it can be shown (see Exercise 3.12) that a graph with no odd cycles is bipartite which means $\mathcal{N}$ is partitioned into subsets $\mathcal{N}_1$ and $\mathcal{N}_2$ and all edges are of the form $\langle i,j\rangle$ for $i\in\mathcal{N}_1, j\in\mathcal{N}_2$. For such problems, the rows of $\mathbf{A}^0$ corresponding to $\mathcal{N}_2$ can be multiplied by -1 converting $\mathbf{A}^0$ to a network optimization problem. Thus, we will consider in the remainder of this section only graphs with odd cycles.

As a result of the possibility of a cycle in the basis, linear programming basic solutions to the graph optimization problem with integer right-hand-side vector $\mathbf{b}$ may be fractional, unlike the network optimization problems. A typical basis $\mathbf{B}$ is of the form

$$\mathbf{B}=\begin{Bmatrix} \mathbf{Q}_1 & & & \\ & \mathbf{Q}_2 & & \\ & & \ddots & \\ & & & \mathbf{Q}_K \end{Bmatrix}$$

where each submatrix $\mathbf{Q}_i$ is nonsingular and corresponds to a rooted tree or a component containing a cycle. The submatrices containing a cycle cause the difficulty.

THEOREM 3.5. A $t \times t$ nonsingular matrix $\mathbf{Q}$ corresponding to an odd cycle in the graph G, satisfies $|\det \mathbf{Q}| = 2$.

PROOF. Let $i(j)$ for $j=1,\ldots,t$, denote a permutation of $\{1,\ldots,t\}$. The determinant of the $t \times t$ matrix $\mathbf{Q}=(q_{ij})$ is calculated by the formula

$$\det \mathbf{Q} = \sum_{\text{all permutations}} (-1)^{N[i(1),\ldots,i(t)]} q_{i(1),1} \cdots q_{i(t),t}$$

where $N[i(1),\ldots,i(t)]$ counts the number of inversions in the permutation. For the matrix depicted in (3.9), but with an odd number of rows and columns, there are two nonzero terms in the summation. One of these terms corresponds to the identity permutation $i(j)=j$ and gives the product of the elements along the diagonal which is $+1$. The second nonzero term corresponds to the permutation $i(j)=j+1$ for $j=1,\ldots,t-1$, $i(t)=1$, which has $t-1$ inversions. Since t is odd, $t-1$ is even and the second nonzero term is also $+1$. Thus $\det \mathbf{Q}=2$ when $\mathbf{Q}$ has the form (3.9). Finally, if $\mathbf{Q}$ corresponds to a cycle, but not in the form (3.9), then it can be put into that form by interchanging rows and columns without changing the magnitude of the determinant. ■

With this background, we can now consider two related graph optimization problems, the maximal matching and minimal covering problems, which have the added restriction that the optimal solutions are required to be integer. Theorem 3.5 tells us that this added restriction is nontrivial, but as we shall see, the difficulty can be overcome by reformulations of these problems.

Let S be a subset of $\mathcal{E}$ and let $d_S(i)$ denote the degree of node i with respect to S; that is, $d_S(i)$ equals the number of edges of S incident to i. A subset M of $\mathcal{E}$ is called a *matching* of G if $d_M(i) \le 1$ for all $i \in \mathcal{N}$. A matching M^* is a *maximal matching* if its cardinality $|M^*|$ is maximal among all matchings. The maximal matching problem has the following formulation

$$\begin{aligned} &\max \sum_{j=1}^{n} x_j \\ &\text{s.t.} \sum_{j=1}^{n} a_{ij} x_j \le 1 \qquad \text{for } i=1,\ldots,m \\ &\qquad x_j = 0 \text{ or } 1 \qquad \text{for } j=1,\ldots,n \end{aligned} \tag{3.10}$$

A subset of edges C is called a *cover* of G if $d_C(i) \geq 1$ for all nodes i. A cover C^* is a *minimal cover* if its cardinality $|C^*|$ is minimal among all covers. The mathematical formulation of the minimal cover problem is

$$\begin{aligned} \min \sum_{i=1}^{m} x_j & \\ \text{s.t.} \sum_{j=1}^{n} a_{ij} x_j \geq 1 & \quad \text{for } i=1,\ldots,m \\ x_j = 0 \text{ or } 1 & \quad \text{for } j=1,\ldots,n \end{aligned} \tag{3.11}$$

There are a number of generalizations of the maximal matching and minimum covering problems including problems of similar form with cost coefficients c_j and/or right-hand-sides b_i not equal to one. An important special integer programming problem that arises in many scheduling applications is the covering problem for which there can be more than two elements in each column equal to $+1$. Usually the covering problem has nonunit cost coefficients as well. Exercises 3.5 and 8.8 address these generalizations.

There is a symmetry between the covering and matching problems as the following two theorems demonstrate.

THEOREM 3.6. Suppose C^* is a minimal cover of the graph G. For each node i such that $d_{C^*}(i) > 1$, remove $d_{C^*}(i) - 1$ edges incident to it. The resulting set of edges M^* is a maximal matching.

PROOF. By construction $d_{M^*}(i) \leq 1$ for all nodes i and by definition, M^* is a matching. It remains to prove that M^* is maximal. To this end, let p be the number of edges deleted from C^* to form M^*; that is, $|M^*| = |C^*| - p$. Each deleted edge $e = \langle i,j \rangle$ such that $d_{C^*}(j) > 1$, must also satisfy $d_{C^*}(i) = 1$ because otherwise we could delete e from C^* and have a cover. Thus, $d_{C^* - \{e\}}(i) = 0$ and there are exactly p nodes such that $d_{M^*}(i) = 0$.

Suppose that M^* is not maximal implying that there is a matching M' with $|M'| = |M^*| + 1 = |C^*| - p + 1$. Each edge in a matching spans two nodes and since $|M'|$ has one more edge than M^*, we can conclude that there are exactly $p-2$ nodes such that $d_{M'}(i) = 0$. The matching M' can be extended to a cover by adding edges to span these nodes. The maximal number of edges required to do this is $p-2$, and therefore the cover C' derived from M' satisfies $|C'| \leq |M'| + p - 2 = |C^*| - 1$ contradicting the minimality of C^*. ■

THEOREM 3.7. Suppose M^* is a maximal matching of the graph G. Add one edge incident to each node i such that $d_{M^*}(i) = 0$. The resulting set of edges is a minimum cover.

PROOF. The proof is similar to the proof of Theorem 3.6 and is left to the reader as an exercise (see Exercise 3.13). ■

With these results, it suffices to study only the maximal matching problem. The difficulty due to fractional linear programming solutions to the matching problem can be overcome by adding constraints to the problem which exclude cycles for which the variables will equal $\frac{1}{2}$. This is our second exposure to the idea of adding constraints, otherwise known as *cuts*, to a linear programming approximation to an integer or nonlinear programming problem, in order to make the approximation more exact. Unlike the cuts derived in Section 2.7, cuts for the maximal matching problem (3.10) can be added before any computation is done to give an exact linear programming approximation to it. For an arbitrary integer programming problem, this is not possible in a practical constructive sense, and a linear programming representation of the integer programming problem is not obtainable.

The fractions in a linear programming basic feasible solution are due to the presence of odd cycles in the graph of the basis. Let $\langle i_1, i_2, \ldots, i_{2k+1}, i_1\rangle$ be such an odd cycle consisting of $2k+1$ edges $e_l = \langle i_l, i_{l+1}\rangle$ for $l = 1,2,\ldots,2k+1$, with $i_{2k+2} = i_1$. A matching can include at most k of these edges implying the constraint $\sum_{l=1}^{2k+1} x_l \leq k$ can be added without eliminating any feasible solutions to the matching problem. The general rule is obtained by considering every subset $\mathcal{N}'$ of the node set with the properties $|\mathcal{N}'| \geq 3$ and odd. Let $\mathcal{E}'$ denote the subset of the edge set with both ends in $\mathcal{N}'$. The constraints to add to (3.10) are

$$\sum_{e_j \in \mathcal{E}'} x_j \leq \frac{|\mathcal{N}'| - 1}{2} \tag{3.12}$$

for every $\mathcal{N}' \subseteq \mathcal{N}$ satisfying $|\mathcal{N}'| \geq 3$ and $|\mathcal{N}'|$ odd.

An algorithm can be derived from the simplex algorithm for solving the matching problem augmented by these valid constraints. We omit the algorithm but give the following result, which is the principle on which solution of the maximal matching problem is achieved.

Relative to a subset S of the edge set $\mathcal{E}$, an *alternating path* in G is a simple path (i.e., no cycles) whose edges alternate between S and $\mathcal{E} - S$. Relative to a matching M, an *augmenting path* is an alternating path connecting vertices i and k such that $d_M(i) = d_M(k) = 0$.

THEOREM 3.8. A matching M is maximum if and only if there is no augmenting path relative to M.

PROOF

1. If there is an augmenting path relative to $M \Rightarrow M$ is not a maximal matching. Let P denote the set of edges in the augmenting path and let

$$M' = (M \cup P) - (M \cap P)$$

Since the path was augmenting, M' is a matching and moreover, $|M'| = |M| + 1$.

2. If there is no augmenting path relative to $M \Rightarrow M$ is a maximal matching. Suppose M is not maximal and there exists a matching M^* such that $|M^*| = |M| + 1$. Define the set of edges

$$D = (M \cup M^*) - (M \cap M^*)$$

We will establish a contradiction by showing that D contains an augmenting path relative to M. Since $|D| = |M| + |M^*| - 2|M \cap M^*|$ and $|M^*| = |M| + 1$, we can conclude that D is odd. Let G^* be the subgraph of G composed of the edges in D and the nodes incident to these edges.

The subgraph G^* may not be connected, and we consider each connected component G' with node set $\mathcal{N}'$ and edge set D'. We show that G' is either a simple path or a simple cycle. Since G' is connected, there is a path from each node to every other node. Every such path must alternate between M and M^* because M and M^* are matchings. For the same reason, the degree at each node of G' must be either one or two. If the degree at each node is two, then G' consists of a simple cycle. The cycle must be of even length because it is alternating between M and M^*. If there is a node with degree one, that is, an end, consider the simple path generated by starting at this node and continuing until another node of degree one is reached. This path must connect all the nodes in G', because otherwise a node not visited by the path must be connected to the nodes on the path at an intermediate node implying that the intermediate node has a degree of at least three.

Considering the entire subgraph $G^* = (\mathcal{N}^*, D)$ again, since $|D|$ is odd and every connected component of G^* which is a cycle is even, there must be a connected component that is a simple path of odd length $\langle i_1, \ldots, i_{2k} \rangle$ alternating between M and M^*. At least one of these simple paths of odd length must have $d_M(i) = d_M(i_{2k}) = 0$ since $|M^*| > |M|$ and such a path is an augmenting path. ■

The final optimization problem that we study is the easy to solve but important minimum spanning tree problem. Let c_k be a weight or length associated with each edge e_k. The problem is to find the tree spanning all the nodes of G such that the sum of the weights is minimal. If the graph is a communications network, finding a spanning tree of minimal length may be a good approximation to minimizing construction costs so that each node can communicate with every other node. More generally, the minimum spanning tree problem arises as a substructure in more complex optimization problems such as the traveling salesman problem studied in Chapter 8.

The minimum spanning tree algorithm is very simply stated. Assume the edges $e_1,\ldots,e_n$ of the graph are ordered so that $c_1<c_2<\cdots<c_n$. We can, without loss of generality, assume all the c_k are distinct because if not, say $c_1=c_2=c_3<c_4$, then we could replace c_2 by $c_2+\varepsilon$, c_3 by $c_3+2\varepsilon$, for ε sufficiently small, without affecting the solution.

Minimum Spanning Tree (Greedy) Algorithm

Step 1. Include edge e_1 in the minimum spanning tree.

Step 2. If $m-1$ edges have been included in the tree, stop with the minimal spanning tree.

Step 3. Add edge e_k to the tree unless it forms a cycle in which case discard the edge. Return to Step 2.

THEOREM 3.9. The greedy algorithm solves the minimal spanning tree problem.

PROOF. Suppose the spanning tree T^* found by the algorithm is not minimal and let $T \neq T^*$ be a spanning tree with total length $l(T)$ less than the total length $l(T^*)$ or T^*. Suppose further that $T^*=\{e_{j_1},\ldots,e_{j_{n-1}}\}$ and that e_{j_k} is the first edge in T^* not in T. Then $T\cup\{e_{j_k}\}$ has exactly one cycle, and this cycle contains an edge $e_{j_0}\notin T^*$. Let $T'=(T\cup\{e_{j_k}\})-\{e_{j_0}\}$; T' contains $m-1$ edges and no cycles implying it is a spanning tree. Now $\{e_{j_1},\ldots,e_{j_{k-1}}\}\cup\{e_{j_0}\}$ contains no cycles because it is contained in T and therefore, by the greedy algorithm, we can conclude that $c_{j_0}>c_{j_k}$ implying $l(T')=l(T)+c_{j_k}-c_{j_0}<l(T)$, which is impossible since T is assumed to be minimal. ■

The minimum spanning tree algorithm is called a greedy algorithm because it follows the myopic rule of selecting at each iteration the least-cost available edge to try to add to the spanning tree under construction. An algorithm for solving shortest-route problems to be presented in Chapter 4 exhibits the same type of greedy behavior in the selection of arcs for a shortest-route chain. In general, greedy algorithms are highly desirable because of their empirical efficiency. Moreover, polynomial bounds can be derived on the number of arithmetic operations required by these algorithms to find optimal solutions. Exercise 8.10 presents and discusses a greedy heuristic algorithm for finding good feasible solutions to a class of integer programming problems.

3.9 EXERCISES

3.1 Solve the network optimization example given in Table 3.1 where the capacity on arc (2,6) is 4 instead of 5.

3.2 Solve the network optimization example given in Table 3.1 where the cost on arc (1,5) is 7 instead of 4.

3.3 Suppose the arc capacities of a transshipment problem are infinite. How would you transform the transshipment problem to a pure transportation problem; that is, a network optimization problem with only origin and destination nodes and no transshipment nodes?

3.4 (E. L. Johnson, 1975) Consider a scheduling problem consisting of five jobs, A, B, C, D, E such that A must be completed before C or D can begin, and B and C must be completed before E begins. The precedence relation among jobs is depicted in Figure 3.26 where the nodes represent events of completing jobs and the arcs represent performing the jobs. The normal completion times for A, B, C, D, E are 30, 25, 1, 20, and 22, respectively, but these times can be decreased by spending more money on the jobs. Specifically, job C cannot be shortened but jobs A, B, D, and E can be at costs of \$1, \$2, \$3, \$4 for each unit of time eliminated, respectively. The problem is to complete all jobs in a time of 40 units while minimizing the total cost of shortening the jobs.

(1) Let the variables t_A, t_B, t_D, t_E represent the decrease in time for jobs A, B, D, E so that their actual completion times are $30 - t_A, 25 - t_B, 20 - t_D, 22 - t_E$. Let the variables t_1, t_2, t_3, t_4 represent the event times

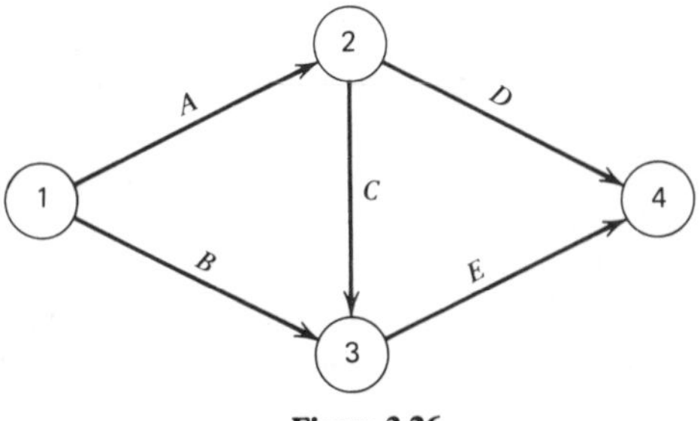

Figure 3.26.

associated with the nodes. Formulate the problem as a linear programming problem.

(2) Formulate and interpret the dual to the problem of part 1 as a network optimization problem of the form (3.1). Find an optimal solution to this problem.

3.5 Consider the set partitioning problem

$$\begin{aligned} z^* = \min \sum_{j=1}^{n} c_j x_j & \\ \text{s.t.} \sum_{j=1}^{n} a_{ij} x_j = 1 & \qquad \text{for } i = 1, \ldots, m \\ x_j = 0 \text{ or } 1 & \qquad \text{for } j = 1, \ldots, n \end{aligned}$$

where the coefficients a_{ij} are all 0 or 1. Give a network optimization problem whose minimal cost is always a lower bound on z^*. Compare the lower bound with the lower bound provided by the linear programming relaxation ($0 \leq x_j \leq 1$ instead of $x_j = 0$ or 1).

3.6 The Caterer's Problem. A caterer wishes to minimize the total cost of using fresh napkins over T days when the requirements for each day are r_t. He can send his soiled napkins to the laundry where there are three speeds of service, $i = 1, 2, 3$, with a cost of c_i per napkin laundered. He can also purchase new napkins at a price of c_0 per napkin. His initial stock of napkins is S. Formulate this problem as a network optimization problem.

3.7 Prove that a necessary and sufficient condition that the circulation problem (3.4) has a feasible solution is that

$$q(P, \bar{P}) \geq l(\bar{P}, P)$$

holds for all $P \subseteq \mathfrak{N}$, where

$$\begin{aligned} \mathfrak{N} &= \text{node set of the network} \\ \bar{P} &= \mathfrak{N} - P \\ q(P, \bar{P}) &= \sum_{(i,j) \in (P, \bar{P})} q_{ij} \\ l(\bar{P}, P) &= \sum_{(i,j) \in (\bar{P}, P)} l_{ij} \end{aligned}$$

3.8 The phase one network optimization problem is

$$\min \sum_{i=1}^{m} (y_i^+ + y_i^-)$$

$$\begin{aligned} \text{s.t. } & \mathbf{Ax} + \mathbf{Us} + \mathbf{Iy}^+ - \mathbf{Iy}^- = \mathbf{b} \\ & \mathbf{0} \leq \mathbf{x} \leq \mathbf{q}, \qquad \mathbf{0} \leq \mathbf{s} \leq \mathbf{p}, \qquad \mathbf{0} \leq \mathbf{y}^+, \qquad \mathbf{0} \leq \mathbf{y}^- \end{aligned}$$

Show how to convert this problem to a maximal flow problem from a single source to a single sink.

3.9 Demonstrate, using the min-cut max-flow theorem, that the maximal flow labeling algorithm has found a maximal flow solution when no more labels can be assigned and the sink is unlabeled in Step 1.

3.10 (Heller and Tompkins, 1956) Suppose **A** is a matrix with elements equal to 0, +1, −1, and such that each column has at most two nonzero elements. Prove that **A** is totally unimodular if and only if its rows can be divided into two disjoint sets I_1 and I_2 satisfying the following:

(1) If the two nonzero elements of a column have the same sign, one of them is in I_1 and the other is in I_2.

(2) If the two nonzero elements of a column have the opposite sign, both of them are in I_1 or both of them are in I_2.

3.11 (Veinott and Dantzig, 1968) Suppose **A** is an integer matrix. Prove that the following statements are equivalent:

(1) **A** is totally unimodular.

(2) The extreme points of $\{\mathbf{x}|\mathbf{A}\mathbf{x} \le \mathbf{b},\ \mathbf{x} \ge \mathbf{0}\}$ are integer for arbitrary integer **b**.

(3) Every nonsingular submatrix of **A** has an integer inverse.

3.12 Prove that a graph with no odd cycles is bipartite; that is, the node set $\mathfrak{N}$ can be partitioned into subsets $\mathfrak{N}_1$ and $\mathfrak{N}_2$ such that all edges are of the form $\langle i,j \rangle$ with $i \in \mathfrak{N}_1$ and $j \in \mathfrak{N}_2$.

3.13 Prove Theorem 3.7 using the proof of Theorem 3.6 as a model.

3.14 (Gale, 1968) Let X and Y be finite sets and X is totally ordered with order relation "$\ge$." Let ϕ be a function from X to subsets of Y. A subset $A \subset X$ is *assignable* if there is a function α, called an *assignment*, from A to Y such that

$$\alpha(x) \ne \alpha(y) \text{ for } x,y \in A,\ x \ne y$$

$$\alpha(x) \in \phi(x) \text{ for all } x \in A$$

Let P denote the family of all assignable subsets of X. Let F denote any arbitrary family of subsets of X. A set B of F is optimal if for any other C of F there is a mapping f from C to B such that $f(x) \ne f(y)$ for $x,y \in C$, $x \ne y$, and $f(x) \ge x$ for all $x \in C$.

(1) A family F of subsets of X is called a *matroid* provided that

  (i) if $A \in F$ and $B \subset A$, then $B \in F$

  (ii) for any $X' \subset X$, all elements of F which are maximal in X' have the same cardinality.

  Prove: The family of assignable subsets P is a matroid.

(2) Prove: If F is a matroid on X, then F contains an optimal element.

(3) Prove: The family P of assignable subsets of X contains an optimal element.

(4) Devise a "greedy" or myopic algorithm for the problem of finding an optimal assignment.

3.15 (Edmonds, 1968) Consider an arbitrary network $G=[\mathcal{N},\mathcal{Q}]$ and let x_{ij} denote the variable associated with the arc $(i,j)\in\mathcal{Q}$. A branching of the network G is a subnetwork with two properties:

(1) Every arc is directed toward a different node.

(2) If the directions of the arcs are ignored, the subnetwork is a forest on the node set G.

For $X\subseteq\mathcal{N}$, $Y\subseteq\mathcal{N}$, define

$$x(X,Y)=\sum_{\substack{i\in X\\ j\in Y\\ (i,j)\in\mathcal{Q}}} x_{ij}$$

Prove that the extreme points of the polyhedron defined by the linear inequalities

$$\begin{aligned} &x_{ij}\geq 0 && \text{for all } (i,j)\in\mathcal{Q}\\ &x(\mathcal{N},j)\leq 1 && \text{for all } j\in\mathcal{N}\\ &x(X,X)\leq |X|-1 && \text{for all nonempty } X\subseteq\mathcal{N} \end{aligned}$$

are precisely the incidence vectors of all branchings in the network G.

3.10 NOTES

SECTIONS 3.1 AND 3.2. The development of this chapter is motivated in large part by Johnson (1965). Johnson (1975) gives an up-to-date survey of network optimization techniques. General references containing the application of linear programming to network optimization problems include Dantzig (1963), Simonnard (1966), Christofides (1975), and Bazaraa and Jarvis (1976). Ford and Fulkerson (1962) treat many of the problems found in this chapter, but from a somewhat different point of view emphasizing ideas derived from maximal flow analysis. The books by Berge (1962) and Busacker and Saaty (1965) are mainly concerned with graph theory but some mathematical programming problems are analyzed. The book by Frank and Frisch (1971) is oriented mainly toward the properties of communication networks, but nevertheless it contains a good amount of mathematical programming. The book by Elmaghraby (1970) contains discussions of many network optimization applications. A survey of network optimization in transportation is given by Magnanti and

Golden (1978). Lawler (1976) treats network optimization and related topics from a somewhat different viewpoint using recently developed methods of combinatorics such as matroid theory. Boesch (1975) has edited a survey of recent papers on the theory and application of network optimization.

SECTIONS 3.3 AND 3.4. These sections are derived in large part from the development given by Johnson (1965). Cunningham (1976) gives a new method to resolve degeneracy in network simplex methods.

SECTION 3.5. The classic reference for the maximal flow problem is Ford and Fulkerson (1962) who do not rely on the simplex method to derive the max-flow min-cut theorem and the labeling algorithm. Frank and Frisch (1971) give several alternate proofs of this theorem. The use of Dilworth's theorem in machine scheduling is given by Ford and Fulkerson (1962). Edmonds and Karp (1972) show how the labeling algorithm can be made into a "good" algorithm by the first scanned-first labeled rule. Zadeh (1973) gives some pathological network optimization samples.

SECTION 3.6. The out-of-kilter algorithm is due to Fulkerson (1961). This algorithm has incorrectly been called a primal-dual algorithm that, as we have seen, is distinctly different in algorithmic strategy (Shapiro, 1977). The choice of a particular algorithm, primal, out-of-kilter, primal-dual, or some other, depends to a large extent on problem structure and the goals for the analysis. Barr et al. (1974) give experimental results with different types of algorithms and algorithmic strategies. Bradley et al. (1977) give details about the implementation of a primal network optimization code.

SECTION 3.7. The basic reference on flows with gains is Jewell (1962) who gives algorithms and discusses applications. A more recent reference is Maurras (1973).

SECTION 3.8. The development in this section follows closely the development of Garfinkel and Nemhauser (1972). Theorems 3.6 to 3.8 can be found in Norman and Rabin (1959). The fact that the convex hull of the feasible (integer) solutions to the matching problem (3.10) can be constructed by the addition of constraints (3.12) to (3.10) was discovered by Edmonds (1965a). Edmonds (1965b) gives an algorithm for the matching problem based on Theorem 3.7. The minimum spanning tree algorithm is given by Kruskal (1956).

4
SHORTEST ROUTE AND DISCRETE DYNAMIC PROGRAMMING PROBLEMS

4.1 INTRODUCTION

The focus of the previous chapter was on network optimization problems that could be solved as linear programming problems by specialized variants of the simplex algorithm. The central constructs used by the algorithms were trees connecting subsets of nodes indicating how the required flow between the nodes could be met. In this chapter, we begin by studying shortest route network problems where the central constructs are simple (circuitless) chains connecting specified pairs of nodes. These chains can be found by recursive algorithms that are more direct and more efficient than the simplex algorithm and its variants. As in Chapter 3, we use concepts of graph theory given in Appendix B.

A number of important discrete dynamic programming models can be viewed as generalizations of the shortest route problem. Specifically, these problems can be represented as networks where the nodes are the states of a dynamic system and the arcs correspond to decisions. A decision to make the transition from node s to node t along the arc (s,t) causes an immediate cost c_{st} to be incurred and involves the passage of one period of time. The networks are therefore acyclic since time is acyclic, and the number of arcs in an optimal chain is constrained to equal the number of periods in the planning horizon. Finally, we study a class of stochastic models with the property that the transition to any specific node t is not certain but occurs with a specified probability. These stochastic problems are called Markovian decision problems because the transitions are governed by a Markov chain.

4.2 SHORTEST ROUTE PROBLEMS

As in Chapter 3, we consider a network with node set $\mathcal{N} = \{1,\ldots,m\}$ and arc set $\mathcal{A}$. In addition, there is a length c_{ij} associated with each $(i,j) \in \mathcal{A}$. Shortest route problems are concerned with finding chains of minimal length between specified pairs of nodes. For convenience, we assume that the network contains a chain from each node to all other nodes and also that it does not contain any circuits of negative length. This ensures that shortest routes exist. We study two types of shortest route problems.

Shortest Route Problem 1. Find the shortest route chain from node 1 to all other nodes.

Shortest Route Problem 2. Find the shortest route from each node to all other nodes.

Shortest Route Problem 1 can be formulated as a network optimization problem of the form (3.1) although it can be more efficiently solved by the following recursion.

THEOREM 4.1. Let y_j denote the length of a shortest route chain from node 1 to node j. The y_j satisfy the recursion

$$y_j = \underset{(i,j)\in\mathcal{A}}{\text{minimum}} \{c_{ij} + y_i\} \qquad j = 2,\ldots,m \tag{4.1}$$

$$y_1 = 0$$

PROOF. (1) $y_j \geq \min_{(i,j)\in\mathcal{A}}\{c_{ij} + y_i\}$

Let $(1, i_2, i_3, \ldots, i_{k-1}, i_k, j)$ denote a shortest route chain from node 1 to node j; such a chain exists by assumption. Thus,

$$y_j = c_{1,i_2} + c_{i_2,i_3} + \cdots + c_{i_{k-1},i_k} + c_{i_k,j}$$

It must be that

$$y_{i_k} = c_{1,i_2} + c_{i_2,i_3} + \cdots + c_{i_{k-1},i_k}$$

because if y_{i_k} were strictly less than the sum on the right, we could replace the chain $(1, i_2, i_3, \ldots, i_k)$ by a shortest route chain to i_k and realize a shorter chain to j by extending the latter chain by (i_k, j). Thus, we can conclude

$$y_j = c_{i_k,j} + y_{i_k} \geq \min_{(i,j)\in\mathcal{A}} \{c_{ij} + y_i\}.$$

(2) $y_j \leq \min_{(i,j)\in\mathcal{A}}\{c_{ij} + y_i\}$

For some node l

$$\min_{(i,j)\in\mathcal{A}} \{c_{ij} + y_i\} = c_{il} + y_l$$

Let $(1, i_1, i_2, \ldots, i_k, l)$ denote a shortest route chain with length y_l from node

1 to node l. The extended chain $(1, i_1, i_2, \ldots, i_k, l, j)$ to node j is a candidate for shortest route chain to that node implying

$$y_j \leq \min_{(i,j) \in \mathcal{A}} \{c_{ij} + y_i\} \quad \blacksquare$$

Before showing how this recursion can be used to solve Shortest Route Problem 1, we show how it relates to a network optimization problem. Specifically, it can be shown (see Exercise 4.1) that the optimal u_j in the following linear programming problem are unique and satisfy the recursion (4.1):

$$\begin{aligned} \max\ & -mu_1 + u_2 + \cdots + u_m \\ \text{s.t.}\ & -u_i + u_j \leq c_{ij} \qquad \text{for all } (i,j) \in \mathcal{A} \\ & u_1 \geq 0 \end{aligned} \tag{4.2}$$

Problem (4.2) is the dual to the network optimization problem

$$\begin{aligned} \min\ & \sum_{(i,j) \in \mathcal{A}} c_{ij} x_{ij} \\ \text{s.t.}\ & \sum_{(1,j) \in \mathcal{A}} x_{1j} - \sum_{(k,1) \in \mathcal{A}} x_{k1} \leq m \\ & \sum_{(i,j) \in \mathcal{A}} x_{ij} - \sum_{(k,i) \in \mathcal{A}} x_{ki} = -1 \qquad i = 2, \ldots, m \\ & x_{ij} \geq 0 \qquad \text{for all } (i,j) \in \mathcal{A} \end{aligned} \tag{4.3}$$

By conservation of flow, the net flow out of node 1 for any feasible solution to (4.3) will be $m-1$ implying that the inequality constraint in (4.3) is not binding. In words, problem (4.3) minimizes the total length of the chains used in sending 1 unit of flow from node 1 to each of the other nodes. It is somewhat surprising that such a formulation yields simultaneously the shortest route chain from node 1 to all other nodes.

Although Theorem 4.1 provides us with a recursive characterization of Shortest Route Problem 1, it does not tell us specifically how to compute the y_j and the corresponding chains. Under the additional assumptions that $(i,j) \in \mathcal{A}$ for all $i, j \in \mathcal{N}$, $i \neq j$ [take $c_{ij} = +\infty$ if necessary to add (i,j) to the arc set] and that $c_{ij} \geq 0$ for all $(i,j) \in \mathcal{A}$, this is accomplished by the following algorithm which requires in the worst case on the order of m^2 additions and comparisons. The exact number of additions and comparisons depends on the data of the problem. Note that, in the very worst case, the simplex algorithm applied to (4.3) conceivably could require on the order of $\binom{n}{m}$ iterations, where n is the number of arcs in the network. This is a much larger number.

The validity of the following algorithm can be established by showing that recursion (4.1) holds after the algorithm has terminated.

Algorithm for Shortest Route Problem 1

Initialization. Affix a temporary label $l_1 = 0$ to node 1 and a temporary label $l_j = +\infty$ to all other nodes j.

Step 1. Select any node j with label l_j that is minimal among nodes that are temporarily labeled. Make l_j a permanent label and replace any temporary label l_k by $l_j + c_{jk}$ if $l_j + c_{jk} < l_k$.

Step 2. Stop if all nodes have permanent labels; then the value of the permanent label $= y_j$. (If only the shortest route chain to node k is desired, stop when node k is permanently labeled.) Otherwise return to Step 1.

The shortest route chains with lengths y_j can be discovered by *backtracking* from node j. Specifically, we can conclude that (i_1, j) is the last arc in a shortest route chain for any node i_1 such that $y_j = c_{i_1, j} + y_i$. Similarly, (i_2, i_1) is the next to last arc in a shortest route chain if $y_{i_1} = c_{i_2, i_1} + y_{i_2}$. Continuing in this fashion, we ultimately backtrack to node 1 as long as there are no circuits in the network with zero length to cause cycling. Note that the information required for backtracking can be recorded as the algorithm is run. The difficulty due to zero length circuits can be overcome by assigning a small positive length ε to zero length arcs. If $(1, i_k, i_{k-1}, \ldots, i_1, j)$ is the shortest route chain to node j, then as we saw in the proof of Theorem 4.1 $(1, i_k, \ldots, i_l)$ is the shortest route chain to node i_l. The implication is that we can find an *arborescence* of shortest route chains from node 1 to all other nodes.

The maximal flow problem (3.2) discussed in Section 3.4 can be solved by solving a sequence of shortest route problems, each of which selects a new simple chain along which to send flow from the source node 1 to the sink node m. The reader is asked in Exercise 4.5 to show that any maximal flow can be decomposed into flows along simple chains from source to sink. To simplify the notation, let $\mathcal{R} = \{r_1, \ldots, r_n\}$ be the arc set, and let q_k denote the capacity of arc r_k. Let $\xi_1, \xi_2, \ldots, \xi_T$ denote an arbitrary set of simple chains connecting node 1 to node m; the set of all such chains can be enormous, and we work with a small subset. Let x_t denote the flow in ξ_t, a variable to be determined, and define the n-dimensional vector $\mathbf{a}^t$ for which component a_k^t is equal to one if arc r_k is in chain t; a_k^t equals zero otherwise. A feasible but not necessarily maximal flow solution results if we solve the linear programming problem

$$\begin{aligned} &\max \sum_{t=1}^{T} x_t \\ &\text{s.t.} \sum_{t=1}^{T} a_k^t x_t \le q_k \qquad k=1,\ldots,n \\ &\qquad\qquad x_t \ge 0 \end{aligned} \tag{4.4}$$

Problem (4.4) is not a network optimization problem because the $n \times T$ matrix $(\mathbf{a}^1,\ldots,\mathbf{a}^T)$ is an arc-chain rather than a node-arc incidence matrix of a network.

Problem (4.4) is an approximation to the maximal flow problem because it does not contain a column for each simple chain from source to sink. After (4.4) is maximized, however, we can make a test to find a flow augmenting chain if such a chain exists. To see how this is done, let $u_1^T,\ldots,u_n^T$ denote the optimal dual variables on the constraints in (4.4) found by the simplex method. According to the theory of the simplex method, a new chain ξ with column representation $\mathbf{a}=(a_1,\ldots,a_n)$ can be used to increase the maximal flow in (4.4) if $1-\Sigma_{k=1}^n u_k^T a_k > 0$. In this case, $\mathbf{a}$ is added as a nonbasic column which we pivot into the basis because it prices out positively. Whether or not such a new column exists can be determined by maximizing $1-\Sigma_{k=1}^n u_k^T a_k$ over all columns corresponding to simple chains from node 1 to node m. This is achieved by solving the subproblem

$$\begin{aligned} &\min \sum_{k=1}^{n} u_k^T a_k \\ &\text{s.t. } (a_1,\ldots,a_n) \text{ corresponds to a simple chain from node 1} \\ &\qquad \text{to node } m \end{aligned}$$

The subproblem is precisely the problem of finding a shortest route from node 1 to node m where the arc "lengths" are the nonnegative values u_k^T. Note that the variables in the subproblem are the coefficients a_k for the linear programming problem (4.4).

If the length of the shortest route chain ξ_{T+1} from node 1 to node m in the network with lengths u_k^T on the arcs r_k is less than one, then we add the column $\mathbf{a}^{T+1}$ to (4.4), reoptimize it, and repeat the shortest route column generating procedure. The other possibility is that the shortest route chain from node 1 to node m has length equal to one. This length is realized by any chain ξ_t for which x_t is basic (we can assume that there is at least one) because basic variables price out zero; that is, $1-\Sigma_{k=1}^{n}u_k^T a_k^t=0$. If the latter case occurs, then the optimal solution to (4.4) is a maximal flow solution. Convergence in a finite number of solutions of (4.4) to a maximal flow solution is ensured because a new simple chain is generated each time the subproblem is solved, and there are only a finite number of simple chains. This column generation procedure for the maximal flow problem is an example of an important method called *generalized linear programming* that we discuss in detail in Chapter 5.

The algorithm stated above for Shortest Route Problem 1 could be used to solve Shortest Route Problem 2 by solving m versions of Problem 1, one for each node. As we saw in the proof of Theorem 4.1, however, shortest route problems have the property that every subchain of a shortest route chain is itself a shortest route chain for the nodes it connects. Thus, solving m versions of Shortest Route Problem 1 would involve a great deal of redundant calculation. The following algorithm, whose validity is left to the reader as an exercise, exploits this property. As with the algorithm for Shortest Route Problem 1, the number of calculations required by this algorithm is on the order of m^3, but with the difference that there is little possibility that substantially fewer calculations will actually be required for a given problem. Again we require nonnegative arc lengths c_{ij}.

Algorithm for Shortest Route Problem 2

Initialization. Define the $m\times m$ matrix $\mathbf{C}^{(0)}=[c_{ij}^{(0)}]$ where

$$c_{ij}^{(0)}=\begin{cases}\text{arc length} & \text{if } (i,j)\in\mathcal{Q}\\ \infty & \text{if } (i,j)\notin\mathcal{Q}\\ 0 & \text{if } i=j\end{cases} \tag{4.5}$$

Go to Step 1 with $k=0$.

Step 1. Construct the matrix $\mathbf{C}^{(k+1)}=[c_{ij}^{(k+1)}]$ by the operation

$$c_{ij}^{(k+1)}=\min\left\{c_{ij}^{(k)},c_{i,k+1}^{(k)}+c_{k+1,j}^{(k)}\right\}.$$

Step 2. If $k=m-1$, stop; otherwise, return to Step 1 with k incremented by 1.

We present an alternative and less efficient method for solving Shortest Route Problem 2 because it involves an operation to be used in the study of discrete dynamic programming in the next section. Let $\mathbf{Q}=(q_{ij})$ and $\mathbf{P}=(p_{jh})$ be $m\times k$ and $k\times l$ matrices, respectively. Define the min-addition operator $\otimes$ on these matrices by $\mathbf{R}=\mathbf{Q}\otimes\mathbf{P}$ where $\mathbf{R}$ is the $m\times l$ matrix with components r_{ih} given by

$$r_{ih} = \min_{j=1,\ldots,k} \{q_{ij} + p_{jh}\}$$

Shortest Route Problem 2 can be solved by factoring the $m \times m$ matrix $\mathbf{C}$ of arcs lengths given in (4.5) with respect to the min-addition operator. Specifically, the coefficients c_{ij}^2 of the matrix

$$\mathbf{C}^2 = \mathbf{C} \otimes \mathbf{C}$$

are the lengths of shortest route chains of two arcs or less from node i to node j. Similarly, $\mathbf{C}^k = \mathbf{C} \otimes \mathbf{C}^{k-1} = \mathbf{C}^{k-1} \otimes \mathbf{C}$ contains the lengths of shortest route chains of no more than k arcs between all pairs of nodes. Shortest Route Problem 2 is solved by computing $\mathbf{C}^m$ since we can limit attention to circuitless shortest route chains. As a final point, note that we need only $\{\log_2(m-1)\} + 1$ min-addition calculations $\mathbf{C}^{2^k} = \mathbf{C}^{2^{k-1}} \otimes \mathbf{C}^{2^{k-1}}$, $k = 1, 2, \ldots, \{\log_2(m-1)\} + 1$.

4.3 DISCRETE DYNAMIC PROGRAMMING

The mathematical programming problems considered previously have all been static in that the passage of time was ignored. Many mathematical programming problems involve sequences of decisions to be made dynamically over time and dynamic programming is a method for the efficient solution of such problems. Moreover, dynamic programming can be used to solve other decision problems where the sequential property is induced for computational convenience.

We begin with a discussion of a specific class of dynamic programming problems and afterward we discuss extensions. A manager wishes to control a dynamic system by making an optimal decision at each of a sequence of points in time. The interval between successive decision points is called a *period* and we assume that a decision is made at the beginning of a period. We will use the term *stage* as a synonym for period when the sequential aspect of a dynamic programming problem is not relative to time. Let N denote the total number of periods or the *planning horizon* of the problem.

At the beginning of each period before a decision is made, the manager finds the system in any one of several possible *states*. We let S_i denote the finite set of elements each of which describes a possible state at the start of period i. For example, S_i can be a subset of R^k denoting inventory levels of k items. Let $D_i(s)$ denote the finite set of possible (or feasible) decisions that can be made at the start of period i and the system is in state s before this decision is made. For example, $D_i(s)$ can consist of nonnegative vectors from R^k denoting production of each of k items.

As a result of the decision $d \in D_i(s)$ made at the start of period i, an immediate cost is incurred and a transition to a new state is made, both functions of s as well as d. Specifically, let $c_i(s,d)$ denote the immediate cost and let $T_i(s,d) \in S_{i+1}$ denote the new state. A *policy* is a sequence of decisions $[d_1, d_2, \ldots, d_N]$, with $d_i \in D_i(s_i)$, $i=1,2,\ldots,N$, such that $s_{i+1} = T_i(s_i, d_i)$, $i=1,\ldots,N$ with the initial state s_1 given. An optimal policy is one such that the total cost $\Sigma_{i=1}^N c_i(s,d)$ is minimal.

Since S_i and $D_i(s)$ are finite for all i, we could compute an optimal policy by enumerating all policies $[d_1, d_2, \ldots, d_N]$, costing them out, and then selecting a minimal one. It is easy to see that such a procedure can quickly become computationally excessive. Moreover, it does not exploit the property of optimal decisions embodied in The Principle of Optimality:

> An optimal policy has the property that whatever the initial state and decision are, the remaining decisions must constitute an optimal policy with regard to the state resulting from these decisions.

The Principle of Optimality is a self-evident statement of a property of our dynamic programming problem. More generally, when we are confronted with a decision problem expressed in sequential terms, we seek a dynamic programming representation of it which is consistent in the sense that the costs and transition functions depend only on the period i, the state $s \in S_i$ and the decision $d \in D_i(s)$, and moreover, for which the Principle of Optimality holds. Although it is possible to formalize mathematically the construction of dynamic programming problems to describe sequential decision problems, in practice, the construction is a less logical, more artistic intellectual activity depending in part on the specific structure of the sequential decision problem.

The Principle of Optimality permits us to develop recursive relationships for computing optimal policies for N period problems by building them up from optimal policies for shorter period problems. This is done in much the same way that the Shortest Route Algorithms discussed in the previous section find shortest route chains in a network. A frequent approach is to develop the recursions by working backward in time. Let f_s^n denote the minimal total cost over all feasible decision sequences when n periods remain, given that just before the immediate decision is made, the system is in state s. The dynamic programming problem is solved by finding $f_{s_1}^N$ where s_1 is the initial state of the dynamic system. This value is computed by working backward with the recursions

$$f_s^n = \min_{d \in D_{N-n+1}(s)} \left\{ c_{N-n+1}(s,d) + f_{T_{N-n+1}(s,d)}^{n-1} \right\} \qquad \text{for all } s \in S_{N-n+1}$$

$$n = 1,2,\ldots,N \quad (4.6)$$

$$f_s^0 = 0 \qquad \text{for all } s \in S_{N+1}$$

The terminal function f_0 is given as identically zero for all $s \in S_{N+1}$ because we assume no salvage value for these states; other terminal functions are permitted. The calculations proceed by first calculating for all $s \in S_N$

$$f_s^1 = \min_{d \in D_N(s)} c_N(s,d) \tag{4.7}$$

This permits us to calculate f_s^2 for all $s \in S_{N-1}$ since we know the numerical values $f^1_{T_{N-1}(s,d)}$ for all possible states $T_{N-1}(s,d) \in S_N$. The backward iterative process continues until we have calculated $f_{s_1}^N$, which is the minimal cost of making N decisions over N periods starting in the current state s_1 which we are given. An optimal policy $[d_1^*, d_2^*, \ldots, d_N^*]$ with minimal cost $f_{s_1}^N$ is found by selecting any $d_i^* \in D_i(s_i)$ such that $f_{s_i}^{N-i+1} = c_i(s_i, d_i^*) + f_{s_{i+1}}^{N-i}$ where $s_{i+1} = T_i(s_i, d_i^*)$, $i = 1, 2, \ldots, N$.

EXAMPLE 4.1. We consider a single item production/inventory system. The state space $S = \{0, 1, 2, \ldots, m\}$, where $s \in S$ is the number of units of inventory on hand at the beginning of a stage which in this case is a production period. The quantity m can be either a physical upper bound due to storage limitations, or an economic upper bound above which it would be too costly, or nonoptimal, to store inventory. In each period, a nonnegative integer demand r occurs which must be met from inventory or production. The nonnegative integer level of production d is the decision variable. The set $D(s) = \{\max(0, r-s), \ldots, m+r-s\}$ describes the feasible values of d where the lower bound ensures that demand is met if inventory is too low, and the upper bound ensures that inventory never goes above m. As a result of $d \in D(s)$, the system makes the transition to the new state $s+d-r \in S$ with cost $c(s,d) = p(d) + h(s+d-r)$ where the nonnegative quantity $p(d)$ is the immediate production cost and the nonnegative quantity $h(s+d-r)$ is the inventory holding cost as a function of inventory level $s+d-r$ at the end of the period. This ending inventory becomes beginning inventory at the start of the next period. The objective is to minimize production and inventory holding costs over N periods starting in state s.

This problem can readily be translated to network terms. The node set is S and for all $d \in D(s)$, there are arcs (s,t) for all $t \in S$ such that $t = s+d-r$. Associated with (s,t) is the arc length $c_{st} = p(t+r-s) + h(t)$. Since we permit $d = r$, we must include the arcs (s,s) with cost $p(r) + h(s)$. These arcs drawn from nodes to themselves do not cause any difficulty in our analysis, and we mention them simply because they are a departure from our usual network structure. The dynamic programming problem in network terms is to find the shortest route chain in the network of length N starting in state s, the given inventory level.

Consider the following numerical example. Suppose $r = 4$ and $m = 5$ and we have the additional restriction that $d \leq 6$ due to production limitations.

The production cost function is

$$p(d)=\begin{cases}0 & \text{if } d=0\\ 7+d & \text{if } d>0\end{cases}$$

The inventory holding function is $h(s+d-r)=1\cdot(s+d-r)$. With this information, we can compute the 6×6 matrix $\mathbf{C}$ of immediate transition costs

$$\mathbf{C}=\begin{bmatrix} 11 & 13 & 15 & +\infty & +\infty & +\infty \\ 10 & 12 & 14 & 16 & +\infty & +\infty \\ 9 & 11 & 13 & 15 & 17 & +\infty \\ 8 & 10 & 12 & 14 & 16 & 18 \\ 0 & 9 & 11 & 13 & 15 & 17 \\ +\infty & 1 & 10 & 12 & 14 & 16 \end{bmatrix}$$

For example, $c_{02}=p(6)+h(2)=(7+6)+2=15$. The $+\infty$ entries correspond to s,t combinations that cannot occur. For illustrative purposes, suppose we wish to compute an optimal production/inventory policy over five periods when current inventory is 3.

Figure 4.1 gives a network representation of the calculation with the recursion (4.6) at stage 5 when starting inventory is 3 units. This network is

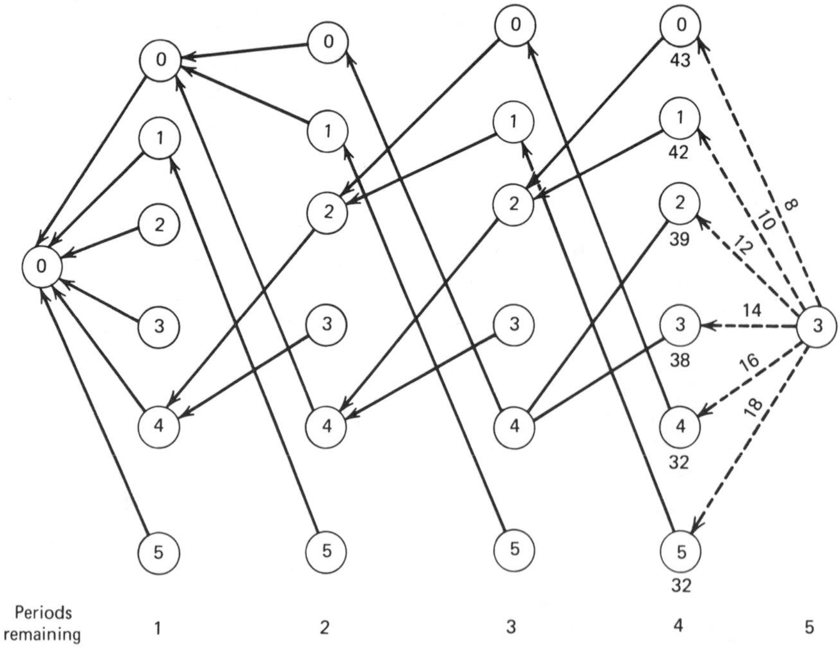

Figure 4.1. Dynamic programming example of production/inventory control.

obtained by replicating in time the nodes of the network originally used to describe the problem and by adding a stage dimension to the arcs. The solid lines from stage 4 onward correspond to optimal decisions. The dotted lines represent the feasible decisions at the given state and stage. For example, the arc from state 3 at stage 5 to state 0 at stage 4 corresponds to the decision to produce one unit. The arc from state 3 at stage 5 to state 5 at stage 4 corresponds to the decision to produce 6 units. The numbers below the nodes at stage 4 are the quantities f_j^4 and the numbers above the dotted arcs are the arc costs. The minimum of $c_{3,j}+f_j^4$ for $j=0,1,2,3,4,5$ is $c_{34}+f_4^4$, and the indicated optimal policy for a 5-period planning horizon, starting with 3 units of inventory is $(d_1^*,d_2^*,d_3^*,d_4^*,d_5^*)=(5,0,6,6,0)$.

If we continued backward with the calculations shown in Figure 4.1, we would discover that the immediate decisions that are optimal when 4 periods remain are also optimal immediate decisions for any n greater than 4. This set of decisions in the original network used to describe the problem is shown in Figure 4.2. We see it consists of the circuit $(0,2,4)$ and paths leading to the circuit from the other states $1,3,5$. For long planning horizons and states on the circuit, the optimal policy is to cycle for most of the periods in the horizon, with a transient set of decisions during the final three periods. For this reason, the circuit is called a *turnpike* circuit. For states off the turnpike circuit, the optimal policy is to go to it immediately and then proceed optimally as just described. The turnpike circuit $(0,2,4)$ is most attractive over the long run because its average cost

$$10\frac{2}{3}=\frac{c_{02}+c_{24}+c_{40}}{3}$$

is minimal among all the circuits. This type of asymptotic result is not particular to our numerical example and will be established for a much

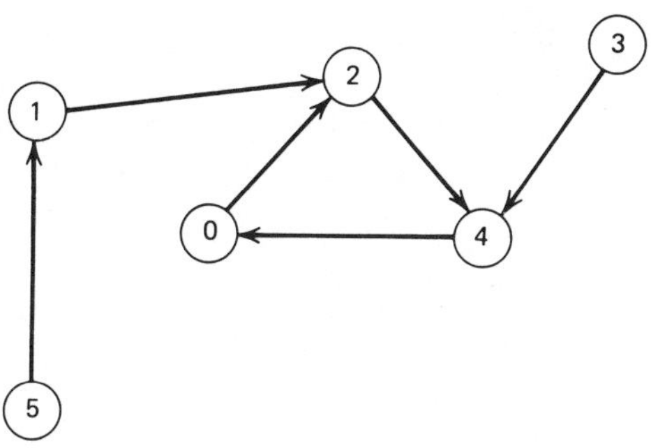

Figure 4.2. Steady-state optimal decisions for production/inventory control example.

more general model in the next section. The result implies that the computational effort required to compute optimal policies over long planning horizons can be greatly reduced from the effort implied by backward iteration with (4.6).

For this problem, the recursions (4.6) can be expressed in terms of the min-addition operator defined at the end of the previous section

$$\mathbf{f}^n = \mathbf{C} \otimes \mathbf{f}^{n-1} \qquad n = 1, 2, \ldots, N$$

where $\mathbf{f}^n$ is the m-vector with components f_s^n and $\mathbf{f}^0$ is the m vector of zeroes. There is, however, a difference in the matrix $\mathbf{C}$ of this example; namely, the diagonal terms are not zero. ▲

EXAMPLE 4.2. One of the simplest mathematical programming models is the *knapsack problem*. It derives its name from the following situation. A person is going on a picnic. He or she will carry a knapsack of b pounds filled with a combination of n items, each weighing a_j pounds and with value c_j. The items are indivisible and can be included any number of times in the knapsack. The person wishes to maximize the value of the knapsack. The knapsack problem has no inherent practical importance, but it has properties similar to those of more complex and important mathematical programming problems.

The knapsack problem can be written as the one constraint integer programming problem

$$\begin{aligned} f^b = \max \sum_{j=1}^{n} c_j x_j& \\ \text{s.t.} \sum_{j=1}^{n} a_j x_j = b& \qquad (4.8) \\ x_j \text{ nonnegative integer}& \end{aligned}$$

where b and the a_j are positive integers, and we assume $a_1 = 1$. The knapsack problem can also be solved by dynamic programming. The problem is so simple that it can be viewed as having one state and many stages. Stage l corresponds to having l pounds remaining in the knapsack to be filled. The decision at each stage is to choose any one of the n items to put into the knapsack and use a_j of the remaining l pounds. The only restriction is that the chosen item must satisfy $a_j \leq l$. Thus, we have

$$\begin{aligned} f^l &= \max_{\substack{j=1,\ldots,n \\ \text{and} \\ a_j \leq l}} \left\{ c_j + f^{l-a_j} \right\} \qquad l = 1, 2, \ldots, b \qquad (4.9) \\ f^0 &= 0 \end{aligned}$$

Suppose that the n items are ordered so that

$$\frac{c_1}{a_1} \leq \frac{c_2}{a_2} \leq \cdots \leq \frac{c_n}{a_n}$$

The linear programming solution to the knapsack problem in the form (4.8) without the integrality constraint is to set $x_n = b/a_n$, $x_j = 0$ for $j = 1,\ldots,n-1$. The optimal linear programming activity a_n is analogous to the minimal cost turnpike circuit in the previous example and plays a central role in asymptotic optimal policies for the knapsack problem. Specifically, it can be shown that there exists a positive b^* such that for all $b \geq b^*$, $f^b = c_n + f^{b-a_n}$. Moreover, an *a priori* upper bound on b^* can be computed (see Exercise 4.10). Thus, for any $b \geq b^*$, the optimal policy is to use the turnpike activity k times where k is the smallest integer such that $b - ka_n < b^*$, and proceed to use up the remaining $b - ka_n$ pounds according to the recursion (4.9).

The integer programming problem (4.8) has only one constraint, but the recursion (4.9) is perfectly valid for an arbitrary integer programming problem where $\mathbf{b}$, $\mathbf{l}$, and the $\mathbf{a}_j$ are integer m-vectors. However, unless there is a constraint $\Sigma_{j=1}^{n} a_{ij} x_j \leq b_i$ of the integer programming problem with all $a_{ij} > 0$, $b_i > 0$, there may be no systematic method for moving from one stage to another. If such a constraint exists, then there is a monotone decrease in the stage as the result of each decision made. This is discussed further in Exercise 4.12 at the end of the chapter.

Table 4.1 gives the data for a knapsack problem with four items.

TABLE 4.1

j	1	2	3	4
c_j	0	8	5	12
a_j	1	5	3	7

Figure 4.3 shows the network representation of calculation with the recursion (4.9) at stage 11. The solid arcs correspond to optimal decisions previously calculated and the dotted arcs correspond to the immediate decisions at that stage. The optimal immediate decision at stage 11 is to include one unit of item 3. The optimal policy for a knapsack of 11 pounds is to include one unit of item 2 weighing 5 pounds and two units of item 3 weighing 3 pounds each. ▲

EXAMPLE 4.3. Let $G = (\mathfrak{N}, \mathfrak{A})$ be a complete network with m nodes and lengths $c_{ij} \geq 0$ associated with each arc (i,j). The *traveling salesman problem*

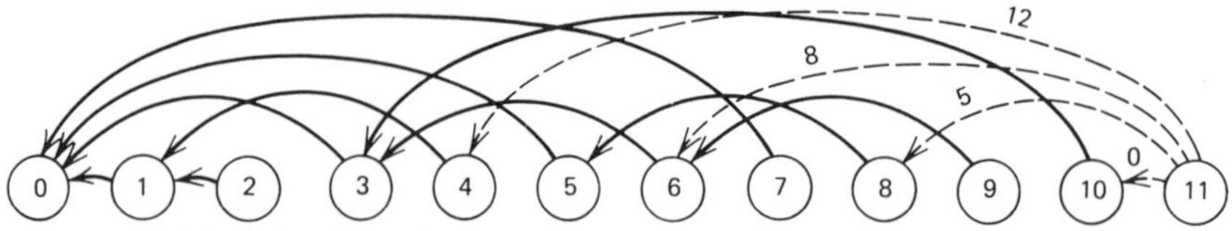

Figure 4.3. Dynamic programming example of the knapsack problem.

is one of finding a minimal length elementary circuit that visits all the nodes in the network. In this chapter we give for illustrative purposes a dynamic programming recursion for solving the traveling salesman problem. When it is reasonably large, say $m \geq 20$, dynamic programming is not an effective technique for the problem because of the large number of states. In Chapter 8, we give a more efficient method which exploits some special structure of the problem.

The states in the dynamic programming recursion for the traveling salesman problem are denoted by (S,j) where S is a subset of $\mathfrak{N}' = \{2,\ldots,m\}$ and $j \in S$ is an identified node. Each (S,j) corresponds to the chains in the network from node 1 visiting the nodes in S and terminating at node j. Let $f_{S,j}$ denote the length of the minimal length path among these paths. The $f_{S,j}$ satisfy the recursion

$$\begin{aligned} f_{S,j} &= \underset{i \in S-j}{\text{minimum}} \{c_{ij} + f_{S-j,i}\} \qquad \text{for } S \neq \phi \\ f_{\phi,-} &= 0 \end{aligned} \tag{4.10}$$

The traveling salesman problem is solved by computing

$$\min_{i=2,\ldots,m} \{c_{i,1} + f_{\mathfrak{N}',i}\}$$

which gives the length of the minimal length circuit and the circuit itself is determined by backtracking.

Figure 4.4 is the network representation of the recursion (4.10) for the case $m=5$. The nodes in the figure are the states S,j and the arcs between the state T, i and $TU\{j\}$, j, correspond to the arcs (i,j) with length c_{ij} in the original network. For convenience, only a few arcs are depicted in the figure. The network is constructed from left to right and we can think of the computation as having m stages. Note that unlike the previous two examples where we worked backward from the terminal state and stage, this recursion works forward from the initial state and stage. The traveling salesman problem is solved by finding the shortest route chain from node ϕ to node 1. Such a chain is depicted by heavy lines in Figure 4.4 and the corresponding minimal length circuit is (1,5), (5,4), (4,2), (2,3), (3,1). The so-called "curse of dimensionality" of dynamic programming is illustrated in Figure 4.4 by the build-up of states in the middle of the network.

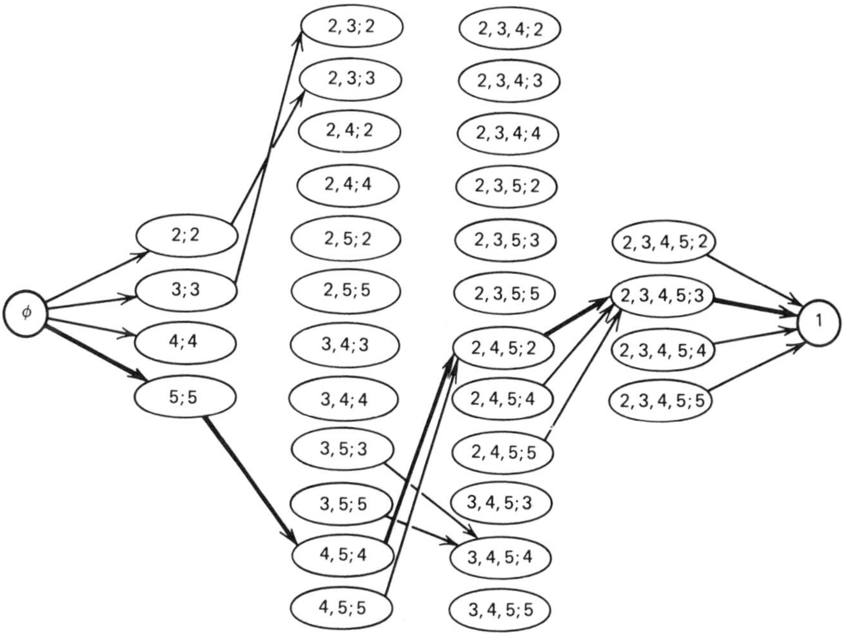

Figure 4.4. Dynamic programming example of the traveling salesman problem.

Specifically, there are $k\binom{m-1}{k}$ nodes at stage k consisting of the states S,j satisfying $|S|=k$. ▲

The following example generalizes in one respect the discrete dynamic programming problems studied above. Specifically, we consider a dynamic programming problem where the state space S_i is a continuum and a subset of R^m rather than a finite set. The set of available decisions for each $s \in S_i$ remains finite because the problems are linear programming indicating the decisions can be limited to extreme points.

EXAMPLE 4.4. Many linear programming applications are multistage of the form

$$\begin{aligned}
\min \quad & \mathbf{c}^1\mathbf{x}^1 + \mathbf{c}^2\mathbf{x}^2 + \mathbf{c}^3\mathbf{x}^3 + \cdots + \mathbf{c}^N\mathbf{x}^N \\
\text{s.t.} \quad & \mathbf{A}^1\mathbf{x}^1 = \mathbf{b}^1 \\
& -\mathbf{K}^1\mathbf{x}^1 + \mathbf{A}^2\mathbf{x}^2 = \mathbf{b}^2 \\
& -\mathbf{K}^2\mathbf{x}^2 + \mathbf{A}^3\mathbf{x}^3 = \mathbf{b}^3 \\
& \quad \vdots \qquad\qquad\qquad \vdots \\
& -\mathbf{K}^{N-1}\mathbf{x}^{N-1} + \mathbf{A}^N\mathbf{x}^N = \mathbf{b}^N \\
& \mathbf{x}^1 \geq 0 \qquad \mathbf{x}^2 \geq 0, \ldots, \mathbf{x}^N \geq 0
\end{aligned}$$

We assume that $\mathbf{A}^i$ is $m \times q$ in each period i. In period i, we must choose $\mathbf{x}^i \geq 0$ so that $\mathbf{A}^i\mathbf{x}^i = \mathbf{b}^i + \mathbf{K}^{i-1}\mathbf{x}^{i-1}$, and at the same time, take into account the future consequences of this selection. The multistage problem can be formulated as a dynamic linear programming problem by taking all $\mathbf{s} \in R^m$ as the state space. Letting $f_\mathbf{s}^n$ denote the minimal cost when the state is $\mathbf{s}$ and n periods remain, we have as before the recursion for all $\mathbf{s} \in R^m$

$$
\begin{aligned}
f_\mathbf{s}^n = \min\{&\mathbf{c}^{N-n+1}\mathbf{x}^{N-n+1} + f_\mathbf{t}^{n-1}\} \\
\text{s.t. } &\mathbf{A}^{N-n+1}\mathbf{x}^{N-n+1} = \mathbf{s} \\
\mathbf{b}^{N-n} + &\mathbf{K}^{N-n+1}\mathbf{x}^{N-n+1} = \mathbf{t} \\
&\mathbf{x}^{N-n+1} \geq 0 \\
&n = 1, 2, \ldots, N
\end{aligned}
\tag{4.11}
$$

In (4.11), we take $f_\mathbf{s}^n = +\infty$ if there is no nonnegative $\mathbf{x}^{N-n+1}$ satisfying $\mathbf{A}^{N-n+1}\mathbf{x}^{N-n+1} = \mathbf{s}$.

It is somewhat misleading to suggest that the dynamic programming recursion (4.11) solves the multistage problem because it is virtually impossible to calculate $f_\mathbf{s}^n$ for all $\mathbf{s} \in R^m$ for m larger than 3 or 4. The situation is not completely hopeless, however, because the function f^n is a piecewise linear convex function which can be approximated. This property of f^n can be shown using linear programming duality theory. In Chapter 6, we consider again multistage problems and give different methods for solving it. ▲

A number of other extensions to our dynamic programming model are possible. First, there may be stochastic behavior of the costs $c_n(s,d)$ and/or the transitions $T_n(s,d)$ where these quantities are random variables that are functions of n, s, and d. If $c_n(s,d)$ is such a random variable, and if the optimality criterion is the natural one of minimizing total expected cost, then it is easy to show that the above analysis is correct if the expected value of $c_n(s,d)$ is used in place of $c_n(s,d)$ in the calculation. The situation is somewhat more complicated if the transition $T_n(s,d)$ is a random variable because then we do not know with certainty the sequence of states to be visited by a policy. The finite state space Markovian decision model discussed in the following section treats this problem.

Another generalization is to investigate the structure of optimal policies as n grows large. Meaningful models and insightful asymptotic results can be constructed and obtained if we assume stationarity of the state and action spaces, and the cost and transition functions; that is, if we assume the data of the problem does not depend on time. These models and results are also discussed in detail in the following section.

A final type of extension of the models discussed above is to permit S_n and $D_n(s)$ to become countably or uncountably infinite as we saw in Example 4.4. Similarly, there are a wide variety of dynamic programming problems called optimal control problems, where decisions are made in continuous rather than discrete time and for which the dynamic transitions are described by differential equations.

4.4 MARKOVIAN DECISION THEORY

In this section, we examine in detail the asymptotic behavior as the planning horizon grows large of an important stochastic dynamic programming problem called the Markovian decision model with finite state and action spaces. This model has the same data in each period. As in the previous section, the state space is $S=\{1,\ldots,m\}$. At the start of each decision period, the state is observed and a decision d is selected from a nonempty finite decision set $D(s)$. As a result of the current state s and the chosen decision d, an immediate expected cost $c(s,d)$ is incurred and the system makes the transition to the new state t with probability $p(t|s,d)$; the probabilities satisfy $\sum_{t=1}^{m} p(t|s,d)=1$. Our development will focus on the case when there is a discount factor α, $0<\alpha<1$, associated with the stream of costs generated by the dynamic process; that is, one unit of cost experienced at the start of period i has a present value equal to α^{i-1}. The objective is to minimize the total expected discounted cost over a finite or infinite horizon. An alternative objective is to minimize total expected average cost over an infinite horizon. This objective can be viewed as the limiting case of discounting when α converges to one from below. We discuss briefly the average cost criterion at the end of this section after we establish our main results for the discounted cost criterion.

Let f_s^n denote the minimal total expected discounted cost starting in state s when there are n periods remaining in the planning horizon. The quantities f_s^n can be calculated recursively by

$$f_s^n = \min_{d \in D(s)} \left\{ c(s,d) + \alpha \sum_{t=1}^{m} p(t|s,d) f_t^{n-1} \right\} \qquad s=1,\ldots,m \qquad (4.12)$$

$$n=1,2,\ldots,N$$

$$f_s^0 = 0 \qquad s=1,\ldots,m$$

As n grows large, we show that the numbers f_s^n converge to f_s^∞ equaling the minimal total expected discounted cost of starting in state s and optimizing the dynamic system for a countably infinite number of periods. Moreover, the scalars f_s^∞ satisfy the functional equation

$$f_s^\infty = \min_{d \in D(s)} \left\{ c(s,d) + \alpha \sum_{t=1}^{m} p(t|s,d) f_t^\infty \right\} \qquad s=1,\dots,m \qquad (4.13)$$

The system (4.13) is a functional equation because the f_s^∞ appear on both sides. Let $\mathbf{f}^n$ denote the $m \times 1$ vector with components f_s^n and let $\mathbf{f}^\infty$ denote the $m \times 1$ vector with components f_s^∞.

Finding a solution to (4.13) is a fixed point problem because it requires the determination of a vector $\mathbf{f}^\infty$ that is mapped into itself by the expectation and minimization operators of (4.13). We will not explicitly make use of fixed point theory in our subsequent development, but our analytic methods will implicitly draw on it. Exercises 4.14 and 4.18 consider directly the role of fixed point theory in infinite horizon dynamic programming. The relationship of fixed point theory to other mathematical programming problems will be discussed again in Chapter 7.

It is necessary for our analysis to define decision functions δ from the state space to the decision space. For $s=1,\dots,m$, $\delta(s)$ is an element of $D(s)$. The set Δ of all such functions is finite and corresponds to the finite set of vectors $(d_1, d_2, \dots, d_m) \in X_{s=1}^m D(s)$. Thus, there are $\Pi_{s=1}^m |D(s)|$ distinct decision functions. Let $\mathbf{c}_\delta$ denote the $m \times 1$ vector with elements $c[s,\delta(s)]$, and let $\mathbf{P}_\delta$ denote the $m \times m$ matrix with the element $p[t|s,\delta(s)]$ in row s and column t. The matrix $\mathbf{P}_\delta$ is a Markov probability or transition matrix.

A *policy* $\pi = [\delta_1, \delta_2, \dots]$ is a countably infinite sequence of decison functions. At the start of period i, if the system is observed to be in state s, then the decision $\delta_i(s)$ is made. Note that, in general, there is a positive probability of being in any one of a number of states at the start of period i as the result of following π. Thus, a decision function δ_i defined on all states is required to specify the appropriate decision at the start of period i when the specific state is observed. A policy using the same decision function δ in every period is called a *stationary policy*, and is denoted by δ^∞. We let $[\rho, \delta^\infty]$ denote the policy of using ρ in the first period and δ in every period thereafter.

An arbitrary policy $\pi = [\delta_1, \delta_2, \delta_3, \dots]$ has associated with it an $m \times 1$ vector of total expected discounted costs over the infinite horizon, denoted by $\mathbf{f}(\pi)$. This vector is given by

$$\mathbf{f}(\pi) = \mathbf{c}_{\delta_1} + \mathbf{P}_{\delta_1}\mathbf{c}_{\delta_2} + \alpha^2 \mathbf{P}_{\delta_1}\mathbf{P}_{\delta_2}\mathbf{c}_{\delta_3} + \dots \qquad (4.14)$$

Note that for any i the matrix $\mathbf{P}_{\delta_1}\mathbf{P}_{\delta_2}\dots\mathbf{P}_{\delta_i}$ is an $m \times m$ Markov transition matrix where the element $p_{st}^i(\pi)$ in row s and column t of the matrix is the probability that the system is in state t at the start of period $i+1$ (i.e., after i transitions), given that it started in state s and we follow the policy π. The infinite vector sum (4.14) converges to $\mathbf{f}(\pi)$ because each component

$$f_s(\pi) = \sum_{i=0}^{\infty} \alpha^i \sum_{t=1}^{m} p_{st}^i(\pi) c_t^{i+1}(\pi)$$

where $c_t^{i+1}(\pi)$ is the cost incurred at the start of the $i+1$st period if the system is in state t as the result of following the policy π. Since for all i and t we have

$$|c_t^{i+1}(\pi)| \leq L = \max_{s,d} |c(s,d)|$$

and

$$\sum_{t=1}^{m} p_{st}^i(\pi) = 1$$

it is easy to show that the infinite sums converge.

A policy π^* is called *optimal* (minimal) if $\mathbf{f}(\pi^*) \leq \mathbf{f}(\pi)$ for all policies π. Although we will show optimal policies exist, moreover, optimal policies that are stationary, their existence is not immediately obvious because there are an infinite number of different policies possible over the infinite horizon. Note also that optimal policies cannot be computed by backward iteration with the recursions (4.12) because there are always an infinite number of periods remaining in an infinite horizon.

The total expected discounted cost vector for a stationary strategy is particularly simple to state using the result of the following lemma. The reader is asked to prove the lemma in Exercise 2.21.

LEMMA 4.1. Let $\mathbf{A}$ be a matrix with the property that $\mathbf{A}^n$ tends to $\mathbf{0}$ (zero matrix) as n tends to infinity. Then $\mathbf{I}-\mathbf{A}$ has an inverse and

$$(\mathbf{I}-\mathbf{A})^{-1} = \sum_{i=0}^{\infty} \mathbf{A}^i$$

Thus, we can compute

$$\begin{aligned} \mathbf{f}(\delta^{\infty}) &= \mathbf{c}_\delta + \alpha \mathbf{P}_\delta \mathbf{c}_\delta + \alpha^2 \mathbf{P}_\delta^2 \mathbf{c}_\delta + \ldots \\ &= \left(\sum_{i=0}^{\infty} \alpha^i \mathbf{P}_\delta^i\right)\mathbf{c}_\delta = (\mathbf{I} - \alpha \mathbf{P}_\delta)^{-1} \mathbf{c}_\delta \end{aligned}$$

where the last equality follows from Lemma 4.1.

EXAMPLE 4.5. An item is sold for \$10/unit. In each period there is random demand for it with a probability of 0.1 that demand $=0$ units, a probability of 0.7 that demand $=1$ unit and a probability of 0.2 that demand $=2$ units. An inventory system for the item is maintained with possible starting inventory levels of $0, 1, 2$. There is a holding cost of

\$3/unit of starting inventory. The items can be purchased at the start of each period at a cost of \$5/unit. Delivery of the item is immediate. If demand exceeds starting inventory plus the number of items purchased, the excess demand is lost. If ending inventory equal to starting inventory plus the number of items purchased minus demand exceeds two units, then the excess ending inventory must be discarded with no salvage value. Finally, there is a discount rate of 0.8. For this example, the state space is $\{0,1,2\}$ corresponding to the levels of starting inventory, and we seek a policy that maximizes the expected net profit starting in any one of these states. For computational convenience, we will actually minimize the negative of the net profit.

There are four reasonable decision functions to be considered for this problem. When starting inventory is zero, we have two reasonable decisions: purchase one item, or purchase two items. The probability of zero demand is low enough to make it unprofitable to purchase zero items. When starting inventory is one, we also have two possible decisions: purchase no items, or purchase one item. There is no profit in purchasing two items when starting inventory is one because then we will incur a holding cost of at least \$3 at the start of the next period that could be avoided. When starting inventory is two units, it is clearly optimal to purchase no units since starting inventory is sufficient to meet demand in the period in any case. Thus, the four decision functions of interest are $(d_0, d_1, d_2) = (1,0,0)$, $(1,1,0)$, $(2,0,0)$, $(2,1,0)$ where d_i is the number of units purchased when the system is in state i, $i = 0,1,2$.

Consider, for example, the stationary policy of using $\delta_1 = (d_{01}, d_{11}, d_{21}) = (1,0,0)$ in every period. The immediate cost vector for this policy is

$$\mathbf{c}_{\delta_1} = \begin{bmatrix} c(0,1) \\ c(1,0) \\ c(2,0) \end{bmatrix} = \begin{bmatrix} -4 \\ -6 \\ -5 \end{bmatrix}$$

These costs are computed as follows: $c(0,1)$ equals the purchase cost of \$5 for the one unit purchased, minus the expected income of $(-10)(0.9) = -\$9$; $c(1,0)$ equals the holding cost of \$3 minus the expected income of $-\$9$; $c(2,0)$ equals the holding cost of \$6 minus the expected income of $(0)(0.1) + (10)(0.7) + 20(0.2) = -\$11$. The transition matrix for δ_1 is

$$\mathbf{P}_{\delta_1} = \begin{bmatrix} 0.9 & 0.1 & 0 \\ 0.9 & 0.1 & 0 \\ 0.2 & 0.7 & 0.1 \end{bmatrix}$$

The total expected discounted cost of the stationary policy δ_1^∞ is

$$(\mathbf{I} - \alpha \mathbf{P}_{\delta_1})^{-1} \mathbf{c}_{\delta_1} = \begin{bmatrix} 4.6 & 0.40 & 0 \\ 3.6 & 1.40 & 0 \\ 2.99 & 0.92 & 1.09 \end{bmatrix} \begin{bmatrix} -4 \\ -6 \\ -5 \end{bmatrix} = \begin{bmatrix} -20.8 \\ -22.8 \\ -22.9 \end{bmatrix} \quad \blacktriangle$$

The connection between the functional equations (4.13) and optimal policies is given in the following theorem. Methods for computing optimal policies are treated afterward.

THEOREM 4.2. Suppose a solution, denoted by y^*, is assumed to exist for the functional equation (4.13). Define a decision function ρ by selecting for each s any decision $\rho(s) \in D(s)$ satisfying

$$c[s,\rho(s)] + \alpha \sum_{t=1}^{m} p[t|s,\rho(s)]y_t^*$$
$$= \min_{d \in D(s)} \left\{ c(s,d) + \alpha \sum_{t=1}^{m} p(t|s,d)y_t^* \right\} = y_s^* \quad (4.15)$$

Then $\mathbf{y}^* = \mathbf{f}(\rho^\infty) = (\mathbf{I} - \alpha \mathbf{P}_\rho)^{-1}\mathbf{c}_\rho$ and ρ^∞ is an optimal policy.

PROOF. By construction of ρ, we have from (4.15) that $\mathbf{y}^* = \mathbf{c}_\rho + \alpha \mathbf{P}_\rho \mathbf{y}^*$ implying $\mathbf{y}^* = (\mathbf{I} - \alpha \mathbf{P}_\rho)^{-1}\mathbf{c}_\rho$, which is the total expected discounted cost vector $\mathbf{f}(\rho^\infty)$ for the policy ρ^∞. To see that ρ^∞ is optimal, that is, that $\mathbf{f}(\rho^\infty) \le \mathbf{f}(\pi)$ for any policy $\pi = [\delta_1, \delta_2, \delta_3, \ldots]$, we use an induction argument to show that $\mathbf{f}(\delta_1, \ldots, \delta_N, \rho^\infty) \ge \mathbf{f}(\rho^\infty)$ for all N. This implies $\mathbf{f}(\pi) = \lim_{N\to\infty} \mathbf{f}(\delta_1, \ldots, \delta_N, \rho^\infty) \ge \mathbf{f}(\rho^\infty)$.

INDUCTION PROOF. The relationships (4.15) for all s can be written in vector form as

$$\mathbf{f}(\rho^\infty) = \mathbf{y}^* \le \mathbf{c}_\delta + \alpha \mathbf{P}_\delta \mathbf{y}^* = \mathbf{f}(\delta, \rho^\infty) \quad (4.16)$$

for any $\delta \in \Delta$, where the inequality follows from the minimum in (4.15). Thus, $\mathbf{f}(\rho^\infty) \le \mathbf{f}(\delta_1, \rho^\infty)$. Our induction hypothesis is $\mathbf{f}(\rho^\infty) \le \mathbf{f}(\delta_1, \ldots, \delta_{N-1}, \rho^\infty)$ and we will show this imples $\mathbf{f}(\rho^\infty) \le \mathbf{f}(\delta_1, \ldots, \delta_N, \rho^\infty)$ to complete the proof.

Let $\delta_1, \ldots, \delta_{N-1}$ be used to define a mapping L from R^m to R^m by

$$L(\mathbf{w}) = \mathbf{c}_{\delta_1} + \alpha \mathbf{P}_{\delta_1}\mathbf{c}_{\delta_2} + \cdots + \alpha^{N-1}\mathbf{P}_{\delta_1} \ldots \mathbf{P}_{\delta_{N-1}}\mathbf{w}$$

This mapping is monotone in the sense that $\mathbf{w}^1 \ge \mathbf{w}^2$ implies $L(\mathbf{w}^1) \ge L(\mathbf{w}^2)$. Monotonicity is a consequence of the fact that the $m \times m$ matrix $\mathbf{P}_{\delta_1} \ldots \mathbf{P}_{\delta_{N-1}}$ has nonnegative coefficients. We apply the monotone mapping L to the vector inequality $\mathbf{f}(\delta_N, \rho^\infty) \ge \mathbf{f}(\rho^\infty)$ from (4.16), which can be written as $\mathbf{c}_{\delta_N} + \alpha \mathbf{P}_{\delta_N} y^* \ge \mathbf{y}^*$. The result is

$$\mathbf{c}_{\delta_1} + \alpha \mathbf{P}_{\delta_1}\mathbf{c}_{\delta_2} + \ldots + \alpha^{N-1}\mathbf{P}_{\delta_1} \ldots \mathbf{P}_{\delta_{N-1}}\mathbf{c}_{\delta_N} + \alpha^N \mathbf{P}_{\delta_1} \ldots \mathbf{P}_{\delta_N}\mathbf{y}^*$$
$$\ge \mathbf{c}_{\delta_1} + \alpha \mathbf{P}_{\delta_1}\mathbf{c}_{\delta_2} + \ldots + \alpha^{N-1}\mathbf{P}_{\delta_1} \ldots \mathbf{P}_{\delta_{N-1}}\mathbf{y}^*$$

which can be written as $\mathbf{f}(\delta_1, \ldots, \delta_N, \rho^\infty) \ge \mathbf{f}(\delta_1, \ldots, \delta_{N-1}, \rho^\infty)$. The right-hand vector satisfies $\mathbf{f}(\delta_1, \ldots, \delta_{N-1}, \rho^\infty) \ge \mathbf{f}(\rho^\infty)$ by the induction hypothesis thereby completing the proof. ■

COROLLARY 4.1. The solution to the functional equation (4.13), if it exists, is unique.

PROOF. The solution $\mathbf{y}^*$ to (4.13), if it exists, equals minimum $\mathbf{f}(\pi)$, where the minimum is the vector minimum over all policies π. Such a minimum is by definition unique. ■

We now show how linear programming can be used to prove the existence of the solution to the functional equations (4.13) by computing it. Equivalently, an optimal stationary policy is determined and the simplex method in this context can be interpreted as a *policy iteration* procedure. At each iteration, a trial stationary policy is selected and tested for optimality. If the policy is not optimal, the stationary policy is changed to a strictly better one.

The linear programming problem related to (4.13) is

$$\begin{aligned} &\max \sum_{s=1}^{m} \frac{(1-\alpha)}{m} y_s \\ &\text{s.t. } y_s - \alpha \sum_{t=1}^{m} p(t|s,d) y_t \le c(s,d) \qquad \text{for all } d \in D(s) \qquad (4.17) \\ &\qquad\qquad\qquad\qquad\qquad\qquad s = 1,\ldots,m \end{aligned}$$

We could choose any positive coefficients in the objective function to derive our main results, but the choice of $(1-\alpha)/m > 0$ is useful for subsequent interpretation and extensions. The dual to (4.17) is

$$\begin{aligned} &\min \sum_{s=1}^{m} \sum_{d \in D(s)} c(s,d) x(s,d) \\ &\text{s.t. } \sum_{d \in D(s)} x(s,d) - \alpha \sum_{u=1}^{m} \sum_{d \in D(u)} p(s|u,d) x(u,d) = \frac{1-\alpha}{m} \qquad (4.18) \\ &\qquad\qquad\qquad\qquad\qquad\qquad s = 1,\ldots,m \\ &\qquad x(s,d) \ge 0 \qquad \text{for all } s,d \end{aligned}$$

It is easy to show that both (4.17) and (4.18) have feasible and therefore optimal solutions. For (4.17), the solution $y_s = l = \min_{s,d} c(s,d)$ for all s, is feasible since $l \le c(s,d) \le c(s,d) + \alpha \Sigma_{t=1}^{m} p(t|s,d) l$ for all s,d. For (4.18), let each decision function δ correspond to the $1 \times m$ vector $\mathbf{x}_\delta = \{x[1,\delta(1)],\ldots,x[m,\delta(m)]\}$. Associated with these variables is the $m \times m$ nonsingular basis matrix $\mathbf{I} - \alpha \mathbf{P}_\delta$, and partitioning $\mathbf{x}_\delta$ into basic variables $(\mathbf{x}_\delta)_\mathbf{B}$ and nonbasic variables $(\mathbf{x}_\delta)_\mathbf{N}$, the corresponding basic solution is $(\mathbf{x}_\delta)_\mathbf{B} = (1-\alpha)/m \cdot \mathbf{1}_m (\mathbf{I} - \alpha \mathbf{P}_\delta)^{-1}$, where $\mathbf{1}_m$ is the $1 \times m$ vector of ones. For any decision function δ, $\mathbf{x}_\delta \ge \mathbf{0}$ since $(\mathbf{I} - \alpha \mathbf{P}_\delta)^{-1} \ge 0$ by Lemma 4.1. The dual solution induced by the basis corresponding to δ is given by $\mathbf{y} = \mathbf{f}(\delta^\infty) = (\mathbf{I} - \alpha \mathbf{P}_\delta)^{-1} \mathbf{c}_\delta$, where $\mathbf{c}_\delta^T$ is the $1 \times m$ vector $\{c[1,\delta(1)],\ldots,c[m,\delta(m)]\}$.

Conversely, any basic feasible solution to (4.18) must have exactly one basic variable $x(s,d)$ for each s, $s=1,\dots,m$. If not, since there are m basic variables, then some row v must have no basic variables $x(v,d)$. But this is impossible because it implies $-\alpha\Sigma_{u=1}^{m}\Sigma_{d\in D(u)}p(v|u,d)x(u,d)=(1-\alpha)/m>0$ although the terms in the sum are nonnegative.

The simplex test for optimality (see Step 1 of the simplex algorithm in Chapter 1) of the basic feasible solution $\mathbf{x}_\delta$ to (4.18) translates here into the requirement that the dual solution $\mathbf{y}=(\mathbf{I}-\alpha\mathbf{P}_\delta)^{-1}\mathbf{c}_\delta$ satisfies the functional equations (4.13). From Chapter 2, we know that the optimality test is for feasibility in the dual (4.17) to (4.18); that is, for all s and d,

$$y_s \le c(s,d)+\alpha\sum_{t=1}^{m}p(t|s,d)y_t$$

For these particular linear programming problems, however, there is for each s at least one $\delta(s)\in D(s)$ such that $x[s,\delta(s)]$ is basic implying

$$y_s = c[s,\delta(s)]+\alpha\sum_{t=1}^{m}p[t|s,\delta(s)]y_t$$

which in turn implies $\mathbf{y}$ satisfies (4.13).

The simplex algorithm applied to (4.18) will produce in a finite number of iterations a dual solution $\mathbf{y}$ optimal in (4.17) because (4.18) has an optimal solution. This optimal dual solution will be unique because it solves (4.13), which has a unique solution. Thus, we have proved the following theorem.

THEOREM 4.3. The unique solution to the functional equation (4.13) is the unique optimal solution to the linear programming problem (4.17).

If the simplex test for optimality of the basic feasible solution $\mathbf{x}_\delta$ fails, then the stationary policy δ^∞ is not optimal because for some s and some $d_s\in D(s)$,

$$c(s,d_s)+\alpha\sum_{t=1}^{m}p(t|s,d_s)y_t<y_s \tag{4.19}$$

indicating that the variable $x(s,d_s)$ wants to enter the basis. Since there is exactly one basic variable for each s, the leaving basic variable is $x[s,\delta(s)]$. In other words, the decision d_s is preferred to $\delta(s)$ when the system is in state s. The optimality test suggests that it is preferable to use the new function ρ obtained from δ by changing $\delta(s)$ to $d_s=\rho(s)$ for all states s such that (4.19) holds. The reader is asked in Exercise 4.17 to prove this assertion about policy iteration; namely, that $\mathbf{f}(\rho^\infty)\le\mathbf{f}(\delta^\infty)$ and $\mathbf{f}(\rho^\infty)\ne\mathbf{f}(\delta^\infty)$. The simplex algorithm proceeds at each iteration by changing the decision for only *one* state s such that (4.19) holds. Nevertheless, the policy iteration interpretation of the simplex algorithm applied to the infinite horizon Markovian decision model remains valid.

The linear programming problems (4.17) and (4.18) for Example 4.5 are

$$\begin{aligned} &\max 0.067y_0+0.067y_1+0.067y_2 \\ \text{s.t.}\quad &0.28y_0-0.08y_1 \leq -4 \\ &0.84y_0-0.56y_1-0.08y_2 \leq -1 \\ &-0.72y_0+0.92y_1 \leq -6 \\ &-0.16y_0+0.44y_1-0.08y_2 \leq -3 \\ &-0.16y_0+0.56y_1+0.92y_2 \leq -5 \end{aligned}$$

and

$$\begin{aligned} \min -4x(0,1)- \quad 1x(0,2)- \quad 6x(1,0)- \quad 3x(1,1)- \quad 5x(2,0) \\ 0.28x(0,1)+0.84x(0,2)-0.72x(1,0)-0.16x(1,1)-0.16x(2,0)=0.067 \\ -0.08x(0,1)-0.56x(0,2)+0.92x(1,0)+0.44x(1,1)-0.56x(2,0)=0.067 \\ -0.08x(0,2) \qquad\qquad -0.08x(1,1)+0.92x(2,0)=0.067 \end{aligned}$$

$$x(0,1)\geq 0 \qquad x(0,2)\geq 0 \qquad x(1,0)\geq 0 \qquad x(1,1)\geq 0 \qquad x(2,0)\geq 0$$

The optimal solutions to this pair of problems are $(y_1, y_2, y_3) = (-20.8, -22.8, -22.9)$ and $[x(0,1),\ x(0,2),\ x(1,0),\ x(1,1),\ x(2,0)] = (0.746, 0, 0.181, 0, 0.073)$. The optimal stationary policy is δ_1^{∞}, evaluated in Example 4.5.

Thus far, we have studied the infinite horizon planning problem implied by (4.13) without relating it to the finite horizon problem embodied in recursions (4.12). We return to this relationship to establish the convergence properties of the finite horizon model as the horizon becomes infinite.

THEOREM 4.4. (Principle of Successive Approximation). For all s,

$$\lim_{n\to\infty} f_s^n = f_s^{\infty}$$

where the f_s^n are the minimal total expected discounted n-period costs computed from the recursions (4.12), and f_s^{∞} are the minimal total expected discounted infinite horizon costs from the functional equations (4.13).

PROOF. Define

$$r_n = \max_{s=1,\ldots,m} |f_s^n - f_s^{\infty}| \qquad n=1,2,\ldots$$

Let δ^{∞} be any optimal stationary policy. For any s, we have

$$f_s^{\infty} = c[s,\delta(s)] + \alpha \sum_{t=1}^{m} p[t|s,\delta(s)] f_t^{\infty}$$

but for the n-period problem,

$$f_s^n \leq c[s,\delta(s)] + \alpha \sum_{t=1}^{m} p[t|s,\delta(s)] f_t^{n-1}$$

since $\delta(s)$ may not be the optimal immediate decision when n periods remain. Subtracting the equality from the inequality, we obtain

$$\begin{aligned} f_s^n - f_s^\infty &\leq \alpha \sum_{t=1}^{m} p[t|s,\delta(s)](f_t^{n-1} - f_t^\infty) \\ &\leq \alpha \sum_{t=1}^{m} p[t|s,\delta(s)] |f_t^{n-1} - f_t^\infty| \\ &\leq \alpha \sum_{t=1}^{m} p[t|s,\delta(s)] r_{n-1} \\ &= \alpha r_{n-1} \end{aligned} \tag{4.20}$$

The third inequality follows from the definition of r_n.

In a similar fashion, we have for some $d_s \in D(s)$ that

$$f_s^n = c(s,d_s) + \alpha \sum_{t=1}^{m} p(t|s,d_s) f_t^{n-1}$$

but

$$f_s^\infty \leq c(s,d_s) + \alpha \sum_{t=1}^{m} p(t|s,d_s) f_t^\infty$$

By the same reasoning that led to (4.20), subtracting the equality from the inequality gives us

$$f_s^\infty - f_s^n \leq \alpha r_{n-1} \tag{4.21}$$

Comparing (4.20) and (4.21), we can conclude

$$|f_s^\infty - f_s^n| \leq \alpha r_{n-1}$$

and since s was arbitrary, this implies

$$r_n = \max_{s=1,\ldots,m} |f_s^\infty - f_s^n| \leq \alpha r_{n-1}$$

This result implies directly that $r_n \leq \alpha^{n-1} r_1$, and since $0<\alpha<1$, we have $\lim_{n\to\infty} r_n = 0$ establishing the theorem. ■

The principle of successive approximation relates to the convergence of the cost vectors. The following result says that if the planning horizon is large, but finite, the optimal immediate decision is the same as if the planning horizon were infinite. Let

$$\Delta^\infty = \{\delta \in \Delta : \mathbf{f}^\infty = (\mathbf{I} - \alpha \mathbf{P}_\delta)^{-1} \mathbf{c}_\delta\}$$

Δ^∞ is the set of decision functions leading to optimal stationary policies. Similarly, let

$$\Delta^n = \{\delta \in \Delta: \mathbf{f}^n = \mathbf{c}_\delta + \alpha \mathbf{P}_\delta \mathbf{f}^{n-1}\} \qquad n = 1, 2, \ldots$$

THEOREM 4.5. (Turnpike Planning Horizon Theorem). There exists an N^* such that for all $n \geq N^*$,

$$\Delta^n \subseteq \Delta^\infty$$

PROOF. Suppose the contrary; that is, suppose there exists no N^* such that for all $n \geq N^*$, $\Delta^n \subseteq \Delta^\infty$. This implies there is a $\delta \notin \Delta^\infty$, such that for an infinite sequence of planning horizons $\{n_i\}_{i=1}^\infty$, $\mathbf{f}^{n_i} = \mathbf{c}_\delta + \alpha \mathbf{P}_\delta \mathbf{f}^{n_i - 1}$. By Theorem 4.4, we have $\lim_{i \to \infty} \mathbf{f}^{n_i} = \lim_{i \to \infty} \mathbf{f}^{n_i - 1} = \mathbf{f}^\infty$, or $\mathbf{f}^\infty = \mathbf{c}_\delta + \alpha \mathbf{P}_\delta \mathbf{f}^\infty$ implying $\mathbf{f}^\infty = (\mathbf{I} - \alpha \mathbf{P}_\delta)^{-1} \mathbf{c}_\delta$, or $\delta \in \Delta^\infty$, a contradiction. ■

According to Theorem 4.5, if the set Δ^∞ contains only one decision function ρ, then $\Delta^n = \{\rho\}$ for all $n \geq N^*$. In this case, an optimal N period policy for $N > N^*$ can be given as $\Pi^N = [\rho, \ldots, \rho, \delta_{N-N^*+1}, \ldots, \delta_N]$; that is, use the decision function ρ at the start of each of the first $N - N^*$ periods, and use δ_i at the start of period i for $i = N - N^* + 1, \ldots, N$. A difficulty with this analysis, however, is that Theorem 4.5 establishes the existence of an N^* with the stated property, but it does not suggest how to compute one. Although it is possible to compute such an N^* using bounds suggested by successive approximation (see Exercise 4.16 at the end of the chapter), the minimal N^* with the desired property is difficult to determine in general. Ideally, we would like to discover when the minimal or near minimal N^* has been reached when doing backward iteration with the recursions (4.12). Unfortunately, the stochastic behavior appears too complex to permit this. If the set Δ^∞ contains two or more decision functions, then the situation is somewhat more complicated because it is possible that $\Delta^n \subsetneq \Delta^\infty$ for all $n \geq N^*$. This is a fine point that does not merit further discussion here.

We discuss briefly results analogous to the ones above for the infinite horizon, finite state, and action space Markovian decision model when the discount rate $\alpha = 1$ and the optimality criterion is a minimal expected average cost. This criterion is ambiguous because it permits optimal policies that are extremely costly in the short run but that are optimal on average. The example in Figure 4.5 illustrates this point. Starting in state 1, any policy that ultimately goes to state 2 has a minimal average cost over an infinite horizon equal to 0. The difficulty can be overcome by the use of secondary optimality criteria involving the behavior of policies for discount rates near to but less than one. In the example of Figure 4.5, the intuitively correct policy when starting in state 1 of going to state 2 at the first transition is the unique optimal stationary decision for all $\alpha < 1$.

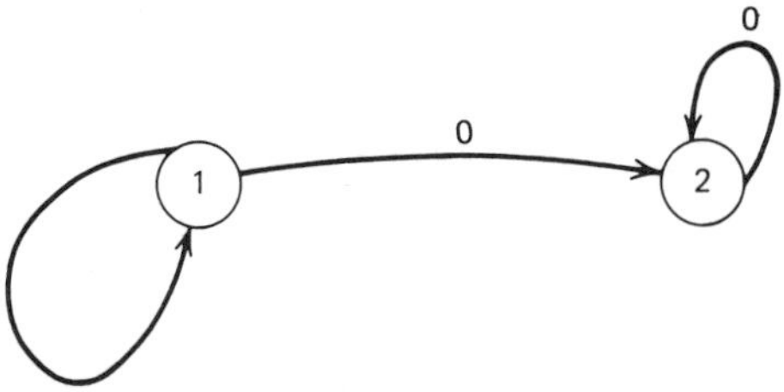

Figure 4.5. Infinite horizon average cost example.

For finite planning horizons and $\alpha = 1$, minimizing f_s^n is equivalent to the stated criterion of minimizing f_s^n/n. It is not clear, however, that $\lim_{n\to\infty} f_s^n/n = \bar{f}_s^\infty$ exists, and that there are optimal stationary policies with this minimal expected average cost. Although we do not prove it here, it can be shown that the limit $\bar{f}_s^\infty$ does exist, and moreover that $\bar{f}_s^\infty = \lim_{\alpha\to 1^-}(1-\alpha)f_{s,\alpha}^n$ where the $f_{s,\alpha}^n$ are the unique solutions to (4.13) as a function of α.

Under a somewhat restrictive assumption, a stationary policy achieving the minimal expected average cost $\bar{f}_s^\infty$ can be found by adding the equation

$$\sum_{s=1}^{m} \sum_{d\in D(s)} x(s,d) = 1$$

to the linear programming problem (4.18). For $\alpha < 1$, this is a redundant equation satisfied by the $x(s,d)$ satisfying the other equations in (4.18). Then, letting $\alpha = 1$, there results the linear programming problem

$$\begin{aligned}
\min \sum_{s=1}^{m} \sum_{d\in D(s)} & c(s,d)x(s,d) \\
\text{s.t.} \quad \sum_{d\in D(s)} x(s,d) - \sum_{u=1}^{m}\sum_{d\in D(u)} & p(s|u,d)x(u,d) = 0 \\
& s = 1,\ldots,m \\
\sum_{s=1}^{m}\sum_{d\in D(s)} x(s,d) &= 1 \\
x(s,d) \geq 0 & \qquad \text{for all } s,d
\end{aligned} \tag{4.22}$$

The additional assumption that must be made to ensure that (4.22) is well defined is that, for any decision function δ, the Markov matrix $\mathbf{P}_\delta$ has a single ergodic class of m states. In other words, starting in an arbitrary state and after a sufficiently large number of state transitions according to the transition matrix $\mathbf{P}_\delta$, there is a positive probability that the system is any state. This assumption ensures that the rank of (4.22) is m implying the selection of a stationary policy at each iteration. For this model, the $x(s,d)$ can be interpreted as the probability that the system is in state s and decision d is taken at the start of a random period.

Letting $\alpha \to 1^-$ and adding the constraint

$$\sum_{s=1}^{m} \sum_{d \in D(s)} x(s,d) = 1$$

can be interpreted in a formulation similar to (4.17) by considering the linear programming dual to (4.22)

$$\begin{aligned} &\max k \\ &\text{s.t. } y_s + k - \alpha \sum_{t=1}^{m} p(t|s,d) y_t \le c(s,d) \qquad \text{for all } d \in D(s) \\ &\qquad\qquad\qquad\qquad\qquad\qquad\qquad\qquad s = 1,\ldots,m \end{aligned}$$

where k is the dual variable on the added constraint. This linear programming problem solves the functional equation analogous to (4.13)

$$z_s + k = \min_{d \in D(s)} \left\{ c(s,d) + \alpha \sum_{t=1}^{m} p(t|s,d) z_t \right\} \qquad s = 1,\ldots,m$$

By duality, we know

$$k = \min \sum_{s=1}^{m} \sum_{d \in D(s)} c(s,d) x(s,d)$$

which is the minimal expected average cost of following an optimal stationary policy starting in any state and z_s is a total expected cost correction due to starting in the specific state s. In other words, $k = \lim_{n\to\infty} f_s^n / n$ for any state s because all states are in the same ergodic class.

4.5 EXERCISES

4.1 Show that the u_j optimal in the linear programming problem (4.2) satisfy the recursion (4.1) and that the x_{ij} optimal in the linear programming problem (4.3) give the shortest routes.

4.2 Suppose we do not make any assumptions about Shortest Route Problem 1 ensuring the existence of optimal solutions for it and the linear programming problems (4.2) and (4.3). Show that (4.2) cannot have a feasible solution if the network has a circuit with negative length. Show that (4.2) has an unbounded solution if there is no chain from node 1 to node k for $k \ge 2$.

4.3 Prove the validity of the stated Algorithm for Shortest Route Problem 2.

4.4 Give a procedure for recording the shortest route chains at the same time their lengths are calculated by the Algorithm for Shortest Route Problem 2.

4.5 For the maximal flow problem (3.2), show that any maximal flow can be represented as flows along simple chains (i.e., circuitless chains) connecting the source to the sink.

4.6 Solve the maximal flow problem of Figure 3.12 using the generalized linear programming approach given in Section 4.2. Start with the chains (1,2,4,6), (1,3,5,6), and (1,4,6) in the linear programming problem (4.4).

4.7 Show that the min-addition operator $\otimes$ is associative $[\mathbf{A}\otimes(\mathbf{B}\otimes\mathbf{C})=(\mathbf{A}\otimes\mathbf{B})\otimes\mathbf{C}]$ and, therefore, it is correct to compute powers of a matrix $\mathbf{C}$ by $\mathbf{C}^k=\mathbf{C}\otimes\mathbf{C}^{k-1}$.

4.8 Solve the numerical example 4.1 when the demand $r=3$, the production restriction is $d\leq 5$, and the production cost function is

$$p(d)=\begin{cases}0 & \text{if } d=0\\ 8+d & \text{if } d>0\end{cases}$$

All other data for the problem is unchanged. Specifically, find an optimal strategy for a 5-period planning horizon starting with 3 units of inventory. Also, determine a least average cost or turnpike circuit for the network with this new data.

4.9 (Gilmore and Gomory, 1965) The *cutting stock problem* is one in which rolls of paper of width W are cut up into smaller rolls of width w_i, $i=1,\ldots,m$ to meet demands of b_i, $i=1,\ldots,m$. The activities or columns $\mathbf{a}_j$ in this problem correspond to feasible cutting patterns satisfying

$$\sum_{i=1}^{m} w_i a_{ij} \leq W$$

and a_{ij} nonnegative integer. The objective is to meet demand while holding waste to a minimum, or equivalently, to meet demand using a minimal number of rolls. The problem can be expressed as

$$\begin{aligned}\min \ & \sum_{j=1}^{n} x_j \\ \text{s.t.} \ & \sum_{j=1}^{n} a_{ij}x_j \geq b_i \qquad i=1,\ldots,m \\ & x_j \geq 0\end{aligned}$$

where x_j is the number of times the cutting pattern a_j is to be used. The difficulty with this formulation for practical applications is the extremely large number n of possible cutting patterns. This suggests the use of a column generation procedure similar to the one used for the maximal flow problem with the node arc formulation (4.4).

(1) Suppose the cutting stock problem is solved using a feasible but incomplete subset of the total number of possible patterns or columns. Let π denote the m-vector of optimal shadow prices for the incomplete problem, and consider the knapsack problem

$$v(\pi) = \max \sum_{i=1}^{m} \pi_i a_i$$

$$\text{s.t.} \sum_{i=1}^{m} w_i a_i \leq W$$

$$a_i \text{ nonnegative integer}$$

Show that $v(\pi)$ is always greater than or equal to one. In addition, show that the incomplete problem has produced an optimal solution to the problem with all columns included if $v(\pi)$ equals one, whereas a new column has been generated to be added to the incomplete problem if $v(\pi)$ is greater than one.

(2) Use the column generation procedure to solve the following cutting stock problem: $W=17$; $w_1=3$, $w_2=9$, $w_3=4$; $b_1=22$, $b_2=18$, $b_3=19$. Start with the feasible set of columns $(5,0,0)$, $(0,1,0)$, $(0,0,4)$.

4.10 (Shapiro and Wagner, 1967)

(1) Assume for the knapsack problem (4.9) that for all $j<n$

$$\frac{c_j}{a_j} < \frac{c_n}{a_n}$$

Show that

$$f^b = c_n + f^{b-a_n}$$

for all integer $b \geq b^*$, where

$$b^* = \frac{c_n \dfrac{a_{n-1}}{a_n}}{\dfrac{c_n}{a_n} - \dfrac{c_{n-1}}{a_{n-1}}}$$

(2) Extend the result of part (1) to include the case when

$$\frac{c_{n-1}}{a_{n-1}} = \frac{c_n}{a_n}$$

by exhibiting a larger b^*.

4.11 (Shapiro and Wagner, 1967) Let α be a discount factor, $0<\alpha<1$, and consider the discounted knapsack problem analogous to (4.9)

$$f^l = \max_{\substack{j=1,\ldots,n \\ \text{and} \\ a_j \le l}} \left\{ c_j + f^{l-a_j} \right\} \qquad l=1,2,\ldots,b$$

$$f^0 = 0$$

Here each item represents an investment alternative with return c_j consuming a_j time periods before another investment can be made. The stage l is the number of time periods remaining in the planning horizon. For convenience, assume the n items have distinct equivalent average costs $c_j/(1-\alpha^{a_j})$ and that they are ordered by

$$\frac{c_1}{1-\alpha^{a_1}} < \frac{c_2}{1-\alpha^{a_2}} < \cdots < \frac{c_n}{1-\alpha^{a_n}}.$$

(1) Prove that an optimal policy has the form that if alternatives j and k ($j<k$) are used x_j^* and x_k^* times, respectively, then alternative k must be used x_k^* times before alternative j is used at all.

(2) Prove that

$$\lim_{l\to\infty} \frac{f^l}{1-\alpha^l} = \frac{c_n}{1-\alpha^{a_n}}$$

(3) What happens to the results of (1) and (2) as α approaches one from below?

4.12 Suppose (4.8) is a general integer programming problem where the activities $\mathbf{a}_j$ and the right-hand-side vector $\mathbf{b}$ are arbitrary integer m vectors. Show that the recursion (4.9) is still valid in this case. What additional assumption about the integer programming problem is required to ensure that the f^l can be systematically computed? For example, use the recursion (4.9) to solve the problem

$$\begin{aligned}
\max\quad & 5x_1+2x_2+4x_3 \\
\text{s.t.}\quad & -1x_1+8x_2+3x_3+x_4 = 9 \\
& 4x_1-1x_2-1x_3 \qquad +x_5 = 6 \\
& 2x_1+3x_2+5x_3 \qquad\qquad +x_6 = 29
\end{aligned}$$

$$x_1, x_2, x_3, x_4, x_5, x_6 \text{ nonnegative integers}$$

4.13 Use an induction argument to show that the function $f_{\mathbf{s}}^n$ in the recursion (4.11) is convex in $\mathbf{s}$ for all n.

4.14 (Shapiro, 1976) For $\mathbf{y} \in R^m$, consider the mapping $\mathbf{y} \to \mathbf{F}(\mathbf{y}) = [F_1(\mathbf{y}), \ldots, F_m(\mathbf{y})] \in R^m$ derived from the Markovian decision model functional equation (4.13)

$$F_s(\mathbf{y}) = \min_{d \in D(S)} \left\{ c(s,d) + \alpha \sum_{t=1}^{m} p(t|s,d) y_t \right\} \qquad s = 1, \ldots, m$$

The solutions $\mathbf{y}^*$, if any, to (4.13) are the fixed points of F; namely, $\mathbf{y}^* = F(\mathbf{y}^*)$. Show that a solution to (4.13) *exists* by verifying that F satisfies the conditions of *Brouwer's fixed point theorem*:

> Let G be a continuous function which maps the compact convex set $X \subseteq R^m$ into itself. There is an $\mathbf{x}^* \in X$ with the property that $G(\mathbf{x}^*) = \mathbf{x}^*$.

4.15 (1) Solve the numerical Example 4.5 when the probabilities of demand are

probability (demand = 0) = 0.1
probability (demand = 1) = 0.7
probability (demand = 2) = 0.1
probability (demand = 3) = 0.1

All other data for the problem is unchanged.

(2) Solve the problem of part (1) when the discount factor $\alpha = 1$ and the optimality criterion is to minimize expected average cost.

4.16 (Shapiro, 1968) show that an upper bound on N^* in the turnpike planning horizon theorem 4.5 is

$$\max_{\delta \notin \Delta^\infty} \min \left\{ n\text{: } n \text{ integer and } \frac{\alpha^n M}{1-\alpha} < \frac{\varepsilon_\delta}{2} \right\}$$

where

$$M = \max_{s,d} |c,(s,d)|$$

and

$$\varepsilon_\delta = \max_{s=1,\ldots,m} \left\{ \left(c[s,\delta(s)] + \alpha \sum_{t=1}^{m} p[t|s,\delta(s)] f_t^\infty \right) - f_s^\infty \right\} > 0.$$

We know $\varepsilon_\delta > 0$ because $\delta \notin \Delta^\infty$.

4.17 Consider the stationary policy δ^∞ for the infinite horizon Markovian decision model. Let $\mathbf{y} = (\mathbf{I} - \alpha \mathbf{P}_\delta)^{-1} \mathbf{c}_\delta$ and suppose δ^∞ is not optimal. Define a new decision function ρ as follows:

(1) For any state s such that

$$\min_{d \in D(s)} \left\{ c(s,d) + \alpha \sum_{t=1}^{m} p(t|s,d) y_t \right\} < y_s$$

let $\rho(s) = d_s$ for any $d_s \in D(s)$ achieving the minimum.

(2) For all other state s, let $\rho(s) = \delta(s)$.

Prove that $\mathbf{f}(\rho^\infty) \le \mathbf{f}(\delta^\infty)$ and $\mathbf{f}(\rho^\infty) \ne \mathbf{f}(\delta^\infty)$.

4.18 (Denardo, 1967) The infinite horizon Markovian decision problem described in Section 4.4 can be generalized using notions of functional analysis. A basic result is a fixed point theorem implicit in the proof of Theorem 4.4.

(1) Consider an arbitrary metric space X with metric ρ. A mapping T of X into itself is called a *contraction mapping* if there exists a positive real number α, $0 \le \alpha < 1$ with the property $\rho(Tx, Ty) \le \alpha\rho(x,y)$ for all $x,y \in X$. Prove the contraction mapping fixed point theorem: If T is a contraction mapping on a complete metric space X, then T has a unique fixed point; that is, there is a unique $x^* \in X$ such $T(x^*) = x^*$.

(2) Let Ω be an arbitrary set representing the state space of a dynamic system. The dynamic programming analysis involves the set V of all bounded functions from Ω to the reals; that is,

$$\sup_{x \in \Omega} |v(x)| < \infty$$

For arbitrary $u, v \in V$, define

$$\rho(u,v) = \sup_{x \in \Omega} |u(x) - v(x)|$$

Prove that the function ρ is a metric on the space V and the space is complete in this metric.

(3) Associated with each point $x \in \Omega$ is a decision set D_x; an element of D_x is denoted by d_x. The set Δ of all decision functions is given by $\Delta = \times_{x \in \Omega} D_X$ and δ denotes a generic element of this set. The decision selected by δ at x is denoted by δ_x. There is a cost function c that assigns a real number to each triplet (x, d_x, v) with $x \in \Omega$, $d_x \in D_s$, and $v \in V$. The number $c(x, d_x, v)$ is the cost of starting at the state $x \in \Omega$, making decision $d_x \in D_x$, and incurring a subsequent cost determined by the transition, possibly probabilistic, to the new state $z \in \Omega$ for which a cost $v(z)$ is incurred. The transition to the new state depends on x and d_x. For each $\delta \in \Delta$, we define a function C_δ mapping V into itself by

$$[C_\delta(v)](x) = c(x, \delta_x, v)$$

where $C_\delta(v)$ is the function in V produced by this mapping and the value of this function at x is denoted by $[C_\delta(v)](x)$ and equals $c(x,\delta_x,v)$.

CONTRACTION ASSUMPTION. For some α satisfying $0 \le \alpha < 1$, $|c(x,d_x,u)-c(x,d_x,v)| \le \alpha\rho(u,v)$ for all u, $v \in V$, $x \in \Omega$, $d_x \in D_x$.

Prove that for each decision function δ, there exists a unique element $v_\delta \in V$ such that

$$v_\delta(x) = c(x,\delta_x,v_\delta) \qquad \text{for all } x \in \Omega$$

HINT. Show that C_δ is a contraction mapping on the complete metric space V under the contraction assumption.

The function v_δ is the total cost function for the decision function δ. It represents a generalization of the vector $\mathbf{f}(\delta^\infty)$ given in Section 4.4 whose components are the total expected discounted costs of starting in each of the states and following the stationary strategy δ^∞. The optimal total cost function is defined by

$$f(x) = \inf_{\delta \in \Delta} v_\delta(x)$$

(4) We define a maximization operator A which is a mapping on the domain V by

$$(Av)(x) = \sup_{d \in D_x} c(x,d_x,v)$$

for each $v \in V$ and $x \in \Omega$. Assume the range of A is contained in V. Prove that the mapping A is a contraction mapping on V and thus there exists a unique element $v^* \in V$ such that

$$v^*(x) = \sup_{d_x \in D_x} c(x,d_x,v^*) \qquad \text{for each } x \in \Omega$$

(5) This last equation is the generalization of the functional equation (4.13). Two questions about our development is whether v^* is approximated or attained by the total cost function v_δ of some $\delta \in \Delta$ and whether v^* is the optimal total cost function; that is, whether $v^* = f$. The following result addresses the former question. The latter question is addressed in the next part below.

Prove that for $\varepsilon > 0$, there exists a decision function δ such that $\rho[C_\delta(v^*),v^*] \le \varepsilon(1-\alpha)$, and any such function δ satisfies $\rho(v_\delta,v^*) \le \varepsilon$. If $\rho[C_\delta(v^*),v^*]=0$, then $v_\delta = v^*$.

(6) For u, $v \in V$, we write $u \ge v$ if $u(x) \ge v(x)$ for all $x \in \Omega$, and $u > v$, if $u \ge v$ and $u \ne v$.

MONOTONICITY ASSUMPTION. If $u \ge v$, then $C_\delta(u) \ge C_\delta(v)$ for all $\delta \in \Delta$.

Prove that if the contraction and monotonicity assumptions are satisfied, then $v^* = f$.

(7) Verify that the contraction and monotonicity assumptions are satisfied by the finite state and action space Markovian decision model of Section 4.4.

4.19 (Nemhauser, 1966) Consider the recursion

$$f_x^n = \underset{0 \le d \le x}{\text{maximum}} \{ c_1 d + c_2(x-d) + f_{\alpha_1 d + \alpha_2(x-d)}^{n-1} \} \qquad n = 1, 2, \ldots$$

where $0 \le \alpha_1 < 1$, $0 \le \alpha_2 < 1$. Let d_x^n denote an optimal solution at stage n to this recursion. Give expressions for $\lim_{n\to\infty} f_x^n$ and $\lim_{n\to\infty} d_x^n$ as functions of $c_1, c_2, \alpha_1, \alpha_2$.

4.20 Suppose at the start of every period of an infinite horizon, the existing stock of capital goods can be allocated between the production of consumption goods and that of new capital goods. Both production processes take one time period and for each the output is proportional to the amount of capital goods invested in it. Of the capital goods used in production, a fixed proportion depreciates, the same proportion for both processes and the amount of capital goods available in the next period is the remainder plus the new production of capital goods. Suppose that consumption goods must be consumed immediately on production, that the utility of a given volume of consumption goods is measured by the logarithm, and that the criterion is the maximization of the sum of discounted utilities.

This problem can be formulated in the following way as an infinite horizon dynamic programming problem. Let

c_1 = proportion of consumption goods produced from one unit of capital stock

c_2 = proportion of new capital goods produced from one unit of capital stock

K = proportion of capital goods in production that depreciate in one period

α = discount factor ($0 < \alpha < 1$)

x = initial stock of capital goods

y = amount allocated to the production of consumption goods ($y \le x$)

$r = y/x$

Determine the optimal strategy for the allocation of capital and the maximum value to be derived from any given stock of capital goods.

HINT. Show that the functional equation describing this model is

$$f(x) = \max_{0 \le r \le 1} \left[\log c_1 r x + \alpha f\{ x[c_2(1-r) + (1-K)] \} \right]$$

4.6 NOTES

SECTION 4.1 Dynamic programming has been treated extensively in a number of books including Bellman (1957), Bellman and Dreyfus (1962), and Nemhauser (1966). Our development here has much in common with Wagner (1975) in the treatment of sequential discrete optimization problems.

SECTION 4.2 The stated algorithm for the Shortest Route Problem 1 was first given by Dijkstra (1959). The algorithm for the Shortest Route Problem 2 is found in Hu (1969); see also Floyd (1962) and Murchland (1967). A survey of shortest route methods is given by Dreyfus (1969) and some computational experimentation is given by Golden (1976). Tomlin (1966) discusses arc-chain and node-arc formulations of the multicommodity flow problem. Lawler (1976) contains a complete treatment of shortest route algorithms including computational complexity considerations.

SECTION 4.3 Wagner (1975) in Chapter 8 discusses some discrete dynamic programming problems in a similar fashion to the way they are presented in this section. Karp and Held (1967) use formalisms from automata theory to describe discrete, deterministic dynamic programming problems and to derive recursions for solving them. The knapsack problem has received considerable attention because its analysis is relevant to the study of more complex dynamic and integer programming problems; see Gilmore and Gomory (1963), Shapiro and Wagner (1967), Garfinkel and Nemhauser (1972). The dynamic programming approach to the traveling salesman problem is due to Held and Karp (1962); see Morin and Marsten (1976) for a more flexible way of doing dynamic programming calculations for the traveling salesman and other problems.

SECTION 4.4 Markovian decision theory is discussed in Bellman (1957) and more fully developed by Howard (1960). Mine and Osaki (1970) give an extensive treatment of the subject to that date, including an extensive bibliography. Ross (1970) treats some of the same topics presented in Section 4.4. Bertsekas (1976) relates Markovian decision models to optimal control problems. The relationship of the finite state and action space Markovian decision model to linear programming is presented by d'Epenoux (1960), (1963); see also Wolfe and Dantzig (1962). Denardo and Fox (1968) and Schweitzer (1971) address the undiscounted linear programming model (4.22) in the case when there are multiple ergodic classes. Analysis of optimal policies as the discount factor approaches one from below is given by Blackwell (1962), Veinott (1966), and Denardo and Miller (1968). Extensions of the finite state and action space model to more general Markovian decision problems with continuous time, state, and action spaces are found in Blackwell (1965), Fox (1966), Grinold (1974), Jewell (1963a, b) and Miller (1968a, b). The turnpike theorem 4.5 is due to Shapiro (1968). Denardo (1967) presents some general theoretical results relating to existence and uniqueness of solutions to infinite horizon dynamic programming problems (see Exercise 4.18).

5 MATHEMATICAL PROGRAMMING DUALITY THEORY AND ITS RELATIONSHIP TO CONVEXITY

5.1 INTRODUCTION

In Chapter 2 we saw that duality theory plays a central role in the solution of linear programming problems. In this chapter we study a related but more general duality theory that is used in the solution of all mathematical programming problems.

Duality theory is concerned with global optimality conditions involving dual variables (also called generalized Lagrange multipliers) that a solution is optimal in a given primal minimization problem. The dual variables in these optimality conditions are optimal in a concave maximization problem that is dual to the given primal problem. If the primal problem is convex and obeys some regularity condition, then an optimal dual solution can be used to find an optimal solution to the primal. For an arbitrary primal problem, however, there is no guarantee that solution of the dual will yield an optimal solution to the primal. We show in later chapters how dual problems are useful nevertheless in the solution of nonconvex problems such as the integer programming problem and the traveling salesman problem.

A central result of this chapter is a demonstration of the mathematical equivalence that exists between convexification and dualization of an arbitrary mathematical programming problem. The well-known generalized linear programming algorithm that constructs linear programming approximations to an arbitrary primal minimization problem is a mechanization of this result. We present this algorithm and demonstrate that any limit point of the sequence of linear programming dual prices produced by the algorithm is optimal in the concave maximization problem that is dual to the primal problem. This result holds even if the generalized linear programming algorithm does not solve the primal problem.

We begin our development of duality theory by considering the arbitrary problem

$$\begin{aligned} v = \min & f(\mathbf{x}) \\ \text{s.t. } & g(\mathbf{x}) \le \mathbf{0} \\ & \mathbf{x} \in X \subseteq R^n \end{aligned} \tag{5.1}$$

where f is a real valued function, $g(\mathbf{x}) = [g_1(\mathbf{x}), \ldots, g_m(\mathbf{x})]$ is a function from R^n to R^m, and X is a nonempty set. Strictly speaking, we should write inf instead of min in problem (5.1) because the minimum may not be attained. Since nonattainment of minima or maxima is very rare in actual mathematical programming problems, we adopt the convention of (5.1). If (5.1) does not have a feasible solution (that is, there is no $\mathbf{x} \in X$ such that $g(\mathbf{x}) \le \mathbf{0}$), then we take $v = +\infty$. Henceforth, we call (5.1) the *primal problem*.

The constraints in problem (5.1) are of two types: The *explicit* constraints $g(x) \le 0$ and the *implicit* constraints $\mathbf{x} \in X$. The problem has been written in this way to emphasize that the explicit constraints are the difficult ones and that the problem would be much easier to solve without them. For example, g may be a nonlinear function from R^n to R^m, whereas X may be a convex polyhedron, perhaps one as simple as $\{\mathbf{x} | \mathbf{x} \ge \mathbf{0}\}$. We will study a number of more detailed examples of this partitioning of the constraint set in later sections of this chapter, and in Chapters 6, 7, and 8.

An intuitive approach to try to avoid the difficult constraints $g(\mathbf{x}) \le \mathbf{0}$ is to price them out by placing the term $\mathbf{u}g(\mathbf{x})$ in the objective function where $\mathbf{u}$ is a nonnegative m-vector of associated dual variables. The result is the *Lagrangean* function

$$L(\mathbf{u}) = \min_{\mathbf{x} \in X} \{ f(\mathbf{x}) + \mathbf{u}g(\mathbf{x}) \} \tag{5.2}$$

For future reference, we also define

$$L(\mathbf{x}, \mathbf{u}) = f(\mathbf{x}) + \mathbf{u}g(\mathbf{x}) \tag{5.3}$$

Let $\bar{\mathbf{x}} \in X$ denote a specific primal solution computed from (5.2) for a specific $\bar{\mathbf{u}} \ge \mathbf{0}$; that is, $L(\bar{\mathbf{u}}) = L(\bar{\mathbf{x}}, \bar{\mathbf{u}})$. There is no guarantee that $\bar{\mathbf{x}}$ is optimal or even feasible in (5.1), but the rationale for selecting $\bar{\mathbf{u}}$ is embodied in the following conditions.

Definition 5.1. A pair $(\bar{\mathbf{x}}, \bar{\mathbf{u}})$ with $\bar{\mathbf{x}} \in X$ and $\bar{\mathbf{u}} \geq \mathbf{0}$ satisfies the *global optimality conditions* for the primal problem (5.1) if

(1) $f(\bar{\mathbf{x}}) + \bar{\mathbf{u}}g(\bar{\mathbf{x}}) = \min_{\mathbf{x} \in X}\{f(\mathbf{x}) + \bar{\mathbf{u}}g(\mathbf{x})\}$
(2) $\bar{\mathbf{u}}g(\bar{\mathbf{x}}) = 0$
(3) $g(\bar{\mathbf{x}}) \leq \mathbf{0}$

THEOREM 5.1. If $(\bar{\mathbf{x}}, \bar{\mathbf{u}})$ satisfies the global optimality conditions, then $\bar{\mathbf{x}}$ is optimal in the primal problem (5.1).

PROOF. Clearly, the solution $\bar{\mathbf{x}}$ is feasible in (5.1) [$\bar{\mathbf{x}} \in X$ and $g(\bar{\mathbf{x}}) \leq \mathbf{0}$]. Let $\tilde{\mathbf{x}}$ be any feasible solution of (5.1). We have

$$f(\bar{\mathbf{x}}) = f(\bar{\mathbf{x}}) + \bar{\mathbf{u}}g(\bar{\mathbf{x}}) \leq f(\tilde{\mathbf{x}}) + \bar{\mathbf{u}}g(\mathbf{x}) \leq f(\tilde{\mathbf{x}})$$

where the equality follows from (2), the left-most inequality follows from (1), and the right-most inequality follows because $\bar{\mathbf{u}} \geq \mathbf{0}$, $g(\tilde{\mathbf{x}}) \leq \mathbf{0}$ implies $\bar{\mathbf{u}}g(\tilde{\mathbf{x}}) \leq 0$. Thus, $f(\bar{\mathbf{x}}) \leq f(\tilde{\mathbf{x}})$ for any feasible solution $\tilde{\mathbf{x}}$ in (5.1) and $\bar{\mathbf{x}}$ is optimal. ■

The conditions given in Definition 5.1 are sufficient conditions that $\bar{\mathbf{x}}$ is optimal in the primal problem (5.1). They may not be necessary conditions in the sense that if $\bar{\mathbf{x}} \in X$ is optimal in (5.1), there may not be a $\bar{\mathbf{u}} \geq \mathbf{0}$ such that $(\bar{\mathbf{x}}, \bar{\mathbf{u}})$ satisfy optimality conditions (1) or (2). This point is illustrated in Example 5.2 in the next section. In general, we cannot expect the conditions to be necessary but a frequent strategy in trying to solve the primal problem (5.1) is to assume a pair $(\bar{\mathbf{x}}, \bar{\mathbf{u}})$ satisfying the conditions can be found.

The necessary existence of a primal-dual pair satisfying the global optimality conditions can be established if f and g are convex functions, X is a convex set, and some regularity condition is satisfied. The special case when the primal problem is a linear programming problem has already been treated in Chapter 2 [see (2.5) and (2.6)]. Details for general convex primal problems are given in Sections 5.4 and 5.5. If the primal problem (5.1) is not convex, and a primal-dual pair satisfying the global optimality conditions is not found, then additional means are needed to solve (5.1) such as changing its representation to make it more likely that the optimality conditions can be established, or devising a systematic search of the set X to find an optimal solution.

If $X = R^n$, and the functions f and g are differentiable, then a necessary condition that $\bar{\mathbf{x}}$ is minimal in optimality condition (1) is $\nabla f(\bar{\mathbf{x}}) + \bar{\mathbf{u}}\nabla g(\bar{\mathbf{x}}) = \mathbf{0}$. The global optimality conditions with (1) replaced by this relation on the gradients are the Kuhn-Tucker necessary conditions for optimality. These conditions are presented and discussed in Section 5.5.

The question remains: What are constructive procedures for trying to find an $(\bar{\mathbf{x}}, \bar{\mathbf{u}})$ pair satisfying the global optimality conditions? The answer is that we must seek an optimal solution to a problem that is dual to (5.1). The next section is concerned with a statement of the dual problem and an analysis of some of its properties. Throughout the chapter, we will use properties of convex sets and functions given in Appendix A.

5.2 PROPERTIES OF THE DUAL PROBLEM

In stating and analyzing the dual problem, we make the additional assumptions that the functions f and g are continuous and that the nonempty set X is compact. These assumptions will remain in effect throughout the remainder of this section and the next two. They are made in order to simplify the mathematical analysis thereby permitting an uncluttered demonstration of our results. The assumptions can be relaxed and most of the results will remain valid in modified form (see Exercise 5.8). Moreover, the vast majority of real-life mathematical programming applications satisfy these assumptions.

THEOREM 5.2. (weak duality). For any $\mathbf{u} \geq \mathbf{0}$,

$$L(\mathbf{u}) \leq v$$

where v is the minimal objective function value of the primal problem (5.1).

PROOF. If problem (5.1) has no feasible solution, then $v = +\infty$ and there is nothing to prove. Otherwise, let $\tilde{\mathbf{x}}$ be any feasible solution to the primal problem (5.1). Since $g(\tilde{\mathbf{x}}) \leq \mathbf{0}$ and $\mathbf{u} \geq \mathbf{0}$, we have

$$L(\mathbf{u}) \leq f(\tilde{\mathbf{x}}) + \mathbf{u}g(\tilde{\mathbf{x}}) \leq f(\tilde{\mathbf{x}})$$

which establishes the theorem by the definition of v. ■

The dual problem to the primal problem (5.1) is obtained by finding the greatest lower bound to v; namely

$$\begin{aligned} d = \sup\, & L(\mathbf{u}) \\ \text{s.t. } & \mathbf{u} \geq \mathbf{0} \end{aligned} \tag{5.4}$$

Clearly $d \leq v$ and without further assumptions on the primal problem (5.1), it is possible that $d < v$ in which case we say there is a *duality gap*.

THEOREM 5.3. If $(\bar{\mathbf{x}}, \bar{\mathbf{u}})$ satisfies the global optimality conditions (see Definition 5.1), then $\bar{\mathbf{u}}$ is optimal in the dual problem (5.4). Moreover, the primal and dual problems have equal optimal objective function values; namely $d=v$.

PROOF. The solution $\bar{\mathbf{u}}$ is nonnegative and thus feasible in the dual problem (5.4). By the definition of the dual problem, $L(\bar{\mathbf{u}}) \leq d$. By optimality condition (1),

$$\begin{aligned} L(\bar{\mathbf{u}}) &= f(\bar{\mathbf{x}}) + \bar{\mathbf{u}} g(\bar{\mathbf{x}}) \\ &= f(\bar{\mathbf{x}}) \end{aligned}$$

where the second equality follows from $\bar{\mathbf{u}} g(\bar{\mathbf{x}}) = 0$ [condition (2)]. The point $\bar{\mathbf{x}} \in X$ is feasible in the primal problem (5.1) since $g(\bar{\mathbf{x}}) \leq \mathbf{0}$ by condition (3), and therefore $L(\bar{\mathbf{u}}) = f(\bar{\mathbf{x}}) \geq d$ by the definition of d. Thus, $L(\bar{\mathbf{u}}) = f(\bar{\mathbf{x}}) = d$, and since $f(\bar{\mathbf{x}}) = v$ by Theorem 5.1 the theorem is proven. ■

The implication of Theorem 5.3 is that we need only consider optimal dual solutions in seeking to establish the global optimality conditions. The indicated solution strategy is to find an optimal solution $\bar{\mathbf{u}}$ to the dual, and then attempt to find a complementary $\mathbf{x}$, which satisfies the global optimality conditions. Specifically, we can compute one or more optimal solutions $\mathbf{x} \in X$ in the Lagrangean (evaluated at $\bar{\mathbf{u}}$) and check each one to see if it verifies conditions (2) and (3). The approach is complicated by the fact that there may be alternative optimal dual solutions as well as alternative $\mathbf{x} \in X$ that are optimal in the Lagrangean for each optimal dual solution. Thus, the specific dual solution and specific complementary primal solution satisfying the optimality conditions can be ambiguous to compute.

The following lemma gives us the property of L to be exploited in methods for solving the dual problem.

LEMMA 5.1. The Lagrangean function L is finite and concave.

PROOF. The function L is finite on R^m by our assumptions that f and g are continuous and X compact. To show L is concave, consider the points $\mathbf{u}^1$ and $\mathbf{u}^2$. For α satisfying $0 \leq \alpha \leq 1$, let $\mathbf{x}^0$ satisfy

$$L\left[\alpha \mathbf{u}^1 + (1-\alpha)\mathbf{u}^2\right] = f(\mathbf{x}^0) + \left\{\alpha \mathbf{u}^1 + (1-\alpha)\mathbf{u}^2\right\} g(\mathbf{x}^0)$$

By the definition of L,

$$L(\mathbf{u}^1) \leq f(\mathbf{x}^0) + \mathbf{u}^1 g(\mathbf{x}^0)$$

and

$$L(\mathbf{u}^2) \leq f(\mathbf{x}^0) + \mathbf{u}^2 g(\mathbf{x}^0)$$

Multiplying the first inequality by α and the second by $(1-\alpha)$, we have

$$\alpha L(\mathbf{u}^1) + (1-\alpha) L(\mathbf{u}^2) \leq f(\mathbf{x}^0) + \left\{\alpha \mathbf{u}^1 + (1-\alpha)\mathbf{u}^2\right\} g(\mathbf{x}^0) \quad ■$$

There are several remarks to be made about Lemma 5.1 and the dual problem (5.4). First, despite the arbitrary structure of the primal problem (5.1), problem (5.4) is a problem of maximizing a concave objective function over a convex feasible region. Moreover, this function is continuous because it is concave and finite on R^m. Because L is concave and continuous, we can derive necessary and sufficient conditions that a solution $\bar{\mathbf{u}}$ is optimal in the dual problem. As a consequence, ascent algorithms can be constructed for finding a global optimum to the dual although such an algorithm may not exist for the primal problem from which it is derived. Optimality conditions for the dual problem, and ascent algorithms for solving it are discussed in detail in Chapters 6 and 7. The dual problem is a well-behaved mathematical programming problem because of the mathematical equivalence between convexification and dualization of an arbitrary primal problem, an equivalence that is demonstrated in the next section.

Second, concavity of $L(\mathbf{u})$ is a property of the function defined for all $\mathbf{u} \in R^m$. The constraints $\mathbf{u} \geq \mathbf{0}$ are imposed in order that $L(\mathbf{u}) \leq v$ and thus that $d \leq v$. If $d < v$, one mathematical procedure for closing the duality gap is to relax the nonnegativity constraints on the dual. Another procedure is to generalize the dual vector to a nonlinear dual function; this is discussed in Section 7.4.

Finally, Theorem 5.3 remains true in modified form without the assumptions of continuity on f and g and the compactness of X in the primal problem (5.1). The proof that L is concave is a relatively straightforward modification of the proof given. However, the Lagrangean function in this case may not be continuous everywhere and this can make computational procedures more difficult.

EXAMPLE 5.1. Consider the quadratic programming problem

$$\begin{aligned} v = \min\ & 4(x_1)^2 + 2x_1x_2 + (x_2)^2 \\ \text{s.t. } & 3x_1 + 1x_2 \geq 6 \\ & x_1 \geq 0 \qquad x_2 \geq 0 \end{aligned} \tag{5.5}$$

Problem (5.5) is a simple example of a general class of quadratic programming problems that we study in greater generality and detail in Chapter 7. The Lagrangean derived from (5.5) is

$$\begin{aligned} L(u) = 6u + \min\{&4(x_1)^2 + 2x_1x_2 + (x_2)^2 - 3ux_1 - 1ux_2\} \\ \text{s.t. } & x_1 \geq 0 \\ & x_2 \geq 0 \end{aligned}$$

The function $L(\mathbf{x}, u)$ is convex in $\mathbf{x}$ because its Hessian matrix is positive definite, and therefore a minimum is attained either at the point $\tilde{\mathbf{x}} = (\tilde{x}_1, \tilde{x}_2)$ satisfying $\nabla_{\mathbf{x}} L(\tilde{\mathbf{x}}, u) = 0$ or along the boundaries $x_1 \geq 0$, $x_2 \geq 0$ if $\tilde{\mathbf{x}}$ has negative components. The condition $\nabla_{\mathbf{x}} L(\mathbf{x}, u) = 0$ produces a linear system in the variable u

$$8x_1 + 2x_2 = 3u$$

$$2x_1 + 2x_2 = 1u$$

which we can invert to give the unique solution

$$\begin{bmatrix} x_1 \\ x_2 \end{bmatrix} = \begin{bmatrix} \dfrac{4u}{12} \\ \dfrac{2u}{12} \end{bmatrix}$$

This solution is nonnegative for all $u \geq 0$ and therefore it minimizes $L(\mathbf{x}, u)$. Substituting in the Lagrangean, we obtain for all $u \geq 0$ that

$$L(u) = 6u - \frac{7u^2}{12}$$

The reader can easily verify that this function is concave. The unique optimal solution to the dual problem of maximizing $L(u)$ subject to $u \geq 0$ is obtained by setting $dL(u)/du = 0$; this gives us $\bar{u} = 36/7$.

If possible, we wish to use the optimal dual solution to solve the original primal problem (5.5) from which it was derived. According to the global optimality conditions, we attempt to do this by computing the primal solution $\bar{\mathbf{x}}$, which is optimal in $L(\mathbf{x}, \bar{u})$. By the formula above, this gives us $\bar{x}_1 = 12/7$ and $\bar{x}_2 = 6/7$ with objective function value equal to $108/7$. This solution, along with the optimal dual solution, satisfy the global optimality conditions (2) and (3), which establishes its optimality in (5.5). The optimal primal solution is unique because the quadratic objective function has a positive definite Hessian; that is, it is strictly convex. The successful optimization of the primal problem (5.5) by the dual approach is no accident because the objective function of the primal minimization problem is convex. ▲

EXAMPLE 5.2. The duality theory in the previous example worked perfectly in the sense that there was a unique optimal solution to the dual, which immediately produced the unique optimal solution to the primal. We consider now an integer programming example for which the obvious dual problem does not produce an optimal solution to the primal. In Chapter 8, we show how the duality theory can be refined to produce a sequence of dual problems with increasing lower bounds until an optimal integer solution is obtained. The problem is

$$\begin{aligned} v = \min 3x_1 + 7x_2 + 10x_3 & \\ \text{s.t. } 1x_1 + 3x_2 + 5x_3 \ge 7 & \\ x_1, x_2, x_3 = 0 \text{ or } 1 & \end{aligned} \tag{5.6}$$

The Lagrangean function derived from (5.6) is

$$L(u) = 7u + \min\{(3-u)x_1 + (7-3u)x_2 + (10-5u)x_3\}$$

$$x_1, x_2, x_3 = 0 \text{ or } 1$$

The minimum is calculated simply by setting $x_j = 0$ if its reduced cost coefficient is positive and setting $x_j = 1$ if its reduced cost coefficient is negative. The result is the family of solutions

	x_1	x_2	x_3
$0 \le u \le 2$	0	0	0
$2 \le u \le 7/3$	0	0	1
$7/3 \le u \le 3$	0	1	1
$3 \le u$	1	1	1

The corresponding Lagrangean is

$$L(u) = \begin{cases} 7u & \text{for } 0 \le u \le 2 \\ 2u + 10 & \text{for } 2 \le u \le 7/3 \\ -u + 17 & \text{for } 7/3 \le u \le 3 \\ -2u + 20 & \text{for } 3 \le u \end{cases}$$

for which it is clear that the optimal $\bar{u} = 7/3$. This is because L increases to that value and decreases after it.

At $\bar{u}$, there are two optimal solutions in the Lagrangean; $\bar{x}_1 = \bar{x}_2 = 0$, $\bar{x}_3 = 1$ and $\bar{x}_1 = 0$, $\bar{x}_2 = \bar{x}_3 = 1$. The former solution is infeasible and the latter is feasible, but the complementary slackness condition (2) does not hold since $3\bar{x}_2 + 5\bar{x}_3 = 8 > 7$ and $\bar{u} > 0$. The max ıal dual objective function value $d = 44/3$ which is less than the minimal primal objective function value $v = 17$. In fact, the solution $\bar{\mathbf{x}}$ is optimal, but we failed to establish its optimality because of the duality gap. It can be shown that this dual approach to the integer programming problem is equivalent to solving the linear programming relaxation which results when $x_j = 0$ or 1 is replaced by $0 \le x_j \le 1$ (see Exercise 5.5). ▲

5.3 EQUIVALENCE OF CONVEXIFICATION AND DUALIZATION

In this section, we consider one of the most important theoretical results of mathematical programming. The minimal objective function value that can be achieved if the primal problem (5.1) is convexified equals the maximal dual objective value. Thus, if the primal problem (5.1) is convex and obeys some regularity condition, then the global optimality conditions are necessary as well as sufficient. In any case, the convexification of a primal problem is naturally a *relaxation* of it; that is, every feasible solution to the primal problem is feasible in the convexification. The special nature of this relaxation is sometimes explicitly mentioned by calling it a *Lagrangean relaxation*.

If the primal problem (5.1) is not convex, a common algorithmic approach is to replace it by its convexified relaxation. The equivalence between convexification and dualization indicates that when this is done, it is the dual problem (5.3) that is actually being solved. The specific algorithm used should take into account the structure of the dual problem and its relation to the primal problem. We remind the reader that we have assumed for convenience that f and g are continuous and X is compact.

By its definition, the Lagrangean function $L(\mathbf{u}) \leq f(\mathbf{x}) + \mathbf{u}g(\mathbf{x})$ for every $\mathbf{x} \in X$. Thus if we plot (see Figure 5.1) the values $[f(\mathbf{x}), g(\mathbf{x})]$ in R^{m+1} the hyperplane $L(\mathbf{u}) = y_0 + \mathbf{uy}$ lies below the resulting set; that is, substituting $\mathbf{y} = g(\mathbf{x})$ gives $y_0 = L(\mathbf{u}) - \mathbf{u}g(\mathbf{x}) \leq f(\mathbf{x})$. Also $L(\mathbf{u})$ is the intercept of this hyperplane with $\mathbf{y} = \mathbf{0}$. Letting

$$[f,g] = \bigcup_{\mathbf{x} \in X} \{(\eta, \xi) \mid \eta \geq f(\mathbf{x}), \xi \geq g(\mathbf{x})\}$$

and $[f,g]^c$ be the convex hull of $[f,g]$, we easily see that the $L(\mathbf{u}) = y_0 + \mathbf{uy}$ must be a supporting hyperplane for $[f,g]^c$ as well. We formally record this result as the following lemma.

LEMMA 5.2. For any $\mathbf{u} \geq \mathbf{0}$, the hyperplane $L(\mathbf{u}) = y_0 + \mathbf{uy}$ supports the set $[f,g]^c$.

PROOF. If $(\eta, \xi) \in [f,g]^c$, then there exist $(\eta^1, \xi^1), \ldots, (\eta^{m+2}, \xi^{m+2}) \in [f,g]$ and nonnegative weights $\lambda_1, \ldots, \lambda_{m+2}$ satisfying

$$\sum_{k=1}^{m+2} \lambda_k = 1 \qquad \sum_{k=1}^{m+2} \eta^k \lambda_k = \eta \qquad \sum_{k=1}^{m+2} \xi^k \lambda_k = \xi$$

(this is Caratheodory's Theorem in Appendix A). By the definition of $[f,g]$ there must exist $\mathbf{x}^k \in X$ satisfying $\eta^k \geq f(\mathbf{x}^k)$, $\xi^k \geq g(\mathbf{x}^k)$ for $k = 1, \ldots, m+2$. These inequalities imply that for any $\mathbf{u} \geq \mathbf{0}$,

$$\sum_{k=1}^{m+2} f(\mathbf{x}^k)\lambda_k + \mathbf{u} \sum_{k=1}^{m+2} g(\mathbf{x}^k)\lambda_k \leq \eta + \mathbf{u}\xi$$

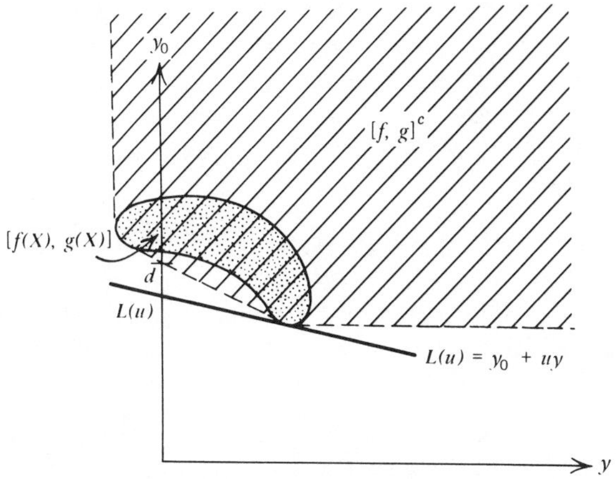

Figure 5.1. Convexified primal problem and dual supporting hyperplane.

But then, since

$$L(\mathbf{u}) \le f(\mathbf{x}^k) + \mathbf{u}g(\mathbf{x}^k)$$

we have

$$L(\mathbf{u}) = \sum_{k=1}^{m+2} L(\mathbf{u})\lambda_k \le \eta + \mathbf{u}\boldsymbol{\xi}$$

that is, $y_0 = L(\mathbf{u}) - \mathbf{u}\boldsymbol{\xi} \le \eta$. Finally, if $L(\mathbf{u}) = f(\overline{\mathbf{x}}) + \mathbf{u}g(\overline{\mathbf{x}})$ then the hyperplane $L(\mathbf{u}) = y_0 + \mathbf{u}\mathbf{y}$ supports $[f,g]^c$ at $[f(\overline{\mathbf{x}}), g(\overline{\mathbf{x}})] \in [f,g]$. ■

Next we define

$$v^c(\boldsymbol{\xi}) = \inf\{\eta \mid (\eta, \boldsymbol{\xi}) \in [f,g]^c\}$$

which is taken to be $+\infty$ if there is no $(\eta, \boldsymbol{\xi}) \in [f,g]^c$; $v^c \equiv v^c(\mathbf{0})$ is the convexified value of the primal problem. We are now in a position to prove the basic result establishing the equivalence of convexification and dualization of the primal problem (5.1).

THEOREM 5.4. The optimal dual objective function value equals the optimal objective function value of the convexified primal; namely

$$v^c = d$$

PROOF. ($d \le v^c$). If $v^c = +\infty$, there is nothing to prove; otherwise, select an arbitrary $(\eta, \mathbf{0}) \in [f,g]^c$. Then from Lemma 5.2, for any $\mathbf{u} \ge \mathbf{0}$, $L(\mathbf{u}) \le \eta + \mathbf{u} \cdot \mathbf{0} = \eta$. Thus,

$$\eta \ge d = \sup L(\mathbf{u}) \quad \text{s.t. } \mathbf{u} \ge \mathbf{0}$$

and since $(\eta, \mathbf{0})$ was chosen arbitrarily from $[f,g]^c$, we can conclude $d \le v^c$.

($v^c \le d$). If $v^c = -\infty$, there is nothing to prove; otherwise, let $r < v^c$ be an arbitrary real number. Then $(r, \mathbf{0}) \notin [f,g]^c$. Since $[f,g]^c$ is a closed convex set (see Exercise 5.7), there is a hyperplane $u_0 y_0 + \mathbf{u}\mathbf{y} = \beta$ strictly separating $(r, \mathbf{0})$ and $[f,g]^c$; namely the nonzero vector $(u_0, \mathbf{u}) \in R^{m+1}$ and real number β satisfy

$$u_0 r + \mathbf{u} \cdot \mathbf{0} < \beta \le u_0 \eta + \mathbf{u}\boldsymbol{\xi} \qquad \text{for all } (\eta, \boldsymbol{\xi}) \in [f,g]^c \tag{5.7}$$

Since η and each component ξ_j of $\boldsymbol{\xi}$ are unbounded from above over $[f,g]$, the rightmost inequality in (5.7) implies that $u_i \ge 0$ for $i = 0, \ldots, m$.

To complete the proof that $v^c \le d$, we distinguish two cases.

Case 1. There exists a point $(\eta, \mathbf{0}) \in [f,g]^c$ for some $\eta \in R$. Then it follows from (5.7) that $u_0 \ne 0$; otherwise $0 < \beta \le 0$. By scaling u_0, $\mathbf{u}$, and β, we may assume that $u_0 = 1$. But then since by definition $[f(\mathbf{x}), g(\mathbf{x})] \in [f,g]$ for every $\mathbf{x} \in X$, (5.7) implies that

$$L(\mathbf{u}) = \min_{\mathbf{x} \in X} \{ f(\mathbf{x}) + \mathbf{u}g(\mathbf{x}) \} \ge \beta > r$$

and therefore since $r < v^c$ was arbitrary

$$\begin{aligned} v^c \ge d = \sup\ & L(\mathbf{u}) \\ \text{s.t. } & \mathbf{u} \ge \mathbf{0} \end{aligned}$$

Case 2. There does not exist a point $(\eta, \mathbf{0}) \in [f,g]^c$ implying $v^c = +\infty$. Then the sets $\{(r, 0) | r \in R\}$ and $[f,g]^c$ are disjoint closed convex sets implying the existence[1] of $(u_0, \mathbf{u}) \in R^{m+1}$ and a scalar β such that

$$u_0 r + \mathbf{u} \cdot \mathbf{0} < \beta \le u_0 \eta + \mathbf{u}\boldsymbol{\xi}$$

for all $r \in R$ and $(\eta, \boldsymbol{\xi}) \in [f,g]$. As before, we have $(u_0, \mathbf{u}) \ge \mathbf{0}$. Letting $r \to +\infty$, the leftmost inequality implies that $u_0 = 0$ and thus that $\beta > 0$. Thus, $\mathbf{u}\boldsymbol{\xi} \ge \beta$ for any $(\eta, \boldsymbol{\xi}) \in [f,g]$ and for any $K > 0$, we have

$$(K\mathbf{u})\boldsymbol{\xi} \ge K\beta$$

Letting $\boldsymbol{\xi} = g(\mathbf{x})$, this inequality implies that

$$\min_{\mathbf{x} \in X} \left[(K\mathbf{u})g(\mathbf{x}) \right] \ge K\beta$$

Thus, if $l = \min_{\mathbf{x} \in X} f(\mathbf{x})$, we have

$$L(K\mathbf{u}) = \min_{\mathbf{x} \in X} \left[f(\mathbf{x}) + (K\mathbf{u})g(\mathbf{x}) \right] \ge l + K\beta$$

[1]Since the sets are not compact, the existence of a strictly separating hyperplane needs additional proof. Such a proof uses the properties f, g continuous, X compact.

Letting K go to $+\infty$ demonstrates that

$$v^c = d = \sup_{\text{s.t. } \mathbf{u} \geq \mathbf{0}} L(\mathbf{u}) = +\infty$$

which completes the proof. ■

5.4 GENERALIZED LINEAR PROGRAMMING

Generalized linear programming is a method for trying to solve the primal problem (5.1) by solving a series of more and more refined linear programming approximations. These approximations are the convexifications of the previous section and the implication of Theorem 5.4 is that convergence, if it occurs, is to an optimal solution to the dual (5.4). If the primal problem (5.1) has convex structure, then, in addition, a primal feasible solution complementary to this dual solution is found so that the primal-dual pair satisfies the global optimality conditions. By Theorems 5.1 and 5.3, both the primal and dual solutions are optimal in their respective problems.

At iteration K, the generalized linear programming algorithm uses the simplex method to solve the *Master LP*

$$\begin{aligned} d^K = \min \sum_{k=1}^{K} f(\mathbf{x}^k)\lambda_k & \\ \text{s.t.} \sum_{k=1}^{K} g(\mathbf{x}^k)\lambda_k \leq 0 & \\ \sum_{k=1}^{K} \lambda_k = 1 & \\ \lambda_k \geq 0 \qquad k = 1,\ldots,K & \end{aligned} \tag{5.8}$$

where the solutions $\mathbf{x}^k \in X$ for $k = 1,\ldots,K$ have been previously generated. We assume (5.8) has a feasible solution; a phase one procedure for finding a feasible solution is discussed in Exercise 5.10. The LP dual to (5.8) is

$$\begin{aligned} d^K = \max w & \\ \text{s.t. } w \leq f(\mathbf{x}^k) + \mathbf{u}g(\mathbf{x}^k) \qquad & k = 1,\ldots,K \\ \mathbf{u} \geq \mathbf{0} & \end{aligned} \tag{5.9}$$

Let λ_k^K, $k = 1,\ldots,K$, and the m-vector $\mathbf{u}^K$, denote optimal solutions to the LP primal problem (5.8) and the LP dual problem (5.9), respectively. Notice that d^K is the optimal dual variable on the convexity row $\Sigma_{k=1}^K \lambda_k = 1$ in the primal (5.8). The generalized linear programming algorithm proceeds by solving the Lagrangean

$$L(\mathbf{u}^K) = \min_{x \in X} \{ f(\mathbf{x}) + \mathbf{u}^K g(\mathbf{x}) \}$$
$$= f(\mathbf{x}^{K+1}) + \mathbf{u}^K g(\mathbf{x}^{K+1}) \tag{5.10}$$

It is common terminology to refer to the Lagrangean calculation as the *subproblem*. By its definition, the dual problem (5.4) can be written as

$$\begin{aligned} d = &\sup w \\ &\text{s.t. } w \le f(\mathbf{x}) + \mathbf{u}g(\mathbf{x}) \qquad \text{for all } \mathbf{x} \in X \\ &\qquad \mathbf{u} \ge \mathbf{0} \end{aligned}$$

Since this problem has at least as many constraints as (5.9), $d^K \ge d$; also by definition, $L(\mathbf{u}^K) \le d$. These inequalities give us immediately the following result.

LEMMA 5.3. At iteration K of the generalized linear programming algorithm

1. $d^{K-1} \ge d^K \ge d$.
2. If $L(\mathbf{u}^K) = d^K$, then $L(\mathbf{u}^K) = d$; that is, $\mathbf{u}^K$ is optimal in the dual (5.4).

Thus, the generalized linear programming algorithm terminates with an optimal solution $\mathbf{u}^K$ to the dual problem (5.4) if $L(\mathbf{u}^K) = d^K$. Note that it is impossible for $L(\mathbf{u}^K)$ to be greater than d^K since there is at least one λ_k, say λ_l, positive in (5.8) implying

$$f(\mathbf{x}^l) + \mathbf{u}^K g(\mathbf{x}^l) - d^K = 0$$

by complementary slackness (that is, the reduced cost coefficient of λ_l must be zero). If $L(\mathbf{u}^K) < d^K$, the algorithm proceeds by adding a column corresponding to $\mathbf{x}^{K+1}$ to (5.8) or equivalently, a row corresponding to $\mathbf{x}^{K+1}$ to (5.9). This column is introduced as a nonbasic column in (5.8) with negative reduced cost coefficient. The simplex method is reinitiated from the previously optimal basic solution to (5.8), and computation continues by pivoting column λ_{K+1} into the basis.

We consider the convergence properties of the generalized linear programming algorithm when $L(\mathbf{u}^K) < d^K$ for all K.

THEOREM 5.5. If there exists an index set $\mathcal{K} \subseteq \{1, 2, \ldots\}$ such that the subsequence of dual solutions $\{\mathbf{u}^K\}_{K \in \mathcal{K}}$ produced by the generalized linear programming algorithm is convergent, say to the limit point $\mathbf{u}^*$, then

1. $\mathbf{u}^*$ is optimal in the dual problem (5.4), and
2. $\lim_K d^K = d = L(\mathbf{u}^*)$.

PROOF. By the definition of problem (5.9), we have for all $k = 1, 2, \ldots, K$

$$f(\mathbf{x}^k) + \mathbf{u}^K g(\mathbf{x}^k) \ge d^K \ge d \tag{5.11}$$

where the right inequality is from Lemma 5.3. Let $d^\infty = \lim d^K$; this limit exists because the d^K are monotonically decreasing and bounded from below by $L(\mathbf{u})$ for any $\mathbf{u} \geq \mathbf{0}$. Taking the limit in (5.11) for $K \in \mathcal{K}$, we obtain

$$f(\mathbf{x}^k) + \mathbf{u}^* g(\mathbf{x}^k) \geq d^\infty \geq d \quad \text{for each } k = 1, 2, \ldots \tag{5.12}$$

Since $g(\cdot)$ is continuous and X is compact, there is a real number B such that $|g_i(\mathbf{x})| \leq B$ for all $\mathbf{x} \in X$ and $i = 1, 2, \ldots, m$. Then

$$\begin{aligned} |L(\mathbf{x}^{K+1}, \mathbf{u}^K) - L(\mathbf{x}^{K+1}, \mathbf{u}^*)| &= |(\mathbf{u}^K - \mathbf{u}^*) g(\mathbf{x}^{K+1})| \\ &\leq B \sum_{i=1}^{m} |u_i^K - u_i^*| \end{aligned}$$

Consequently, given $\varepsilon > 0$ there is a $K_1 \in \mathcal{K}$ such that for all $K \in \mathcal{K}$, $K \geq K_1$, the right-hand side is bounded by ε and therefore

$$\begin{aligned} L(\mathbf{u}^K) = L(\mathbf{x}^{K+1}, \mathbf{u}^K) &\geq L(\mathbf{x}^{K+1}, \mathbf{u}^*) - \varepsilon \\ &= f(\mathbf{x}^{K+1}) + \mathbf{u}^* g(\mathbf{x}^{K+1}) - \varepsilon \end{aligned}$$

Thus, from (5.12) and the definition of d,

$$d \leq d^\infty \leq f(\mathbf{x}^{K+1}) + \mathbf{u}^* g(\mathbf{x}^{K+1}) \leq L(\mathbf{u}^K) + \varepsilon \leq d + \varepsilon$$

Since $\varepsilon > 0$ was arbitrary, we can conclude that $d^\infty = \lim d^K = d$. This also implies that

$$\lim_{K \in \mathcal{K}} L(\mathbf{u}^K) = L\Big(\lim_{K \in \mathcal{K}} \mathbf{u}^K\Big) = L(\mathbf{u}^*) = d$$

where the first equality follows from the continuity of L. ■

The following lemma gives a sufficient condition that there be a converging subsequence of the $\mathbf{u}^K$ generated by the generalized linear programming algorithm. The condition is a familiar regularity condition for the Kuhn-Tucker optimality conditions studied in the next section.

LEMMA 5.4. A sufficient condition that the generalized linear programming algorithm produces a converging subsequence of dual variables is that there exists $\mathbf{x}^1 \in X$ satisfying the *constraint qualification*

$$g_i(\mathbf{x}^1) < 0 \qquad i = 1, \ldots, m \tag{5.13}$$

PROOF. Consider the constraint written with respect to $\mathbf{x}^1$ in problem (5.9). At any iteration K of the generalized linear programming algorithm, we have

$$\sum_{i=1}^{m} u_i^K[-g_i(\mathbf{x}^1)] \leq f(\mathbf{x}^1) - d^K \leq f(\mathbf{x}^1) - d$$

where the left-hand inequality follows because $\mathbf{x}^1$ is represented by a column in (5.8), which must have nonnegative reduced cost, and the right-hand inequality follows from Lemma 5.3. Since $-g_i(\mathbf{x}^1)>0$ for all i, $M=\min_{i=1,\ldots,m} -g_i(\mathbf{x}^1)>0$ and we can divide by L

$$\sum_{i=1}^{m} u_i^K \leq \sum_{i=1}^{m} u_i^K \frac{-g_i(\mathbf{x}^1)}{M} \leq \frac{f(\mathbf{x}^1)-d}{M}$$

Thus, the $\mathbf{u}^K$ are bounded for all K and there exists a converging subsequence. ■

Note that under the hypothesis of Theorem 5.5, the generalized linear programming algorithm mechanizes the duality result of the previous section. The algorithm establishes $v^c \leq d$ which is the reverse of the easily established weak duality result $d \leq v^c$. To see this, observe that at each step of the algorithm,

$$d^K = \sum_{k=1}^{K} f(\mathbf{x}^k)\lambda_k^K$$

and

$$\sum_{k=1}^{K} g(\mathbf{x}^k)\lambda_k^K \leq \mathbf{0}$$

Thus, $(d^K,\mathbf{0})\in[f,g]^c$ implying $v^c \leq d^K$ and since $\lim d^K = d$, we can conclude $v^c \leq d$ and therefore $v^c = d$. There is, however, a subtle distinction between the results of Theorem 5.4 and Theorem 5.5. When the generalized linear programming algorithm converges, we not only have $d=v^c$, but we have found a $\bar{\mathbf{u}}\geq\mathbf{0}$ such that $L(\bar{\mathbf{u}})=d$; namely, we have attainment of the dual supremum objective function value in (5.4).

Thus far, we have concentrated on properties of the dual problem (5.4) and how it can be optimized using generalized linear programming. We have not addressed the question of the conditions on the primal problem (5.1) under which we can be assured that knowledge of an optimal dual solution can be guaranteed to produce an optimal primal solution via the global optimality conditions. The following result tells us when we can necessarily expect the global optimality conditions to exist for some primal-dual pair. The properties of the generalized linear programming algorithm demonstrated above provide a constructive proof of the existence of the conditions in the special case when the primal problem (5.1) is convex.

THEOREM 5.6. If the constraint qualification (5.13) holds, and if f, g, and X in the primal problem (5.1) are convex, then there exists a primal-dual pair $(\mathbf{x}^*,\mathbf{u}^*)$ satisfying the global optimality conditions.

PROOF. First we establish that there exists an $\mathbf{x}^* \in X$ such that $g(\mathbf{x}^*) \le \mathbf{0}$ and $f(\mathbf{x}^*) = d \le v$; that is, that $\mathbf{x}^*$ is optimal in (5.1) and there is no duality gap. Then we show that the global optimality conditions must hold. At iteration K of the algorithm, the convexity assumptions imply that the solution $\mathbf{y}^K = \sum_{k=1}^{K} \lambda_k^K \mathbf{x}^k$ satisfies

$$f(\mathbf{y}^K) \le \sum_{k=1}^{K} \lambda_k^K f(\mathbf{x}^k) = d^K$$

$$g(\mathbf{y}^K) \le \sum_{k=1}^{K} \lambda_k^K g(\mathbf{x}^k) \le \mathbf{0}$$

$$\mathbf{y}^K \in X$$

Since X is compact, there is a subsequence of the $\mathbf{y}^K$ converging to $\mathbf{x}^* \in X$. Since the constraint qualification holds, there is a convergent subsequence of the dual solutions $\mathbf{u}^K$ and without loss of generality, we can assume $\mathbf{y}^K$ and $\mathbf{u}^K$ converge for K in the same index set $\mathcal{K} \subseteq \{1, 2, \ldots\}$. Let $\mathbf{u}^*$ denote the limit point of the $\mathbf{u}^K$. Since f and g are continuous, we clearly have

$$f(\mathbf{x}^*) = f\Big(\lim_{K \in \mathcal{K}} \mathbf{y}^K\Big) = \lim_{K \in \mathcal{K}} f(\mathbf{y}^K) \le d = \lim_{K \in \mathcal{K}} d^K \qquad (5.14)$$

$$g(\mathbf{x}^*) = g\Big(\lim_{K \in \mathcal{K}} \mathbf{y}^K\Big) = \lim_{K \in \mathcal{K}} g(\mathbf{y}^K) \le \mathbf{0} \qquad (5.15)$$

where $d = \lim_{K \in \mathcal{K}} d^K$ from Theorem 5.5. In other words, the solution $\mathbf{x}^*$ is feasible in the primal problem (5.1) and $f(\mathbf{x}^*) \le d$ implying $\mathbf{x}^*$ is optimal in (5.1) and $v = f(\mathbf{x}^*) = d$.

To establish that $(\mathbf{x}^*, \mathbf{u}^*)$ satisfy the global optimality conditions, we use Theorem 5.5 which tells us that $\mathbf{u}^*$ is optimal in the dual problem (5.4). This implies

$$d = L(\mathbf{u}^*) \le f(\mathbf{x}^*) + \mathbf{u}^* g(\mathbf{x}^*) \le f(\mathbf{x}^*) = d$$

where the first inequality follows from the definition of L and the second inequality holds because $\mathbf{u}^* \ge \mathbf{0}$ and $g(\mathbf{x}^*) \le \mathbf{0}$. Thus, we can conclude

$$L(\mathbf{u}^*) = f(\mathbf{x}^*) + \mathbf{u}^* g(\mathbf{x}^*)$$

implying

$$\mathbf{u}^* g(\mathbf{x}^*) = 0$$

which, along with $g(\mathbf{x}^*) \le 0$, are the global optimality conditions for problem (5.1). ■

EXAMPLE 5.3. We apply the generalized linear programming algorithm to the Example 5.1. The initial points are $(x_1^1, x_2^1) = (0, 0)$ and $(x_1^2, x_2^2) = (2, 2)$ and the resulting Master LP is

$$\begin{aligned} d^2 = \min\ & 0\lambda_1 + 28\lambda_2 \\ \text{s.t. } & 6\lambda_1 - 2\lambda_2 + s = 0 \\ & \lambda_1 + \lambda_2 = 1 \\ & \lambda_1 \ge 0 \quad \lambda_2 \ge 0 \quad s \ge 0 \end{aligned}$$

The optimal solution to this problem is $(\lambda_1^2, \lambda_2^2) = (1/4, 3/4)$ and the optimal dual variable $u^2 = 7/2$. Given these weights on $\mathbf{x}^1$ and $\mathbf{x}^2$, the indicated primal solution is $\mathbf{y}^2 = 1/4(0,0) + 3/4(2,2) = (3/4, 3/2)$. Solution of the Lagrangean subproblem gives the point $(x_1^3, x_2^3) = (7/6, 7/12)$ which has reduced cost equal to $f(\mathbf{x}^3) + \mathbf{u}^2 g(\mathbf{x}^3) - \theta = -343/144$, where $\theta = 21$ is the shadow price on the convexity row. The feasible solution $\mathbf{y}^2$ has objective function value equal to 15.75 and the dual lower bound value is $L(u^2) =$ 13.85. These figures bracket the minimal objective function value of 15.42 calculated previously in Example 5.1. The new Master LP is

$$
\begin{aligned}
d^3 = \min 0\lambda_1 + 28\lambda_2 + \frac{1029}{144}\lambda_3 \\
\text{s.t. } 6\lambda_1 - 2\lambda_2 + \frac{23}{12}\lambda_3 + s = 0 \\
\lambda_1 + \lambda_2 + \lambda_3 = 1 \\
\lambda_1 \geq 0 \qquad \lambda_2 \geq 0 \qquad \lambda_3 \geq 0 \qquad s \geq 0
\end{aligned}
$$

The optimal weights in this LP are $(\lambda_1, \lambda_2, \lambda_3) = (0, 47/96, 49/96)$ and the indicated primal feasible solution is $\mathbf{y}^3 = (74/47, 60/47)$ with objective function value equal to 15.56. The dual solution $\mathbf{u}^3 = 5.326$ and the dual lower bound value is 15.41. Thus, we have closely bracketed the minimal objective function value and could terminate with $\mathbf{y}^3$ as an approximately optimal solution. ▲

5.5 KUHN-TUCKER CONDITIONS

In Section 5.1, we stated the global optimality conditions and established their validity as sufficient conditions for solving the primal problem (5.1) without assuming any properties for the functions f and g_i. For a large class of mathematical programming problems that we study in this section and in Chapter 7, it is appropriate to assume additional properties of differentiability. Specifically, we consider

$$
\begin{aligned}
v = \min f(\mathbf{x}) \\
\text{s.t. } g_i(\mathbf{x}) \leq 0 \qquad i = 1, \ldots, m \\
\mathbf{x} \in R^n
\end{aligned}
\tag{5.16}
$$

where f and the g_i are differentiable. With this structure for the primal problem, a *necessary* condition for $\mathbf{x}$ to minimize $f(\mathbf{x}) + \mathbf{u}g(\mathbf{x})$ for a fixed $\mathbf{u}$ is that $\nabla f(\mathbf{x}) + \mathbf{u}\nabla g(\mathbf{x}) = \mathbf{0}$. Our interest here is to investigate the consequences of replacing global optimality condition (1) by this necessary condition.

Definition 5.2. For $\bar{\mathbf{x}} \in R^n$ and $\bar{\mathbf{u}} \geq \mathbf{0}$, the *Kuhn-Tucker conditions* are

(1) $\nabla f(\bar{\mathbf{x}}) + \bar{\mathbf{u}} \nabla g(\bar{\mathbf{x}}) = \mathbf{0}$,

(2) $\bar{\mathbf{u}} g(\bar{\mathbf{x}}) = 0$,

(3) $g(\bar{\mathbf{x}}) \leq \mathbf{0}$.

THEOREM 5.7. Suppose f and the g_i in problem (5.16) are convex. If $(\bar{\mathbf{x}}, \bar{\mathbf{u}})$ satisfy the Kuhn-Tucker conditions, then $\bar{\mathbf{x}}$ is optimal in (5.16).

PROOF. The function $L(\mathbf{x}, \bar{\mathbf{u}}) = f(\mathbf{x}) + \bar{\mathbf{u}} g(\mathbf{x})$ is convex since f and g are convex and $\bar{\mathbf{u}} \geq \mathbf{0}$. Thus, Kuhn-Tucker condition (1) is sufficient as well as necessary for $\bar{\mathbf{x}}$ to minimize $L(\mathbf{x}, \bar{\mathbf{u}})$ over $\mathbf{x} \in R^n$. In other words, $L(\bar{\mathbf{u}}) = f(\bar{\mathbf{x}}) + \bar{\mathbf{u}} g(\bar{\mathbf{x}})$ and the optimality of $\bar{\mathbf{x}}$ in (5.16) follows immediately from Theorem 5.1. ∎

The Kuhn-Tucker conditions are clearly just a restatement of the global optimality conditions for (5.16) in the special case when f and the g_i are convex. Our main concern in this section is to investigate the relationship between these two sets of conditions when convexity is not assumed. Without further analysis, the situation is as depicted in Figure 5.2. Figure 5.2 shows that the set of $\mathbf{x} \in R^n$, which satisfies the global optimality conditions for any $\mathbf{u} \geq \mathbf{0}$ are a subset, possibly empty, of the set of optimal solutions to (5.16). This was established by Theorem 5.1. Figure 5.2 also shows that the set of $\mathbf{x} \in R^n$ which satisfies the global optimality conditions for any $\mathbf{u} \geq \mathbf{0}$ is a subset of the set of solutions to (5.16) which satisfies the Kuhn-Tucker conditions. This is because $\nabla f(\mathbf{x}) + \mathbf{u} \nabla g(\mathbf{x}) = \mathbf{0}$ is a necessary condition for minimizing $f(\mathbf{x}) + \mathbf{u} g(\mathbf{x})$.

Thus, without further analysis, we do not know the relationship, if any, between the set of optimal solutions to (5.16) and the set of solutions to (5.16) which satisfies the Kuhn-Tucker conditions for any $\mathbf{u} \geq \mathbf{0}$. It turns out that if appropriate regularity assumptions on (5.16) are satisfied, then the situation is as depicted in Figure 5.3; namely, the Kuhn-Tucker conditions are *necessary* optimality conditions for problem (5.16).

The study of necessary optimality conditions naturally leads us to consider solutions that satisfy the conditions but are not optimal.

Definition 5.3. A feasible solution $\tilde{\mathbf{x}}$ to (5.16) is a *local minimum* if there exists a nonempty neighborhood $N(\tilde{\mathbf{x}}; \boldsymbol{\delta})$ such that $f(\mathbf{x}) \geq f(\tilde{\mathbf{x}})$ for all feasible $\mathbf{x} \in N(\mathbf{x}; \boldsymbol{\delta})$. A local minimum is a *strict local minimum* if $f(\mathbf{x}) > f(\tilde{\mathbf{x}})$ for all feasible $\mathbf{x} \neq \tilde{\mathbf{x}}$ in $N(\tilde{\mathbf{x}}; \boldsymbol{\delta})$.

In these terms, an optimal solution to (5.16) is a global as well as local minimum, but there can be local minima that are not global. The presence

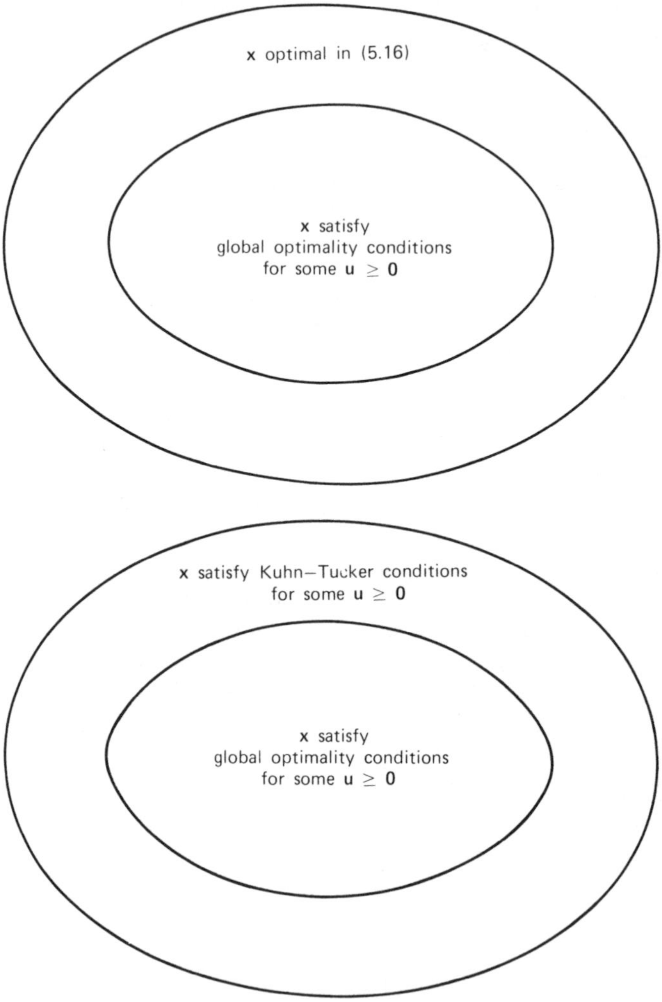

Figure 5.2. (a) Global optimality conditions are sufficient for optimality. (b) Global optimality conditions are sufficient for Kuhn-Tucker conditions.

of the gradients ∇f and ∇g_i in the Kuhn-Tucker conditions suggests that they are local conditions based on the behavior of the functions in the neighborhood of a local minimal solution to (5.16). Thus, we will attempt to show that if $\mathbf{x}^*$ is a local minimum for problem (5.16), then there exists a dual solution $\mathbf{u}^*$ such that the Kuhn-Tucker conditions hold for $(\mathbf{x}^*, \mathbf{u}^*)$. As we shall see, this statement is true, but only with some additional

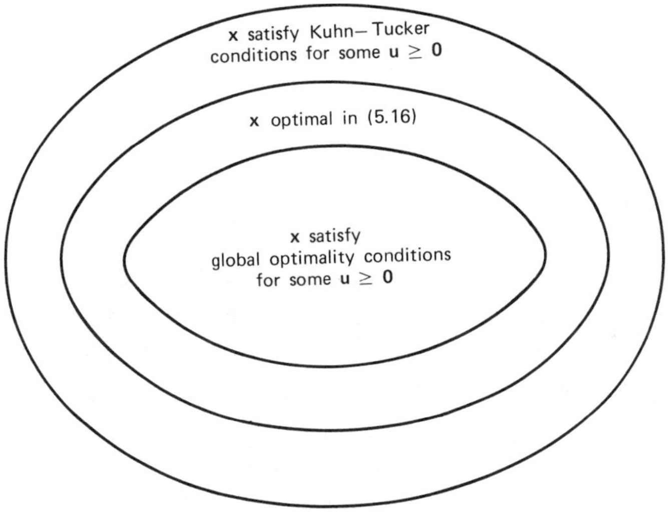

Figure 5.3. Kuhn-Tucker conditions are necessary for optimality.

assumptions about the regularity of the constraint region of (5.16) in the neighborhood of local minima.

Our analysis begins by considering linearizations of problem (5.16) in neighborhoods of an arbitrary feasible solution $\tilde{\mathbf{x}}$. First, we need to make precise the notion of feasible directions of movement from $\tilde{\mathbf{x}}$.

Definition 5.4. Let $\tilde{\mathbf{x}}$ be a feasible solution to (5.16). A *feasible direction* $\mathbf{z}\neq 0$ from $\tilde{\mathbf{x}}$ is a vector with the property that there exists a $\theta_1 > 0$ such that $\tilde{\mathbf{x}}+\theta\mathbf{z}$ is feasible for all θ satisfying $0 \leq \theta \leq \theta_1$.

Problem (5.16) is linearized about $\tilde{\mathbf{x}}$ by considering the sets

$$Z^1(\tilde{\mathbf{x}}) = \{\mathbf{z} | \mathbf{z}^T \nabla g_i(\tilde{\mathbf{x}}) \leq 0 \qquad \text{for all } i \in I(\tilde{\mathbf{x}})\}$$

where

$$I(\tilde{\mathbf{x}}) = \{i | g_i(\tilde{\mathbf{x}}) = 0\}$$

and

$$Z^2(\tilde{\mathbf{x}}) = \{\mathbf{z} | \mathbf{z}^T \nabla f(\tilde{\mathbf{x}}) < 0\}$$

The set $I(\tilde{\mathbf{x}})$ gives the indices of the binding constraints at $\tilde{\mathbf{x}}$ and the set $Z^1(\tilde{\mathbf{x}})$ gives a linearized approximation to the feasible region in a neighborhood of the feasible solution $\tilde{\mathbf{x}}$. To see this, consider any $\mathbf{z} \notin Z^1(\tilde{\mathbf{x}})$; say $\mathbf{z}^T \nabla g_k(\tilde{\mathbf{x}}) > 0$ for some k such that $g_k(\tilde{\mathbf{x}}) = 0$. The differentiability of g_k

implies for $\theta \geq 0$ that

$$g_k(\tilde{\mathbf{x}}+\theta\mathbf{z}) = g_k(\tilde{\mathbf{x}}) + \theta\mathbf{z}^T\nabla g_k(\tilde{\mathbf{x}}) + \theta\xi_k(\theta)$$

where $\xi_k(\theta)$ tends to zero as $\theta \to 0^+$. If θ is sufficiently small, then

$$\theta\mathbf{z}^T\nabla g_k(\tilde{\mathbf{x}}) + \theta\xi_k(\theta) > 0$$

implying that

$$g_k(\tilde{\mathbf{x}}+\theta\mathbf{z}) > 0$$

or in other words that $\tilde{\mathbf{x}}+\theta\mathbf{z}$ is infeasible. Thus, the set $Z^1(\tilde{\mathbf{x}})$ contains all the feasible directions from $\tilde{\mathbf{x}}$. Similarly, the set $Z^2(\tilde{\mathbf{x}})$ contains all the directions in which the objective function strictly decreases from $\tilde{\mathbf{x}}$.

THEOREM 5.8. Suppose $\tilde{\mathbf{x}}$ is a feasible solution to problem (5.16). Then $Z^1(\tilde{\mathbf{x}}) \cap Z^2(\tilde{\mathbf{x}})$ is empty if and only if there exists a $\tilde{\mathbf{u}} \geq \mathbf{0}$ such that

$$\nabla f(\tilde{\mathbf{x}}) + \tilde{\mathbf{u}}g(\tilde{\mathbf{x}}) = \mathbf{0}$$
$$\tilde{\mathbf{u}}g(\tilde{\mathbf{x}}) = 0$$

PROOF. The set $Z^1(\tilde{\mathbf{x}})$ is not empty for any $\tilde{\mathbf{x}}$ because it always contains the zero vector. Thus, the set $Z^1(\tilde{\mathbf{x}}) \cap Z^2(\tilde{\mathbf{x}})$ is empty if and only if

$$\mathbf{z}^T\nabla f(\tilde{\mathbf{x}}) \geq 0$$

for all $\mathbf{z}$ satisfying

$$\mathbf{z}^T\nabla g_i(\tilde{\mathbf{x}}) \leq 0 \qquad \text{for all } i \in I(\tilde{\mathbf{x}})$$

By Farkas' lemma 2.2, these inequalities hold if and only if there exists $\tilde{u}_i \geq 0$, $i \in I(\tilde{\mathbf{x}})$, such that

$$\nabla f(\tilde{\mathbf{x}}) + \sum_{i \in I(\tilde{\mathbf{x}})} \tilde{u}_i \nabla g_i(\tilde{\mathbf{x}}) = \mathbf{0}$$

We expand the $\tilde{u}_i$ to an m-vector $\tilde{\mathbf{u}}$ by taking $\tilde{u}_i = 0$ for $i \notin I(\tilde{\mathbf{x}})$. By this construction, we have $\nabla f(\tilde{\mathbf{x}}) + \tilde{\mathbf{u}}\nabla g(\tilde{\mathbf{x}}) = \mathbf{0}$ and moreover, $\sum_{i=1}^m \tilde{u}_i g_i(\tilde{\mathbf{x}}) = 0$ since at least one of the terms $\tilde{u}_i$ and $g_i(\tilde{x})$ is zero for each i. ■

We might expect that if $Z^1(\tilde{\mathbf{x}}) \cap Z^2(\tilde{\mathbf{x}})$ is nonempty then $\tilde{\mathbf{x}}$ is not a local minimum because then there is a feasible direction $\mathbf{z} \in Z^1(\tilde{\mathbf{x}}) \cap Z(\tilde{\mathbf{x}})$ from $\tilde{\mathbf{x}}$ in which f strictly decreases. By contraposition, we are saying that we might expect $Z^1(\tilde{\mathbf{x}}) \cap Z^2(\mathbf{x})$ to be empty if $\tilde{\mathbf{x}}$ is a local minimum. If this were so, then Theorem 5.8 would provide us with the characterization we seek of local minima in terms of the Kuhn-Tucker conditions; that is, if $\tilde{\mathbf{x}}$ is a local minimum, then there exists a $\tilde{\mathbf{u}} \geq \mathbf{0}$ such that the Kuhn-Tucker conditions hold for $(\tilde{\mathbf{x}}, \tilde{\mathbf{u}})$. Unfortunately, the situation is more complicated as illustrated by the following example.

EXAMPLE 5.4. Consider the problem

$$\min f(x_1,x_2) = -x_1$$

$$\text{s.t. } g_1(x_1,x_2) = -(1-x_1)^3 + x_2 \leq 0$$

$$g_2(x_1,x_2) = -x_1 \leq 0$$

$$g_3(x_1,x_2) = -x_2 \leq 0$$

The constraint region is depicted in Figure 5.4 and we can see that the solution $\tilde{\mathbf{x}}=(1,0)$ is a global as well as a local minimum. The binding constraints at $\tilde{\mathbf{x}}$ are g_1 and g_3 and using the facts that $\nabla g_1(\tilde{\mathbf{x}})=(0,1)$ and $\nabla g_3(\tilde{\mathbf{x}})=(0,-1)$ we can easily compute that

$$Z^1(\tilde{\mathbf{x}}) = \{(z_1,z_2) | z_2 = 0\}$$

Similarly, we have

$$Z^2(\tilde{\mathbf{x}}) = \{(z_1,z_2) | z_1 > 0\}$$

and therefore the set $Z^1(\tilde{\mathbf{x}}) \cap Z^2(\tilde{\mathbf{x}}) = \{(z_1,z_2) | z_1 > 0, z_2 = 0\}$ is not empty at the local (global) minimum $\tilde{x}$. ▲

The difficulty illustrated by the last example is a serious one for the application of the Kuhn-Tucker conditions to the practical solution of problem (5.16) in its general form as stated. We discuss below some qualifications on the constraint region $\{\mathbf{x} | g_i(\mathbf{x}) \leq 0 \text{ for } i=1,\ldots,m\}$ which eliminate the difficulty, but these qualifications are mainly theoretical and can be easily verified only in the special case when the functions g_i are linear, and to a lesser extent, when the functions g_i are convex.

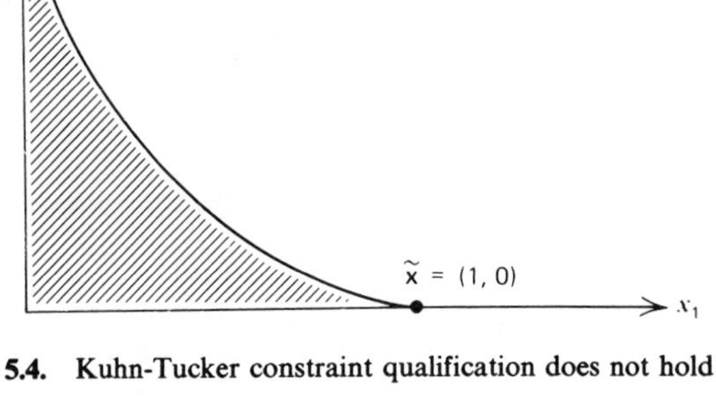

Figure 5.4. Kuhn-Tucker constraint qualification does not hold at $\tilde{x}$.

The constraint qualification we use is the one originally proposed by Kuhn and Tucker. In words, the qualification requires for each point $\tilde{\mathbf{x}}$ that any $\mathbf{z} \in Z^1(\tilde{\mathbf{x}})$ is tangent to a differentiable arc in the feasible region. Specifically, the *Kuhn-Tucker constraint qualification* at a point $\tilde{\mathbf{x}}$ is that for each $\mathbf{z} \in Z^1(\tilde{\mathbf{x}})$, there exists a function σ defined on $[0, \varepsilon]$ for some $\varepsilon > 0$ and taking values in R^n such that

(1) $\sigma(0) = \tilde{\mathbf{x}}$
(2) $\sigma(\theta)$ is feasible for all $\theta \in [0, \varepsilon]$
(3) $\sigma(\theta)$ is differentiable at $\theta = 0$

and moreover,

$$\frac{d\sigma(0)}{d\theta} = \lambda \mathbf{z} \qquad \text{for some } \lambda > 0$$

THEOREM 5.9. Suppose the objective function f in (5.16) is continuously differentiable. If $\tilde{\mathbf{x}}$ is a local minimum of (5.16) and if the Kuhn-Tucker constraint qualification holds at $\tilde{\mathbf{x}}$, then $Z^1(\tilde{\mathbf{x}}) \cap Z^2(\tilde{\mathbf{x}})$ is empty implying the Kuhn-Tucker conditions hold for some $\tilde{\mathbf{u}} \geq \mathbf{0}$ at $\tilde{\mathbf{x}}$.

PROOF. We show that if $Z^1(\tilde{\mathbf{x}}) \cap Z^2(\tilde{\mathbf{x}})$ is not empty, then $\tilde{\mathbf{x}}$ is not a local minimum. Let $\mathbf{z}$ be any point in $Z^1(\tilde{\mathbf{x}}) \cap Z^2(\tilde{\mathbf{x}})$ and consider the function σ corresponding to $\mathbf{z}$ in the constraint qualification. For $\theta \in [0, \varepsilon]$, $\sigma(\theta)$ is a feasible solution to (5.16) by property (2) and we can show that for $\theta = \tilde{\theta}$ sufficiently small, $f[\sigma(\tilde{\theta})] < f(\tilde{\mathbf{x}})$ implying $\tilde{\mathbf{x}}$ is not a local minimum. This is accomplished by considering the expansion

$$f\left[\sigma(\tilde{\theta})\right] = f(\tilde{\mathbf{x}}) + \left[\sigma(\tilde{\theta}) - \tilde{\mathbf{x}}\right]^T \nabla f\left\{\tilde{\mathbf{x}} + \xi(\tilde{\theta})\left[\sigma(\tilde{\theta}) - \tilde{\mathbf{x}}\right]\right\} \tag{5.17}$$

where $0 \leq \xi(\tilde{\theta}) \leq 1$.

Properties 1 and 3 of the function σ can be combined to give us

$$\lim_{\theta \to 0^+} \frac{\sigma(\theta) - \tilde{\mathbf{x}}}{\theta} = \lambda \mathbf{z} \qquad \text{for some } \lambda > 0$$

This limit, plus the continuity of ∇f, implies for $\theta = \tilde{\theta}$ sufficiently small and $\delta = \lambda \mathbf{z}^T \nabla f(\tilde{\mathbf{x}})/2 > 0$ that

$$\left| \frac{(\sigma(\theta) - \tilde{\mathbf{x}})^T}{\tilde{\theta}} \nabla f\left\{\tilde{\mathbf{x}} + \xi(\tilde{\theta})\left[\sigma(\tilde{\theta}) - \tilde{\mathbf{x}}\right]\right\} - \lambda \mathbf{z}^T \nabla f(\tilde{\mathbf{x}}) \right| < \delta$$

Thus, we have from (5.17) that

$$\begin{aligned} f\left[\sigma(\tilde{\theta})\right] &\leq f(\tilde{\mathbf{x}}) + \tilde{\theta}\left[\lambda \mathbf{z}^T \nabla f(\tilde{\mathbf{x}}) + \delta\right] \\ &< f(\tilde{\mathbf{x}}) \end{aligned}$$

where the last inequality follows from the definition of δ. This establishes that $\tilde{\mathbf{x}}$ is not a local minimum if $Z^1(\tilde{\mathbf{x}}) \cap Z^2(\tilde{\mathbf{x}})$ is not empty, or by contraposition, $Z^1(\tilde{\mathbf{x}}) \cap Z^2(\tilde{\mathbf{x}})$ is empty if $\tilde{\mathbf{x}}$ is a local minimum. Thus, we have by Theorem 5.8 that the Kuhn-Tucker conditions hold for some $\tilde{\mathbf{u}} \geq \mathbf{0}$ at $\tilde{\mathbf{x}}$. ■

COROLLARY 5.1. Suppose the objective function f in (5.16) is continuously differentiable. Suppose further that the constraint functions g_i are linear; that is, $g(\mathbf{x}) = \mathbf{A}\mathbf{x} - \mathbf{b}$ for some matrix $\mathbf{A}$ of rank m. If $\tilde{\mathbf{x}}$ is a local minimum, then the Kuhn-Tucker conditions hold for some $\tilde{\mathbf{u}} \geq \mathbf{0}$ at $\tilde{\mathbf{x}}$.

PROOF. The proof is left as an exercise for the reader. ■

The Kuhn-Tucker constraint qualification is clearly not verifiable for mathematical programming problems of the general form (5.16). Other constraint qualifications have been proposed, but none permit easier verification for the general problem (5.16). Exercise 5.18 discusses more recent efforts at mathematical characterizations of constraint qualifications.

The difficulties with the constraint qualification should not make us lose sight of the fact that the Kuhn-Tucker conditions are necessary conditions that can be satisfied by feasible solutions to (5.16) other than local minima. This point is illustrated in the following example.

EXAMPLE 5.5. Consider the problem

$$
\begin{aligned}
\min f(x_1, x_2) &= x_1 \\
\text{s.t. } g_1(x_1, x_2) &= (x_1 - 1)^2 + (x_2 + 2)^2 - 16 \leq 0 \\
g_2(x_1, x_2) &= -x_1^2 \quad - x_2^2 \qquad + 13 \leq 0
\end{aligned}
$$

This problem is depicted in Figure 5.5 where the shaded area is the nonconvex feasible region. The Kuhn-Tucker conditions hold for the three points indicated in the figure. The point $\mathbf{x}^1$ is the global minimum and the point $\mathbf{x}^2$ is a local minimum. The point $\mathbf{x}^3$ is neither a local minimum nor maximum, but simply one at which the Kuhn-Tucker conditions hold. ▲

We complete our discussion of the Kuhn-Tucker conditions and the constraint qualification by considering (5.16) again in the special case when f and the g_i are convex as well as differentiable. We have already observed that the Kuhn-Tucker conditions are sufficient for global optimality in this case. We wish to show that they are necessary as well under the constraint qualification (5.13) that ensured convergence of the generalized linear programming algorithm. This double use of the constraint qualification (5.13) is more than a coincidence because the generalized linear programming algorithm could be used to prove the following theorem (see Exercise 5.17).

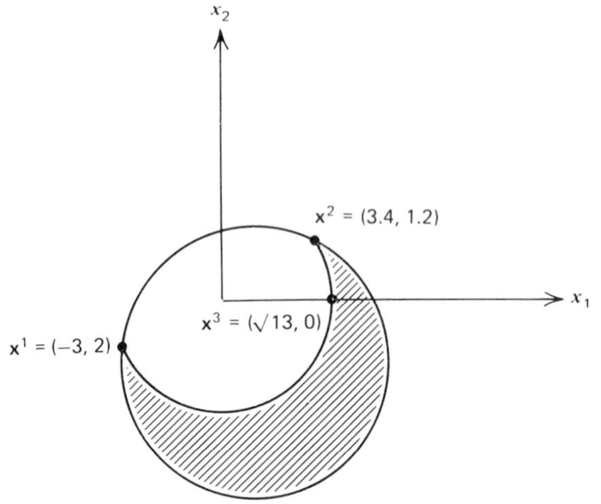

Figure 5.5. Kuhn-Tucker solutions for a nonconvex problem.

THEOREM 5.10. Suppose f, g_i are convex in (5.16) and suppose further that there exists an $\mathbf{x}^1 \in R^n$ such that

$$g_i(\mathbf{x}^1) < 0 \qquad i = 1, \ldots, m$$

Let $\bar{\mathbf{x}}$ denote any optimal solution to (5.16). Then there exists a $\bar{\mathbf{u}} \geq \mathbf{0}$ such that the Kuhn-Tucker optimality conditions hold.

PROOF. We prove this result using Theorem 5.8 and the approach used in Theorem 5.9. Specifically, we show that if there exists a $\mathbf{z} \in Z^1(\bar{\mathbf{x}}) \cap Z^2(\bar{\mathbf{x}})$, then $\bar{\mathbf{x}}$ is not minimal. By contraposition, if $\bar{\mathbf{x}}$ is minimal, then $Z^1(\bar{\mathbf{x}}) \cap Z^2(\bar{\mathbf{x}})$ is empty implying by Theorem 5.8 that the Kuhn-Tucker conditions hold for $\bar{\mathbf{x}}$ and some $\bar{\mathbf{u}} \geq \mathbf{0}$.

Thus, suppose we have a vector $\mathbf{z} \in Z^1(\bar{\mathbf{x}}) \cap Z^2(\bar{\mathbf{x}})$. For appropriate values of $\theta > 0$ and $\lambda > 0$, we will show that the vector

$$y = \bar{\mathbf{x}} + \theta \mathbf{z} + \lambda(\mathbf{x}^1 - \bar{\mathbf{x}} - \theta \mathbf{z})$$

satisfies $g_i(\mathbf{y}) \leq 0$ for $i = 1, \ldots, m$, and $f(\mathbf{y}) < f(\bar{\mathbf{x}})$. Note first that since $\mathbf{z}^T \nabla f(\bar{\mathbf{x}}) < 0$, we have for small values of θ that

$$f(\bar{\mathbf{x}} + \theta \mathbf{z}) < f(\bar{\mathbf{x}}) \tag{5.18}$$

because $f(\bar{\mathbf{x}} + \theta \mathbf{z}) = f(\bar{\mathbf{x}}) + \theta \mathbf{z}^T \nabla f[\bar{\mathbf{x}} + \xi(\theta)\theta \mathbf{z}]$, where $0 \leq \xi(\theta) \leq 1$, and ∇f is continuous because f is convex and differentiable on R^n (see Theorem A.16 in Appendix A). We can select small values $\theta \geq 0$ and $\lambda \geq 0$ such that for all $\theta \in (0, \theta_1]$ and all $\lambda \in (0, \lambda_1]$,

$$f\left[\bar{\mathbf{x}} + \theta \mathbf{z} + \lambda(\mathbf{x}^1 - \bar{\mathbf{x}} - \theta \mathbf{z})\right] \leq f(\bar{\mathbf{x}} + \theta \mathbf{z}) + \lambda\{f(\mathbf{x}^1) - f(\bar{\mathbf{x}} + \theta \mathbf{z})\} < f(\bar{\mathbf{x}})$$

The first inequality follows from the convexity of f and the second inequality from (5.18) and the smallness of θ_1 and λ_1.

Select $\lambda = \tilde{\lambda} \in (0, \lambda_1]$ and consider any g_i. Since $g_i(\bar{\mathbf{x}}) \leq 0$, g_i is continuous and $g_i(\mathbf{x}^1) < 0$, we can select $\tilde{\theta}_i \in (0, \theta_1]$ such that for all $\theta \in (0, \tilde{\theta}_i]$,

$$g_i(\bar{\mathbf{x}} + \theta \mathbf{z}) \leq -\frac{\tilde{\lambda} g_i(\mathbf{x}^1)}{1 - \tilde{\lambda}}$$

Thus, we have for all $\theta \in (0, \tilde{\theta}_i]$ that

$$g_i\left[\bar{\mathbf{x}} + \theta\mathbf{z} + \tilde{\lambda}(\mathbf{x}^1 - \bar{\mathbf{x}} - \theta\mathbf{z})\right] \leq g_i(\bar{\mathbf{x}} + \theta\mathbf{z}) + \tilde{\lambda}\left[\, g_i(\mathbf{x}^1) - g_i(\bar{\mathbf{x}} + \theta\mathbf{z})\right] \leq 0$$

where the first inequality follows from the convexity of g_i and the second from our selection of $\tilde{\theta}_i$. Let $\tilde{\theta} = \min\{\tilde{\theta}_i | i = 1, \ldots, m\}$; we have by construction that for any $\theta \in (0, \tilde{\theta}]$, the vector $\mathbf{y} = \bar{\mathbf{x}} + \theta\mathbf{z} + \tilde{\lambda}(\mathbf{x}^1 - \bar{\mathbf{x}} - \theta\mathbf{z})$ is feasible and moreover $f(\mathbf{y}) < f(\bar{\mathbf{x}})$. ■

5.6 EXERCISES

5.1 Construct an example of a primal problem (5.1) for which v is finite, $v = d$, but there is no optimal solution to the dual (5.4); that is, the supremum value d of $L(\mathbf{u})$ is not attained.

5.2 Consider the function

$$\begin{aligned} v(\mathbf{y}) = \min\, & f(\mathbf{x}) \\ \text{s.t. } & g(\mathbf{x}) \leq \mathbf{y} \\ & \mathbf{x} \in X \end{aligned}$$

Prove that if $v(\bar{\mathbf{y}}) = -\infty$ for some $\bar{\mathbf{y}}$, then $L(\mathbf{u}) = \min_{\mathbf{x} \in X}\{f(\mathbf{x}) + \mathbf{u}g(\mathbf{x})\}$ equals $-\infty$ for all $\mathbf{u} \geq \mathbf{0}$.

5.3 Give the Kuhn-Tucker conditions for the problem

$$\begin{aligned} & \min f(\mathbf{x}) \\ & \text{s.t. } g_i(\mathbf{x}) \leq \mathbf{0} \qquad i = 1, \ldots, m_1 \\ & \qquad h_k(\mathbf{x}) = \mathbf{0} \qquad k = 1, \ldots, m_2 \\ & \qquad \mathbf{x} \geq \mathbf{0} \end{aligned}$$

where f, g_i, and h_k are differentiable.

5.4 State the Kuhn-Tucker conditions for the problem in Example 5.5 and find all the primal-dual solution pairs satisfying these conditions. Construct and solve the dual (5.3) for the same problem. Compare your results.

5.5 (Nemhauser and Ullman, 1968) Consider the zero-one integer programming problem

$$v = \min \mathbf{cx}$$
$$\text{s.t. } \mathbf{Ax} \geq \mathbf{b}$$
$$x_j = 0 \text{ or } 1$$

where $\mathbf{A}$ is $m \times n$. For $\mathbf{u} \geq \mathbf{0}$, define the Lagrangean

$$L(\mathbf{u}) = \mathbf{ub} + \min_{x_j = 0 \text{ or } 1} (\mathbf{c} - \mathbf{uA})\mathbf{x}$$

As in numerical Example 5.2, the minimum is calculated by setting $x_j = 0$ if $c_j + \mathbf{ua}_j \geq 0$ and $x_j = 1$ if $c_j + \mathbf{ua}_j \leq 0$. Suppose $\bar{\mathbf{u}}$ is optimal in the dual problem of maximizing L over non-negative $\mathbf{u}$. Show that $\bar{\mathbf{u}}$ is optimal in the dual to the linear programming relaxation of this problem, which results when $x_j = 0$ or 1 is replaced by $0 \leq x_j \leq 1$.

5.6 Consider the quadratic programming problem

$$v = \min \tfrac{1}{2}\mathbf{xQx} - \mathbf{cx}$$
$$\text{s.t. } \mathbf{Ax} \leq \mathbf{b}$$

where $\mathbf{Q}$ is symmetric and positive semidefinite. Assume the problem has an optimal solution. Show that the dual to this problem is

$$d = \max -\mathbf{ub} - \tfrac{1}{2}\mathbf{xQx}$$
$$\text{s.t. } \mathbf{xQ} + \mathbf{uA} - \mathbf{c} = \mathbf{0}$$
$$\mathbf{u} \geq \mathbf{0}$$

in the sense that if $\bar{\mathbf{x}}$ is optimal in the primal, then there exists a $\bar{\mathbf{u}}$ such that $(\bar{\mathbf{x}}, \bar{\mathbf{u}})$ is optimal in the dual and moreover, $d = v$.

5.7 Prove that the set $[f,g]^c$ defined in Section 5.3 is a closed set.

5.8 Consider the arbitrary mathematical programming problem

$$v = \sup f(\mathbf{x})$$
$$\text{s.t. } g(\mathbf{x}) \leq \mathbf{0}$$
$$\mathbf{x} \in X$$

For $\mathbf{u} \geq \mathbf{0}$ define the Lagrangean

$$L(\mathbf{u}) = \inf_{\mathbf{x} \in X} \{f(\mathbf{x}) + \mathbf{u}g(\mathbf{x})\}$$

and define the dual problem in the usual way as

$$d = \sup L(\mathbf{u})$$
$$\text{s.t. } \mathbf{u} \geq \mathbf{0}$$

Define $v^c(\xi)=\inf\{\eta|(\eta,\xi)\in\text{cl}([f,g]^c)\}$, where cl( ) denotes closure. Prove Theorem 5.4 in this more general case under the additional assumption that at least one of the numbers v and d is finite.

5.9 Consider the function h from R^n to R^1 defined by

$$h(\mathbf{w})=\inf_{k\in K}\{a^k+b^k\mathbf{w}\}$$

where $a^k\in R^1$, $\mathbf{b}^k\in R^n$, and K is an arbitrary index set. Show that h is concave.

5.10 (Magnanti et al., 1976) A phase one Master LP for the generalized linear programming algorithm is

$$\begin{aligned} \sigma^K &= \min \sigma \\ \text{s.t.} \quad & \sum_{k=1}^{K} g_i(\mathbf{x}^k)\lambda_k - \sigma \le 0 \qquad i=1,\ldots,m \\ & \sum_{k=1}^{K} \lambda_k = 1 \\ & \lambda_k \ge 0 \qquad \sigma \ge 0 \end{aligned}$$

where the g_i are continuous functions and the $\mathbf{x}^k$ are taken from the compact set X. Let $-\mathbf{u}^K\ge\mathbf{0}$ denote the optimal dual solution on the first m rows of this problem. For $\mathbf{u}\ge\mathbf{0}$, define

$$P(\mathbf{u})=\min_{\mathbf{x}\in X}\mathbf{u}g(\mathbf{x})$$

and the phase one dual

$$\begin{aligned} \sigma &= \sup P(\mathbf{u}) \\ \text{s.t.} \quad & \sum_{i=1}^{m} u_i \le 1 \\ & \mathbf{u}\ge\mathbf{0} \end{aligned}$$

(1) Prove:

(i) $\sigma^K\ge\sigma$

(ii) if $P(\mathbf{u}^K)\ge\sigma^K$, then $P(\mathbf{u}^K)=\sigma=\sigma^K$,

(iii) any limit point $\mathbf{u}^*$ of the sequence $\{\mathbf{u}^K\}_{K=1}^{\infty}$ is optimal in the phase one dual,

(iv) $\lim\sigma^K=\sigma=P(\mathbf{u}^*)$ where $\mathbf{u}^*$ is any limit from (iii).

(2) Prove that exactly one of the following alternatives is valid:

(i) There is a $\mathbf{u}\ge\mathbf{0}$ such that $\min_{\mathbf{x}\in X}[\mathbf{u}g(\mathbf{x})]>0$

(ii) The vector $\mathbf{0}$ is in the convex hull of G, where

$$G=\bigcup_{\mathbf{x}\in X}\{\boldsymbol{\eta}\in R^m;\ \boldsymbol{\eta}\ge g(\mathbf{x})\}$$

5.11 A population is divided into m subgroups (called strata). A sample of n_i items for $i=1,2,\ldots,m$, is to be selected from each strata in order to estimate the mean μ of the total population. The estimator is

$$\hat{\mu}=\sum_{i=1}^{m} w_i \bar{x}_i$$

where $\bar{x}_i$ is the mean of the items sampled from the ith stratum and w_i is the fraction of the total number of items in the population that are in the ith stratum (i.e., the "weight" of the ith stratum). The variance of this estimator is

$$\sigma_{\hat{\mu}}^2=\sum_{i=1}^{m} w_i^2 \frac{\sigma_i^2}{n_i}$$

where σ_i^2 is the variance of the items in the ith stratum and n_i (as above) is the sample size in the ith stratum.

Suppose that there are C dollars available to make the survey and that it costs c_i dollars for each item sampled in the ith stratum. Use the Kuhn-Tucker conditions to show that the sample size that minimizes the variance is to choose

$$n_i=\left(\sum_{i=1}^{m} n_i\right)\frac{\dfrac{w_i\sigma_i}{\sqrt{c_i}}}{\displaystyle\sum_{i=1}^{m}\frac{w_i\sigma_i}{\sqrt{c_i}}}$$

and the optimal total sample size is

$$\sum_{i=1}^{m} n_i=C\frac{\displaystyle\sum_{i=1}^{m}\frac{w_i\sigma_i}{\sqrt{c_i}}}{\displaystyle\sum_{i=1}^{m} w_i\sigma_i\sqrt{c_i}}$$

5.12 (1) Use the Kuhn-Tucker conditions to solve the optimization problem

$$\max \prod_{j=1}^{m} x_j$$
$$\text{s.t.} \sum_{j=1}^{m} x_j=k$$

(2) Use (1) to prove the inequality: Given $a_j \geq 0$ for $j=1,\ldots,m$

$$\frac{\sum_{j=1}^{m} a_j}{m} \geq \left(\prod_{j=1}^{m} a_j\right)^{1/m}$$

5.13 Use the generalized linear programming algorithm to solve the dual (5.3) for the problem of Example 5.2.

5.14 The *saddlepoint problem* for problem (5.1) is to find $\bar{\mathbf{x}} \in X \subseteq R^n$, $\bar{\mathbf{u}} \geq \mathbf{0}$ such that

$$L(\bar{\mathbf{x}},\mathbf{u}) \leq L(\bar{\mathbf{x}},\bar{\mathbf{u}}) \leq L(\mathbf{x},\bar{\mathbf{u}})$$

for all $\mathbf{x} \in X$, $\mathbf{u} \geq \mathbf{0}$. Such a pair $(\bar{\mathbf{x}},\bar{\mathbf{u}})$ is called a *saddlepoint solution*.

(1) Prove that the pair $(\bar{\mathbf{x}},\bar{\mathbf{u}})$ is a saddlepoint solution if $(\bar{\mathbf{x}},\bar{\mathbf{u}})$ satisfies the global optimality conditions.
(2) Prove that the vector $\bar{\mathbf{u}} \geq \mathbf{0}$ is an optimal solution to the dual problem (5.4) if $(\bar{\mathbf{x}},\bar{\mathbf{u}})$ is a saddlepoint solution.
(3) Construct a mathematical programming problem for which all saddlepoint solutions do not satisfy the global optimality conditions.

5.15 Prove Corollary 5.1 by showing that the Kuhn-Tucker constraint qualification holds at each feasible solution $\mathbf{x}$.

5.16 (Fritz John, 1948) Suppose the functions f and the g_i in (5.16) are continuously differentiable. Prove that, if $\bar{\mathbf{x}}$ is a feasible solution, then there exists $\bar{u}_i$ for $i=0,1,2,\ldots,m$, such that

$$\bar{u}_0 \nabla f(\bar{\mathbf{x}}) + \sum_{i=1}^{m} \bar{u}_i \nabla g_i(\bar{\mathbf{x}}) = \mathbf{0}$$

$$\bar{u}_i g_i(\bar{\mathbf{x}}) = 0 \qquad i=1,\ldots,m$$

HINT. Use the theorem of the alternative (Exercise 2.8) and the fact that $\mathbf{x}$ cannot be minimal if there exists a $\mathbf{z}$ with the property that $\mathbf{z}^T \nabla g_i(\bar{\mathbf{x}}) < 0$ for all $i \in I(\bar{\mathbf{x}})$ and $\mathbf{z}^T \nabla f(\bar{\mathbf{x}}) < 0$.

5.17 Prove Theorem 5.10 by applying the generalized linear programming algorithm. Assume

$$\nabla f(\mathbf{x}) + \mathbf{u} \nabla g(\mathbf{x}) = \mathbf{0}$$

has a solution for all $\mathbf{u} \geq \mathbf{0}$.

5.18 (Avriel (1976))

(1) Consider an arbitrary set $A \subset R^n$ and a point $\mathbf{x} \in A$. Let $S(A,\mathbf{x})$ denote the intersection of all closed cones containing the set $\{\mathbf{a}-\mathbf{x}|\mathbf{a} \in A\}$. The *closed cone of tangents* of the set A at $\mathbf{x}$, denoted by $S(A,\mathbf{x})$, is given by

$$S(A,\mathbf{x}) = \bigcap_{k=1}^{\infty} \tilde{S}(A \cap N_{1/k}(\mathbf{x}), \mathbf{x})$$

where $N_{1/k}(\mathbf{x})$ is a spherical neighborhood of $\mathbf{x}$. Prove that a vector $\mathbf{z}$ is contained in $S(A,\mathbf{x})$ if and only if there exists a sequence of vectors $\{\mathbf{x}^k\} \subset A$ converging to $\mathbf{x}$ and a sequence of nonnegative numbers $\{\alpha^k\}$ such that the sequence $\{\alpha^k(\mathbf{x}^k - \mathbf{x})\}$ converges to $\mathbf{z}$.

(2) The *positively normal cone* to a set $A \subset R^n$, denoted by A', is the set consisting of all $\mathbf{x} \in R^n$ such that $\mathbf{x}^T\mathbf{y} \geq 0$ for all $\mathbf{y} \in A$. Prove that given two sets $A_1 \subset R^n$ and $A_2 \subset R^n$, then $A_1 \subset A_2$ implies $A_2' \subset A_1'$.

(3) Suppose $\bar{\mathbf{x}}$ is a feasible solution to (5.16). Prove that the set $Z^1(\bar{\mathbf{x}}) \cap Z^2(\bar{\mathbf{x}})$ is empty if and only if $\nabla f(\mathbf{x}) \in [Z^1(\bar{\mathbf{x}})]'$.

(4) Suppose that $\bar{\mathbf{x}}$ is a local minimum of problem (5.16). Prove that

$$\nabla f(\bar{\mathbf{x}}) \in [\, S(F, \bar{\mathbf{x}})\,]'$$

where

$$F = \{\mathbf{x} |\, g_i(\mathbf{x}) \leq 0 \quad \text{for } 1 = j, \ldots, m\}$$

(5) Prove that, if we let $\bar{\mathbf{x}}$ be a local minimum to problem (5.16) and supposing $[Z^1(\bar{\mathbf{x}})] = [S(F, \bar{\mathbf{x}})]'$, then the Kuhn-Tucker conditions hold at $\bar{\mathbf{x}}$ for some $\bar{\mathbf{u}} \geq \mathbf{0}$.

HINT. Use parts (3) and (4) and Theorem 5.8.

5.19 Suppose that $\bar{x}_{ij}$ for $i = 1, \ldots, m$; $j = 1, \ldots, n$, are optimal in the problem

$$\max \sum_{i=1}^{m} \lambda_i \sum_{j=1}^{n} p_{ij} U_i(x_{ij})$$

$$\text{s.t.} \sum_{i=1}^{m} x_{ij} \leq c_j \qquad j = 1, \ldots, m$$

where

$$\lambda_i > 0, p_{ij} > 0, U_i' > 0, U_i'' \text{ exists}, U_i'' < 0 \text{ and}$$

$$\frac{U_i'(x)}{U_i''(x)} = ax + b_i$$

(1) State the Kuhn-Tucker conditions for this problem.

(2) Show that the $\bar{x}_{ij}$ are linear functions of the c_j.

5.20 Consider the problem

$$v(b) = \max \sum_{j=1}^{n} a_j \log(x_j + c_j)$$
$$\sum_{j=1}^{n} x_j \leq b$$
$$x_j \leq 0$$

where $a_j > 0$, $c_j > 0$, $b > 0$.

(1) Show that the objective function is concave.

(2) Use the Kuhn-Tucker conditions to find optimal values of $x_1, \ldots, x_n$ and give $v(b)$ as a function of the a_j and the c_j. You may assume the variables are indexed in decreasing order of the a_j / c_j.

5.21 (Hurter and Wendell, 1972) We wish to locate a plant in R^2 so as to minimize the costs of supply and transportation of variable inputs q_i from m supply locations $\mathbf{s}_i \in R^2$ for $i = 1, \ldots, m$, subject to a constraint that the plant meet a specified level of output $q^0 \in R^1$. We let $\mathbf{q} \in R^m$ denote the vector of inputs. An arbitrary vector of inputs $\mathbf{q}$ is converted to the scalar output quantity $g(\mathbf{q})$. A mathematical statement of the location and supply problem is

$$\min \sum_{i=1}^{m} \{t_i \rho_i(\mathbf{x}) + p_i\} q_i$$
$$\text{s.t. } g(\mathbf{q}) = q^0$$
$$\mathbf{x} \in R^2 \qquad \mathbf{q} \in R^m$$

where t_i is a unit transportation cost from location i, $\rho_i(\mathbf{x}) = \|\mathbf{x} - \mathbf{s}_i\|$ is the Euclidean distance from $\mathbf{x}$ to $\mathbf{s}_i$ and p_i is the unit cost of the input from location q_i.

(1) Suppose the production function g has the *Cobb-Douglas* form

$$g(\mathbf{q}) = \exp\left[a_0 + \sum_{i=1}^{m} a_i \ln q_i\right]$$

where $a_i > 0$ for $i = 1, \ldots, m$. Show that the location and supply problem can be solved by solving the pure location problem

$$\min \sum_{i=1}^{m} \omega_i(\mathbf{x})$$
$$\text{s.t. } \mathbf{x} \in R^2$$

where

$$\omega_i(\mathbf{x}) = a_i \ln\left[\rho_i(\mathbf{x}) + \frac{p_i}{t_i}\right]$$

Are the functions ω_i convex?

(2) Suppose the production function g has the *constant elasticity of substitution* form

$$g(q) = \sum_{i=1}^{m} (b_i q_i^{-c})^{-1/c}$$

where $b_i > 0$ for $i = 1, \ldots, m$, and $c \neq 0$ satisfies $c > -1$. Show that the location and supply problem can be solved by solving

$$\min \sum_{i=1}^{m} \sigma_i(\mathbf{x})$$
$$\text{s.t.} \quad \mathbf{x} \in R^2$$

where

$$\sigma_i(\mathbf{x}) = c_i \left[\rho_i(\mathbf{x}) + \frac{p_i}{t_i}\right]^{c/(1+c)}$$

and

$$c_i = b_i^{1/(c+1)} t_i^{c/(c+1)}$$

Are the functions σ_i convex?

5.7 NOTES

SECTION 5.1. The orientation of this chapter is different from the usual treatment by most authors of duality in mathematical programming. A dominant theme since Kuhn and Tucker's pioneering work in 1951 has been the Kuhn-Tucker conditions for optimality involving dual variables. A great deal of analysis has been given to the constraint qualifications required to make the conditions necessary for problems with differentiable functions and also sufficient for convex programming problems. We have taken the point of view that the global optimality conditions are the fundamental constructive tool in the analysis of mathematical programming problems. In this regard, Everett (1963) was perhaps the first to take a constructive rather than existential approach to Lagrange multipliers. Geoffrion (1970) took this same point of view with respect to convex programming, but much of the theory is applicable to arbitrary mathematical programming problems. Rockafellar (1970) treats in depth many of the mathematical issues underlying duality theory. Luenberger (1969) addresses optimization in infinite dimensional spaces using dual concepts.

SECTION 5.2. Dual problems are discussed by Fisher et al. (1975), Geoffrion (1970), Varaiya (1972), and Whittle (1971). Rockafellar (1970) gives the mathematical properties of general convex and concave functions such as the Lagrangean L.

SECTION 5.3. The development in this section is taken from Magnanti et al. (1976). Whittle (1971) gives a similar result for the primal problem (5.1) with equality constraints. Geoffrion (1974) coined the term Lagrangean relaxation.

SECTION 5.4. Generalized linear programming is also known as Dantzig-Wolfe decomposition in part because it was first proposed for decomposing large-scale linear programming problems (Dantzig and Wolfe, 1961). The convergence proof of Theorem 5.5 can be found in a different form in Dantzig (1963). See Lasdon (1970) for additional discussion of the method applied to a variety of problems.

SECTION 5.5. The central role played by the sets $Z^1(\mathbf{x})$ and $Z^2(\mathbf{x})$ in the study of the Kuhn-Tucker conditions is due to Fiacco and McCormick (1968). We follow more closely Avriel (1976) who gives an extensive development of the conditions and various constraint qualifications. Avriel also gives a survey of the literature on these subjects.

6 NONDIFFERENTIABLE OPTIMIZATION AND LARGE-SCALE LINEAR PROGRAMMING

6.1 INTRODUCTION

Large-scale linear programming problems often possess special structures permitting them to be decomposed into a set of smaller problems each of which can be solved somewhat separately. These decomposition methods are intimately related to nondifferentiable optimization techniques and also to the dual methods discussed in Chapter 5. To illustrate these relationships, let us consider

$$\begin{aligned} v &= \min \mathbf{cx} \\ &\text{s.t. } \mathbf{Ax} - \mathbf{b} \leq \mathbf{0} \\ \mathbf{x} &\in X = \{\mathbf{x} | \mathbf{Tx} = \mathbf{q},\ \mathbf{x} \geq \mathbf{0}\} \end{aligned} \tag{6.1}$$

where $\mathbf{T}$ is the node-arc incidence matrix of a large network. The problem has been written with the constraints in this decomposed form to emphasize that it would be much easier to solve using one of the specialized network algorithms of Chapter 3 if the complicating constraints $\mathbf{Ax} - \mathbf{b} \leq \mathbf{0}$ were not present. We try to exploit the special structure of X and deal implicitly with the constraints $\mathbf{Ax} - \mathbf{b} \leq \mathbf{0}$, by solving the dual problem

$$\begin{aligned} \mathbf{d} &= \max L(\mathbf{u}) \\ &\text{s.t. } \mathbf{u} \geq \mathbf{0} \end{aligned}$$

where

$$L(\mathbf{u}) = -\mathbf{ub} + \min_{\mathbf{x} \in X} (\mathbf{c} + \mathbf{uA})\mathbf{x}$$

Note that since (6.1) is a linear programming problem, we will have $d = v$ and (6.1) can be optimized by first finding an optimal dual solution $\bar{\mathbf{u}}$ and then a complementary $\bar{\mathbf{x}} \in X$ satisfying the global optimality conditions.

The Lagrangean calculation is the solution of a linear programming problem, although a special one defined over a network, and we can expect for some $\mathbf{u} \geq \mathbf{0}$ that there will be alternative optima; that is, distinct $\mathbf{x}^1, \mathbf{x}^2 \in X$ satisfying

$$L(\mathbf{u}) = \mathbf{c}\mathbf{x}^1 + \mathbf{u}(\mathbf{A}\mathbf{x}^1 - \mathbf{b}) = \mathbf{c}\mathbf{x}^2 + \mathbf{u}(\mathbf{A}\mathbf{x}^2 - \mathbf{b})$$

If we try to differentiate L in the obvious way, we come to the conclusion that the gradient of L at $\mathbf{u}$ probably does not exist since $\mathbf{x}^1 \neq \mathbf{x}^2$ in all likelihood implies $\mathbf{A}\mathbf{x}^1 - \mathbf{b} \neq \mathbf{A}\mathbf{x}^2 - \mathbf{b}$. The nondifferentiability of L at those points $\mathbf{u}$ where there are alternative optima might appear unimportant because continuous, concave functions are differentiable almost everywhere; that is, $\nabla L(\mathbf{u})$ exists except on a set of u with measure zero. However, the methods we discuss in this chapter for computing an optimal dual solution naturally produce dual solutions for which there are alternative optima to the Lagrangean. Thus, we must be able to deal directly with the nondifferentiability of L.

The previous example illustrates how dual methods can be used to exploit special structures arising in linear programming problems, and how nondifferentiable Lagrangeans can result. In general, decomposition schemes depend on the structures and they may be effected by the construction of dual problems, or by the construction of other nondifferentiable concave and convex optimization problems. In later chapters, we approximate mathematical programming problems that are not linear and may not be large scale by large-scale linear programming problems. The same ideas and methods developed in this chapter are applicable to these induced large-scale linear programming problems, including the nonlinear differentiable convex programming problem in Chapter 7, and the integer programming and the traveling salesman problems in Chapter 8.

We have already seen in Chapter 5 that the generalized linear programming algorithm solves dual problems. The algorithm was originally proposed to decompose large-scale linear programming problems, but there are other generalizations of the simplex algorithm that can be used. The three decomposition algorithms to be discussed in this chapter are generalizations of the primal, dual, and primal-dual simplex algorithms. To be precise, the generalized linear programming algorithm is a generalization of the primal simplex algorithm because its subproblems generate columns for a master problem that is iteratively resolved by the primal simplex algorithm. Benders' algorithm, to be discussed for the first time in this chapter, generalizes the dual simplex algorithm because its subproblems generate rows for a master problem that is iteratively resolved by the dual simplex algorithm. These two algorithms can also be contrasted by their economic interpretation as decentralization mechanisms for decomposing large organizations: generalized linear programming is viewed as a method of price directive decomposition and Benders' algorithm as a resource directive decomposition.

The third decomposition algorithm, the generalized primal-dual algorithm, is an alternative to the other two, generalized linear programming and Benders', for both price and resource directive decomposition. However, our main purpose in presenting it is to show how the constructs of convex analysis given in Appendix A can be mechanized into an ascent algorithm for solving nondifferentiable optimization problems. We also present an approximate ascent algorithm, called the subgradient optimization algorithm, with interesting analytic properties that is essentially a heuristic simplification of the generalized primal-dual algorithm. We assume here that the reader is familiar with the properties of convex and concave nondifferentiable functions given in Appendix A.

6.2 NECESSARY AND SUFFICIENT OPTIMALITY CONDITIONS FOR NONDIFFERENTIABLE CONCAVE PROGRAMMING

The basic ideas of nondifferentiable optimization will be illustrated by considering the unconstrained optimization problem

$$\begin{aligned} w &= \max f(\mathbf{y}) \\ &\text{s.t. } \mathbf{y} \in R^m \end{aligned} \tag{6.2}$$

where f is a concave, real-valued (finite) function defined for all $\mathbf{y} \in R^m$. We assume w is finite. The concave function f is necessarily continuous on R^m, but as we have seen, it need not be differentiable everywhere. Once we have developed the necessary and sufficient optimality conditions for the unconstrained problem (6.2), we extend the results to linearly constrained nondifferentiable concave programming problems. Finally, note that unlike almost all of the mathematical programming problems discussed in this book, we have chosen to consider a maximization problem rather than a minimization problem. This is because the most common nondifferentiable optimization problem to be encountered in this and later chapters is the maximization problem (5.4) that is dual to the generic minimization problem (5.1).

Although a continuous, concave function may not have a gradient everywhere, it does have everywhere a generalization of the gradient. Recall that a vector $\bar{\gamma}$ is called a *subgradient* of f at $\bar{\mathbf{y}}$ if

$$f(\mathbf{y}) \leq f(\bar{\mathbf{y}}) + (\mathbf{y} - \bar{\mathbf{y}})\bar{\gamma} \qquad \text{for all } \mathbf{y}$$

If there is a unique subgradient of f at $\bar{\mathbf{y}}$, then the subgradient is the gradient.

LEMMA 6.1. Let $\bar{\mathbf{x}} \in X$ denote an optimal solution for the Lagrangean function (5.2) at $\bar{\mathbf{u}}$; that is,

$$L(\bar{\mathbf{u}}) = f(\bar{\mathbf{x}}) + \bar{\mathbf{u}}g(\bar{\mathbf{x}}) = \min_{\mathbf{x} \in X} \{ f(\mathbf{x}) + \bar{\mathbf{u}}g(\mathbf{x}) \}$$

The vector $\bar{\gamma} = g(\bar{\mathbf{x}})$ is a subgradient of L at $\bar{\mathbf{u}}$.

PROOF. For any $\mathbf{u}$, we have by the definition of L that

$$L(\mathbf{u}) \leq f(\bar{\mathbf{x}}) + \mathbf{u}g(\bar{\mathbf{x}}) = L(\bar{\mathbf{u}}) - \bar{\mathbf{u}}g(\bar{\mathbf{x}}) + \mathbf{u}g(\bar{\mathbf{x}})$$

or

$$L(\mathbf{u}) \leq L(\bar{\mathbf{u}}) + (\mathbf{u} - \bar{\mathbf{u}})\bar{\gamma}. \quad \blacksquare$$

LEMMA 6.2. Let $\bar{\gamma}$ be any subgradient of the concave function f at $\bar{\mathbf{y}}$. Then the set $\{\mathbf{y} | (\mathbf{y} - \bar{\mathbf{y}})\bar{\gamma} \geq 0\}$ contains all optimal solutions to the concave maximization problem (6.2).

PROOF. Since $\bar{\gamma}$ is a subgradient, we have for all $\mathbf{y}$ that

$$(\mathbf{y} - \bar{\mathbf{y}})\bar{\gamma} \geq f(\mathbf{y}) - f(\bar{\mathbf{y}})$$

If $\mathbf{y}$ is optimal in (6.2), then $f(\mathbf{y}) - f(\bar{\mathbf{y}}) \geq 0$ implying the desired result. $\blacksquare$

Lemma 6.1 says, in effect, that any subgradient $\bar{\gamma}$ of the concave function f at a point $\bar{\mathbf{y}}$ points into a half space containing all optimal solutions to problem (6.2). In this sense, nondifferentiable optimization is no different than differentiable optimization. The difficulty is that if the function f is not differentiable at $\bar{\mathbf{y}}$, then it may not actually increase in the direction $\bar{\gamma}$ although $\bar{\mathbf{y}}$ is not optimal. This situation is depicted in Figure 6.1 for the Lagrangean function of Example 6.1 discussed in Section 6.4. The Lagrangean does not increase in the direction of either of the subgradients $\bar{\gamma}^0$ and $\bar{\gamma}^1$ at the point $\bar{\mathbf{u}} = (2,2)$, but it does increase in any direction lying in the intersection of the two half spaces defined by $\bar{\gamma}^0$ and $\bar{\gamma}^1$. Note that if the intersection of the two half spaces defined by $\bar{\gamma}^0$ and $\bar{\gamma}^1$ were to consist of only the zero vector, then we could conclude that $\bar{\mathbf{u}}$ is optimal.

The implication of the preceding discussion is that the construction of an ascent algorithm for problem (6.2) depends upon knowledge of the set $\partial f(\bar{\mathbf{y}})$ of all subgradients of f at $\bar{\mathbf{y}}$. This set is called the *subdifferential* and it is a closed, convex subset of R^m. Moreover, $\partial f(\bar{\mathbf{y}})$ is nonempty for all $\bar{\mathbf{y}} \in R^m$ because of our assumption that f is finite everywhere. In Section 6.5, we show how $\partial f(\bar{\mathbf{y}})$ can be successively built up from its elements until either a direction of ascent is discovered or $\bar{\mathbf{y}}$ is proven optimal. For our purposes at the moment, however, we take $\partial f(\bar{\mathbf{y}})$ as given.

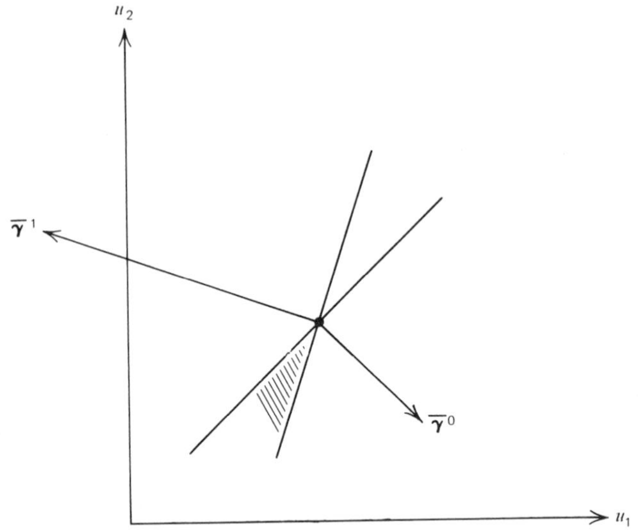

Figure 6.1. Ascent directions of a concave function at a nondifferentiable point.

The fundamental result from Appendix A that we use in our analysis and algorithm construction is that the directional derivative of f at any point $\bar{\mathbf{y}}$ in any direction $\mathbf{d}$ exists and is given by

$$\nabla f(\bar{\mathbf{y}};\mathbf{d}) = \min_{\gamma \in \partial f(\bar{\mathbf{y}})} \mathbf{d}\gamma \tag{6.3}$$

If a point $\bar{\mathbf{y}}$ is optimal in (6.2), we would expect $\nabla f(\bar{\mathbf{y}};\mathbf{d}) \leq 0$ in all directions $\mathbf{d}$; that is, the function decreases in all directions from the maximum. Since f is concave, we can make a stronger statement: The point $\bar{\mathbf{y}}$ is optimal in (6.2) if and only if $\nabla f(\bar{\mathbf{y}};\mathbf{d}) \leq 0$ for all directions $\mathbf{d}$. The following theorem is a restatement of this idea in constructive terms using (6.3) to characterize directional derivatives.

THEOREM 6.1. The solution $\bar{\mathbf{y}}$ is optimal in (6.2) if and only if

$$\begin{aligned} &\max_{\mathbf{d}} \ \underset{\gamma \in \partial f(\bar{\mathbf{y}})}{\text{minimum}} \ \mathbf{d}\gamma = 0 \\ &\text{s.t.} \ -1 \leq d_i \leq 1 \qquad i = 1,\ldots,m \end{aligned} \tag{6.4}$$

PROOF.
(1)

$$\max_{\mathbf{d}} \ \min_{\gamma \in \partial f(\bar{\mathbf{y}})} \mathbf{d}\gamma = 0 \Rightarrow \bar{\mathbf{y}} \text{ optimal}$$

Let $\mathbf{y}$ be any other solution in the maximization problem (6.2). Then

$\mathbf{y}-\bar{\mathbf{y}}=\theta\mathbf{d}$ for $\mathbf{d}$ satisfying the constraints of (6.4) and some $\theta>0$. For any subgradient $\bar{\gamma}\in\partial f(\bar{\mathbf{y}})$, we have

$$\begin{aligned} f(\mathbf{y}) &\leq f(\bar{\mathbf{y}})+(\mathbf{y}-\bar{\mathbf{y}})\bar{\gamma} \\ &= f(\bar{\mathbf{y}})+\theta\mathbf{d}\bar{\gamma} \\ &\leq f(\bar{\mathbf{y}}) \end{aligned}$$

where the last inequality follows because $\theta>0$ and $\mathbf{d}\bar{\gamma}\leq 0$ by hypothesis. Since $\mathbf{y}$ was an arbitrary solution, this permits us to conclude that $\bar{\mathbf{y}}$ is optimal.

(2)

$$\max_{\mathbf{d}} \min_{\gamma\in\partial f(\bar{\mathbf{y}})} \mathbf{d}\gamma>0 \Rightarrow \bar{\mathbf{y}} \text{ not optimal}$$

Note that the maximum in (6.4) must be greater than or equal to 0 since $\mathbf{d}=0$ is feasible in (6.4). By hypothesis, there exists according to (6.3) a direction $\mathbf{d}$ and a subgradient $\bar{\gamma}\in\partial f(\bar{\mathbf{y}})$ such that

$$\nabla f(\bar{\mathbf{y}};\mathbf{d})=\mathbf{d}\bar{\gamma}>0$$

By definition,

$$\nabla f(\bar{\mathbf{y}};\mathbf{d})=\lim_{\theta\to 0^+}\frac{f(\bar{\mathbf{y}}+\theta\mathbf{d})-f(\bar{\mathbf{y}})}{\theta}$$

which implies for some $\bar{\theta}>0$ that $f(\bar{\mathbf{y}}+\theta\mathbf{d})>f(\bar{\mathbf{y}})$ establishing the second part of the theorem. ∎

The following corollary gives another way of stating the necessary and sufficient optimality conditions of Theorem 6.1. The proof is left as an exercise for the reader (see Exercise 6.1).

COROLLARY 6.1. The solution $\bar{\mathbf{y}}$ is optimal in (6.2) if and only if $\mathbf{0}\in\partial f(\bar{\mathbf{y}})$.

These results can be readily extended to nondifferentiable optimization problems with linear constraints; namely

$$\begin{aligned} &\max f(\mathbf{y}) \\ &\text{s.t. } \mathbf{A}\mathbf{y}\leq\mathbf{b} \\ &\qquad \mathbf{y}\geq\mathbf{0} \end{aligned} \tag{6.5}$$

Consider a point $\bar{\mathbf{y}}$ feasible in (6.5). A direction $\mathbf{d}\neq 0$ is a feasible direction at $\bar{\mathbf{y}}$ if $\bar{\mathbf{y}}+\theta\mathbf{d}$ is feasible for some $\theta>0$. Feasible directions can be characterized by reference to the index sets of the binding constraints

$$K(\bar{\mathbf{y}})=\left\{k\,\middle|\,\sum_{i=1}^{m}a_{ki}\bar{y}_i=b_k\right\}$$

and (6.6)

$$I(\bar{\mathbf{y}})=\{i\,|\,\bar{y}_i=0\}$$

The feasible directions from $\bar{\mathbf{y}}$ are those $\mathbf{d} \in R^m$ satisfying $\sum_{i=1}^m a_{ki} d_i \leq 0$, $k \in K(\bar{\mathbf{y}})$, and $d_i \geq 0$, $i \in I(\bar{\mathbf{y}})$. The analogue to Theorem 6.1 that can be proved in much the same way is as follows: The solution $\bar{\mathbf{y}}$ is optimal in (6.5) if and only if

$$\begin{aligned} \max_{\mathbf{d}} \ \min_{\gamma \in \partial f(\bar{\mathbf{y}})} \ & \mathbf{d}\gamma = 0 \\ \text{s.t.} \ \sum_{i=1}^{m} a_{ki} d_i \leq 0 & \qquad k \in K(\bar{\mathbf{y}}) \\ 0 \leq d_i \leq 1 & \qquad i \in I(\bar{\mathbf{y}}) \\ -1 \leq d_i \leq 1 & \qquad i \in I^c(\bar{\mathbf{y}}) \end{aligned} \tag{6.7}$$

We conclude this section by relating the constructs and analyses discussed above to the dual problem (5.4), which we rewrite here for completeness:

$$\begin{aligned} d = \sup & \ L(\mathbf{u}) \\ \text{s.t.} & \ \mathbf{u} \geq \mathbf{0} \end{aligned}$$

where

$$L(\mathbf{u}) = \min_{\mathbf{x} \in X} \{ f(\mathbf{x}) + \mathbf{u}\mathbf{g}(\mathbf{x}) \}$$

and f, g are continuous functions and X is a compact set. For $\bar{\mathbf{u}} \geq \mathbf{0}$, let

$$X(\bar{\mathbf{u}}) = \{ \mathbf{x} \in X \mid L(\bar{\mathbf{u}}) = f(\mathbf{x}) + \bar{\mathbf{u}} g(\mathbf{x}) \}$$

denote the set of optimal solutions to the Lagrangean at $\bar{\mathbf{u}}$, and let

$$g[X(\bar{\mathbf{u}})] = \{ \gamma \in R^m | \gamma = g(\mathbf{x}) \qquad \text{for } \mathbf{x} \in X(\bar{\mathbf{u}}) \} \tag{6.8}$$

denote the image set of these solutions under the mapping g.

We have already seen in Lemma 6.1 that any $\gamma \in g[X(\bar{\mathbf{u}})]$ is a subgradient, or in other words, $g[X(\bar{\mathbf{u}})] \subseteq \partial L(\bar{\mathbf{u}})$, the subdifferential of L at $\bar{\mathbf{u}}$. It can be shown, moreover, that the convex hull of $g[X(\mathbf{u})]$ equals $\partial L(\mathbf{u})$ (see Exercise 6.2). These properties of the Lagrangean can be used to establish necessary and sufficient optimality conditions for the dual problem: The vector $\bar{\mathbf{u}} \geq \mathbf{0}$ is optimal in the dual problem if and only if

$$\begin{aligned} \max_{\mathbf{d}} \ \min_{\mathbf{x} \in X(\bar{\mathbf{u}})} \ & \mathbf{d}g(\mathbf{x}) = 0 \\ \text{s.t.} \ 0 \leq d_i \leq 1 & \qquad i \in I(\bar{\mathbf{u}}) \\ -1 \leq d_i \leq 1 & \qquad i \in I^c(\bar{\mathbf{u}}) \end{aligned} \tag{6.9}$$

where $I(\bar{\mathbf{u}}) = \{ i | \bar{u}_i = 0 \}$.

6.3 ASCENT ALGORITHMS

Theorem 6.1 in the previous section provides the fundamental procedure for testing a solution for optimality in a nondifferentiable concave programming problem. If the solution is not optimal, the procedure indicates a direction of ascent or increase of the concave objective function. In this section we give a formal definition of the general class of ascent algorithms. In the following two sections we show how the procedure of Theorem 6.1 is mechanized into specific ascent algorithms for nondifferentiable concave programming problems.

Consider the problem

$$v = \max f(\mathbf{y}) \qquad (6.10)$$
$$\mathbf{y} \in Y \subseteq R^m$$

where we assume that Y is convex and nonempty and that the directional derivative $\nabla f(\mathbf{y}; \mathbf{d})$ exists at all points $\mathbf{y}$ and in all directions $\mathbf{d}$, implying that f is continuous. Recall that continuous, nondifferentiable functions can possess directional derivatives even at a point where the gradient does not exist. An ascent algorithm for (6.10) generates a finite or infinite sequence $\{\mathbf{y}^l\}$ of solutions satisfying for all l

$$\mathbf{y}^l \in Y$$
$$\mathbf{y}^{l+1} = \mathbf{y}^l + \theta_l \mathbf{d}^l \qquad (6.11)$$

and

$$f(\mathbf{y}^{l+1}) > f(\mathbf{y}^l)$$

where $\mathbf{d}^l \in R^m$ ($\mathbf{d}^l \neq \mathbf{0}$) is a *feasible direction of ascent* from $\mathbf{y}^l$, and the scalar $\theta_l > 0$ is the *step length*. Formally, a direction $\mathbf{d}$ is a *feasible direction* at a feasible point $\mathbf{y}$ only if $\mathbf{y} + \theta\mathbf{d}$ is feasible for all $\theta \in [0, \varepsilon)$ for some $\varepsilon > 0$. A feasible direction $\mathbf{d}$ is a *feasible direction of ascent* at $\mathbf{y}$ if $\nabla f(\mathbf{y}; \mathbf{d}) > 0$.

A point $\mathbf{y}^* \in Y$ is called a *global maximum* in (6.10) if $f(\mathbf{y}^*) \geq f(\mathbf{y})$ for all $\mathbf{y} \in Y$. By necessity, we must have $\nabla f(\mathbf{y}^*; \mathbf{d}) \leq 0$ for all feasible directions $\mathbf{d}$ at a global maximum $\mathbf{y}^*$. Without additional assumptions about f, however, a point $\mathbf{y}$ satisfying $\nabla f(\mathbf{y}; \mathbf{d}) \leq 0$ for all feasible directions $\mathbf{d}$, called a *local maximum*, may not be a global maximum. The goal of any ascent algorithm is to find a local maximum. If an ascent algorithm produces a local maximum $\mathbf{y}^l$, then it terminates finitely with this $\mathbf{y}^l$. If none of the $\mathbf{y}^l$ in (6.11) are local maxima, then it is desired that the ascent algorithm generate one or more convergent subsequences $\{\mathbf{y}^{l_i}\}$ with limit points that are local maxima.

An iteration of an ascent algorithm at a point $\mathbf{y}$ consists of two procedures. The first is the *direction finding procedure* that determines a feasible direction of ascent $\mathbf{d}$ at $\mathbf{y}$, or ascertains that $\mathbf{y}$ is a local maximum. The second is the *step length procedure* that determines the scalar $\theta_l > 0$ thereby determining the next point $\mathbf{y}^{l+1} = \mathbf{y}^l + \theta_l \mathbf{d}^l$ with higher objective function value.

6.4 SUBGRADIENT OPTIMIZATION

In this section we present a very simple, approximately ascending algorithm for the unconstrained nondifferentiable concave programming problem (6.2) called the *subgradient optimization algorithm.* We also discuss how the method can be extended to problems with constraints. If certain assumptions are satisfied, this algorithm converges to an optimal solution to (6.2). The assumptions are of two types. First, there is the mathematical assumption that the norms of the subgradients encountered by the algorithm are uniformly bounded. This is a fairly weak assumption satisfied by all the linear programming decomposition problems we study in this chapter and later chapters. The other assumption is that we know the value w of the maximal objective function in (6.2) and use it in determining the step length of each iteration. This we cannot know, except in very special cases, and therefore the theoretical convergence of the algorithm cannot be guaranteed in practice. We present the algorithm because of its mathematical interest and also because it has worked well in practice and can be integrated with the generalized primal-dual ascent algorithm of the next section. Convergence of the generalized primal-dual ascent algorithm does not depend upon knowledge we cannot have.

The subgradient optimization algorithm generates the solutions $\{\mathbf{y}^l\}$ by the rule

$$\mathbf{y}^{l+1} = \mathbf{y}^l + \theta_l \gamma^l \qquad l = 0, 1, 2, \ldots \tag{6.12}$$

where γ^l is any subgradient of f at $\mathbf{y}^l$, and

$$\theta_l = \frac{\lambda_l [w - f(\mathbf{y}^l)]}{\|\gamma^l\|^2} \tag{6.13}$$

where $0 < \varepsilon_1 \leq \lambda_l \leq 2 - \varepsilon_2$ with $\varepsilon_2 > 0$. Note that the direction finding procedure of this algorithm is to select any subgradient and no effort is made to discover if f actually increases in the subgradient direction. Similarly, the step length θ_l is determined solely by the formula (6.13) and no further effort is made to try to achieve $f(\mathbf{y}^{l+1}) > f(\mathbf{y}^l)$. For these reasons, the subgradient optimization algorithm is only approximately ascending.

The following convergence theorem is the main result of subgradient optimization. We exclude the unlikely and trivial, but fortunate case when one of the $\mathbf{y}^l$ generated is found to be optimal in (6.2) because the corresponding subgradient $\gamma^l = 0$.

THEOREM 6.2. Consider the concave maximization problem (6.2) and suppose an optimal solution $\mathbf{y}^*$ exists. Suppose further that we apply the subgradient optimization algorithm with the additional assumption that there exists an $M > 0$ such that $\|\gamma\|^2 \le M$ for all $\gamma \in \partial f(\mathbf{y})$ and any $\mathbf{y}$ in the set $\{\mathbf{y} \mid \|\mathbf{y} - \mathbf{y}^*\| \le \|\mathbf{y}^0 - \mathbf{y}^*\|\}$. Then $\lim f(\mathbf{y}^l) = w$ and any limit point of the sequence $\{\mathbf{y}^l\}$ is an optimal solution.

PROOF. First we show that $\|\mathbf{y}^l - \mathbf{y}^*\|$ is monotonically decreasing. Starting with (6.12), we have

$$\begin{aligned} \|\mathbf{y}^{l+1} - \mathbf{y}^*\|^2 &= \|\mathbf{y}^l + \theta_l \gamma^l - \mathbf{y}^*\|^2 \\ &= \|\mathbf{y}^l - \mathbf{y}^*\|^2 + \theta_l^2 \|\gamma^l\|^2 + 2\theta_l (\mathbf{y}^l - \mathbf{y}^*)\gamma^l \\ &\le \|\mathbf{y}^l - \mathbf{y}^*\|^2 + \theta_l^2 \|\gamma^l\|^2 - 2\theta_l [f(\mathbf{y}^*) - f(\mathbf{y}^l)] \\ &= \|\mathbf{y}^l - \mathbf{y}^*\|^2 + \frac{(\lambda_l^2 - 2\lambda_l)[f(\mathbf{y}^*) - f(\mathbf{y}^l)]^2}{\|\gamma^l\|^2} \end{aligned} \tag{6.14}$$

where the inequality follows because γ^l is a subgradient, and the final equality is obtained from the formula (6.13). Letting

$$\Delta = \frac{[f(\mathbf{y}^*) - f(\mathbf{y}^l)]^2}{\|\gamma^l\|^2} \ge 0$$

we have

$$(\lambda_l^2 - 2\lambda_l)\Delta = \lambda_l(\lambda_l - 2)\Delta \le -\varepsilon_1 \varepsilon_2 \Delta \tag{6.15}$$

since $\lambda_l \ge \varepsilon_1$ and $\lambda_l \le 2 - \varepsilon_2$ from (6.13). Combining (6.14) and (6.15), we have

$$\|\mathbf{y}^{l+1} - \mathbf{y}^*\|^2 \le \|\mathbf{y}^l - \mathbf{y}^*\|^2 - \frac{\varepsilon_1 \varepsilon_2 [f(\mathbf{y}^*) - f(\mathbf{y}^l)]}{\|\gamma^l\|^2}. \tag{6.16}$$

The direct implication of (6.16) is that the sequence of nonnegative numbers $\|\mathbf{y}^l - \mathbf{y}^*\|^2$ is monotonically decreasing, which in turn implies $\lim_{l \to \infty} \|\mathbf{y}^l - \mathbf{y}^*\|^2$ exists. The existence of this limit implies from (6.16) that

$$\lim_{l \to \infty} \frac{[f(\mathbf{y}^*) - f(\mathbf{y}^l)]}{\|\gamma^l\|^2} = 0$$

because otherwise there would be an infinite subsequence of the

$\|\mathbf{y}^{l+1}-\mathbf{y}^*\|^2$ decreasing at each step by at least some $\varepsilon_3>0$ which is clearly impossible. Since $\|\gamma^l\|^2\leq M$ for all l, we can conclude that

$$\lim_{l\to\infty} f(\mathbf{y}^l)=f(\mathbf{y}^*)=w$$

Finally, the sequence $\{\mathbf{y}^l\}$ must have at least one converging subsequence because the $\mathbf{y}^l$ are restricted to the bounded set $\{y|\,\|\mathbf{y}^l-\mathbf{y}^*\|\leq\|\mathbf{y}^0-\mathbf{y}^*\|\}$. If $\{\mathbf{y}^{l_i}\}$ is such a subsequence converging to $\mathbf{y}^{**}$, we have since f is continuous that

$$\lim_{i\to\infty} f(\mathbf{y}^{l_i})=f\Big(\lim_{i\to\infty}\mathbf{y}^{l_i}\Big)=f(\mathbf{y}^{**})=f(\mathbf{y}^*)=w$$

and thus $\mathbf{y}^{**}$ is optimal in (6.2). ■

COROLLARY 6.2. Under the assumptions of Theorem 6.1, the subgradient optimization algorithm generates a sequence of points converging to an optimal solution.

PROOF. Suppose the sequence $\{\mathbf{y}^l\}$ has two limit points, say $\{\mathbf{y}^{l_i}\}$ converges to $\mathbf{z}^1$ and $\{\mathbf{y}^{l_j}\}$ converges to $\mathbf{z}^2$ and $\mathbf{z}^1\neq\mathbf{z}^2$. By Theorem 6.2, both $\mathbf{z}^1$ and $\mathbf{z}^2$ are optimal solutions implying from the proof of the theorem that

$$\lim_{l\to\infty}\|\mathbf{y}^l-\mathbf{z}^1\|=\lim_{j\to\infty}\|y^{l_j}-\mathbf{z}^1\|=\lim_{i\to\infty}\|\mathbf{y}^{l_i}-\mathbf{z}^1\|=p_1$$

and

$$\lim_{l\to\infty}\|\mathbf{y}^l-\mathbf{z}^2\|=\lim_{i\to\infty}\|\mathbf{y}^{l_i}-\mathbf{z}^2\|=\lim_{j\to\infty}\|\mathbf{y}^{l_j}-\mathbf{z}^2\|=p_2$$

Without loss of generality, we assume $p_1\geq p_2$. Then

$$\begin{aligned}\|\mathbf{y}^{l_2}-\mathbf{z}^1\|^2&=\|\mathbf{y}^{l_j}-\mathbf{z}^2+\mathbf{z}^2-\mathbf{z}^1\|^2\\&=\|\mathbf{y}^{l_j}-\mathbf{z}^2\|^2+\|\mathbf{z}^2-\mathbf{z}^1\|^2+2(\mathbf{y}^{l_j}-\mathbf{z}^2)(\mathbf{z}^2-\mathbf{z}^1)\end{aligned}$$

The rightmost term in the last equation goes to zero in the limit as $j\to\infty$ implying that

$$\lim_{j\to\infty}\|\mathbf{y}^{l_j}-\mathbf{z}^1\|^2=p_2^2+\|\mathbf{z}^2-\mathbf{z}^1\|^2>p_1^2$$

which is impossible. ■

An important special application of subgradient optimization is to the solution of systems of linear equations. In this context, the methods are known as *linear relaxation methods*. Consider the linear inequality system

$$\boldsymbol{\alpha}^k\mathbf{y}-b_k\geq 0 \qquad k=1,\ldots,K \tag{6.17}$$

where each $\boldsymbol{\alpha}^k\in R^m$. We wish to find a $\mathbf{y}$ satisfying (6.17). We could of course, use a phase one simplex procedure, but linear relaxation methods may be more efficient when the number of constraints in (6.17) is large and most of them are not binding.

Let $\mathbf{y}$ denote a trial solution to (6.17) and define the function

$$f(\mathbf{y}) = \min_{k=1,\ldots,K} \{\boldsymbol{\alpha}^k \mathbf{y} - b_k\}$$

The function f is concave on R^m because it is the minimum of a collection of linear functions (see Exercise 5.9). The idea is to max$f(\mathbf{y})$ for $\mathbf{y} \in R^m$ by the subgradient optimization algorithm. If the maximum value $w \geq 0$, this implies there is a $\bar{\mathbf{y}}$ satisfying $f(\bar{\mathbf{y}}) \geq 0$ and such a $\bar{\mathbf{y}}$ is a solution to (6.17) by the definition of f. The application of subgradient optimization to this problem is straightforward. Specifically, we generate the sequence $\{\mathbf{y}^l\}$ by the rules (6.12) and (6.13), which become in this context

$$\mathbf{y}^{l+1} = \mathbf{y}^l + \theta_l \boldsymbol{\alpha}^l$$

where

$$f(\mathbf{y}^l) = \boldsymbol{\alpha}^l \mathbf{y}^l + b_l \tag{6.18}$$

and

$$\theta_l = \frac{-\lambda_l f(\mathbf{y}^l)}{\|\boldsymbol{\alpha}^l\|^2}$$

where $0 < \varepsilon_1 \leq \lambda_l \leq 2 - \varepsilon_2$ with $\varepsilon_2 > 0$. If the system (6.17) does not have a solution, then the subgradient optimization algorithm will oscillate in the sense that the values $f(\mathbf{y}^l)$ will decrease as often as they increase. This is a drawback of the method because oscillations can also be the result of poor performance of the algorithm due to a poor starting solution $\mathbf{y}^0$ and/or poor choices of the λ_l.

The subgradient optimization algorithm can be extended to nondifferentiable problems with constraints such as (6.10) with f concave by projecting the vector $\mathbf{y}^{l+1}$ given by (6.12) onto the feasible region Y if it is infeasible (see Exercise 6.5). This can be difficult to accomplish in the general case when Y is nonlinear. It is very easy to do, however, when the constraints added to (6.2) are $\mathbf{y} \geq 0$ in which case (6.12) is replaced by the rule

$$\begin{aligned} y_i^{l+1} &= \max\{0, y_i^l + \theta_l \gamma_i^l\} \qquad i = 1, \ldots, m \\ l &= 0, 1, 2, \ldots \end{aligned} \tag{6.19}$$

Thus, subgradient optimization is well suited as a method for solving the dual problem (5.4).

EXAMPLE 6.1. Consider the network optimization problem of Example 3.1 with side constraints that (1) the total flow in and out of the transshipment node 3 cannot exceed 7 and (2) the total flow between the factories at nodes 5 and 6 cannot exceed 1. A mathematical statement of the resulting problem is

$$\begin{aligned} v = \min\ & 4x_{13} + 5x_{15} + 3x_{24} + 6x_{26} + 1x_{34} + 2x_{35} + 7x_{43} + 5x_{46} + 8x_{56} \\ & + 1x_{65} + 3s_1^+ + 4s_2^+ \\ \text{s.t.}\quad & x_{13} + x_{34} + x_{35} + x_{43} \leq 7 \\ & x_{56} + x_{65} \leq 1 \\ & \text{network constraints} \end{aligned} \tag{6.20}$$

where the network constraints on the x_{ij}, s_1^+, and s_2^+ are the ones depicted in Figure 3.1. Note that we have denoted the variable flows in the arcs $r_l = (i,j)$ by x_{ij} following the convention of Section 3.3. The optimal solution to the original network optimization problem depicted in Figure 3.11 does not satisfy the first of the side constraints and we construct a dual problem to exploit the network structure in solving problem (6.20).

For $u_1 \geq 0$, $u_2 \geq 0$, the Lagrangean for the dual problem is

$$\begin{aligned} L(u_1, u_2) = -7u_1 - 1u_2 + \min(4 + u_1)x_{13} + 5x_{15} + 3x_{24} \\ + 6x_{26} + (1 + u_1)x_{34} + (2 + u_1)x_{35} + (7 + u_1)x_{43} \\ + 5x_{46} + (8 + u_2)x_{56} + (1 + u_2)x_{65} + 3s_1^+ + 4s_2^+ \\ \text{s.t. network constraints} \end{aligned} \tag{6.21}$$

Since (6.20) is a linear programming problem, we can be assured that the global optimality conditions of Chapter 5 will hold for some x_{ij}, s_1^+, and s_2^+ satisfying the network constraints and some $\mathbf{u} \geq \mathbf{0}$ which is optimal in the dual problem $d = \max L(\mathbf{u})$, s.t. $\mathbf{u} \geq \mathbf{0}$; moreover, $d = v$.

Our strategy is to maximize L using the subgradient optimization algorithm starting at the initial dual solution $\mathbf{u}^0 = (0,0)$. To do this we need to specify in (6.13) a target value and values of the λ_l. Our selections for these quantities can only be educated guesses based on previous experience with the algorithm, preferably on similar problems. Since the minimal objective function value in (6.20) without the side constraints is 95, we take 98 as our target value. In addition, we take

$$\lambda_0 = 2 \qquad \lambda_1 = 1 \qquad \lambda_2 = 3/4 \qquad \text{and} \qquad \lambda_l = 1/2 \qquad \text{for } l \geq 3$$

The solution depicted in Figure 3.11 is optimal in the Lagrangean at $\mathbf{u}^0$ with value $L(\mathbf{u}^0) = 95$. The indicated subgradient γ^0 is therefore

$$\gamma_1^0 = x_{13}(\mathbf{u}^0) + x_{34}(\mathbf{u}^0) + x_{35}(\mathbf{u}^0) + x_{43}(\mathbf{u}^0) - 7 = 1$$
$$\gamma_2^0 = x_{56}(\mathbf{u}^0) + x_{65}(\mathbf{u}^0) = -1$$

where $x_{ij}(\mathbf{u})$ denotes the optimal values of x_{ij} as a function of $\mathbf{u}$. We use the

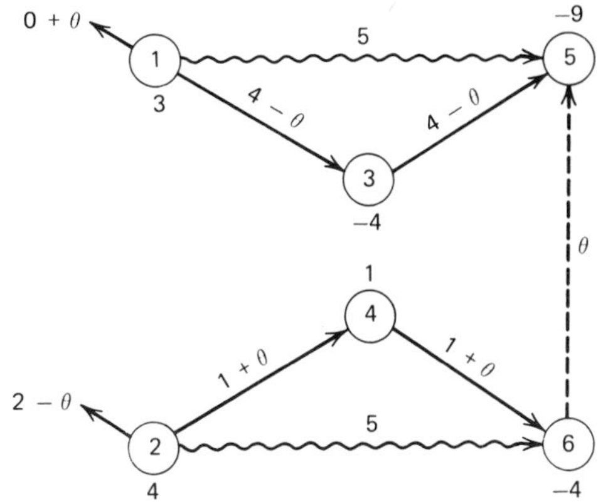

Figure 6.2. Reoptimizing network problem to compute Lagrangean at a dual solution.

rule (6.19) to compute the new dual solution u^1; that is,

$$u_1^1 = \max\left\{0, 0+2\frac{(98-95)}{2}(1)\right\} = 3.0$$

$$u_2^1 = \max\left\{0, 0+2\frac{(98-95)}{2}(-1)\right\} = 0$$

The spanning forest of Figure 3.11 shown in Figure 6.2 is feasible but not optimal in the Lagrangean (6.20) evaluated at $\mathbf{u}^1$. The numbers π_i beside each node i are the linear programming shadow prices when the arc and slack activity costs are as given in (6.20). These shadow prices are computed by Step 3 of the network simplex algorithm given in Chapter 3. Note that the shadow prices are the linear programming dual variables on the original constraints of the network optimization problem, whereas the dual variables explicitly stated in (6.21) are on the side constraints.

Relative to these shadow prices, the variable x_{65} has a negative reduced cost; that is, $c_{65} + u_2^1 - \pi_6 + \pi_5 = 1 + 0 + 4 - 9 = -4$. Thus, we increase the flow in the arc $(6,5)$ by the quantity $\theta \geq 0$. Following the principles of the network simplex algorithm, the maximal value of θ preserving feasibility is $\theta = 2$. At this value of θ, the variable s_2^+ is driven out of the basis. Figure 6.3 gives the resulting spanning forest and the new shadow prices are the numbers beside the nodes. This spanning forest is optimal in the Lagrangean evaluated at $\mathbf{u}^1 = (3,0)$ and therefore $L(\mathbf{u}^1) = 91$. Note that the step from $\mathbf{u}^0$ to $\mathbf{u}^1$ has resulted in a *decrease* in the value of L.

We continue with the subgradient optimization algorithm by computing the new subgradient of L at $\mathbf{u}^1$ corresponding to the optimal spanning forest shown in Figure 6.3. This new subgradient is $\gamma^1 = (-3, 1)$ and the

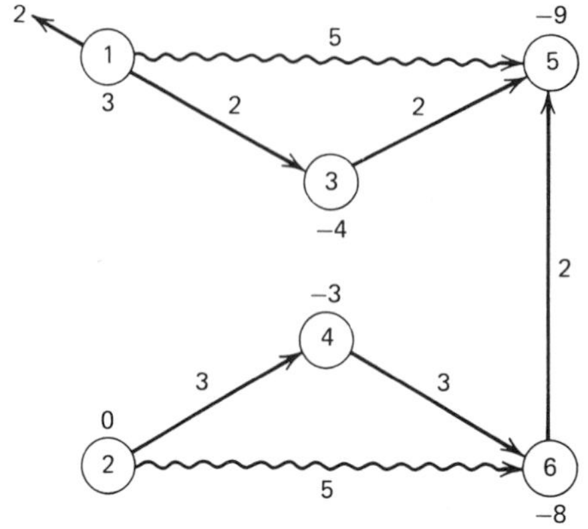

Figure 6.3. Optimal network problem at a dual solution.

new dual solution computed by (6.19) is $\mathbf{u}^2 = (0.3, 0.9)$. The optimal spanning forest at $\mathbf{u}^2$ turns out to be the same as the one in Figure 6.2, and in fact, the subgradient optimization algorithm continues to oscillate back and forth between the two spanning forests of Figures 6.2 and 6.3. The situation is depicted in Figure 6.4 where the nonnegative orthant is split into two segments corresponding to the sets of dual solutions $\mathbf{u}$ such that

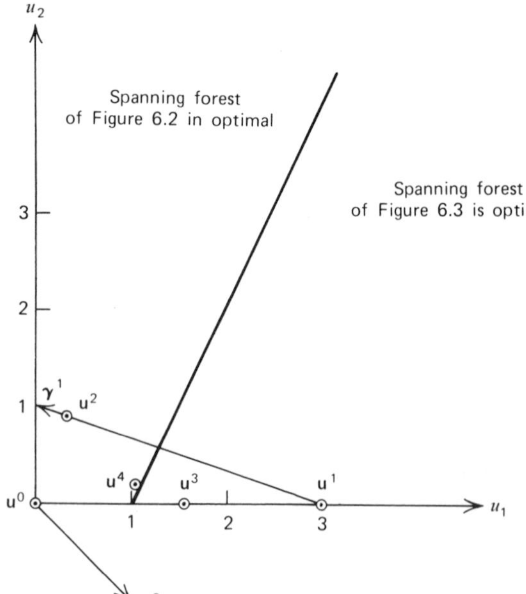

Figure 6.4. Subgradient optimization applied to a dual problem.

the spanning forests in Figures 6.2 and 6.3 are optimal, respectively, in the Lagrangean (6.20). The optimal solution to the dual problem is the solution $\bar{\mathbf{u}}=(1,0)$ with $v=d=96$, and the optimal solution to the linear programming problem (6.19) is to take $\frac{3}{4}$ of the solution of Figure 6.2 and $\frac{1}{4}$ of the solution of Figure 6.3. Note that the subgradient optimization algorithm produces a solution $\mathbf{u}^4$ that is fairly close to the optimal solution $\bar{\mathbf{u}}$. The next solution would be further away because the target value of 98 is above 96 which is the true maximal value of L. ▲

6.5 GENERALIZED PRIMAL-DUAL ASCENT ALGORITHM

In this section we present a generalized version of the primal-dual algorithm of Section 2.8 and show that it can be interpreted as an ascent algorithm for nondifferentiable concave programming problems. As we shall see, the direction-finding step of the algorithm is a mechanization of the minimization problem (6.3) to compute directional derivations at each point tested during the ascent. The essence of the method is illustrated by the application of the algorithm to the unconstrained problem (6.2) and in the special case when the concave function f is piecewise linear; namely

$$f(\mathbf{y}) = \min_{t=1,\ldots,T} \{p^t + \mathbf{q}^t\mathbf{y}\}$$

Once the special case is developed in detail, we discuss how the algorithm can be extended to handle more general concave functions and to incorporate constraints on $\mathbf{y}$.

The piecewise linear function f is the type we encounter in the decomposition of large-scale linear programming problems in Section 6.6. However, the p^t and $\mathbf{q}^t$ will not be given explicitly and will be generated as needed by the simplex algorithm. Since there are only a finite number of pieces to generate, the generalized primal-dual ascent algorithm converges in a finite number of arithmetic steps to an optimal solution to (6.2). Nevertheless, the number T of possible linear segments for f can be enormous in applications, and the effectiveness of an algorithm for maximizing f depends on how many linear segments it generates, and how close they are to being optimal. Sometimes the number of segments near the maximum is too large for exact optimization of f, and truncation of an algorithm before convergence is practical and necessary.

The piecewise linear form of f permits us to write (6.2) as the linear programming problem

$$\begin{aligned} w &= \max v \\ v &\le p^t + q^t\mathbf{y} \qquad t=1,\ldots,T \end{aligned} \tag{6.22}$$

For each $\mathbf{y} \in R^m$, there is a maximal associated value

$$\nu(\mathbf{y}) = \min_{t=1,\ldots,T} \{p^t + q^t\mathbf{y}\} = f(\mathbf{y})$$

Thus (6.22) and (6.2) have the same maximal objective function values, and the set of optimal $\mathbf{y}$ are identical. The linear programming dual to (6.22) is

$$\begin{aligned} w = \min \sum_{t=1}^{T} p^t \lambda_t & \\ \text{s.t.} \sum_{t=1}^{T} \mathbf{q}^t \lambda_t = \mathbf{0} & \\ \sum_{t=1}^{T} \lambda_t = 1 & \\ \lambda_t \geq 0 \qquad t = 1, \ldots, T & \end{aligned} \tag{6.23}$$

Our goal is to apply the primal-dual simplex algorithm to the pair (6.22), (6.23) with (6.22) as the primal problem. Let $\mathbf{y}$ and $\nu = f(\mathbf{y})$ be the given values of the variables in (6.22), and let

$$T(\mathbf{y}) = \{t \mid f(\mathbf{y}) = p^t + \mathbf{q}^t\mathbf{y}\} \tag{6.24}$$

The primal-dual simplex algorithm is based on the test: The solution $(\mathbf{y}, \nu) = [\bar{\mathbf{y}}, f(\bar{\mathbf{y}})]$ is optimal in the linear programming problem (6.22) if and only if there is a solution to the system

$$\begin{aligned} \sum_{t \in T(\bar{\mathbf{y}})} \mathbf{q}^t \lambda_t = \mathbf{0} & \\ \sum_{t \in T(\bar{y})} \lambda_t = 1 & \\ \lambda_t \geq 0 \qquad t \in T(\bar{\mathbf{y}}) & \end{aligned} \tag{6.25}$$

The validity of (6.25) is a direct consequence of Corollary 2.2 pertaining to linear programming complementary slackness. It is also a restatement of Corollary 6.1 that $\bar{\mathbf{y}}$ is optimal in the unconstrained problem (6.2) if and only if $\mathbf{0} \in \partial f(\bar{\mathbf{y}})$ (in the case when the concave function f is piecewise linear). To see why this last statement is so, note that $\mathbf{q}^t$ for any $t \in T(\bar{\mathbf{y}})$ is a subgradient of f at $\bar{\mathbf{y}}$. This is because $t \in T(\bar{\mathbf{y}})$ implies

$$f(\bar{\mathbf{y}}) = p^t + \mathbf{q}^t\bar{\mathbf{y}}$$

whereas by definition

$$f(\mathbf{y}) \leq p^t + \mathbf{q}^t\mathbf{y} \qquad \text{for all } \mathbf{y}$$

implying

$$f(\mathbf{y}) \le f(\bar{\mathbf{y}}) + \mathbf{q}^t(\mathbf{y} - \bar{\mathbf{y}}) \qquad \text{for all } \mathbf{y}$$

Moreover, any convex combination of subgradients is a subgradient as can easily be seen from the definition of a subgradient. Thus every vector

$$\mathbf{q} = \sum_{t \in T(\bar{\mathbf{y}})} \mathbf{q}^t \lambda_t$$

from (6.25) is a subgradient of f at $\bar{\mathbf{y}}$. Moreover, the set of vectors $\mathbf{q}$ generated by (6.25) is precisely the set of *all* subgradients of f at $\bar{\mathbf{y}}$ [i.e., the subdifferential $\partial f(\mathbf{y})$]. The equivalence between Corollary 6.1 and (6.25) is now clear because (6.25) states, in effect, that $\bar{\mathbf{y}}$ is optimal in (6.2) if and only if $\mathbf{0} \in \partial f(\bar{\mathbf{y}})$. If this test fails, then $\bar{\mathbf{y}}$ is not optimal and, as we shall see, a direction of ascent from $\bar{\mathbf{y}}$ is indicated.

Before discussing the ascent step of the primal-dual simplex algorithm, however, we must face up to the serious difficulty due to the possible large size of the index set $T(\mathbf{y})$ and the fact that it often is not explicitly given but needs to be generated. The difficulty is resolved at each point $\bar{\mathbf{y}}$ generated by the algorithm by working with a subset $T'(\bar{\mathbf{y}}) \subseteq T(\bar{\mathbf{y}})$ and iteratively augmenting it until $\bar{\mathbf{y}}$ is proven optimal or an ascent step can be taken.

Thus, suppose we are at an arbitrary point $\bar{\mathbf{y}}$ with a nonempty subset $T'(\bar{\mathbf{y}})$ of the set $T(\bar{\mathbf{y}})$ defined in (6.24). We try to prove $\bar{\mathbf{y}}$ is optimal in (6.22) with value $\nu = f(\mathbf{y})$ by trying to establish that (6.25) has a solution. Operationally, we solve the phase one linear programming problem

$$\begin{aligned} \omega = \min \sum_{i=1}^{m} s_i^+ + s_i^- & \\ \text{s.t.} \quad \sum_{t \in T'(\bar{\mathbf{y}})} q_i^t \lambda_t + s_i^+ - s_i^- = 0 & \qquad i = 1, \ldots, m \\ \sum_{t \in T'(\bar{\mathbf{y}})} \lambda_t = 1 & \\ \lambda_t \ge 0 \qquad t \in T'(\bar{\mathbf{y}}) & \\ s_i^+ \ge 0 \qquad s_i^- \ge 0 \qquad i = 1, \ldots, m & \end{aligned} \tag{6.26}$$

If the minimal objective function value is zero, then $\bar{\mathbf{y}}$ is optimal in (6.22) by our previous arguments. If the minimum in (6.26) is greater than zero, then either $\bar{\mathbf{y}}$ is not optimal, or $\bar{\mathbf{y}}$ is optimal but there are subgradients $\mathbf{q}^t$ for $t \in T(\bar{\mathbf{y}}) - T'(\bar{\mathbf{y}})$ which need to be added to (6.26) to establish optimality. The generalized primal-dual algorithm proceeds by assuming that $\bar{\mathbf{y}}$ is not optimal and that a direction of ascent of the function f has been found.

Insight into the nature of problem (6.26) is gained by looking at its dual

$$\begin{aligned} \omega = \max\ & \nu \\ & \nu \leq \mathbf{q}^t\mathbf{d} \qquad t \in T'(\bar{\mathbf{y}}) \\ & -1 \leq d_i \leq 1 \qquad i = 1,\ldots,m \end{aligned} \tag{6.27}$$

We can argue that (6.27) is a direction-finding procedure because it approximates the more general direction-finding problem (6.4) in the special case when f is piecewise linear. This follows from the fact that the subdifferential $\partial f(\bar{\mathbf{y}})$ is given by the convex polyhedron (6.25) implying

$$\nabla f(\bar{\mathbf{y}};\mathbf{d}) = \min_{t \in T(\bar{\mathbf{y}})} \mathbf{q}^t\mathbf{d}$$

Thus, if $\bar{\mathbf{d}}$ is optimal in (6.27), we have

$$\omega = \underset{t \in T'(\bar{\mathbf{y}})}{\text{minimum}}\, \mathbf{q}^t\bar{\mathbf{d}}$$

and ω is approximating $\nabla f(\bar{\mathbf{y}};\bar{\mathbf{d}})$ in the sense that $\omega \geq \nabla f(\bar{\mathbf{y}};\bar{\mathbf{d}})$ with the possibility of strict inequality because the $\mathbf{q}^t$, $t \in T(\bar{\mathbf{y}}) - T'(\bar{\mathbf{y}})$ are omitted in calculating ω. When $\bar{\mathbf{y}}$ is optimal in (6.22) because $\omega = 0$, then for all directions $\mathbf{d}$

$$\nabla f(\bar{\mathbf{y}};\bar{\mathbf{d}}) \leq \min_{t \in T'(\bar{\mathbf{y}})} \mathbf{q}^t\mathbf{d} \leq 0$$

which is consistent with our previous results. On the other hand, when $\omega > 0$ in (6.26) and (6.27) and an optimal $\bar{\mathbf{d}} \neq 0$ is found, then the generalized primal-dual algorithm tries $\bar{\mathbf{d}}$ as a direction of ascent on the assumption that $\nabla f(\bar{\mathbf{y}};\bar{\mathbf{d}}) = \omega > 0$. If $\nabla f(\bar{\mathbf{y}};\bar{\mathbf{d}})$ is less than ω, we show that the step length procedure will discover it and subsequently the direction-finding problem (6.27) will be rerun with additional constraints.

It might appear reasonable in determining the step length in the direction $\bar{\mathbf{d}} \neq 0$ optimal in (6.27) to solve $\max_{\theta \geq 0} f(\bar{\mathbf{y}} + \theta\bar{\mathbf{d}})$. In practice, this may be effective but to ensure convergence, a different rule is required. The rule is as follows: Select $\theta = \bar{\theta}$ to be the maximal value such that

$$f(\bar{\mathbf{y}} + \theta\bar{\mathbf{d}}) = f(\bar{\mathbf{y}}) + \theta\omega \tag{6.28}$$

Since f is a piecewise linear function, we expect it to increase linearly at the positive rate ω a positive distance in the direction $\bar{\mathbf{d}}$.

We have assumed throughout the chapter that the maximal value of f for $\mathbf{y} \in R^m$ is finite, implying that we can assume there is some $\theta = \theta_1$ such that $\theta \leq \theta_1$ in determining the step. Moreover, it is possible to construct a finite search of the half line $(\bar{\mathbf{y}}, \bar{\mathbf{y}} + \theta\mathbf{d})$ for $\theta \geq 0$ to discover such a θ_1. Alternatively, we may compute θ_1 by using some known or assumed upper bound $\bar{f}$ on the maximal value of f.

Thus, we assume that a θ_1 exists such that $\theta \leq \theta_1$ is valid in the step length procedure. By reference to Figures 6.5 and 6.6, we illustrate this procedure. It begins at $\theta = \theta_1$ where we compute $f(\bar{\mathbf{y}} + \theta_1 \mathbf{d})$ to see if $\bar{\theta}$ in (6.28) is equal to θ_1. In Figure 6.5, we discover that $f(\mathbf{y} + \theta_1 \bar{\mathbf{d}}) = p^1 + \mathbf{q}^1 \bar{\mathbf{y}} + \theta_1 \mathbf{q}^1 \bar{\mathbf{d}} < f(\mathbf{y}) + \theta_1 \omega$ indicating that $\bar{\theta} \leq \theta_1$. Since f is a piecewise linear, concave function, it is clear that $\bar{\theta} \leq \theta_2$ where θ_2 is the value of θ where the two line segments $f(\mathbf{y}) + \theta\omega$ and $p^1 + \mathbf{q}^1\bar{\mathbf{y}} + \theta \mathbf{q}^1 \mathbf{d}$ intersect. As shown in Figure 6.5, the presence of the third line segment $p^2 + \mathbf{q}^2 \bar{\mathbf{y}} + \theta \mathbf{q}^2 \mathbf{d}$, as yet undiscovered, indicates that in fact $\bar{\theta} = \theta_3 < \theta_2$. Nevertheless, the step length calculation proceeds by computation of θ_2 from the following equation describing the intersection of the line segments

$$f(\mathbf{y}) + \theta_2 \omega = p^1 + \mathbf{q}^1 \left(\bar{\mathbf{y}} + \theta_2 \bar{\mathbf{d}} \right)$$

or

$$\theta_2 = \frac{p^1 + \mathbf{q}^1 \bar{\mathbf{y}} - f(\bar{\mathbf{y}})}{\phi - \mathbf{q}^1 \mathbf{d}} \tag{6.29}$$

Note that the numerator in (6.29) must be nonnegative by the definition of $f(\bar{\mathbf{y}})$ as the minimum over k of the piecewise linear segments $p^k + \mathbf{q}^k \bar{\mathbf{y}}$. In

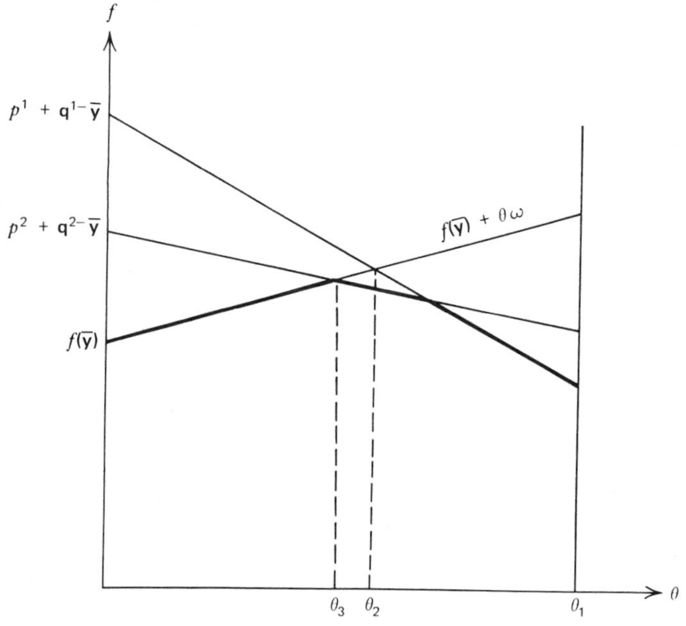

Figure 6.5. Primal-dual ascent algorithm—line search produces a solution with higher objective function value.

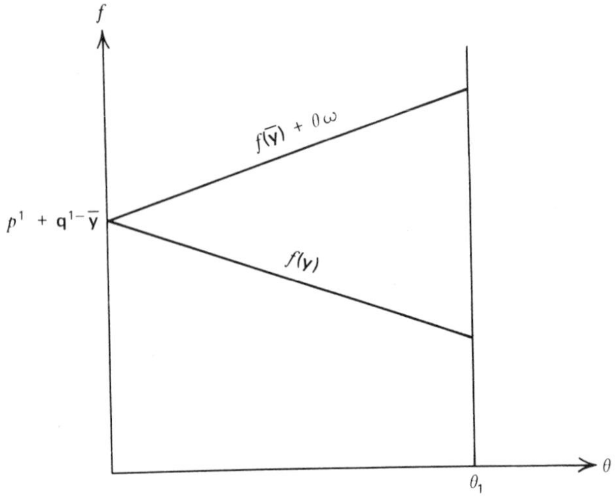

Figure 6.6. Primal-dual ascent algorithm—line search fails to produce a solution with higher objective function value.

addition, the denominator is positive since $p^1+\mathbf{q}^1(\bar{\mathbf{y}}+\theta_1\bar{\mathbf{d}})<f(\mathbf{y})+\theta_1\omega$ and $f(\bar{\mathbf{y}})\leq p^1+\mathbf{q}^1\bar{\mathbf{y}}$. The procedure applied at $\theta=\theta_1$ is then repeated at $\theta=\theta_2$.

Figure 6.6 depicts a typical situation when f does not increase in the direction $\bar{\mathbf{d}}$. The fact that $\bar{\mathbf{d}}$ is not a direction of ascent is passed to (6.27) by recognizing that (1) $\mathbf{q}^1$ is a subgradient of f at $\bar{\mathbf{y}}$ that has not been previously considered and (2) the constraint $\nu\leq\mathbf{q}^1\mathbf{d}$ cuts off the previously optimal solution $\omega,\bar{\mathbf{d}}$ in (6.24) since $\omega>\mathbf{q}^1\bar{\mathbf{d}}$ by our previous discussion of (6.29).

The general rule for searching in the direction $\bar{\mathbf{d}}$ selected by (6.27) is to compute recursively the p^k, $\mathbf{q}^k$, and θ_k by

$$f\left(\bar{\mathbf{y}}+\theta_k\bar{\mathbf{d}}\right)=p^k+\mathbf{q}^k\left(\bar{\mathbf{y}}+\theta_k\bar{\mathbf{d}}\right)$$

and if the right-hand side is less than $f(\bar{\mathbf{y}})+\theta_k\omega$, then we decrease θ to

$$\theta_{k+1}=\frac{p^k+\mathbf{q}^k\bar{\mathbf{y}}-f(\bar{\mathbf{y}})}{\omega-\mathbf{q}^k\bar{\mathbf{d}}}$$

where θ_1 is given. The calculation stops when either $f(\bar{\mathbf{y}})+\theta_k\omega=f(\bar{\mathbf{y}}+\theta_k\bar{\mathbf{d}})=p^k+\mathbf{q}^k(\bar{\mathbf{y}}+\theta_k\mathbf{d})>f(\bar{\mathbf{y}})$ for $\theta_k>0$, or $\theta_{k+1}=0$. In the former case, the new trial solution for problem (6.2) is $\bar{\mathbf{y}}+\theta_k\bar{\mathbf{d}}$ and in the latter case, the constraint $\nu\leq\mathbf{q}^k\mathbf{d}$ is added to (6.27). Since there are only a finite number of $k\in T(\bar{\mathbf{y}})$, the process of adding constraints and resolving (6.27) must terminate finitely either with $\omega=0$ implying $\bar{\mathbf{y}}$ is optimal in (6.2) or with a new solution $\bar{\mathbf{y}}+\theta_k\bar{\mathbf{d}}$ satisfying

$$f(\bar{\mathbf{y}}+\theta_k\bar{\mathbf{d}})>f(\bar{\mathbf{y}})$$

The convergence criterion of the generalized primal-dual simplex algorithm is the same as the criterion for the primal-dual simplex algorithm. When the algorithm moves from $\bar{\mathbf{y}}$ to the new point $\bar{\mathbf{y}}+\theta_k\bar{\mathbf{d}}$, the binding constraints in (6.27) at $\bar{\mathbf{y}}$ remain valid constraints for (6.27) at the new point. In other words, $t\in T(\bar{\mathbf{y}}+\theta_k\bar{\mathbf{d}})$ for all $t\in T(\bar{\mathbf{y}})$ such that $\omega=\mathbf{q}^t\bar{\mathbf{d}}$ in (6.27). Thus, the optimal objective function value ω in (6.26) and (6.27) decreases in moving from $\bar{\mathbf{y}}$ to $\bar{\mathbf{y}}+\theta_k\bar{\mathbf{d}}$, and it must ultimately decrease to zero since there are only a finite number of linear segments p^k, $\mathbf{q}^k$.

The generalized primal-dual algorithm can be easily extended to incorporate linear constraints $\mathbf{Ay}\leq\mathbf{b}$, $\mathbf{y}\geq\mathbf{0}$, on the piecewise linear function. The direction-finding linear programming problem (6.27) becomes

$$\begin{aligned}
\omega=\max \nu& \\
\nu\leq\mathbf{q}^t\mathbf{d} \qquad & t\in T'(\bar{\mathbf{y}})\\
\sum_{i=1}^{m} a_{ki}d_i\leq 0 \qquad & k\in K(\mathbf{y})\\
0\leq d_i\leq 1 \qquad & i\in I(\bar{y})\\
-1\leq d_i\leq 1 \qquad & i\in I^c(\bar{y})
\end{aligned}\tag{6.30}$$

where $K(\bar{\mathbf{y}})$ is the index set of binding constraints at $\bar{\mathbf{y}}$, and $I(\bar{\mathbf{y}})$ is the index set of variables at zero at $\bar{\mathbf{y}}$. If $\omega\geq 0$, then (6.30) has produced an apparent direction of ascent. This rule for calculating the step length in that direction is the same as (6.28), but one must take into account the constraints on (6.30) in doing the step length analysis.

Our development of the generalized primal-dual algorithm and its interpretation as an ascent algorithm has been for the special case when the nondifferentiable concave function f is piecewise linear. The algorithm with suitable modifications can be extended to handle general nondifferentiable concave functions. The primal-dual arguments work in the piecewise linear case because the line searches in trial ascent directions are finite and a positive step of minimal length is made in true directions of ascent. In the general case when the functions are nonlinear, these convenient properties are not available and we need a new construct.

For $\varepsilon>0$, a solution $\bar{\mathbf{y}}$ is called *ε-optimal* in the unconstrained concave maximization problem (6.2) if $f(\bar{\mathbf{y}})>f(\mathbf{y})-\varepsilon$ for all $\mathbf{y}$. The central construct used in finding ε-optimal solutions is the *ε-subdifferential* defined by

$$\partial_\varepsilon f(\bar{\mathbf{y}})=\{\gamma|f(\mathbf{y})\leq f(\bar{\mathbf{y}})+(\mathbf{y}-\bar{\mathbf{y}})\gamma+\varepsilon \qquad \text{for all } \mathbf{y}\}\tag{6.31}$$

The analogue to Theorem 6.1 characterizing ε-optimal solutions is: the solution $\bar{\mathbf{y}}$ is ε-optimal in the unconstrained concave maximization problem (6.2) if and only if

$$\begin{aligned} \max_{\mathbf{d}} \min_{\gamma \in \partial_\varepsilon f(\bar{\mathbf{y}})} \mathbf{d}\gamma = 0 \\ \text{s.t. } -1 \le d_i \le 1 \qquad i=1,\ldots,m \end{aligned} \tag{6.32}$$

This test is used in the following algorithm where we are still assuming for convenience that the concave function f is finite for all $\mathbf{y} \in R^m$.

ε-Subgradient Method

Step 1. Select an initial vector $\mathbf{y}^0$ and scalars $\varepsilon_0 > 0$ and α, $0 < \alpha < 1$.

Step 2. Given $\mathbf{y}^l$ and $\varepsilon_l > 0$, set $\varepsilon_{l+1} = \alpha^k \varepsilon_l$, where k is the smallest integer such that $0 \notin \partial_{\varepsilon_{l+1}} f(\mathbf{y}^l)$. If there is no such k, then $\mathbf{y}^l$ is optimal in (6.2).

Step 3. Find any vector $\mathbf{d}$ such that

$$\min_{\gamma \in \partial_{\varepsilon_{l+1}} f(\mathbf{y}^l)} \mathbf{d}^l \gamma > 0$$

Step 4. Set $\mathbf{y}^{l+1} = \mathbf{y}^l + \theta_l \mathbf{d}^l$ where $\theta_l > 0$ satisfies

$$f(\mathbf{y}^{l+1}) > f(\mathbf{y}^l) + \varepsilon_{l+1}$$

Return to Step 2.

The reader is asked in Exercise 6.6 to show that Step 4 of this algorithm can always be carried out and that the algorithm has the correct convergence properties for solving $\max f(\mathbf{y})$ over $\mathbf{y} \in R^m$. Note, however, that Step 3 assumes complete knowledge of the ε_{l+1} subdifferential of f which in general is not available. We also ask the reader in Exercise 6.6 to provide a method for generating it.

EXAMPLE 6.2. We apply the generalized primal-dual algorithm to the dual problem of Example 6.1. The algorithm is initiated at the dual feasible solution $\mathbf{u}^0 = (0,0)$, where we have as an initial subgradient $\gamma^0 = \begin{pmatrix} 1 \\ -1 \end{pmatrix}$ corresponding to the optimal spanning forest of Figure 6.4. Thus, the direction-finding subproblem (6.30) at $\mathbf{u}^0$ is

$$\begin{aligned} \omega = \max \nu & \\ \nu &\le d_1 - d_2 \\ 0 &\le d_1 \le 1 \\ 0 &\le d_2 \le 1 \end{aligned}$$

The optimal solution to this problem is $\mathbf{d}^0 = (1,0)$ with $\omega = 1$ and we try to use this as a direction of ascent. Actually, unknown to the algorithm, γ^0 is the gradient of L at $\mathbf{u}^0$ and $\mathbf{d}^0$ is a true direction of ascent with $\nabla L(\mathbf{u}^0;\mathbf{d}^0) = 1$.

If we make the assumption that the maximal value of L over the nonnegative orthant is less than or equal to 100, we can limit our search for the first break-point of L along the half line $\{(u_1,u_2)|0\le u_1,\ u_2=0\}$ to the line segment $\{(u_1,u_2)|0\le u_1\le 5,\ u_2=0\}$. We discover at $\mathbf{u}=(5,0)$ that the spanning tree of Figure 6.3 is optimal and $L(5,0)=84<100=L(0,0)+5\omega$. We then use the subgradient $\gamma^1=(-3,1)$ to draw in another line segment of L in the direction $u_1\ge 0$ and $u_2=0$. The two line segments $95+u_1$ and $99-3u_1$ intersect at the solution $(u_1,0)=(1,0)$ and we discover that $L(1,0)=96=L(0,0)+1\omega$ indicating that the solution is the first break point of L in the indicated direction and the next solution to be evaluated.

The generalized primal-dual algorithm proceeds by solving the direction finding problem at the new solution $(1,0)$ using γ^0 and γ^1 as the initial subgradients at that solution. This problem is

$$\begin{aligned}\omega = \max\ & \nu\\ & \nu\le d_1-d_2\\ & \nu\le 3d_1+d_2\\ & -1\le d_1\le 1\\ & 0\le d_2\le 1\end{aligned}\tag{6.33}$$

and $d=(0,0)$ is optimal in it; that is, $\omega=0$ and we have established the optimality of $(1,0)$ in the dual problem. An optimal solution to the primal problem (6.20) is found by observing that the optimal shadow prices $\lambda_1=\frac{3}{4}$, $\lambda_2=\frac{1}{4}$ on the subgradient inequalities in (6.33) are optimal weights for the primal solutions corresponding to Figures 6.2 and 6.3, respectively. ▲

6.6 PRICE DIRECTIVE AND RESOURCE DIRECTIVE DECOMPOSITION METHODS

Decomposition of large-scale linear programming problems historically began with the study of the block diagonal problem

$$\begin{aligned}
v=\min\ & \mathbf{c}^1\mathbf{x}^1+\cdots && +\mathbf{c}^R\mathbf{x}^R && (6.34a)\\
\text{s.t.}\ & \mathbf{Q}^1\mathbf{x}^1+\cdots && +\mathbf{Q}^R\mathbf{x}^R\le\mathbf{q} && (6.34b)\\
& \mathbf{A}^1\mathbf{x}^1 && =\mathbf{b}^1 && \\
& \qquad\ddots && && (6.34c)\\
& && \mathbf{A}^R\mathbf{x}^R=\mathbf{b}^R && \\
& \mathbf{x}^1\ge\mathbf{0} && \mathbf{x}^R\ge\mathbf{0} && (6.34d)
\end{aligned}$$

The idea of the methods is to separate (6.34) into $R+1$ smaller linear programming problems, one coordination problem concerned with the joint constraints (6.34b) and R subproblems, each one using the constraint set $\mathbf{A}^r\mathbf{x}^r=\mathbf{b}^r$, $\mathbf{x}^r \geq \mathbf{0}$. In price directive decomposition, the separation is accomplished by putting prices, or dual variables, on the joint constraints and placing them in the objective function. The price directive coordination problem is concerned with calculating optimal prices on the shared resources to be used in the subproblems so that an optimal solution to the global problem (6.34) is achieved by optimizing separately each of the subproblems. In resource directive decomposition, each of the R subproblems is given a portion of the shared resources $\mathbf{q}$. The resource directive coordination problem is concerned with effecting an apportionment that permits (6.34) to be optimized by optimizing separately each of the subproblems.

It has only recently been fully recognized that price and resource directive methods for (6.34) are intimately involved with methods for nondifferentiable optimization, convex analysis, and duality. The decomposition approach gives rise to problems that are nondifferentiable because a small change in an objective function or right-hand-side coefficient of a linear programming problem can cause a large change in the optimal extreme point solution and the implied dual solution. In this sense, an LP model can be an inaccurate representation of the real world, but its data requirements are usually far simpler than those of a nonlinear model, which might exhibit smoother behavior.

Decomposition methods were originally proposed in order to be able to solve (6.34) on computers with limited core storage capacity which would not permit it to be solved as an ordinary linear programming problem. There are other methods, such as basis partitioning, which can accomplish the same reduction in core storage demands. However, the price and resource directive decomposition methods have an important economic interpretation as coordination mechanisms for decentralization of large-scale organizations. It is for this reason that they have continued to receive a great deal of attention in spite of the fact that the algorithms have not yet been very widely implemented and used. We will develop the decomposition methods using the organizational decentralization point of view. The price directive decentralization is an economic explanation of the generalized linear programming algorithm developed in Section 5.4 and the resource directive decentralization is an economic explanation of Benders' algorithm, which we will be seeing for the first time in this section.

Thus, we consider (6.34) to be a mathematical description of an organization consisting of a headquarters with the constraints (6.34b) and R divisions with the individual constraints (6.34c). We assume (6.34) has an optimal solution. In the price directive approach, the goal of a coordinator at headquarters is to set prices $\mathbf{u} \geq \mathbf{0}$ on the shared resources $\mathbf{q}$ so that each of the R divisions can autonomously optimize their own operations. Mathematically, the prices $\mathbf{u} \geq \mathbf{0}$ are used to construct the Lagrangean

$$\begin{aligned} L(\mathbf{u}) = -\mathbf{u}\mathbf{q} + \min \sum_{r=1}^{R} (\mathbf{c}^r + \mathbf{u}\mathbf{Q}^r)\mathbf{x}^r& \\ \text{s.t. } \mathbf{A}^r\mathbf{x}^r = \mathbf{b}^r \qquad r = 1,\ldots,R& \\ \mathbf{x}^r \geq \mathbf{0}& \end{aligned} \tag{6.35}$$

The Lagrangean clearly separates into

$$L(\mathbf{u}) = \mathbf{u}\mathbf{q} + \sum_{r=1}^{R} L^r(\mathbf{u})$$

where for each r we have the subproblem

$$\begin{aligned} L^r(\mathbf{u}) = \min(\mathbf{c}^r + \mathbf{u}\mathbf{Q}^r)\mathbf{x}^r& \\ \text{s.t. } \mathbf{x}^r \in \mathbf{X}^r = \{\mathbf{x}^r | \mathbf{A}\mathbf{x}^r = \mathbf{b}^r,\ \mathbf{x}^r \geq \mathbf{0}\}& \end{aligned} \tag{6.36}$$

Thus, each division can autonomously optimize its own operations if the prices are given. For expositional simplicity, we assume the sets X^r are nonempty and bounded so that problem (6.36) always has an optimal solution. If X^r were unbounded, it would be possible that $L^r(\mathbf{u}) = -\infty$ for some $\mathbf{u} \geq \mathbf{0}$ and it would be necessary to constrain $\mathbf{u}$ so that this cannot occur (see Exercise 6.7). The rationale for selecting the prices is to satisfy Global Optimality Conditions for Block Diagonal Linear Programming: The solutions $\bar{\mathbf{x}}^r \in X^r$ for $r = 1,\ldots,R$, and $\bar{\mathbf{u}} \geq \mathbf{0}$ satisfy the global optimality conditions for (6.34) if

(1) $L(\bar{\mathbf{u}}) = -\bar{\mathbf{u}}\mathbf{q} + \sum_{r=1}^{R} (\mathbf{c}^r + \bar{\mathbf{u}}\mathbf{Q}^r)\bar{\mathbf{x}}^r$

(2) $\bar{\mathbf{u}}\left(\sum_{r=1}^{R} \mathbf{Q}^r\bar{\mathbf{x}}^r - \mathbf{q}\right) = 0$

(3) $\sum_{r=1}^{R} \mathbf{Q}^r\bar{\mathbf{x}}^r \leq \mathbf{q}$

As we saw in Chapter 5, if $\bar{\mathbf{x}}^r$ for $r = 1,\ldots,R$, and $\bar{\mathbf{u}} \geq \mathbf{0}$ satisfy the global optimality conditions, then the $\bar{\mathbf{x}}^r$ are optimal in (6.34) and $\bar{\mathbf{u}}$ is optimal in the nondifferentiable dual problem called the *price directive coordination problem*

$$\begin{aligned} d = \max -\mathbf{u}\mathbf{q} + \sum_{r=1}^{R} L^r(\mathbf{u})& \\ \text{s.t. } \quad \mathbf{u} \geq \mathbf{0}& \end{aligned} \tag{6.37}$$

The economic interpretation of price directive decomposition is derived from the application of the generalized linear programming algorithm to solving the pair of problems (6.34) and (6.37).

At any arbitrary iteration K of the price directive decentralization of the organization, the coordinator at headquarters has received solutions $\mathbf{x}^{r,k} \in X^r$ for $k=1,\ldots,K$, from the divisions. He uses these solutions to calculate new prices by solving the *price directive master problem*

$$\begin{aligned}
\min \quad & \sum_{r=1}^{R} \sum_{k=1}^{K} (\mathbf{c}^r \mathbf{x}^{r,k})\lambda_{r,k} \\
\text{s.t.} \quad & \sum_{r=1}^{R} \sum_{k=1}^{K} (\mathbf{Q}^r \mathbf{x}^{r,k})\lambda_{r,k} \leq \mathbf{q} \\
& \sum_{k=1}^{K} \lambda_{r,k} = 1 \qquad r=1,\ldots,R \\
& \lambda_{r,k} \geq 0
\end{aligned} \tag{6.38}$$

The interpretation of the master problem is that the coordinator makes the best use he can of the information he has about the operating possibilities of each of the divisions in minimizing total cost; namely, he considers all convex combinations of the solutions $\mathbf{x}^{r,k}$ and this gives him a partial description of the feasible region X^r. For expositional simplicity, we have assumed he has the same number of solutions from each set X^r. The coordinator tests this information by passing to each of the divisions the vector $\mathbf{u}^K \geq \mathbf{0}$, which are the optimal prices on the shared resources $\mathbf{q}$ in (6.34). Each division responds by solving its subproblem (6.36) with $\mathbf{u}=\mathbf{u}^K$ thereby producing a new solution $\mathbf{x}^{r,K+1}$ which is passed to the coordinator along with the objective function value

$$L^r(\mathbf{u}^K) = (\mathbf{c}^r + \mathbf{u}^K \mathbf{Q}^r)\mathbf{x}^{r,K+1}$$

The new solutions $\mathbf{x}^{r,K+1}$ from the divisions are evaluated to see if they can be used to improve the total cost of operating the entire organization. Specifically, for each r, the coordinator compares $\mathbf{x}^{r,K+1}$ to the $\mathbf{x}^{r,k}$ which were previously used; that is, he considers those $\mathbf{x}^{r,k}$ with $\lambda_{r,k}^K > 0$ in the master problem. Such $\mathbf{x}^{r,k}$ have a reduced cost relative to $\mathbf{u}^K$ of

$$\theta^{r,K} = (\mathbf{c}^r + \mathbf{u}^K \mathbf{Q}^r)\mathbf{x}^{r,k}$$

and $\mathbf{x}^{r,K+1}$ can be used to reduce total cost in the master problem (6.38) if $L^r(\mathbf{u}^K) < \theta^{r,K}$. Note that $\theta^{r,K}$ is the optimal shadow price at iteration K on the rth convexity row in (6.38). In linear programming terms, this last statement is equivalent to saying that the new column

$$\begin{bmatrix} \mathbf{c}^r\mathbf{x}^{r,K+1} \\ \mathbf{Q}^r\mathbf{x}^{r,K+1} \\ 1 \end{bmatrix}$$

is a candidate to enter the optimal basis in (6.38) if it prices out negatively.

If $L^r(\mathbf{u}^K)=\theta^{r,K}$ for $r=1,\ldots,R$, then the coordinator can conclude that he has enough information about the feasible regions X^r of the divisions because each division has failed to produce a new solution that can reduce total cost in the master problem. In this case, the solutions

$$\bar{\mathbf{x}}^r=\sum_{k=1}^{K}\lambda_{r,k}^{K}\mathbf{x}^{r,k} \qquad r=1,\ldots,R \tag{6.39}$$

are optimal in the block diagonal problem (6.34). This can be verified by direct appeal to the global optimality conditions for (6.34) using the dual solution $\mathbf{u}^K\geq\mathbf{0}$. Note that the price directive decentralization scheme is imperfect because there will probably be more than one $\lambda_{r,k}^K$ at a positive level in (6.39) indicating that there is more than one $\mathbf{x}^{r,k}$ optimal in the divisional subproblem evaluated at $\mathbf{u}^K$. Thus, the coordinator must tell each division what weights to attach to alternative optima in the divisional subproblem as well as providing the price information. As a final point, note that the iterative scheme between the master problem and the divisional subproblems must converge after a finite number of iterations because each time the problem is resolved we have added at least one new extreme point solution from some division's feasible region X^r to the master problem.

In resource directive decomposition, the coordinator wishes to select the resource vectors $\mathbf{q}^1,\ldots,\mathbf{q}^R$ satisfying $\sum_{r=1}^R\mathbf{q}^r\leq\mathbf{q}$ so that the global problem (6.34) is solved when each division solves its own linear programming problem

$$\begin{aligned} v^r(\mathbf{q}^r)=\min\ & \mathbf{c}^r\mathbf{x}^r && (6.40\text{a})\\ \text{s.t. } & \mathbf{Q}^r\mathbf{x}^r\leq\mathbf{q}^r && (6.40\text{b})\\ & \mathbf{A}^r\mathbf{x}^r=\mathbf{b}^r && (6.40\text{c})\\ & \mathbf{x}^r\geq\mathbf{0} && (6.40\text{d}) \end{aligned}$$

The mechanism for optimally selecting the $\mathbf{q}^r$ is to solve the nondifferentiable convex optimization problem called the *resource directive coordination problem*

$$\begin{aligned} v=\min\ & \sum_{r=1}^{R}\mathbf{v}^r(\mathbf{q}^r) \\ \text{s.t. } & \mathbf{q}^1+\cdots+\mathbf{q}^R\leq\mathbf{q} \end{aligned} \tag{6.41}$$

As we know from Chapter 2, each of the functions v^r is convex, but unlike the price directive coordination problem (6.37), the coordinator must include the possibility that $v^r(\mathbf{q}^r) = +\infty$ when (6.40) is infeasible.

The specific algorithm for (6.41) is derived by expressing v^r in its dual form

$$\begin{aligned} v^r(\mathbf{q}^r) = \max\{-\mathbf{u}^r\mathbf{q}^r + \mathbf{y}^r\mathbf{b}^r\} \\ \text{s.t.} \; -\mathbf{u}^r\mathbf{Q}^r + \mathbf{y}^r\mathbf{A}^r \le \mathbf{c}^r \\ \mathbf{u}^r \ge 0 \end{aligned} \tag{6.42}$$

First, we note that if (6.40) is infeasible, then there must be an extreme ray solution $(\mathbf{u}^{r,l}, \mathbf{y}^{r,l})$ of the dual feasible region satisfying

$$\begin{aligned} -\mathbf{u}^{r,l}\mathbf{Q}^r + \mathbf{y}^{r,l}\mathbf{A}^r \le \mathbf{0} \\ \mathbf{u}^{r,l} \ge 0 \end{aligned} \tag{6.43}$$

and

$$-\mathbf{u}^{r,l}\mathbf{q}^r + \mathbf{y}^{r,l}\mathbf{b}^r > 0$$

These conditions imply $v^r(\mathbf{q}^r) = +\infty$ because for any $\lambda > 0$ and any $(\bar{\mathbf{u}}^r, \bar{\mathbf{y}}^r)$ feasible in (6.42), the solution $(\bar{\mathbf{u}}^r, \bar{\mathbf{y}}^r) + \lambda(\mathbf{u}^{r,l}, \mathbf{y}^{r,l})$ is also feasible in (6.42) and its objective function value goes to $+\infty$ as λ goes to $+\infty$. To prevent such $\mathbf{q}^r$ from being selected in the price directive coordination problem (6.41), it suffices to add to it the constraint

$$-\mathbf{u}^{r,l}\mathbf{q}^r + \mathbf{y}^{r,l}\mathbf{b}^r \le 0 \tag{6.44}$$

Thus, there are a number of implied linear constraints to add to (6.41) to limit the search for an optimal resource directive decomposition to the set of $\mathbf{q}^r$ such that $v^r(\mathbf{q}^r)$ is finite. These constraints are added as they are needed.

On the other hand, if (6.40) has a feasible and therefore an optimal solution (recall that we assumed that $X^r = \{\mathbf{x} | \mathbf{A}^r\mathbf{x}^r = \mathbf{b}^r, \mathbf{x}^r \ge \mathbf{0}\}$ is nonempty and bounded), then

$$v^r(\mathbf{q}^r) = \max_k \{-\mathbf{u}^{r,k}\mathbf{q}^r + u_0^{r,k}\} \tag{6.45}$$

where $\{\mathbf{u}^{r,k}, \mathbf{y}^{r,k}\}$ are the extreme points of the feasible region in (6.42), and $u_0^{r,k} = \mathbf{y}^{r,k}\mathbf{b}^r$. The specific algorithm by which the coordinator effects his resource direction uses the form (6.45) to describe the convex function $v^r(\mathbf{q}^r)$ where it is finite, and adds constraints of the form (6.44) when $v^r(\mathbf{q}^r) = +\infty$ to prevent reoccurrence of the infeasible $\mathbf{q}^r$.

Thus, at iteration K of the resource directive decomposition of problem (6.34), the coordinator has bid prices $\mathbf{u}^{r,k}$ and constants $u_0^{r,k}$ for $k = 1, \ldots, K$, from the divisions that are derived from a subset of the set of all dual extreme points in (6.42). He uses this information to approximate the functions v^r and thereby select resource vectors $\mathbf{q}^r$ to send to the divisions by solving the resource directive master problem

$$
\begin{aligned}
\nu^K = \min \nu^1 + \ldots + \nu^R & \\
\nu^r \geq -u_0^{r,k} + \mathbf{u}^{r,k}\mathbf{q}^r \qquad & k = 1,\ldots,K \\
& r = 1,\ldots,R \\
\mathbf{q}^1 + \ldots + \mathbf{q}^R \leq \mathbf{q} &
\end{aligned}
\tag{6.46}
$$

Let $\mathbf{q}^{1,K},\ldots,\mathbf{q}^{R,K},\nu^{1,K},\ldots,\nu^{R,K}$ denote an optimal solution to problem (6.46).

LEMMA 6.3. The resource directive master problem minimal objective function value is a lower bound on the minimal objective function value of the block diagonal linear programming problem (6.34); that is

$$\nu^K \leq v$$

PROOF. Let $\mathbf{x}^1,\ldots,\mathbf{x}^R$, be an arbitrary feasible solution in (6.34), and let $\mathbf{q}^1 = \mathbf{Q}\mathbf{x}^1,\ldots,\mathbf{q}^R = \mathbf{Q}\mathbf{x}^R$; clearly, $\mathbf{q}^1 + \ldots + \mathbf{q}^R \leq \mathbf{q}$. For any k, we have for the dual feasible solution $(\mathbf{u}^{r,k},\mathbf{y}^{r,k})$ that

$$-\mathbf{u}^{r,k}\mathbf{Q}^r + \mathbf{y}^{r,k}\mathbf{A}^r \leq \mathbf{c}^r \qquad r = 1,\ldots,R$$

and since $\mathbf{x}^r \geq \mathbf{0}$, we can conclude

$$-\mathbf{u}^{r,k}\mathbf{Q}^r\mathbf{x}^r + \mathbf{y}^{r,k}\mathbf{A}^r\mathbf{x}^r \leq \mathbf{c}^r\mathbf{x}^r$$

or

$$-\mathbf{u}^{r,k}\mathbf{q}^r + u_0^{r,k} \leq \mathbf{c}^r\mathbf{x}^r \tag{6.47}$$

where

$$\mathbf{y}^{r,k}\mathbf{A}^r\mathbf{x}^r = \mathbf{y}^{r,k}\mathbf{b}^r = u_0^{r,k}$$

The optimal value $\nu^{r,K}$ is equal to the left-hand side of (6.47) for some k because otherwise it could not be optimal. Therefore

$$\nu^{r,k} \leq \mathbf{c}^r\mathbf{x}^r \qquad r = 1,\ldots,R$$

and we can conclude that

$$\nu^K = \sum_{r=1}^{R} \nu^{r,K} \leq \sum_{r=1}^{R} \mathbf{c}^r\mathbf{x}^r$$

Since $\mathbf{x}^1,\ldots,\mathbf{x}^R$ was an arbitrary feasible solution in (6.34), it follows immediately that $\nu^K \leq v$. ■

Thus, the master problem provides a lower bound on the minimal cost of operating the organization as formulated by (6.34). The coordinator tests the solution to see if in fact he can achieve the lower bound cost by sending the vector $\mathbf{q}^{r,K}$ to each division r and asking the division to send new bid prices on its share of the resources. The division resolves (6.40) with $\mathbf{q}^{r,K}$ as the right-hand side of (6.40b). If (6.40) is infeasible with this allotment of resource, then a constraint of the form (6.44) is by necessity discovered and added to the resource directive master program. If (6.40) is feasible, the division sends to the coordinator an optimal solution $\mathbf{x}^{r,K+1}$, an optimal dual solution $\mathbf{u}^{r,K+1}$ on the constraints (6.40b), and the scalar value $u_0^{r,K+1}=\mathbf{y}^{r,K+1}\mathbf{b}^r$ where $\mathbf{y}^{r,K+1}$ are optimal dual values on the constraints (6.40c). The coordinator checks this information against the quantity $\nu^{r,K}$ to see if the resource directive master problem needs to be updated. Specifically, if $\nu^{r,K}<-u_0^{r,K+1}+\mathbf{u}^{r,K+1}\mathbf{q}^{r,K}$, then the previously obtained bid prices from division r were inaccurate and a constraint of the form

$$\nu^r \geq -u_0^{r,K+1}+\mathbf{u}^{r,K+1}\mathbf{q}^r$$

is added to the master problem.

If the division linear programming problem (6.40) is feasible for all r, then the solutions $\mathbf{x}^{1,K+1},\ldots,\mathbf{x}^{R,K+1}$ are feasible in (6.34). The following theorem establishes that this solution is optimal if for all r there are no new constraints of this form to add to the master problem (6.46).

THEOREM 6.3. Suppose the R subproblems (6.40) have optimal solutions $\mathbf{x}^{r,K+1}$ for $r=1,\ldots,R$, and let $\mathbf{u}^{r,K+1}$, $\mathbf{y}^{r,K+1}$ for $r=1,\ldots,R$ denote optimal solutions in the dual problems (6.42). If for $r=1,\ldots,R$,

$$\mathbf{c}^r\mathbf{x}^{r,K+1}=-\mathbf{u}^{r,K+1}\mathbf{q}^{r,K}+\mathbf{y}^{r,K+1}\mathbf{b}^r \leq \nu^{r,K} \tag{6.48}$$

then $\mathbf{x}^{1,K+1},\ldots,\mathbf{x}^{R,K+1}$ is optimal in the block diagonal linear programming problem (6.34).

PROOF. By linear programming duality for problems (6.40) and (6.42), we have

$$\mathbf{c}^r\mathbf{x}^{r,K+1}=-\mathbf{u}^{r,K+1}\mathbf{q}^{r,K}+\mathbf{y}^{r,K+1}\mathbf{b}^r$$

and this implies from (6.46) and (6.48) that

$$v \leq \sum_{r=1}^{R} \mathbf{c}^r\mathbf{x}^{r,K+1} \leq \sum_{r=1}^{R} \nu^{r,K} = \nu^K$$

By Lemma 6.3, we must also have $v \geq \nu^K$ implying the desired result. ■

The algorithm embodied in the iterative solution of the resource directive master problem (6.46) and the divisional subproblems (6.40) converges finitely to an optimal solution to problem (6.34). This is because there are only a finite number of extreme points and extreme rays to generate constraints for the master problem.

EXAMPLE 6.3. A coal supplier has contracted to sell coal at a fixed price to customers in three separate locations over the next three time periods. The coal is to be supplied from mines at two other locations, each mine with its own extraction costs which are a function of the cumulative supply at the mine. The supplier wishes to meet his contracts at minimal total discounted extraction and transportation costs.

We let the indices 1,2 refer to the locations of the mines and the indices 3,4,5 refer to the locations of the customers. The variable y_{it} denotes the supply of coal measured in tons from the mine at location i in period t. The quantity q_i is an upper bound on the tons of coal that can be extracted from the mine at location i due to capacity limitations. The extraction cost from the mine at location i is given by $f_i(w_i)$ where w_i is cumulative extraction. Thus, if y_{it} is supplied from mine i in period t, the extraction cost for that period is

$$f_i(\mathbf{y}_{i1}+\cdots+\mathbf{y}_{it})-f_i(\mathbf{y}_{i1}+\cdots+\mathbf{y}_{i,t-1})$$

We let the variable x_{ijt} denote the flow of coal in tons from the mine at location i to the customer at location j in period t. The associated transportation unit cost is c_{ij} per ton of flow. The quantity d_{jt} denotes the tons of coal that must be delivered to the customers in location j in period t. The suppliers discount factor (see Section 4.4) is denoted by α, $0<\alpha<1$.

Tables 6.1–6.3 give the data for a sample coal supplier's problem of three periods duration. The extraction cost functions are assumed to be linear with the indicated slopes. The extraction capacities are $q_1=10$, $q_2=8$, and the discount factor $\alpha=0.8$. These data are input to the mathematical programming problem to be solved by the supplier

$$\begin{aligned}
\min \sum_{t=1}^{3} \alpha^{t-1}\Bigg\{ & \sum_{i=1}^{2} f_i\left(\sum_{\tau=1}^{t} y_{i\tau}\right)-f_i\left(\sum_{\tau=1}^{t-1} y_{i\tau}\right) \\
& + \sum_{i=1}^{2}\sum_{j=3}^{5} c_{ij}x_{ijt}\Bigg\} \\
\text{s.t.} \quad & \sum_{j=3}^{5} x_{ijt}-y_{it}=0 \qquad i=1,2 \text{ and } t=1,2,3 \qquad (6.49) \\
& \sum_{i=1}^{2} x_{ijt}=d_{jt} \qquad j=3,4,5 \text{ and } t=1,2,3 \\
& x_{ijt}\geq 0 \qquad q_i\geq y_{it}\geq 0
\end{aligned}$$

The objective function of this problem is time dependent because extraction costs in each period depend on extractions in previous periods.

We effect a resource directive decomposition by fixing the values of the supply variables y_{it}, permitting the residual problem to be solved as three separate transportation problems, one for each time period. The supply

TABLE 6.1 EXAMPLE OF RESOURCE DIRECTIVE DECOMPOSITION EXTRACTION COST FUNCTIONS

	MINE AT LOCATION 1		
Cumulative extraction	0 to 8 tons	8 to 16 tons	16 to 32 tons
Marginal costs	\$1/ton	\$4/ton	\$6/ton
	MINE AT LOCATION 2		
Cumulative extraction	0 to 4 tons	4 to 20 tons	20 to 40 tons
Marginal costs	\$2/ton	\$3/ton	\$4/ton

TABLE 6.2 EXAMPLE OF RESOURCE DIRECTIVE DECOMPOSITION COAL DEMANDS (TONS)

Demand Location	Period 1	Period 2	Period 3
3	2	3	5
4	4	4	4
5	8	5	1

TABLE 6.3 EXAMPLE OF RESOURCE DIRECTIVE DECOMPOSITION TRANSPORTATION COSTS (\$/TON)

From/To	Customer at Location 3	Customer at Location 4	Customer at Location 5
Mine at location 1	2	3	7
Mine at location 2	5	4	2

variables are then adjusted by Benders' algorithm. Specifically, the resource directive coordination problem for $\mathbf{y}=(y_{11},y_{21},y_{12},y_{22},y_{13},y_{23})$ is

$$\min v(\mathbf{y})=\sum_{t=1}^{3}\alpha^{t-1}\sum_{i=1}^{2}\left[f_i\left(\sum_{\tau=1}^{t}y_{i\tau}\right)-f_i\left(\sum_{\tau=1}^{t-1}y_{i\tau}\right)\right]$$

$$+\sum_{t=1}^{3}\alpha^{t-1}z_t(y_{it},y_{2t}) \tag{6.50}$$

$$\text{s.t.}\quad q_i\geq y_{it}\geq 0 \qquad i=1,2;\ t=1,2,3$$

where

$$z_t(y_{1t}, y_{2t}) = \min \sum_{i=1}^{2} \sum_{j=3}^{5} c_{ij} x_{ijt}$$

$$\text{s.t.} \sum_{j=3}^{5} x_{ijt} = y_{it} \qquad i = 1, 2$$

$$\sum_{i=1}^{2} (-x_{ijt}) = -d_{jt} \qquad j = 3, 4, 5$$

$$x_{ijt} \geq 0$$

or equivalently,

$$z_t(y_{1t}, y_{2t}) = \max \sum_{i=1}^{2} y_{it} u_{it} - \sum_{j=3}^{5} d_{jt} u_{jt}$$

$$\text{s.t. } u_{it} - u_{jt} \leq c_{ij} \qquad i = 1, 2; j = 1, 2, 3$$

The resource directive master problem derived from (6.50) is

$$\min \nu \qquad (6.51a)$$

$$\text{s.t. } \nu \geq v(\mathbf{y}^k) - \mathbf{y}^k \boldsymbol{\gamma}^k + \mathbf{y} \boldsymbol{\gamma}^k \qquad k = 1, \ldots, \mathbf{K} \qquad (6.51b)$$

$$\sum_{i=1}^{2} y_{it} = \sum_{j=3}^{5} d_{jt} \qquad t = 1, 2, 3 \qquad (6.51c)$$

$$q_i \geq y_{it} \geq 0 \qquad (6.51d)$$

where the constraints (6.51c) have been added to ensure feasibility of the supplies y_{it} in the transportation subproblems in (6.50). The $\mathbf{y}^k$ are previously generated supply vectors and the $\boldsymbol{\gamma}^k$ are subgradients of v calculated at these points. Specifically, the components of $\boldsymbol{\gamma}$ for $i = 1, 2$ are

$$\gamma_{it} = \alpha^{t-1} \left\{ f_i^* \left(\sum_{\tau=1}^{t} y_{i\tau} \right) + u_{it} \right\} + \sum_{\theta=t+1}^{3} \alpha^{\theta-1} \left\{ f_i^* \left(\sum_{t=1}^{\theta} y_{it} \right) - f_i^* \left(\sum_{t=1}^{\theta-1} y_{it} \right) \right\} \qquad (6.52)$$

where f_i^* indicates a slope of the piecewise linear function f_i.

We begin our calculations by assuming the initial supplies $y_{11}^1 = 7$, $y_{12}^1 = 6$, $y_{13}^1 = 5$, $y_{21}^1 = 7$, $y_{22}^1 = 6$, $y_{23}^1 = 5$. Optimal transportation subproblem solutions and dual solutions corresponding to this supply are shown in Table 6.4. The cost of the corresponding feasible solution to the coal supply problem (6.49) is \$158.28. Benders' algorithm proceeds by calculating the subgradient $\gamma^1 = (\gamma_{11}, \gamma_{21}, \gamma_{12}, \gamma_{13}, \gamma_{23})$ according to the formula (6.53). These values are $\gamma_{11} = 4.68$, $\gamma_{12} = -2.00$, $\gamma_{21} = 4.48$, $\gamma_{22} = 3.20$, $\gamma_{13} = 3.84$, $\gamma_{23} = 2.56$ and they are used to construct the resource directive master problem.

$$\min \nu$$
$$\text{s.t. } \nu \geq 42.44 + 4.68\mathbf{y}_{11} - 2.00\mathbf{y}_{21} + 4.48\mathbf{y}_{12} + 3.20\mathbf{y}_{22} + 3.84\mathbf{y}_{13} + 2.56\mathbf{y}_{23}$$
$$y_{11} + y_{21} = 14$$
$$y_{12} + y_{22} = 12 \tag{6.53}$$
$$y_{13} + y_{23} = 10$$
$$10 \geq y_{11} \geq 0 \qquad 10 \geq y_{12} \geq 0 \qquad 10 \geq y_{13} \geq 0$$
$$8 \geq y_{21} \geq 0 \qquad 8 \geq y_{22} \geq 0 \qquad 8 \geq y_{23} \geq 0$$

An optimal solution to this problem is the new supply vector $y_{11}^2 = 6$, $y_{21}^2 = 8$, $y_{12}^2 = 4$, $y_{22}^2 = 8$, $y_{13}^2 = 2$, $y_{23}^2 = 8$ that gives an optimal value of $\nu =$ \$126.20. This figure is a lower bound on the minimal cost in the coal supply problem (6.49) to be compared with the upper bound of \$158.28 previously computed.

We resolve the transportation problem with the new supply solution $\mathbf{y}^2$. Optimal solutions analogous to Table 6.4 are shown in Table 6.5. The cost of the corresponding feasible solution to (6.49) is \$156.22 indicating the solution is superior in cost to the first feasible solution. The new subgradient γ^2, calculated from (6.52), is $\gamma_{11}^2 = 3.40$, $\gamma_{21}^2 = 3.64$, $\gamma_{12}^2 = 3.20$, $\gamma_{22}^2 = 3.84$, $\gamma_{13}^2 = 2.56$, $\gamma_{23}^2 = 4.48$. The new constraint to add to the Benders' master problem is

$$\nu \geq 22.72 + 3.40\mathbf{y}_{11} + 3.64\mathbf{y}_{21} + 3.20\mathbf{y}_{12} + 3.84\mathbf{y}_{22} + 2.56\mathbf{y}_{13} + 4.48\mathbf{y}_{23} \tag{6.54}$$

TABLE 6.4 RESOURCE DIRECTIVE EXAMPLE OPTIMAL SUBPROBLEM SOLUTIONS ITERATION 1

	$t=1$	$t=2$	$t=3$
$\mathbf{x}_{13t}$	2	3	5
$\mathbf{x}_{14t}$	4	3	0
$\mathbf{x}_{15t}$	1	0	0
$\mathbf{u}_{1t}$	0	0	0
$\mathbf{x}_{23t}$	0	0	0
$\mathbf{x}_{24t}$	0	1	4
$\mathbf{x}_{25t}$	7	5	1
$\mathbf{u}_{2t}$	−5	1	1

TABLE 6.5 RESOURCE DIRECTIVE EXAMPLE OPTIMAL SUBPROBLEM SOLUTIONS ITERATION 2

	$t=1$	$t=2$	$t=3$
$\mathbf{x}_{13t}$	2	3	2
$\mathbf{x}_{14t}$	4	4	0
$\mathbf{x}_{15t}$	0	3	0
$\mathbf{u}_{1t}$	0	0	0
$\mathbf{x}_{23t}$	0	0	3
$\mathbf{x}_{24t}$	0	0	4
$\mathbf{x}_{25t}$	8	2	1
$\mathbf{u}_{2t}$	1	5	3

and it provides a new supply solution $y_{11}^3=10$, $y_{21}^3=4$, $y_{12}^3=10$, $y_{22}^3=2$, $y_{13}^3=10$, $y_{23}^3=0$ with $v=\$136.56$.

We terminate our calculations at this point with a feasible solution to the coal supply problem (6.49) costing \$156.72 and a lower bound on the minimal cost equaling \$136.56. ▲

There are a number of decomposable structures arising in mathematical programming in addition to the block diagonal structure of problem (6.34). We will discuss briefly several models and indicate how generalized linear programming and Benders' algorithm can be used on them. For example, Benders' algorithm has often been proposed and sometimes used to decompose the *mixed integer programming problem*

$$\begin{aligned} v = \min\ & \mathbf{cx} + \mathbf{hy} \\ \text{s.t.}\ & \mathbf{Ax} + \mathbf{Qy} \le \mathbf{b} \\ & x_j = 0 \text{ or } 1 \qquad \mathbf{y} \ge \mathbf{0} \end{aligned} \tag{6.55}$$

The decomposition of (6.55) consists of an integer programming master problem involving the $\mathbf{x}$ variables, and a linear programming subproblem involving the $\mathbf{y}$ variables. The approach is resource directive because for $\tilde{\mathbf{x}}$ fixed at zero-one values, (6.55) reduces to the linear programming subproblem

$$\begin{aligned} v(\tilde{\mathbf{x}}) = \min\ & \mathbf{hy} \\ \text{s.t.}\ & \mathbf{Qy} \le \mathbf{b} - \mathbf{A}\tilde{\mathbf{x}} \\ & \mathbf{y} \ge \mathbf{0} \end{aligned} \tag{6.56}$$

and $\mathbf{b} - \mathbf{A}\tilde{\mathbf{x}}$ is the resource available for this minimization.

Benders' algorithm for mixed integer programming proceeds in much the same manner as it did for the block diagonal linear programming problem. Specifically, we consider the dual to (6.56)

$$\begin{aligned} v(\tilde{\mathbf{x}}) = \max \mathbf{u}(\mathbf{b} - \mathbf{A}\tilde{\mathbf{x}}) \\ \mathbf{uQ} \le \mathbf{h} \\ \mathbf{u} \le \mathbf{0} \end{aligned} \tag{6.57}$$

The function $v(\mathbf{x})$ is a nondifferentiable convex function and the mixed integer programming problem (6.55) can be written as the nonlinear integer programming problem

$$\begin{aligned} v = \min \mathbf{cx} + v(\mathbf{x}) \\ \text{s.t. } x_j = 0 \text{ or } 1 \end{aligned}$$

The piecewise linear nature of v permits us to convert this problem into a linear integer programming problem.

Let $\mathbf{u}^t$ for $t = 1, \ldots, T$, denote the extreme points of the dual feasible region in (6.57) and let $\mathbf{u}^k$ for $k = 1, \ldots, K$, denote the extreme rays. At any arbitrary iteration of the algorithm, we use subsets of the dual extreme points and rays to construct the integer programming master problem

$$\nu' = \min \nu$$

$$\nu \ge \mathbf{cx} + \mathbf{u}^t(\mathbf{b} - \mathbf{Ax}) \qquad t = 1, \ldots, T' \tag{6.58a}$$

$$\mathbf{u}^k(\mathbf{b} - \mathbf{Ax}) \le 0 \qquad k = 1, \ldots, K' \tag{6.58b}$$

$$x_j = 0 \text{ or } 1$$

This problem is essentially a pure integer programming problem that can be solved by the methods discussed in Chapter 8. For our purposes here, it suffices to assume that some algorithm can be used to produce an optimal solution $\tilde{\mathbf{x}}$. The same argument used in Lemma 6.2 can be used here to show that ν' is a lower bound on the minimal mixed integer programming cost v. The solution $\tilde{\mathbf{x}}$ is used in the pair of linear programming problems (6.56) and (6.57) which are optimized by the simplex method. If (6.56) is infeasible, then a new dual extreme ray is discovered and a constraint is added to the set (6.58b). If (6.56) is feasible with right-hand-side $\mathbf{b} - \mathbf{A}\tilde{\mathbf{x}}$, then it has an optimal solution $\tilde{\mathbf{y}}$ and $(\tilde{\mathbf{x}}, \tilde{\mathbf{y}})$ is a feasible solution to the mixed integer programming problem (6.55). Let $\tilde{\mathbf{u}}$ denote the optimal solution to (6.57) found by the simplex method. The solution $(\tilde{\mathbf{x}}, \tilde{\mathbf{y}})$ is optimal in (6.55) if $\nu' \ge \mathbf{c}\tilde{\mathbf{x}} + \mathbf{h}\tilde{\mathbf{y}} = \mathbf{c}\tilde{\mathbf{x}} + \tilde{\mathbf{u}}(\mathbf{b} - \mathbf{A}\tilde{\mathbf{x}})$. If this optimality test fails, then the constraint

$$\nu \ge \mathbf{cx} + \tilde{\mathbf{u}}(\mathbf{b} - \mathbf{Ax})$$

is added to the set (6.58a).

Benders' algorithm for the mixed integer programming problem converges in a finite number of iterations to an optimal solution because each time the integer programming master problem (6.58) is solved, there is a new constraint in (6.58a) or (6.58b), and there are only a finite number of such constraints possible. The algorithm has the desirable feature of producing a feasible solution to (6.55) at each iteration that (6.56) is feasible, and the lower bound ν' on the cost of an optimal solution in (6.55). Moreover, the lower bounds are monotonically increasing with iterative solutions of the master problem.

Another linear programming problem with special structure, which can be usefully decomposed is the *two stage stochastic programming problem with recourse*

$$v = \min \mathbf{cx} + \sum_{r=1}^{R} p_r(\mathbf{h}^r\mathbf{y}^r) \tag{6.59a}$$

$$\text{s.t. } \mathbf{A}^0\mathbf{x} \leq \mathbf{b}^0 \tag{6.59b}$$

$$\mathbf{A}^r\mathbf{x} + \mathbf{Q}^r\mathbf{y}^r \leq \mathbf{b}^r \qquad r = 1, \ldots, R \tag{6.59c}$$

$$\mathbf{x} \geq \mathbf{0} \qquad \mathbf{y}^r \geq \mathbf{0} \qquad r = 1, \ldots, R \tag{6.59d}$$

where the p_r are probabilities; that is, $\sum_{r=1}^{R} p_r = 1$ and $p_r \geq 0$ for $r = 1, \ldots, R$. The variables $\mathbf{x}$ in this model are the first stage decision variables. After they have been selected, there are R possible outcomes of the data $\mathbf{h}^r$, $\mathbf{Q}^r$, and $\mathbf{b}^r$, each outcome occurring with probability p_r. The decision vectors $\mathbf{y}^r$ are used to achieve feasibility on the constraints (6.59c) in the second state of decision making after the random outcomes have occurred. The objective is to minimize the expected cost of the first and second stage decisions.

The stochastic programming problem (6.59) can be solved by resource directive methods using Benders' algorithm. Specifically, we rewrite it in partitioned form as

$$\begin{aligned} v = \min &\left\{ \mathbf{cx} + \sum_{r=1}^{R} v^r(\mathbf{x}) \right\} \\ \text{s.t. } &\mathbf{A}^0\mathbf{x} \leq \mathbf{b}^0 \\ &\mathbf{x} \geq \mathbf{0} \end{aligned} \tag{6.60}$$

where for each v we have the subproblem

$$\begin{aligned} v^r(x) = \min\ & p_r(\mathbf{h}^r\mathbf{y}^r) \\ \text{s.t. } & \mathbf{Q}^r\mathbf{y}^r \leq \mathbf{b}^r - \mathbf{A}^r\mathbf{x} \\ & \mathbf{y}^r \geq \mathbf{0} \end{aligned} \tag{6.61}$$

or equivalently,

$$\begin{aligned} v^r(x) = \max\ & \mathbf{u}^r(\mathbf{b}^r - \mathbf{A}^r\mathbf{x}) \\ \text{s.t. } & \mathbf{u}^r\mathbf{Q}^r \leq p_r\mathbf{h}^r \\ & \mathbf{u}^r \leq \mathbf{0} \end{aligned} \tag{6.62}$$

The representation (6.62) of the piecewise linear convex function v^r is the basis for the construction of a master problem analogous to (6.46) and (6.58). The construction is straightforward and we leave the details to the reader (see Exercise 6.9).

The final decomposable linear programming problem to be discussed is the *multistage linear programming problem*

$$
\begin{aligned}
v = \min\ & \mathbf{c}^1\mathbf{x}^1 + \mathbf{c}^2\mathbf{x}^2 + \ldots + \mathbf{c}^N\mathbf{x}^N \\
\text{s.t. } & \mathbf{A}^1\mathbf{x}^1 && = \mathbf{b}^1 \\
& -\mathbf{K}^1\mathbf{x}^1 + \mathbf{A}^2\mathbf{x}^2 && = \mathbf{b}^2 \\
& \quad \ddots \\
& -\mathbf{K}^{N-1}\mathbf{x}^{N-1} + \mathbf{A}^N\mathbf{x}^N && = \mathbf{b}^N \\
& \mathbf{x}^1 \geq \mathbf{0} \quad \mathbf{x}^2 \geq \mathbf{0}, \ldots, \mathbf{x}^{N-1} \geq \mathbf{0} \quad \mathbf{x}^N \geq \mathbf{0}
\end{aligned}
\tag{6.63}
$$

This class of problems is extremely important because many applications of linear programming involve time-dependent rather than static decisions. Replication over N periods of a large-scale single period model can clearly produce multistage models of great size that can be extremely difficult to solve exactly. Decomposition methods for multistage models such as (6.63) are more difficult to construct and implement because of the complicated interactions over time. We present here one method that illustrates the use of Benders' algorithm. A nested decomposition approach based on generalized linear programming is treated in Exercise 6.12.

A sequence of vectors $\mathbf{z}^1, \mathbf{z}^2, \ldots, \mathbf{z}^{N-1}$, with the same dimensions as $\mathbf{x}^1, \mathbf{x}^2, \ldots, \mathbf{x}^{N-1}$ determines a *trajectory* of right-hand-side vectors $\mathbf{b}^1, \mathbf{b}^2 + \mathbf{K}^1\mathbf{z}^1, \ldots, \mathbf{b}^N + \mathbf{K}^{N-1}\mathbf{z}^{N-1}$. A decision sequence $\mathbf{x}^1, \mathbf{x}^2, \ldots, \mathbf{x}^N$ is said to *follow* this trajectory if

$$\mathbf{A}^1\mathbf{x}^1 = \mathbf{b}^1 \qquad \mathbf{x}^1 \geq \mathbf{0}$$

and

$$\mathbf{A}\mathbf{x}^r = \mathbf{b}^r + \mathbf{K}^{r-1}\mathbf{z}^{r-1} \qquad \mathbf{x}^r \geq \mathbf{0} \qquad \text{for } r = 2, 3, \ldots, N$$

A following sequence of a trajectory may not be feasible in (6.63), but we can enforce feasibility by adding the conditions $\mathbf{K}\mathbf{x}^r = \mathbf{K}\mathbf{z}^r$, $r = 1, \ldots, N-1$. These ideas can be combined in a decomposition approach to (6.63), which separates it into N nondifferentiable optimization problems. Specifically, (6.63) is equivalent to

$$v = \min \sum_{r=1}^{N} v^r(\mathbf{z}^r)$$

where for $r=1,\ldots,N$, we solve the subproblem (6.64)

$$\begin{aligned} v^r(\mathbf{z}^r) = &\min \mathbf{c}^r\mathbf{x}^r \\ \text{s.t. } &\mathbf{A}^r\mathbf{x}^r = \mathbf{b}^r + \mathbf{K}^{r-1}\mathbf{z}^{r-1} \\ &\mathbf{K}^r\mathbf{x}^r = \mathbf{K}^r\mathbf{z}^r \\ &\mathbf{x}^r \geq \mathbf{0} \end{aligned} \tag{6.64}$$

and $\mathbf{K}^0\mathbf{z}^0 = \mathbf{0}$. Each of the functions v^r is nondifferentiable and convex and again we work with the dual representation

$$\begin{aligned} v^r(\mathbf{z}^r) = &\max \mathbf{u}^r(\mathbf{b}^r + \mathbf{K}^{r-1}\mathbf{z}^{r-1}) + \mathbf{y}^r\mathbf{K}^r\mathbf{z}^r \\ \text{s.t. } &\mathbf{u}^r\mathbf{A}^r + \mathbf{y}^r\mathbf{A}^r \leq \mathbf{c}^r \end{aligned} \tag{6.65}$$

Extreme points and rays of (6.65) are used in the construction of a master problem approximating (6.64) which interacts with the subproblems in the usual way. Details are left to the reader in Exercise 6.10.

6.7 EXERCISES

6.1 Prove Corollary 6.1 using the characterization (6.3) for the directional derivative of f.

6.2 (Grinold, 1970) Consider the problem

$$\begin{aligned} v = &\min f(\mathbf{x}) \\ \text{s.t. } &g(\mathbf{x}) \leq \mathbf{0} \\ &\mathbf{x} \in X \subseteq R^n \end{aligned}$$

where f and g are continuous and X is nonempty and compact. The function g maps R^n into R^m. Consider also the dual

$$\begin{aligned} d = &\sup L(\mathbf{u}) \\ &\text{s.t. } \mathbf{u} \geq \mathbf{0} \end{aligned}$$

where

$$L(\mathbf{u}) = \min_{x \in X} f(\mathbf{x}) + \mathbf{u}g(\mathbf{x})$$

Let

$$X(\mathbf{u}) = \{\mathbf{x} \in X \mid L(\mathbf{u}) = f(\mathbf{x}) + \mathbf{u}g(\mathbf{x})\}$$

and

$$g(X(\mathbf{u})) = \{\gamma \in R^m \mid \gamma = g(\mathbf{x}) \text{ for } \mathbf{x} \in X(\mathbf{u})\}$$

(1) Prove that

$$\nabla L(\mathbf{u}; \mathbf{d}) = \min_{\gamma \in \partial L(\mathbf{u})} \mathbf{d}\gamma$$

where $\partial L(\mathbf{u})$ denotes the subdifferential (the set of all subgradients) of L at $\mathbf{u}$.

(2) Prove that the set $\partial L(\mathbf{u})$ equals the convex hull of $g(X(\mathbf{u}))$.

6.3 Consider the linear inequality system

$$\alpha^k \mathbf{y} - b_k \geq 0 \qquad k = 1, \ldots, K$$

where each $\alpha^k \in R^m$. Prove that if there exists a $\bar{\mathbf{y}} \in R^m$ such that $\alpha^k \bar{\mathbf{y}} > b_k$ for all k, then the linear relaxation method given in Section 6.4 [see (6.18)] converges in a finite number of iterations with a $\mathbf{y}^l$ satisfying the linear inequality system.

6.4 Prove convergence of the subgradient optimization algorithm if the θ_l in (6.12) are chosen to satisfy $\theta_l \to 0$ as $l \to \infty$ and $\sum_{l=1}^{L} \theta_l \to \infty$ as $L \to \infty$.

6.5 Suppose we wish to apply the subgradient optimization algorithm to the constrained problem

$$\max f(\mathbf{y})$$
$$\text{s.t. } \mathbf{y} \in Y$$

where f is concave and Y is a closed convex set in R^m. Show that Theorem 6.2 is still true if we change (6.12) to

$$\mathbf{y}^{l+1} \in P(\mathbf{y}^l + \theta_l \gamma^l)$$

where P is the projection of $\mathbf{y}^l + \theta_l \gamma^l$ onto Y; that is, for any $\mathbf{z} \in R^m$, $P(\mathbf{z})$ is the set of $\bar{\mathbf{y}} \in Y$ satisfying

$$\|\bar{\mathbf{y}} - \mathbf{z}\| = \min_{\mathbf{y} \in Y} \|\mathbf{y} - \mathbf{z}\|$$

The norm $\|\cdot\|$ can be any norm in R^m.

6.6 (Bertsekas and Mitter, 1973) This exercise addresses properties of the ε-subgradient method given at the end of Section 6.5. We assume throughout that the nondifferentiable concave function $f(\mathbf{y})$ is finite for all $\mathbf{y} \in R^m$ and that f attains its maximum on R^m.

(1) Prove the following: Suppose $\mathbf{0} \notin \partial_\varepsilon f(\mathbf{y})$. Let $\mathbf{d}$ be any vector satisfying

$$\min_{\gamma \in \partial_\varepsilon f(\mathbf{y})} \mathbf{d}\gamma > 0$$

Then, we have

$$\max_{\theta \geq 0} f(\mathbf{y} + \theta \mathbf{d}) > f(\mathbf{y}) + \varepsilon$$

Thus, step 4 of the ε-subgradient method can always be carried out. Consider the vectors $\mathbf{y}^l$ generated by the ε-subgradient method.

(2) Prove that either

$$f(\mathbf{y}^k) = \max_{\mathbf{y} \in R^m} f(\mathbf{y}) \qquad \text{for some } k$$

or the infinite sequence satisfies

$$\lim_{l \to \infty} f(\mathbf{y}^l) = \max_{\mathbf{y} \in R^m} f(\mathbf{y})$$

(3) Suppose $\{\mathbf{z} | f(\mathbf{z}) = \max f(\mathbf{y})\}$ is bounded. Prove that there is at least one convergent subsequence of the sequence $\{\mathbf{y}^l\}$ and the limit point of the subsequence is optimal.

(4) Show how Step 3 of the ε-subgradient method can be mechanized along the lines of (6.27) when the ε-subdifferentials are not known but need to be generated.

6.7 How should the price directive decomposition scheme continue if $L^r(\mathbf{u}) = -\infty$ for some division r in (6.36)? In other words, what information should be sent to the price directive master problem (6.38) in this case?

6.8 Solve Example 6.3 using price directive rather than resource directive decomposition.

6.9 Give the resource directive master problem for the two stage stochastic programming problem with recourse (6.59).

6.10 Give the resource directive master problem for the multistage linear programming problem (6.63).

6.11 Show how the generalized primal-dual algorithm can be used to effect both price and resource directive decompositions of the block diagonal problem (6.34). Specifically, show how the generalized primal-dual algorithm can be used to ascend in the space of prices on the joint constraints (6.34b), or descend in the space of the resource shared in (6.34b). Give economic interpretations to both realizations of the algorithm.

6.12 Give a generalized linear programming nested decomposition of the multistage linear programming problem (6.63). Specifically, the decomposition consists of $N-2$ linear programming problems that are both price directive master problems and subproblems, one linear programming problem that is only a master problem, and one that is only a subproblem.

6.13 (Dantzig and Van Slyke, 1967) Generalized linear programming can be applied to the continuous-time linear optimal control problem

$$\begin{aligned}
&\min x_0(T) \\
\text{s.t. } &\frac{d\mathbf{x}(t)}{dt} = \mathbf{A}\mathbf{x}(t) + \mathbf{B}\mathbf{y}(t) \\
&\mathbf{y}(t) \in Y(t) \\
&\mathbf{x}(0) = \mathbf{x}^0 \\
&x_i(T) = x_i^T \qquad i = 1, \ldots, n
\end{aligned}$$

where $\mathbf{x}(t) = [x_0(t), \ldots, x_n(t)]$ is the state vector, $\mathbf{y}(t) = [y_1,(t), \ldots, y_r(t)]$ is the control vector, and $Y(t)$ is a closed convex set for all $t \in [0, T]$. This problem states that we wish to find an optimal control vector $\bar{\mathbf{y}}(t)$ such that $x_0(T)$ is a minimum. For a given y, the differential equations can be integrated yielding

$$\mathbf{x}(T)=\boldsymbol{\phi}(T)\mathbf{x}^0+\int_0^T \boldsymbol{\phi}(T)\boldsymbol{\phi}^{-1}(t)\mathbf{B}\mathbf{y}(t)\,dt$$

where $\boldsymbol{\phi}(t)$ is an $(n+1)\times(n+1)$ matrix satisfying

$$\frac{d\boldsymbol{\phi}(t)}{dt}=\mathbf{A}\boldsymbol{\phi}(t)$$
$$\boldsymbol{\phi}(0)=I$$

implying

$$\boldsymbol{\phi}(t)=e^{\mathbf{A}t}=I+At+\tfrac{1}{2}A^2t^2+\ldots$$

(1) Let

$$\mathbf{p}^{\mathbf{y}}=\int_0^T \boldsymbol{\phi}(T)\boldsymbol{\phi}^{-1}(t)\mathbf{B}\mathbf{y}(t)\,dt$$

and define

$$P=\{\mathbf{p}^{y}\,|\,\mathbf{y}(t)\in Y(t),\,0\le t\le T\}$$

Show that P is a convex set.

(2) Show how generalized linear programming can be used to solve the continuous-time linear optimal control problem by generating columns from P. What is the master problem and what is the subproblem used to generate columns for the master problem?

(3) Suppose the generalized linear programming master problem produces an optimal dual vector $\bar{\mathbf{u}}\in R^r$ on the columns $\boldsymbol{\rho}^{\mathbf{y}}$ in the master such that

$$\int_0^T \bar{\mathbf{u}}\boldsymbol{\phi}(T)\boldsymbol{\phi}^{-1}(t)\mathbf{B}[\mathbf{y}(t)-\bar{\mathbf{y}}(t)]\,dt\le 0$$

for all $\mathbf{y}$ satisfying $\mathbf{y}(t)\in Y(t)$, where $\bar{\mathbf{y}}(t)$ is the indicated control from the master; that is,

$$\bar{\mathbf{y}}(t)=\sum_{k=1}^{K}\bar{\lambda}_k^K\mathbf{y}^k(t)\qquad \text{for all } t\in[0,T]$$

where $\bar{\lambda}_k^K$ are the optimal weights in the master problem on the previously generated controls $\mathbf{y}^k(t)$, $t\in[0,T]$. Show that $\bar{\mathbf{y}}(t)$ is an optimal control in the continuous-time linear optimal control problem. This is the integral form of *Pontryagin's maximum principle*.

(4) Suppose there are no control constraints $Y(t)$; that is, $Y(t)=R^r$ for all $t\in[0,T]$. Show that the integral inequality of part (3) must hold

pointwise and therefore $\bar{\mathbf{y}}(t)$ is an optimal control if

$$\bar{\boldsymbol{\Pi}}(t)\mathbf{B}\bar{\mathbf{y}}(t)=\max_{\mathbf{y}(t)\in R^r}\bar{\boldsymbol{\Pi}}(t)\mathbf{B}\mathbf{y}(t)$$

for almost all $t\in[0,T]$ where $\bar{\boldsymbol{\Pi}}(t)=-\bar{\mathbf{u}}\boldsymbol{\phi}(T)\boldsymbol{\phi}^{-1}(t)$. This is *Pontryagin's maximum principle*. Show also that $\bar{\boldsymbol{\Pi}}(t)$ satisfies the *adjoint equation*

$$\frac{d\bar{\boldsymbol{\Pi}}(t)}{dt}=-\bar{\boldsymbol{\Pi}}(t)\mathbf{A}$$

6.14 (Gale, 1967; Geoffrion, 1971) Consider the family of mathematical programming problems defined for all $\mathbf{b}\in R^m$

$$\begin{aligned} v(\mathbf{b})=\min f(\mathbf{x}) & \\ \text{s.t. } g(\mathbf{x})\le\mathbf{b} & \\ \mathbf{x}\in X & \end{aligned}$$

where g is a mapping from R^n to R^m, f,g are convex functions, and X is a nonempty convex set. Let (P) denote the specific problem when $\mathbf{b}=\mathbf{0}$ and define its dual in the usual way; that is

$$\begin{aligned} d=\sup L(\mathbf{u}) & \qquad \text{(D)} \\ \text{s.t. } \mathbf{u}\ge\mathbf{0} & \end{aligned}$$

where

$$L(\mathbf{u})=\min_{\mathbf{x}\in X} f(\mathbf{x})+\mathbf{u}g(\mathbf{x})$$

We say (P) is *stable* if $v(\mathbf{0})$ is finite and there exists a scalar $M>0$ such that

$$\frac{v(\mathbf{0})-v(\mathbf{b})}{\|\mathbf{b}\|}\le M \qquad \text{for all } \mathbf{b}\ne\mathbf{0}$$

where $\|\cdot\|$ is any norm on R^m.

(1) Define the set

$$B=\{\mathbf{b}\in R^m \mid g(\mathbf{x})\le\mathbf{b} \text{ for some } \mathbf{x}\in X\}$$

Prove that B is a convex set and v is a convex function on B.

(2) Suppose h is a convex function on a convex set $Y\subseteq R^m$ and $h(\mathbf{y})<+\infty$ for all $\mathbf{y}\in Y$. Let $\|\cdot\|$ be any norm on R^m, and let $\bar{\mathbf{y}}$ be any point where h is finite. Prove that the convex function h has a subgradient at $\mathbf{y}\in Y$ if and only if there exists a positive scalar M such that

$$\frac{h(\mathbf{y})-h(\bar{\mathbf{y}})}{\|\bar{\mathbf{y}}-\mathbf{y}\|}\le M \qquad \text{for all } \mathbf{y}\in Y \text{ such that } \mathbf{y}\ne\bar{\mathbf{y}}$$

Note that for a convex function h defined on R^m, the vector $\bar{\gamma}$ is a subgradient at $\bar{\mathbf{y}}$ if

$$h(\mathbf{y}) \geq h(\bar{\mathbf{y}}) + (\mathbf{y} - \bar{\mathbf{y}})\bar{\gamma} \qquad \text{for all } \mathbf{y}$$

(3) Prove that if (P) has an optimal solution, then $\mathbf{u}$ is optimal in (D) if and only if $-\mathbf{u}$ is a subgradient of v at $\mathbf{b}=\mathbf{0}$.

(4) Prove that if (P) has an optimal solution, then an optimal dual solution exists if and only if (P) is stable.

6.8 NOTES

SECTION 6.1. Decomposition of large-scale linear programming is covered in detail by Lasdon (1970). That text also has some mathematical developments touching on nondifferentiable optimization; see also Danskin (1967). The constructs of convex analysis used in nondifferentiable optimization methods are treated extensively in the monograph of Rockafellar (1970). Balinski (1975) edited a special issue of Mathematical Programming devoted to nondifferentiable optimization. Recent results are summarized by LeMaréchal (1978).

SECTION 6.2. The necessary and sufficient optimality conditions using constructs of convex analysis can be found in Grinold (1972) and Fisher et al. (1975).

SECTION 6.3. The properties of feasible direction ascent algorithms were addressed directly in an early monograph by Zoutendijk (1960) who included some discussion about the primal-dual simplex algorithm as a feasible direction ascent algorithm.

SECTION 6.4. Subgradient optimization has been studied theoretically by a number of Russian authors including Polyak (1967, 1969) and Demyanov (1966). It was first successfully applied to mathematical programming problems by Held and Karp (1971) when they used it to approximately solve dual problems to the traveling salesman problem. See also Held et al. (1974) for its application to other large-scale linear programming problems such as the multicommodity flow problem. Linear relaxation methods were proposed and studied by Agmon (1954) and Motzkin and Schoenberg (1954). Oettli (1972) has studied the rates of convergence of primal-dual methods for mathematical programming problems.

SECTION 6.5. Grinold (1972) has shown how the primal-dual algorithm could be generalized for solving large-scale linear programming problems. Fisher et al. (1975) discuss how the same methods can be used to solve dual problems arising in the study of discrete optimization problems such as the integer programming and traveling salesman problems to be studied in Chapter 8. The use of ε-optimality in nondifferentiable optimization is given by Bertsekas and Mitter (1973) and a generalized primal-dual algorithm based on ε-optimality ideas is due to LeMaréchal (1974).

SECTION 6.6. The price directive interpretation of generalized linear programming is given by Baumol and Fabian (1964); see also Lasdon (1968). The original reference for Benders' algorithm is Benders (1962); an extension to resource directive decomposition where the subproblem is nonlinear is given by Geoffrion (1972). Grinold (1972) shows how most of the large-scale linear programming problems covered in this section can be solved by the generalized primal-dual algorithm. Geoffrion and Graves (1974) successfully applied Benders' algorithm to a large-scale mixed integer programming model of warehouse location and distribution. The resource directive solution of the multistage linear programming problem is due to Grinold (1972).

7 NONLINEAR PROGRAMMING

7.1 INTRODUCTION

The central concern of this chapter is a study of optimization problems and methods where the objective function and sometimes the constraint functions are differentiable. Numerical methods based on principles of multivariate differential calculus are used to solve these problems, often by trying to establish the Kuhn-Tucker conditions of Section 5.5 for some primal-dual pair. Convexity of the objective and constraint functions is required for the conditions to be sufficient as well as necessary for optimality, and the convergence of nonlinear programming algorithms to a globally optimal solution is also dependent on these properties.

Nonlinear programming problems with nonlinear objective functions but linear constraints are easier to solve than those with nonlinear constraints. We present a classical feasible direction algorithm for linearly constrained problems. For the more difficult nonlinearly constrained problems, we present penalty function methods and show that they are, in effect, derived from generalizations of the mathematical programming duality theory presented in Chapter 5.

The remainder of the chapter is devoted to mathematical programming constructs related to the existence and computation of economic equilibria. Considerable insight into the relationship is gained by our study of the quadratic programming problem and its generalization, the linear complementarity problem. On the one hand, economic equilibrium conditions can often be viewed as the Kuhn-Tucker conditions for an underlying mathematical programming problem implying the existence and computation of equilibria can be established by solving the underlying problem. On the other hand, the *existence* of economic equilibria can be established by fixed point arguments without recourse to mathematical programming, but solution methods for the linear complementarity problem can be extended to the *computation* of fixed points. These computational methods are applicable to more general fixed point problems and not only those arising in the study of economic equilibria.

Nonlinear programming is a vast subject that we can only sample in this chapter. Unfortunately, we will not be able to give sufficient attention to a host of important numerical analysis difficulties that must be overcome in the construction of nonlinear programming algorithms. Of particular importance in this regard are the unconstrained optimization procedures discussed briefly in the next section.

7.2 UNCONSTRAINED OPTIMIZATION

The computation of unconstrained optima is a basic subroutine in constrained nonlinear programming. Moreover, a number of applications of nonlinear programming are unconstrained optimization problems including least-squares analysis in statistical regression and some problems in engineering design.

Our study of unconstrained optimization begins with methods for minimizing a function f of a single variable. For the moment, we consider the specific problem

$$\begin{aligned} &\min f(t) \\ &\text{s.t. } 0 \le t \le L \end{aligned}$$

where f is a continuous function that is not constant on any interval, and $-f$ is a unimodal function; that is, f has a single local minimum. A typical function of this type is depicted in Figure 7.1. Suppose we select arbitrarily the two points $a, b \in [0, L]$ and compute $f(a)$ and $f(b)$. If $f(b) > f(a)$, as depicted in Figure 7.1, then we can conclude that the minimal point $t^* \in [0, b]$ since f could not decrease again in the interval $[b, L]$. If, however, it had been the case that $f(b) < f(a)$, then by the same reasoning we could conclude that $t^* \in [a, L]$. Finally, if $f(b) = f(a)$, then we could conclude that $t^* \in [a, b]$. A *line search algorithm* is one that systematically evaluates f, eliminating subregions of $[0, L]$ until the location of the minimum point t^* is determined to a desired level of accuracy. According to a certain criterion, the line search algorithm we are about to discuss is optimal among all line search algorithms.

For any positive integer n, suppose we wish to determine the largest interval $[0, L_n]$ with the property that a line search algorithm can always locate the minimum of $f(t)$ on a subinterval of unit length by calculating at most n values of $f(t)$. Since the maximum value of L_n may not be attained, we define

$$F_n = \sup L_n$$

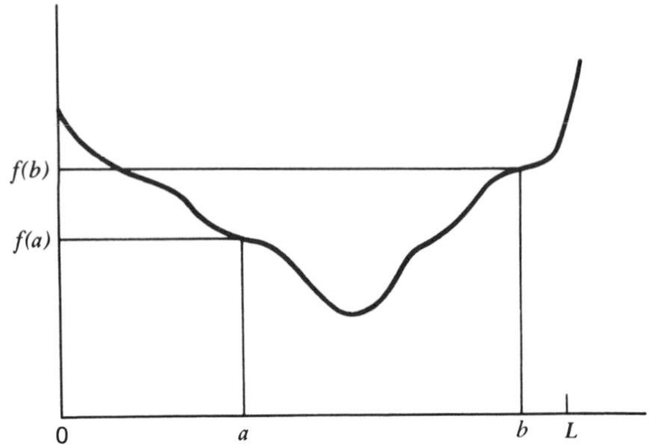

Figure 7.1. Line search for the minimum of a function of a single variable.

THEOREM 7.1. F_n is the nth Fibonacci number; that is, $F_0 = F_1 = 1$, and $F_n = F_{n-1} + F_{n-2}$, $n = 2, 3, \ldots$.

PROOF. The proof is by induction on n. The case $n=1$ is trivial since, with one calculation, $F_n = 1$. Therefore, we make the induction hypothesis

$$F_k = F_{k-1} + F_{k-2} \qquad k = 2, \ldots, n-1$$

We are required to show that $F_n = F_{n-1} + F_{n-2}$.

1. $F_n \leq F_{n-1} + F_{n-2}$. Pick any two points t_1, t_2 in an interval $[0, L_n]$ for which a line search exists yielding a subinterval of one unit or less containing t^*. Assume $0 < t_1 < t_2 < L_n$. We want to show that $L_n \leq F_{n-1} + F_{n-2}$. It could happen that $f(t_1) < f(t_2)$, in which case we would conclude that $t^* \in [0, t_2]$. Since it could also happen that t^* is actually in $[0, t_1]$, it must be that $t_1 \leq F_{n-2}$ because we would have $n-2$ calculations left to find t^* within a subinterval of one unit. On the other hand, it could happen that $f(t_1) > f(t_2)$ implying $t^* \in [t_1, L_n]$ which in turn implies $L_n - t_1 \leq F_{n-1}$ because we would have $n-1$ calculations left to find t^* within a subinterval of one unit. The point t_2 could be anywhere in the interval $[t_1, L_n]$. Combining the two inequalities involving t_1 gives us the desired result.
2. $F_n \geq F_{n-1} + F_{n-2}$. Take $L_n = F_{n-1} + F_{n-2} - \varepsilon$ for arbitrary $\varepsilon > 0$ and consider a line search of the interval $[0, L_n]$. We choose to make two calculations of $f(t)$ at the points $t_1 = F_{n-2} - \varepsilon/2$ and $t_2 = F_{n-1} - \varepsilon/2$. We can verify that this value of L_n admits a line search locating t^* within one unit. If $t^* \in [0, t_1]$, then we can achieve our goal because $t_1 < F_{n-2}$ and we have $n-2$ calculations remaining. The same argument holds if $t^* \in [t_2, L_n]$ because $L_n - t_2 < F_{n-2}$. If $t^* \in [t_1, t_2]$, we can again achieve our goal because $t_2 - t_1 = F_{n-1} - F_{n-2} = F_{n-3} < F_{n-2}$ and $n-2$ calculations remain. ■

The line search algorithm suggested by Theorem 7.1 is clear. Given an interval $[0, L]$ over which we wish to locate to within one unit the t^* minimizing $f(t)$, we begin by making two calculations of f at $t_1 = F_{n-2}$ and $t_2 = F_{n-1}$, where F_n is the smallest Fibonacci number such that $F_n \geq L$. If it is ascertained that $t^* \in [0, F_{n-1}]$, because $f(t_1) < f(t_2)$, then we make another calculation at $t_3 = F_{n-3}$ and repeat the elimination process. If it is ascertained that $t^* \in [F_{n-2}, L]$, because $f(t_1) > f(t_2)$, then we make another calculation at $t_3 = L - F_{n-3}$, and repeat the elimination process. Finally, if it is ascertained that $t \in [F_{n-2}, F_{n-1}]$, because $f(t_1) = f(t_2)$, then we make two more calculations at $t_3 = F_{n-5}$ and $t_4 = F_{n-4}$ because in this case we have reduced the interval of interest to one of length $F_{n-1} - F_{n-2} = F_{n-3}$.

Another approach to one-dimensional minimization when the function f is twice differentiable is curve fitting by Newton's method. We know f achieves its minimum at a stationary point x, that is, at an x satisfying $g(x) = f'(x) = 0$. Newton's method for finding such an x is to approximate f at each x_k in the line search by its second order Taylor expansion $h(x) = f(x_k) + (x - x_k)f'(x_k) + \frac{1}{2}(x - x_k)^2 f''(x_k)$. The next point x_{k+1} is found by minimizing $h(x)$, which is accomplished by setting $h'(x) = 0$. This gives us $x_{k+1} = x_k - [f'(x_k)]/[f''(x_k)]$. Note that the new point x_{k+1} does not depend upon the value of $f(x_k)$. In terms of $g(x)$, the new point satisfies

$$x_{k+1} = x_k - \frac{g(x_k)}{g'(x_k)} \tag{7.1}$$

Newton's method as described can be applied to any one-dimensional optimization when f is twice differentiable. The following theorem gives sufficient conditions under which the method will converge to a stationary point.

THEOREM 7.2. Suppose the function $g(x) = f'(x)$ has a continuous second derivative, and let x^* satisfy $g(x^*) = 0$, $g'(x^*) \neq 0$. Then, if x_1 is sufficiently close to x^*, the sequence $\{x_k\}_{k=1}^{\infty}$ generated by Newton's method (7.1) converges to x^*.

PROOF. For all x in a sufficiently small neighborhood of x^*, there is a k_1 satisfying $|g''(x)| < k_1$ by the continuity of g'', and there is a k_2 satisfying $|g'(x)| > k_2$ since $g'(x^*) \neq 0$. Since g'' is continuous, we have for any x_k in the sequence that $0 = g(x^*) = g(x_k) + (x^* - x_k)g'(x_k) + \frac{1}{2}(x^* - x_k)^2 g''(y)$ for some point y between x^* and x_k. Using (7.1) to express x_{k+1} in terms of x_k and rearranging, this gives us

$$x_{k+1} - x^* = -\frac{1}{2}\frac{g''(y)}{g'(y)}(x_k - x^*)^2$$

Thus, in the sufficiently small neighborhood of x^*,

$$|x_{k+1} - x^*| \leq \frac{k_1}{2k_2}|x_k - x^*|^2$$

and if we choose $|x_1 - x^*| < 2k_2/k_1$, we have by a simple induction argument that $|x_{k+1} - x^*| < \beta^k |x_1 - x^*|$ for $\beta = k_1|x_1 - x^*|/2k_2 < 1$, implying x_{k+1} converges to x^*. ■

This concludes a very brief discussion of one-dimensional optimization. The two procedures presented are representative of two broad classes of one-dimensional procedures, combinatorial and differential.

We now turn to a consideration of methods for the unconstrained minimization of functions defined over R^n. As we have said in Chapter 5, a necessary condition for a differentiable function f to possess a minimum at the point $\mathbf{x}^* \in R^n$ is that $\nabla f(\mathbf{x}^*) = \mathbf{0}$. Such a point is called a *stationary point*. Thus we are faced with the problem of finding a solution to the system of nonlinear equations $\nabla f(\mathbf{x}) = \mathbf{0}$. The ascent algorithm of Chapter 6 for nondifferentiable functions could be applied to this problem, with the minor modification that we would use it to try to descend to a minimum. By assuming continuous differentiability of f, however, we can derive a convergent descent algorithm for this problem that has better convergence properties and does not require, as we did in Chapter 6, knowledge of the minimum value of f.

The optimal gradient descent algorithm generates the sequence $\{\mathbf{x}^k\}_{k=1}^{\infty}$ given by

$$\mathbf{x}^{k+1} = \mathbf{x}^k - t_k \nabla f(\mathbf{x}^k) \tag{7.2}$$

where t_k is the smallest $t \geq 0$ such that

$$f[\mathbf{x}^k - t_k \nabla f(\mathbf{x}^k)] = \min_{t \geq 0} f[\mathbf{x}^k - t\nabla f(\mathbf{x}^k)]$$

Note that the calculation of t_k is a one-dimensional optimization of the type just discussed.

THEOREM 7.3. Suppose f has continuous partial derivatives and also that $\{\mathbf{x} | f(\mathbf{x}) \leq f(\mathbf{x}^1)\}$ is compact. Then any limit point of the sequence generated by the optimal gradient descent algorithm (7.2) is a stationary point.

PROOF. The compactness of $\{\mathbf{x} | f(\mathbf{x}) \leq f(\mathbf{x}^1)\}$ implies the existence of at least one limit point of the sequence generated by the algorithm. Suppose the contrary; that is, suppose there exists a limit point $\mathbf{x}^*$ such that $\nabla f(\mathbf{x}^*) \neq 0$. Let $\{\mathbf{x}^{k_i}\}_{i=1}^{\infty}$ be a subsequence converging to $\mathbf{x}^*$. We have

$$f(\mathbf{x}^{k_i}) > f(\mathbf{x}^{k_i+1}) \geq f(\mathbf{x}^{k_{i+1}}) > f(\mathbf{x}^*) \tag{7.3}$$

for all i because the algorithm either terminates finitely with a stationary point or, since f is differentiable, it finds a strict improvement at each step. Let t^* be chosen such that

$$f[\mathbf{x}^* - t^*\nabla f(\mathbf{x}^*)] = \min_{t \geq 0} f[\mathbf{x}^* - t\nabla f(\mathbf{x}^*)]$$

and let $\mathbf{x}^{**}$ denote the point $\mathbf{x}^* - t^*\nabla f(\mathbf{x}^*)$. Note that $\mathbf{x}^{**} \neq \mathbf{x}^*$ since $\nabla f(\mathbf{x}^*) \neq 0$, and let $\varepsilon = f(\mathbf{x}^*) - f(\mathbf{x}^{**}) > 0$. Since f is continuous, given this $\varepsilon > 0$, there exists a $\delta > 0$ such that $\|\mathbf{x} - \mathbf{x}^{**}\| < \delta$ implies $|f(\mathbf{x}) - f(\mathbf{x}^{**})| < \varepsilon/2$.

Choose an $\mathbf{x}^{k_j}$ satisfying

$$\|\mathbf{x}^{k_j} - \mathbf{x}^*\| < \frac{\delta}{2}$$

and

$$\|\nabla f(\mathbf{x}^*) - \nabla f(\mathbf{x}^{k_j})\| < \frac{\delta}{2t^*}$$

where the first inequality follows from the definition of $\mathbf{x}^*$ as the limit point of the subsequence $\{\mathbf{x}^{k_i}\}$ and the second inequality follows from the continuity of ∇f. Thus,

$$\begin{aligned}
&\|\mathbf{x}^{k_j} - t^*\nabla f(\mathbf{x}^{k_j}) - [\mathbf{x}^* - t^*\nabla f(\mathbf{x}^*)]\| \\
&\leq \|\mathbf{x}^{k_j} - \mathbf{x}^*\| + t^*\|\nabla f(\mathbf{x}^*) - \nabla f(\mathbf{x}^{k_j})\| \\
&\leq \frac{\delta}{2} + \frac{\delta}{2} = \delta
\end{aligned}$$

The implication is that $\mathbf{x}^{k_j} - t^*\nabla f(\mathbf{x}^{k_j})$ is in a δ neighborhood of $\mathbf{x}^{**}$. These conditions permit us to write

$$\begin{aligned}
f(\mathbf{x}^{k_j+1}) &\leq f[\mathbf{x}^{k_j} - t^*\nabla f(\mathbf{x}^{k_j})] \\
&\leq f(\mathbf{x}^{**}) + \frac{\varepsilon}{2} \\
&< f(\mathbf{x}^*) \\
&< f(\mathbf{x}^{k_{j+1}}) \\
&\leq f(\mathbf{x}^{k_j+1})
\end{aligned}$$

where the first inequality follows from the definition of $\mathbf{x}^{k_j+1}$ as a minimizing point in the direction $-\nabla f(\mathbf{x}^{k_j})$, the second inequality follows from the fact that $\mathbf{x}^{k_j} - t^*\nabla f(\mathbf{x}^{k_j})$ is in a δ neighborhood of $\mathbf{x}^{**}$, the first strict inequality by the definition of ε, the second strict inequality and the final inequality by the descent properties of the algorithm. Thus, we have exhibited a contradiction and the theorem is proven. ■

The optimal gradient descent algorithm relies on first-order approximations to the function f at the points $\mathbf{x}^k$. For a highly nonlinear function with continuous second partials, it is preferable to use second-order approximations. In addition to tracking more closely the curvature of f, the second order approximations have better convergence properties. This is the idea behind Newton's method for functions of a vector argument which uses the second-order Taylor series approximation

$$f(\mathbf{x}) \cong g(\mathbf{x}) = f(\mathbf{x}^k) + \nabla f(\mathbf{x}^k)(\mathbf{x}-\mathbf{x}^k) + \tfrac{1}{2}(\mathbf{x}-\mathbf{x}^k)H_f(\mathbf{x}^k)(\mathbf{x}-\mathbf{x}^k)$$

where $H_f(\mathbf{x}^k)$ is the $n \times n$ Hessian matrix. The point $\mathbf{x}^{k+1}$ is chosen to be any point which minimizes the approximation. These points must satisfy $\nabla g(\mathbf{x}^{k+1}) = \mathbf{0}$ or $\nabla f(\mathbf{x}^k) + (\mathbf{x}^{k+1} - \mathbf{x}^k)H_f(\mathbf{x}^k) = 0$. This is a linear system with zero, one, or an infinity of solutions.

Newton's method cannot be expected to converge to a stationary point. It is intended for use primarily in regions of R^n close to a strict local minimum of f. In such a region, it can be shown that the function f is strictly convex, implying that the Hessian matrix is positive definite and invertible and the descent algorithm is given by

$$\mathbf{x}^{k+1} = \mathbf{x}^k - \nabla f(\mathbf{x}^k)\left[H_f(\mathbf{x}^k)\right]^{-1} \tag{7.4}$$

Comparing the optimal gradient descent algorithm (7.2) to (7.4), we see readily that they are both special cases of the general algorithm

$$\mathbf{x}^{k+1} = \mathbf{x}^k - t_k \nabla f(\mathbf{x}_k)\mathbf{M}_k$$

for suitable matrices $\mathbf{M}_k$. In regions distant from a local minimum, the step can be chosen by selecting t_k as we did for the optimal gradient algorithm and $M_k = I$. Near a local minimum, we take $t_k = 1$ and $M_k = [H_f(\mathbf{x}^k)]^{-1}$ to obtain (7.4). Global convergence to a stationary point can be guaranteed by looking at the eigenvalues of $H_f(\mathbf{x}^k)$ in the following manner. At each point $\mathbf{x}^k$, calculate the eigenvalues of $H_f(\mathbf{x}^k)$ and let ε_k be the smallest nonnegative scalar such that the matrix $\varepsilon_k I + H_f(\mathbf{x}^k)$ has eigenvalues greater than or equal to a preselected $\delta > 0$. This implies the matrix is positive definite. The descent step is then given by

$$\mathbf{x}^{k+1} = \mathbf{x}^k - t_k \nabla f(\mathbf{x}^k)\left[\varepsilon_k I + H_f(\mathbf{x}^k)\right]^{-1}$$

where t_k is chosen to minimize over $t \geq 0$, the quantity

$$f\left\{\mathbf{x}^k - t\nabla f(\mathbf{x}^k)\left[\varepsilon_k I + H_f(\mathbf{x}^k)\right]^{-1}\right\}$$

If the algorithm enters a region near a strict local minimum where the smallest eigenvalue of $H_f(\mathbf{x}^k)$ is as great as δ, then this algorithm reduces to Newton's method. Outside of such regions, the function is made to look positive definite. Selection of the proper $\delta > 0$ is an art because a small δ can produce nearly singular matrices, whereas a large δ can delay or eliminate terminal convergence to a stationary point by Newton's method.

An inherent difficulty with the second-order methods described above is the necessity of calculating and inverting the Hessian matrix H_f at each step. Methods have been devised for approximating the Hessian and its inverse by faster iterative schemes. This is the purpose of *conjugate gradient methods*. Space does not permit a development of these methods, but there is an exercise at the end of the chapter presenting some of the central ideas.

7.3 LINEARLY CONSTRAINED NONLINEAR PROGRAMMING PROBLEMS

The specific problem we will address in this section is

$$\begin{aligned} v = \min & f(\mathbf{x}) \\ \text{s.t. } & \mathbf{A}\mathbf{x} = \mathbf{b} \\ & \mathbf{x} \geq \mathbf{0} \end{aligned} \tag{7.5}$$

where f is a differentiable function defined on R^n and the matrix $\mathbf{A}$ is $m \times n$. A number of algorithms have been proposed for this problem. In the special case when f is separable and convex, that is, $f(\mathbf{x}) = \Sigma_{j=1}^{n} f_j(x_j)$ and the f_j are convex, then we have already observed in Chapter 1 [see problem (1.20)] that linear programming can be used to obtain an approximately optimal solution using piecewise linear approximations of the f_j. If f is separable and nonconvex, then mixed integer programming can be used to obtain an approximately optimal solution; the way in which this can be done is discussed in Chapter 8.

Our main concern, therefore, is solving problem (7.5) when f is differentiable and not separable. In general, we can expect an algorithm to converge to an optimal solution to (7.5) only if f is convex. If f is nonconvex, then the best we can hope for is that the algorithm converges to a local minimum.

In this section, we present a typical algorithm for problem (7.5), called the *feasible directions algorithm*, and show that it converges to an optimal solution under the appropriate assumptions. Space does not permit an extensive discussion of other algorithms for (7.5). One alternative is the generalized linear programming algorithm discussed in Section 5.4. Other algorithms are discussed in exercises at the end of the chapter. The method of feasible directions is a generalization of the optimal gradient method for unconstrained optimization discussed in the previous section. The method is also closely related to the ascent (or descent) methods discussed in Chapter 6. The difference is that we make explicit use of the differentiability properties of f in ensuring and establishing convergence of the algorithm.

For convenience, define the set

$$F=\{\mathbf{x}|\mathbf{A}\mathbf{x}=\mathbf{b},\mathbf{x}\geq\mathbf{0}\}$$

The feasible directions algorithm generates the sequence

$$\mathbf{x}^{k+1}=\mathbf{x}^k+t_k\mathbf{r}^k \tag{7.6}$$

where each $\mathbf{x}^k\in F$ and $\mathbf{r}^k$ is a feasible direction of descent at $\mathbf{x}^k$; that is, the directional derivative $\nabla f(\mathbf{x}^k;\mathbf{r}^k)<0$ and $\mathbf{x}^k+\theta\mathbf{r}^k$ is feasible for $\theta>0$ and sufficiently small. The vector $\mathbf{r}^k$ is selected by solving the direction-finding linear programming problem

$$\begin{aligned}\min\ &\nabla f(\mathbf{x}^k)\mathbf{z}\\ \text{s.t. } &\mathbf{A}\mathbf{z}=\mathbf{b}\\ &\mathbf{z}\geq\mathbf{0}\end{aligned} \tag{7.7}$$

THEOREM 7.4. If the minimal objective function value in (7.7) equals $\nabla f(\mathbf{x}^k)\mathbf{x}^k$, then there exists a $\mathbf{u}^k$ such that $\mathbf{x}^k,\mathbf{u}^k$ satisfy the Kuhn-Tucker conditions for (7.7).

PROOF. The Kuhn-Tucker conditions for (7.7) in terms of $\mathbf{x}^k$ and $\mathbf{u}^k$ are

$$\nabla f(\mathbf{x}^k)-\mathbf{u}^k\mathbf{A}\geq\mathbf{0} \qquad \text{and} \qquad (\nabla f(\mathbf{x}^k)-\mathbf{u}^k\mathbf{A})\mathbf{x}^k=0$$

$$\mathbf{A}\mathbf{x}^k=\mathbf{b}$$

$$\mathbf{x}^k\geq\mathbf{0}$$

Thus, the theorem is proven by establishing the first, complementary pair of conditions.

Let $\mathbf{z}^k$ denote an optimal solution to problem (7.7) found by the simplex method and let $\mathbf{u}^k$ denote the optimal dual solution. The vector $\mathbf{u}^k$ satisfies $\nabla f(\mathbf{x}^k)-\mathbf{u}^k\mathbf{A}\geq\mathbf{0}$ and $[\nabla f(\mathbf{x}^k)-\mathbf{u}^k\mathbf{A}]\mathbf{z}^k=0$. Since $\nabla f(\mathbf{x}^k)\mathbf{z}^k=\nabla f(\mathbf{x}^k)\mathbf{x}^k$ by assumption, we have $[\nabla f(\mathbf{x}^k)-\mathbf{u}^k\mathbf{A}]\mathbf{x}^k=0$ which establishes the theorem. ■

COROLLARY 7.1. Suppose the function f in problem (7.5) is convex. If the minimal objective function value in (7.7) equals $f(\mathbf{x}^k)\cdot\mathbf{x}^k$, then $\mathbf{x}^k$ is optimal in (7.5).

PROOF. By Theorem 7.4, there exists a $\mathbf{u}^k$ such that $\mathbf{x}^k,\mathbf{u}^k$ satisfy the Kuhn-Tucker conditions. By Theorem 5.7, these conditions are sufficient that $\mathbf{x}^k$ is optimal in (7.5). ■

Thus, the feasible directions algorithm tests a vector $\mathbf{x}^k$ to see if it satisfies the Kuhn-Tucker conditions in conjunction with some dual vector $\mathbf{u}^k$. If so, then the algorithm terminates with $\mathbf{x}^k$; we can only be sure that $\mathbf{x}^k$ is optimal in (7.5) if f is convex. If the optimal solution $\mathbf{z}^k$ to (7.7) satisfies $\nabla f(\mathbf{x}^k)\mathbf{z}^k < \nabla f(\mathbf{x}^k)\mathbf{x}^k$, then the algorithm proceeds by using $\mathbf{r}^k = \mathbf{z}^k - \mathbf{x}^k$ as a direction of descent. Note that $\mathbf{r}^k$ is a feasible direction of descent because $\nabla f(\mathbf{x}^k; \mathbf{r}^k) = \nabla f(\mathbf{x}^k)\mathbf{r}^k < 0$ and moreover $\mathbf{x}^k + \theta \mathbf{r}^k$ is feasible for $\theta \in [0,1]$. A plausible method for selecting the step length t_k is to choose t_k in a way similar to what we did in the unconstrained case; namely, to choose t_k to satisfy

$$f(\mathbf{x}^k + t_k \mathbf{r}^k) = \min_{0 \le t \le 1} f(\mathbf{x}^k + t\mathbf{r}^k)$$

This choice produces a sequence $\{\mathbf{x}^k\}$ with monotonic strictly decreasing objective function values $\{f(\mathbf{x}^k)\}$, but there is no guarantee that the limit of these values equals v, even if f is assumed to be convex.

In the remainder of this section, we make assumptions about the function f and the feasible region F, which ensure that the feasible directions algorithm when suitably specified, generates a subsequence of feasible solutions converging to an optimal solution to (7.5). Specifically, we assume that f is convex with continuous second partials and also that F is bounded and therefore it is compact. To ensure convergence the step length t_k must be selected on the basis of second-order as well as first-order information about f. The difficulty is that we cannot calculate in general the required second-order information. The convergence results are worth studying nevertheless because they give insight into the method. This situation is not atypical of nonlinear programming algorithms in that it is often necessary to assume information about the functions for which we do not have to ensure convergence. This was also the case for the subgradient optimization algorithm presented in Section 6.4.

The step length choice we need to ensure convergence is

$$t_k = \min\{t'_k, 1\} \tag{7.8}$$

where

$$t'_k = \frac{-\nabla f(\mathbf{x}^k)\cdot \mathbf{r}^k}{L} > 0$$

and where $\mathbf{r}^k = \mathbf{z}^k - \mathbf{x}^k$ for $\mathbf{z}^k$ optimal in (7.7) and L satisfies

$$|(\mathbf{y}-\mathbf{x})H_f(\mathbf{w})(\mathbf{y}-\mathbf{x})| < L \qquad \text{for any } \mathbf{x}, \mathbf{y}, \mathbf{w} \in F$$

where $H_f(\mathbf{w})$ is the Hessian of f. Some intuitive insight into this step length choice can be gained by considering the scalar convex function

$$\phi(t)=f(\mathbf{x}^k)+t\nabla f(\mathbf{x}^k)\mathbf{r}^k+\tfrac{1}{2}t^2L$$

For any $t\geq 0$, since f has continuous second partials

$$f(\mathbf{x}^k+t\mathbf{r}^k)=f(\mathbf{x}^k)+t\nabla f(\mathbf{x}^k)\mathbf{r}^k+\tfrac{1}{2}t^2\mathbf{r}^kH_f(\mathbf{w})\mathbf{r}^k\leq\phi(t) \tag{7.9}$$

where $\mathbf{w}$ is some point on the line segment $[\mathbf{x}^k, \mathbf{x}^k+t\mathbf{r}^k]$. Thus, $\phi(0)=f(\mathbf{x}^k)$ and ϕ lies above f on the half line emanating out of $\mathbf{x}^k$ in the direction $\mathbf{r}^k$. Moreover

$$\left.\frac{d\phi(t)}{dt}\right|_{t=0}=\nabla f(\mathbf{x}^k)\mathbf{r}^k=\left.\frac{df(\mathbf{x}^k+t\mathbf{r}^k)}{dt}\right|_{t=0}$$

Finally, the step length rule (7.8) is selected so as to minimize $\phi(t)$ on the interval $[0, 1]$ because

$$\frac{d\phi(t)}{dt}=0=\nabla f(\mathbf{x}^k)\mathbf{r}^k+tL$$

The following lemma provides a crucial property for convergence of the feasible directions algorithm.

LEMMA 7.1. Let $\{a_k\}_{k=1}^{\infty}$ denote a sequence of nonnegative numbers satisfying the recursion

$$a_{k+1}\leq a_k\max\left(1-a_k, \tfrac{1}{2}\right)$$

then

$$\lim_{k\to\infty} a_k=0$$

PROOF. Since the a_k are monotonically decreasing and bounded from below by zero, we can conclude that the limit exists. Our task is to show it equals zero. First of all, we need to establish that there is a K such that for all $k\geq K$,

$$a_{k+1}\leq a_k(1-a_k)$$

because $a_k\leq\frac{1}{2}$ implying $\max(1-a_k, \frac{1}{2})=1-a_k$. Suppose $a_1>\frac{1}{2}$; then $a_2\leq\frac{1}{2}a_1$ because $\max(1-a_1, \frac{1}{2})=\frac{1}{2}$. If $a_2>\frac{1}{2}$, then $a_3\leq\frac{1}{2}a_2$ which implies $a_3\leq(\frac{1}{2})^2a_1$. In general, we will have $a_k\leq(\frac{1}{2})^{k-1}a_1$ until $a_l\leq\frac{1}{2}$. Once $a_l\leq\frac{1}{2}$, then for all $k\geq l, \max(1-a_k, \frac{1}{2})=1-a_k$ implying $a_{k+1}\leq a_k(1-a_k)$.

Thus, it suffices to show that $\lim_{k\to\infty}a_k=0$, where $a_k\geq 0$ for all k where $a_{k+1}\leq a_k(1-a_k)$. Suppose $\lim_{k\to\infty}a_k=a^*>0$ for some $a^*<\frac{1}{2}$. We show a contradiction by showing there is a k sufficiently large such that $a_{k+1}<a^*$. In particular, pick k large enough that

$$a^*\leq a_k\leq a^*+\frac{(a^*)^2}{1+a^*}$$

then,

$$a_{k+1} \le a_k(1-a_k) \le \left(a^* + \frac{(a^*)^2}{1+a^*}\right)(1-a^*) = a^* - \frac{2(a^*)^3}{1+a^*} < a^* \quad \blacksquare$$

THEOREM 7.5. Suppose that f in (7.5) is convex with continuous second partials and moreover that the feasible region F is compact. Then, the feasible directions algorithm with direction $\mathbf{r}^k = \mathbf{z}^k - \mathbf{x}^k$ for z^k optimal in (7.7), and step length t_k selected according to (7.8), has the following properties:

1. $f(\mathbf{x}^k) > f(\mathbf{x}^{k+1})$ or $\mathbf{x}^k$ is optimal
2. $f(\mathbf{x}^k) \ge v \ge f(\mathbf{x}^k) + \nabla f(\mathbf{x}^k)\mathbf{r}^k$
3. $\lim_{k\to\infty} f(\mathbf{x}^k) = v$
4. $\lim_{k\to\infty} \nabla f(\mathbf{x}^k)\mathbf{r}^k = 0$
5. Any limit point of the sequence $\{\mathbf{x}^k\}_{k=1}^{\infty}$ is optimal

PROOF.

1. If $\nabla f(\mathbf{x}^k)\mathbf{z}^k = \nabla f(\mathbf{x}^k)\mathbf{x}^k$, then $\mathbf{x}^k$ is optimal by Theorem 7.4. Otherwise $\nabla f(\mathbf{x}^k)\mathbf{r}^k < 0$ for $\mathbf{r}^k = \mathbf{z}^k - \mathbf{x}^k$. If we follow the rule (7.8) for selecting the step length t_k, then the minimum of the function ϕ approximating f along the half line $\mathbf{x}^k + t\mathbf{r}^k$ for $t \ge 0$ occurs at t_k and $\phi(t_k) < \phi(0) = f(\mathbf{x}^k)$ since $d\phi(0)/dt = \nabla f(\mathbf{x}^k)\cdot\mathbf{r}^k$. Finally, by (7.9), we have $f(\mathbf{x}^{k+1}) = f(\mathbf{x}^k + t_k\mathbf{r}^k) \le \phi(t_k)$ which establishes 1.

2. For any $y \in F$,

$$f(\mathbf{y}) \ge f(\mathbf{x}^k) + \nabla f(\mathbf{x}^k)(\mathbf{y} - \mathbf{x}^k)$$

because f is convex. Taking the minimum over F on both sides gives us

$$v \ge f(\mathbf{x}^k) + \nabla f(\mathbf{x}^k)\mathbf{r}^k$$

by the definition of $\mathbf{r}^k$; namely, $\mathbf{r}^k = \mathbf{z}^k - \mathbf{x}^k$ where $\mathbf{z}^k$ is optimal in (7.7).

3. By Taylor's theorem,

$$\begin{aligned} f(\mathbf{x}^{k+1}) &= f(\mathbf{x}^k + t_k\mathbf{r}^k) \\ &= f(\mathbf{x}^k) + t_k\nabla f(\mathbf{x}^k)\mathbf{r}^k + \tfrac{1}{2}t_k^2\mathbf{r}^k H_f(\mathbf{w})\mathbf{r}^k \end{aligned} \tag{7.10}$$

for some $\mathbf{w} \in [\mathbf{x}^k, \mathbf{x}^k + t^k\mathbf{r}^k]$. This implies

$$f(\mathbf{x}^{k+1}) - v = f(\mathbf{x}^k) + \tfrac{1}{2}t_k\nabla f(\mathbf{x}^k)\mathbf{r}^k + \tfrac{1}{2}t_k^2\left\{\frac{\nabla f(\mathbf{x}^k)\mathbf{r}^k}{t_k} + \mathbf{r}^k H_f(\mathbf{w})\mathbf{r}^k\right\} - v$$

By the definition of L and the relation (7.8) for selecting t_k, the term in brackets is nonpositive. Thus,

$$0 \leq f(\mathbf{x}^{k+1}) - v \leq f(\mathbf{x}^k) - v + \tfrac{1}{2} t_k \nabla f(\mathbf{x}^k)\mathbf{r}^k$$

and by part 2,

$$0 \leq f(\mathbf{x}^{k+1}) - v \leq \left[f(\mathbf{x}^k) - v \right]\left(1 - \tfrac{1}{2} t_k\right)$$

Dividing by $2L$, and using (7.8) again, there results

$$0 \leq \frac{f(\mathbf{x}^{k+1}) - v}{2L} \leq \frac{f(\mathbf{x}^k) - v}{2L} \max\left\{ 1 - \frac{f(\mathbf{x}^k) - v}{2L}, \tfrac{1}{2} \right\}$$

letting $a_k = [f(\mathbf{x}^k) - v]/2L \geq 0$, we have the recursive relation

$$0 \leq a_{k+1} \leq a_k \max\left\{ 1 - a_k, \tfrac{1}{2} \right\} \qquad k = 1, 2, \cdots$$

By Lemma 7.1, $\lim_{k\to\infty} a_k = 0$ implying part 3.

4. By (7.8) we can establish that $\lim_{k\to\infty} \nabla f(\mathbf{x}^k)\mathbf{r}^k = 0$ if $\lim_{k\to\infty} t_k = 0$. To show this, we have for $\mathbf{w} \in [\mathbf{x}^k, \mathbf{x}^k + t_k\mathbf{r}^k]$ that

$$\begin{aligned} v &\leq f(\mathbf{x}^k + t_k\mathbf{r}^k) \\ &= f(\mathbf{x}^k) + t_k \nabla f(\mathbf{x}^k)\mathbf{r}^k + \tfrac{1}{2} t_k^2 \mathbf{r}^k H_f(\mathbf{w})\mathbf{r}^k \\ &\leq f(\mathbf{x}^k) - t_k^2 L + \tfrac{1}{2} t_k^2 L \\ &= f(\mathbf{x}^k) - \tfrac{1}{2} t_k^2 L \end{aligned}$$

where the first inequality follows because v is the minimum objective function cost, the first equality by Taylor's theorem, and the second inequality by the definitions of t_k and L. Thus, $f(\mathbf{x}^k) - v \geq \frac{1}{2} t_k^2 L \geq 0$, and since $\lim_{k\to\infty} f(\mathbf{x}^k) = v$ by part 3, we must have $\lim_{k\to\infty} t_k = 0$, which is what we needed to show.

5. The sequence $\{\mathbf{x}^k\}_{k=1}^{\infty}$ has at least one limit point in the feasible region F because the set F is assumed to be compact. Any limit point $\mathbf{x}^* = \lim_{i\to\infty} \mathbf{x}^{k_i}$ is optimal because f is continuous implying $v = \lim_{i\to\infty} f(\mathbf{x}^{k_i}) = f(\lim_{i\to\infty} \mathbf{x}^{k_i}) = f(\mathbf{x}^*)$. ■

EXAMPLE 7.1. Let x_1 and x_2 denote the amount of money spent by a company for advertising and production of each of two products. The company wishes to maximize profit by solving the mathematical programming problem

$$\begin{aligned} v = \max\ & 3(1 - e^{-1.2x_1} - 1.2x_1 e^{-1.2x_1}) \\ & + 4(1 - e^{-1.5x_2} - 1.5x_2 e^{-1.5x_2}) \\ & + (1 - e^{-x_1 x_2}) - x_1 - x_2 \\ \text{s.t. } & x_1 + x_2 \leq 3 \\ & -2x_1 + x_2 \leq 1 \\ & x \geq 0 \qquad x_2 \geq 0 \end{aligned}$$

Let X denote the feasible region. We illustrate the feasible directions algorithm by applying it to this problem.

We take as our starting solution the point $(x_1^1, x_2^1) = (1.5, 1.5)$ and solve the linear programming approximation derived from the gradient

$$\frac{\partial f}{\partial x_1} = 4.32 x_1 e^{-1.2x_1} + x_2 e^{-x_1 x_2} - 1$$

$$\frac{\partial f}{\partial x_2} = 9 x_2 e^{-1.5x_2} + x_1 e^{-x_1 x_2} - 1$$

evaluated at that point. The linear programming approximation is

$$\begin{aligned} &\max 0.227 z_1 + 0.575 z_2 \\ &\text{s.t. } z_1 z_2 \in X \end{aligned}$$

and the optimal solution is $z_1^1 = \frac{2}{3}, z_2^1 = \frac{7}{3}$.The indicated direction of ascent is $\mathbf{r}^1 = \mathbf{z}^1 - \mathbf{x}^1 = (-0.833, 0.833)$. The value of the original nonlinear objective function at $(x_1^1, x_2^1) = (1.5, 1.5)$ is 2.14, and according to Theorem 7.5, the upper bound from the linear programming approximation is $2.14 + \nabla f(\mathbf{x}^1) \cdot \mathbf{r}^1 = 2.14 + (0.227, 0.575)(-0.833, 0.833) = 2.43$.

The maximum of f in the direction $\mathbf{r}^1$ occurs approximately at the point $\mathbf{x}^2 = (1.325, 1.675)$ where the objective function value is 2.169. The gradient $\nabla f(\mathbf{x}^2) = (0.349, 0.365)$ and the solution to the linear programming approximation with these objective function coefficients is again $\mathbf{z}^2 = (0.667, 2.133)$. The upper bound on the objective at $\mathbf{x}^2$ is $2.169 + \nabla f(\mathbf{x}^2) \cdot \mathbf{r}^2 = 2.169 + (0.349, 0.365)(-0.658, 0.458) = 2.229$ and we terminate the algorithm because this bound is sufficiently near $f(\mathbf{x}^2) = 2.169$. ▲

7.4 NONLINEARLY CONSTRAINED NONLINEAR PROGRAMMING PROBLEMS: GENERALIZED LAGRANGEAN FUNCTIONS AND PENALTY METHODS

In this section, we consider the nonlinear programming problem

$$\begin{aligned} v = &\min f(\mathbf{x}) \\ &\text{s.t. } g_i(\mathbf{x}) \le 0 \qquad i = 1, \ldots, m \\ &\quad \mathbf{x} \in R^n \end{aligned} \tag{7.11}$$

where, for the moment, we assume only that f and the g_i have continuous second partials. This problem can be quite difficult to solve because of the nonlinear constraints and the nonconvexity of the functions. We could try to overcome the difficulty due to nonlinear constraints by putting nonnegative dual variables (Lagrange multipliers) on the constraints and adding them to the objective function. The resulting Lagrangean calculation is an unconstrained optimization problem, which we hope could be solved more easily than (7.11), for example, by one of the methods given in

Section 7.2. The dual variables in the Lagrangean could be iteratively selected by the generalized linear programming algorithm of Chapter 5 or one of the methods discussed in Chapter 6. As we demonstrated in Chapter 5, however, dualization of (7.11) is equivalent to convexification of it and thus there is no guarantee that this approach will produce an optimal solution. In this section we consider a generalization of the concept of duality that permits the construction of stronger dual problems for analyzing highly nonlinear convex and even nonconvex problems. Penalty methods, which have proven computationally effective on these problems, are closely related to the extended duality theory.

The following definition gives the most general class of multiplier functions we will consider in generalizing the duality theory presented in Chapter 5.

Definition 7.1. A *multiplier function* $U[\mathbf{y};\boldsymbol{\sigma},t]$ is a real valued function defined for $\mathbf{y}\in R^m, \boldsymbol{\sigma}\in R^m, t\in R^1$ and $t>0$, with the properties

(1) it is monotonically nondecreasing in $\mathbf{y}$
(2) $U[\mathbf{0};\boldsymbol{\sigma},t]=0$ for all $\boldsymbol{\sigma}\geq\mathbf{0}, t>0$

If we take $U[\mathbf{y};\boldsymbol{\sigma},t]=\boldsymbol{\sigma}\mathbf{y}$, then we have the ordinary multiplier function discussed extensively in Chapter 5. More generally, the nonnegative vector $\boldsymbol{\sigma}$ plays a role closely related to the ordinary multiplier vector or vector of dual variables, and in certain circumstances, it will equal the dual variables. The scalar t is used in algorithms as a variable that is chosen sequentially to achieve convergence to global or local optimal solutions to (7.11).

We use the multiplier function to define the *generalized Lagrangean function*

$$\mathfrak{L}(\boldsymbol{\sigma},t)=\min_{\mathbf{x}\in R^n}\{f(\mathbf{x})+U[g(\mathbf{x});\boldsymbol{\sigma},t]\} \tag{7.12}$$

and for further reference, we define

$$\mathfrak{L}(\mathbf{x},\boldsymbol{\sigma},t)=f(\mathbf{x})+U[g(\mathbf{x});\boldsymbol{\sigma},t]$$

Figure 7.2 shows why we can expect (7.12) to provide tighter lower bound on v than the conventional (linear) Lagrangean function. At all values of $\mathbf{x}$

$$f(\mathbf{x})+U[g(\mathbf{x});\boldsymbol{\sigma},t]\geq f(\mathbf{x})+\boldsymbol{\sigma}g(\mathbf{x})$$

and therefore the minimum over $\mathbf{x}$ in (7.12) is greater than the minimum over $\mathbf{x}$ in (5.2).

A trivial example of a multiplier function that would permit us to solve (7.11) by one calculation of the generalized Lagrangean is $U^0[\mathbf{y};\boldsymbol{\sigma},t]=0$ for all $\mathbf{y}\leq\mathbf{0}$ and $U^0[\mathbf{y};\boldsymbol{\sigma},t]=+\infty$ for all $\mathbf{y}\nleq\mathbf{0}$. In words, feasible solution

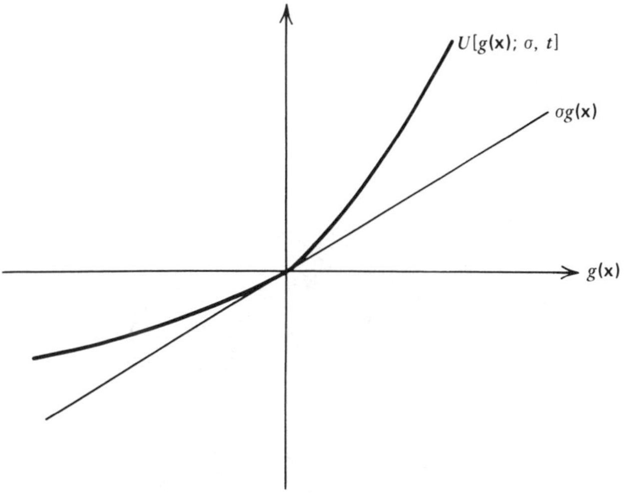

Figure 7.2. Linear and nonlinear multiplier functions.

receive zero penalty under U^0, and infeasible solutions receive an infinite penalty. This multiplier function is clearly nonconstructive, but it can be approximated in **y** space by a continuous and differentiable multiplier function that converges to the discontinuous step function U^0 as t goes to $+\infty$. Two such multiplier functions are

$$U[g(\mathbf{x}); \boldsymbol{\sigma}, t] = \sum_{i=1}^{m} \sigma_i \left\{ \frac{e^{tg_i(\mathbf{x})} - 1}{t} \right\} \tag{7.13}$$

and

$$U[g(\mathbf{x}); \boldsymbol{\sigma}, t] = t \sum_{i=1}^{m} [\max\{0, g_i(\mathbf{x})\}]^2 \tag{7.14}$$

where $\boldsymbol{\sigma} \geq \mathbf{0}$ and $t > 0$.

A generalized duality theory similar to the one in Chapter 5 can be developed. The central result is an extension of the weak duality theorem 5.2 which remains valid because of the way in which we specified the multiplier function U.

THEOREM 7.6. (Generalized Weak Duality). For any multiplier function and any $\boldsymbol{\sigma} \geq \mathbf{0}$, $t > 0$,

$$\mathfrak{L}(\boldsymbol{\sigma}, t) \leq v$$

PROOF. Let $\tilde{\mathbf{x}}$ be any feasible solution to (7.13). Then

$$\mathfrak{L}(\boldsymbol{\sigma}, t) \leq f(\tilde{\mathbf{x}}) + U[g(\tilde{\mathbf{x}}); \boldsymbol{\sigma}, t]$$

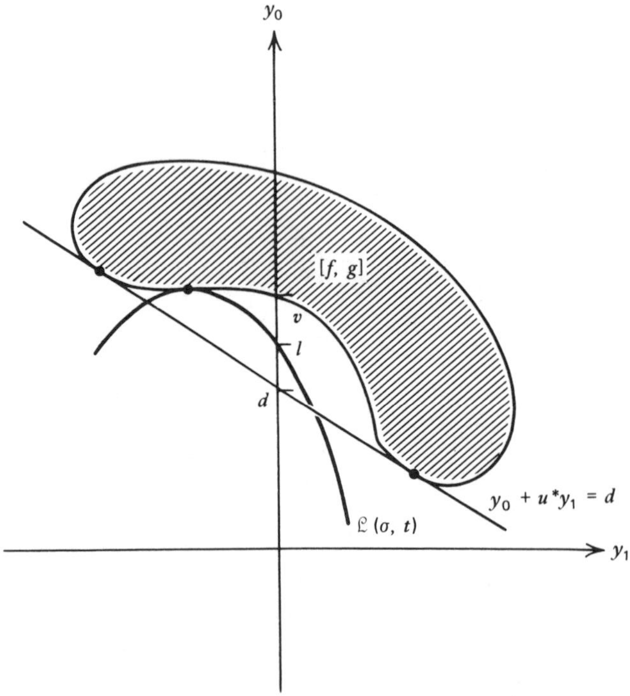

Figure 7.3. Nonlinear dual support provides greater lower bounds.

But $\tilde{\mathbf{x}}$ is feasible implying $g(\tilde{\mathbf{x}}) \leq \mathbf{0}$ and thus $U[g(\tilde{\mathbf{x}}); \boldsymbol{\sigma}, t] \leq U[\mathbf{0}; \boldsymbol{\sigma}, t]$ by the monotonicity of U. Moreover, $U[\mathbf{0}; \boldsymbol{\sigma}, t] = 0$ by property 2 of U implying $\mathcal{L}(\boldsymbol{\sigma}, t) \leq f(\tilde{\mathbf{x}})$. Since $\tilde{\mathbf{x}}$ was an arbitrary feasible solution, the lemma is established. ■

Figure 7.3 reproduces Figure 5.1 to illustrate why we can expect the generalized duality theory to produce stronger lower bounds than the duality theory of Chapter 5, although the calculation of $\mathcal{L}(\boldsymbol{\sigma}, t)$ may require more work than the calculation of the ordinary Lagrangean $L(\mathbf{u})$. The scalar $u^* > 0$ is an optimal dual solution to the dual problem (5.4) yielding the maximal lower bound $d = L(u^*)$ to the minimal objective function value v in (7.11). The hyperplane $y_0 + u^* y_1 = d$ is a linear support of the region $[f, g]$ of attainable values of the objective function f and the constraint function g. As shown, the generalized Lagrangean $\mathcal{L}$ is a nonlinear support of the region $[f, g]$ achieving a higher lower bound l.

Definition 7.2. A solution $\bar{\mathbf{x}} \in R^n$ and a multiplier function $\bar{U}$ defined at specific values $\bar{\boldsymbol{\sigma}} \geq \mathbf{0}$, $\bar{t} > 0$ are said to satisfy the *generalized global optimal-*

ity conditions if

(1) $\mathfrak{L}(\bar{\boldsymbol{\sigma}},\bar{t})=f(\bar{\mathbf{x}})+\bar{U}[g(\bar{\mathbf{x}});\bar{\boldsymbol{\sigma}},\bar{t}]$
(2) $\bar{U}[g(\bar{\mathbf{x}});\bar{\boldsymbol{\sigma}},\bar{t}]=0$
(3) $g(\bar{\mathbf{x}})\leq\mathbf{0}$

THEOREM 7.7. If $\bar{\mathbf{x}}\in R^n$ satisfies the generalized global optimality conditions for some multiplier function $\bar{U}$ defined at specific values $\bar{\boldsymbol{\sigma}}\geq\mathbf{0}$, $\bar{t}>0$, then $\bar{\mathbf{x}}$ is optimal in (7.11).

PROOF. The proof is very similar to the proof of Theorem 5.1 and is omitted. ■

The generalized optimality conditions are stronger sufficient conditions that $\bar{\mathbf{x}}$ is optimal than the optimality conditions of Chapter 5 because the lower bounds are tighter. The nature of typical multiplier functions, such as (7.13), admit the following generalized dual problem analogous to the dual problem (5.4):

$$\bar{d}=\max \mathfrak{L}(\boldsymbol{\sigma},\bar{t})$$
$$\text{s.t. } \boldsymbol{\sigma}\geq\mathbf{0} \tag{7.15}$$

where $\mathfrak{L}$ is defined with respect to a given multiplier function $\bar{U}$ and a given value of $\bar{t}>0$.

THEOREM 7.8. If $\bar{\mathbf{x}}\in R^n$, $\bar{\boldsymbol{\sigma}}\geq\mathbf{0}$ and $\bar{\boldsymbol{\sigma}}\in R^m$, $\bar{t}>0$ and $\bar{t}\in R^1$ satisfy the generalized global optimality conditions, then $\bar{\boldsymbol{\sigma}}$ is optimal in the generalized dual problem (7.15).

PROOF. The proof is very similar to the proof of Theorem 5.2 and is omitted. ■

We see from Theorems 7.7 and 7.8, and problem (7.15) that the generalized duality theory behaves exactly as the duality theory discussed in Chapter 5. For fixed t in the multiplier function U, the nonnegative m-vector $\boldsymbol{\sigma}$ is iteratively selected to optimize the generalized dual problem. This can be done by generalizing the ascent methods of Chapter 6. Further details are given in Exercise 7.10.

Thus, the generalized duality theory just discussed provides stronger dual problems than the dual problems previously discussed in Chapters 5 and 6. Our interest at this point is to investigate conditions on (7.11) ensuring that local minima can be identified and consequently can be found by a specific multiplier function. Roughly speaking, if a point $\bar{\mathbf{x}}$ satisfies the Kuhn-Tucker conditions for some $\bar{\mathbf{u}}$, then sufficient condition for it to be a local minimum is that the function $L(\mathbf{x},\bar{\mathbf{u}})=f(\mathbf{x})+\bar{\mathbf{u}}g(\mathbf{x})$ is convex in a neighborhood of $\bar{\mathbf{x}}$. This idea is made precise in the following theorem.

THEOREM 7.9. Suppose that $(\bar{\mathbf{x}}, \bar{\mathbf{u}})$ satisfy the Kuhn-Tucker conditions for (7.11) and moreover that

$$\mathbf{z}^T\left[\nabla^2 f(\bar{\mathbf{x}}) + \sum_{i=1}^{m} \bar{u}_i \nabla^2 g_i(\bar{\mathbf{x}})\right]\mathbf{z} > 0 \tag{7.16}$$

for all $\mathbf{z}$ satisfying

$$\mathbf{z}^T \nabla g_i(\bar{\mathbf{x}}) = 0 \qquad i \in \tilde{I}(\bar{\mathbf{x}})$$

$$\mathbf{z}^T \nabla g_i(\bar{\mathbf{x}}) \le 0 \qquad i \in I(\bar{\mathbf{x}}) - \tilde{I}(\bar{\mathbf{x}})$$

where

$$I(\bar{\mathbf{x}}) = \{ i \mid g_i(\bar{\mathbf{x}}) = 0 \}$$

and

$$\tilde{I}(\bar{\mathbf{x}}) = \{ i \in I(\bar{\mathbf{x}}) \mid \bar{u}_i > 0 \}$$

Then $\bar{\mathbf{x}}$ is a strict local minimum in problem (7.11).

PROOF. The proof is left to the reader as an exercise (see 7.5). ■

The previous theorem illustrates the importance of second-order information about the functions f and the g_i in identifying local minima in (7.11). Nonlinear multiplier functions produce generalized Lagrangeans that are capable of using such information to find local minima. We show how this can be done by considering the function

$$\mathcal{L}(\mathbf{x}, \boldsymbol{\sigma}, t) = f(\mathbf{x}) + \sum_{i=1}^{m} \sigma_i \left\{ \frac{e^{tg_i(\mathbf{x})} - 1}{t} \right\}$$

Note that

$$\nabla \mathcal{L}(\mathbf{x}, \boldsymbol{\sigma}, t) = \nabla f(\mathbf{x}) + \sum_{i=1}^{m} \sigma_i e^{tg_i(\mathbf{x})} \nabla g_i(\mathbf{x}) \tag{7.17}$$

and therefore for $\bar{\mathbf{x}}, \bar{\mathbf{u}}$ satisfying the Kuhn-Tucker conditions, we have for any $t > 0$

$$\begin{aligned} \nabla \mathcal{L}(\bar{\mathbf{x}}, \bar{\mathbf{u}}, t) &= \nabla f(\bar{\mathbf{x}}) + \sum_{i \in I(\bar{\mathbf{x}})} \bar{u}_i \nabla g_i(\bar{\mathbf{x}}) \\ &= \nabla L(\bar{\mathbf{x}}, \bar{\mathbf{u}}) = 0 \end{aligned}$$

Moreover, the Hessian of $\mathcal{L}$ with respect to $\mathbf{x}$ at the point $\bar{\mathbf{x}}, \bar{\mathbf{u}}, t$ for any $t > 0$ is

$$\begin{aligned} \nabla^2 \mathcal{L}(\bar{\mathbf{x}}, \bar{\mathbf{u}}, t) &= \nabla^2 f(\bar{\mathbf{x}}) + \sum_{i \in I(\bar{\mathbf{x}})} \bar{u}_i \nabla^2 g_i(\bar{\mathbf{x}}) + t \sum_{i \in I(\bar{\mathbf{x}})} \bar{u}_i [\nabla g_i(\bar{\mathbf{x}})]^T \nabla g_i(\bar{\mathbf{x}}) \\ &= \nabla^2 L(\bar{\mathbf{x}}, \bar{\mathbf{u}}) + t \sum_{i \in I(\bar{\mathbf{x}})} \bar{u}_i [\nabla g_i(\bar{\mathbf{x}})]^T \nabla g_i(\bar{\mathbf{x}}) \end{aligned} \tag{7.18}$$

The term $[\nabla g_i(\bar{\mathbf{x}})]^T \nabla g_i(\bar{\mathbf{x}})$ is an $n \times n$ dyadic matrix and it is positive semidefinite because for any $\mathbf{z} \in R^n$, $\mathbf{z}^T[\nabla g_i(\bar{\mathbf{x}})]^T \nabla g_i(\bar{\mathbf{x}})\mathbf{z} = w^2 \geq 0$ where the scalar $w = \nabla g_i(\bar{\mathbf{x}})\mathbf{z}$. We see from (7.18) that if $\nabla^2 L(\bar{\mathbf{x}}, \bar{\mathbf{u}})$ is positive definite, then so is $\nabla^2 \mathfrak{L}(\bar{\mathbf{x}}, \bar{\mathbf{u}}, t)$ for any positive t and minimization of $\mathfrak{L}(\mathbf{x}, \bar{\mathbf{u}}, t)$ will produce the local minimum $\bar{\mathbf{x}}$ in (7.11) for any positive t. Even when $\nabla^2 L(\bar{\mathbf{x}}, \bar{\mathbf{u}})$ is not positive definite, the dyadic terms in (7.18) can cause $\nabla^2 \mathfrak{L}(\bar{\mathbf{x}}, \bar{\mathbf{u}}, t)$ to be positive definite for sufficiently large t under the appropriate conditions.

THEOREM 7.10. Suppose $\bar{\mathbf{x}}$ is a strict local minimum in problem (7.11) satisfying along with some $\bar{\mathbf{u}} \geq \mathbf{0}$ the sufficient conditions of Theorem 7.9. Suppose further that $\bar{\mathbf{x}}, \bar{\mathbf{u}}$ satisfy the condition of strict complementary slackness; that is, $\tilde{I}(\bar{\mathbf{x}}) = I(\bar{\mathbf{x}})$ or in other words, $\bar{u}_i > 0$ for all i such that $g_i(\bar{\mathbf{x}}) = 0$. Then there exists a positive scalar Q such that for all $\bar{t} > Q$, the Hessian $\nabla^2 \mathfrak{L}(\bar{\mathbf{x}}, \bar{\mathbf{u}}, \bar{t})$ is positive definite. Thus, $\bar{\mathbf{x}}$ is a strict local minimum of $\mathfrak{L}$.

PROOF. Let S denote the subspace of R^n spanned by the vectors $\nabla g_i(\bar{\mathbf{x}})$ for $i \in I(\bar{\mathbf{x}})$ and let $S^\perp$ denote the orthogonal complement. Define the function $h: R^n \to R$ by

$$h(y) = \mathbf{y}^T \left\{ \sum_{i \in I(\bar{\mathbf{x}})} \bar{u}_i \left[\nabla g_i(\bar{\mathbf{x}}) \right]^T \nabla g_i(\bar{\mathbf{x}}) \right\} \mathbf{y}$$

$$= \sum_{i \in I(\bar{\mathbf{x}})} \bar{u}_i \left[\mathbf{y}^T \nabla g_i(\bar{\mathbf{x}}) \right]^2$$

The function h is continuous on R^n, it is zero for $\mathbf{y} \in S^\perp$, and it is positive for $\mathbf{y} \in S$, $\mathbf{y} \neq \mathbf{0}$. Let $M = \|\nabla^2 L(\bar{\mathbf{x}}, \bar{\mathbf{u}})\|$ denote the sup norm of the matrix $\nabla^2 L(\bar{\mathbf{x}}, \bar{\mathbf{u}})$. There are three cases to consider.

Case 1. $S = \{\mathbf{0}\}$. In this case, we have $\mathbf{y}^T \nabla g_i(\bar{\mathbf{x}}) = 0$ for all $\mathbf{y} \in R^n$ which implies by (7.16) that $\nabla^2 L(\bar{\mathbf{x}}, \bar{\mathbf{u}})$ is positive definite and hence $\nabla^2 \mathfrak{L}(\bar{\mathbf{x}}, \bar{\mathbf{u}}, t)$ is positive definite for any $t > 0$.

Case 2. $S = R^n$. Let $m_1 > 0$ denote the minimum of $h(\mathbf{y})$ for $\mathbf{y}$ in the compact set $\{\mathbf{y} \in S^\perp \mid \|\mathbf{y}\| = 1\}$. Then we have for all $\mathbf{y} \in S$

$$\mathbf{y}^T \nabla^2 \mathfrak{L}(\bar{\mathbf{x}}, \bar{\mathbf{u}}, t)\mathbf{y} = \mathbf{y}^T \nabla^2 L(\bar{\mathbf{x}}, \bar{\mathbf{u}})\mathbf{y} + th(y)$$

$$\geq -M\|\mathbf{y}\|^2 + tm_1\|\mathbf{y}\|^2$$

Thus, $\nabla^2 \mathfrak{L}(\bar{\mathbf{x}}, \bar{\mathbf{u}}, \bar{t})$ is positive definite for all $\bar{t} > M/m_1$.

Case 3. $\{\mathbf{0}\} \neq S \neq R^n$. Then $S^\perp \neq \{\mathbf{0}\}$ and for any nonzero $\mathbf{y} \in S^\perp$. $\mathbf{y}^T \nabla^2 L(\bar{\mathbf{x}}, \bar{\mathbf{u}})\mathbf{y} > 0$ by (7.16) implying $\mathbf{y}^T \nabla^2 L(\bar{\mathbf{x}}, \bar{\mathbf{u}})\mathbf{y}$ has a minimum $m_2 > 0$ on the compact set $\{\mathbf{y} \in S^\perp \mid \|\mathbf{y}\| = 1\}$. For any $\mathbf{y} \in R^n$, there exists $\mathbf{w} \in S$,

$\mathbf{z} \in S^{\perp}$ such that $\mathbf{y}=\mathbf{w}+\mathbf{z}$. Thus,

$$\begin{aligned}
\mathbf{y}^T \nabla^2 \mathcal{L}(\bar{\mathbf{x}},\bar{\mathbf{u}},t)\mathbf{y} &= (\mathbf{w}+\mathbf{z})^T \nabla^2 L(\bar{\mathbf{x}},\bar{\mathbf{u}})(\mathbf{w}+\mathbf{z}) \\
&\quad + (\mathbf{w}+\mathbf{z})^T \left\{ t \sum_{i\in I(\bar{\mathbf{x}})} \bar{u}_i [\nabla g_i(\bar{\mathbf{x}})]^T \nabla g_i(\bar{\mathbf{x}}) \right\} (\mathbf{w}+\mathbf{z}) \\
&= \mathbf{w}^T \nabla^2 L(\bar{\mathbf{x}},\bar{\mathbf{u}})\mathbf{w} + 2\mathbf{z}^T \nabla^2 L(\bar{\mathbf{x}},\bar{\mathbf{u}})\mathbf{w} \\
&\quad + \mathbf{z}^T \nabla^2 L(\bar{\mathbf{x}},\bar{\mathbf{u}})\mathbf{z}^T + th(\mathbf{w}) \\
&\geq -M\|\mathbf{w}\|^2 - 2M\|\mathbf{w}\|\,\|\mathbf{z}\| + m_2\|\mathbf{z}\|^2 + tm_1\|\mathbf{w}\|^2 \\
&= tm_1\|\mathbf{w}\|^2 + m_2\left\{ \|\mathbf{z}\| - \frac{M}{m_2}\|\mathbf{w}\| \right\}^2 - \frac{M^2}{m_2}\|\mathbf{w}\|^2 - M\|\mathbf{w}\|^2 \\
&= \|\mathbf{w}\|^2 \left\{ tm_1 - \frac{M^2}{m_2} - M \right\} + m_2\left\{ \|\mathbf{z}\| - \frac{M}{m_2}\|\mathbf{w}\| \right\}^2
\end{aligned}$$

The term on the right is positive, and therefore the entire expression is positive and $\nabla^2 \mathcal{L}(\bar{\mathbf{x}},\bar{\mathbf{u}},\bar{t})$ is positive definite if

$$\bar{t} > Q = \frac{M}{m_1}\left(1 + \frac{M}{m_2}\right) \quad \blacksquare$$

An important and computationally successful approach to the solution of problem (7.11) is the Sequential Unconstrained Minimization Technique (SUMT) which uses a penalty function closely related to the multiplier functions discussed above. SUMT generates points in the interior of the feasible region

$$F^0 = \{\mathbf{x} \mid g_i(\mathbf{x}) < 0,\ i=1,\ldots,m\} \tag{7.19}$$

by the unconstrained minimization of the penalty function

$$P(\mathbf{x},t) = f(\mathbf{x}) - t\sum_{i=1}^{m} \frac{1}{g_i(\mathbf{x})} \tag{7.20}$$

where the scalar t is positive. We assume F^0 is nonempty. The penalty function goes to ∞ if any of the boundaries $g_i(\mathbf{x})=0$ is approached from within F^0. The goal of the algorithm is to generate a sequence of points in F^0 converging to an optimal solution to (7.11) by solving iteratively the penalty function for a sequence of values of t going to zero.

The function $-t\Sigma_{i=1}^{m}(1/y_i)$ defined for vectors $\mathbf{y}$ with all negative components is not a multiplier function according to Definition 7.1. It is monotonically increasing for the $\mathbf{y}$'s of interest, namely $\mathbf{y}$ with negative components, but it does not equal zero for $\mathbf{y}=0$. The discrepancy can be partially explained if we consider the closely related multiplier function

$$U[\mathbf{y};\boldsymbol{\sigma},t]=\begin{cases} -t\sum_{i=1}^{m}\left\{\dfrac{1}{y_i-t}\right\}-m & \text{for } \mathbf{y}\le\dfrac{t\mathbf{e}_m}{2}\\ +\infty & \text{for } \mathbf{y}\not\le\dfrac{t\mathbf{e}_m}{2}\end{cases}$$

which is merely a translation of the penalty function term we will use. The situation is depicted for $m=1$ in Figure 7.4. The advantage of the penalty function is that unlike the function $\mathcal{L}(\mathbf{x},\boldsymbol{\sigma},t)$, unconstrained optimization of it by setting $\nabla P(\mathbf{x},t)=0$ always yields a feasible solution to (7.11). The disadvantage is that dual lower bounds on v are not produced except in the special case when f and the g_i are convex.

The following theorem characterizes the convergence properties of SUMT for problem (7.11).

THEOREM 7.11. Let $\{t_k\}_{k=1}^{\infty}$ be a sequence of positive scalars converging strictly monotonically to zero. Suppose the penalty function $P(\mathbf{x},t_k)$ attains its minimum at $\mathbf{x}^k$. Then

$$\lim_{k\to\infty} P(\mathbf{x}^k,t_k)=v$$

where v is the minimal objective function value in (7.11).

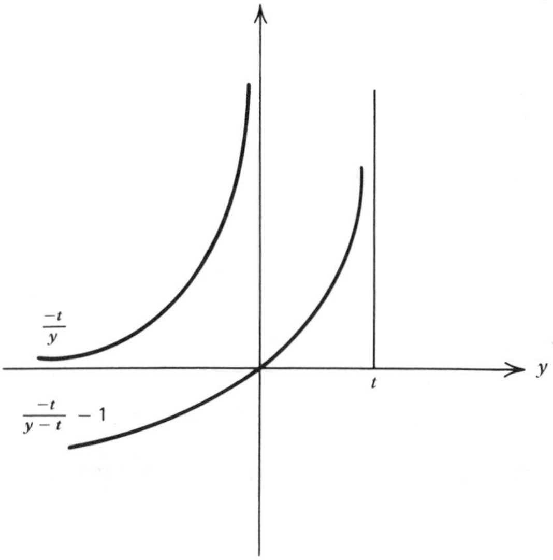

Figure 7.4. Multiplier and penalty functions.

PROOF. For all $\mathbf{x} \in F^0$, we have

$$P(\mathbf{x}, t_k) = f(\mathbf{x}) - t_k \sum_{i=1}^{m} \frac{1}{g_i(\mathbf{x})} > f(\mathbf{x}) \geq v$$

For any $\varepsilon > 0$, since f is continuous we can choose an $\mathbf{x}^* \in F^0$ such that $f(\mathbf{x}^*) < v + \varepsilon/2$. In addition, we can choose an index l such that

$$t_l \leq \left\{ \min_{i=1,\ldots,m} \left[-g_i(\mathbf{x}^*) \right] \right\} \frac{\varepsilon}{2m}$$

Then for $k > l$, we have

$$\begin{aligned} v &\leq \min_{\mathbf{x} \in F^0} P(\mathbf{x}, t_k) = P(\mathbf{x}^k, t_k) \\ &\leq P(\mathbf{x}^l, t_k) \\ &< P(\mathbf{x}^l, t_l) \\ &\leq P(\mathbf{x}^*, t_l) \\ &= f(\mathbf{x}^*) - t_l \sum_{i=1}^{m} \frac{1}{g_i(\mathbf{x}^*)} < v + \frac{\varepsilon}{2} + \frac{\varepsilon}{2} = v - \varepsilon \end{aligned}$$

where the first inequality follows from the fact that $\mathbf{x}^l$ may not be minimal in $P(\mathbf{x}, t_k)$, the first strict inequality because $t_k < t_l$, the second inequality for the same reason as the first, and the final strict inequality from the definitions of $\mathbf{x}^*$ and t_l. Thus, for any $\varepsilon > 0$, there exists an l such that for all $k \geq l + 1$, we have $v \leq P(\mathbf{x}^k, t_k) < v - \varepsilon$ which establishes the theorem. ■

COROLLARY 7.2. Under the assumption of Theorem 7.11, we also have

1. $\lim_{k \to \infty} f(\mathbf{x}^k) = v$
2. $\lim_{k \to \infty} t_k \sum_{i=1}^{m} \frac{1}{g_i(\mathbf{x}^k)} = 0$
3. If the sequence $\{\mathbf{x}^k\}_{k=1}^{\infty}$ has a limit point then any such limit point is optimal in (7.11)

PROOF. The proof of 1 follows immediately from Theorem 7.11 since

$$P(\mathbf{x}^k, t_k) > f(\mathbf{x}^k) > v$$

and $\lim_{k \to \infty} P(\mathbf{x}^k, t_k) = v$. The proof of 2 follows immediately from the equality $-t_k \sum_{i=1}^{m} \frac{1}{g_i(\mathbf{x}^k)} = P(\mathbf{x}^k, t_k) - f(\mathbf{x}^k)$ where the limit of both terms on the right exist. The proof of 3 follows immediately from part 1 and the continuity of f. ■

The results just derived show that SUMT can, in theory, find an optimal solution to a very broad class of mathematical programming problems including (7.11) without our stated assumptions of differentiability on the functions. However, the unconstrained minimization of the penalty function $P(x,t)$ is intended for those problems where the functions are differentiable.

THEOREM 7.12. Suppose that for any $M>0$, the set

$$\{\mathbf{x} \mid f(\mathbf{x}) \le M \text{ and } g_i(\mathbf{x}) \le 0 \text{ for } i=1,\dots,m\}$$

is bounded and therefore compact. Then $P(\mathbf{x},t)$ attains its minimum at a point $\bar{\mathbf{x}} \in F^0$ such that

$$\nabla P(\bar{\mathbf{x}},t)=0$$

PROOF. Pick any $\mathbf{x}^0 \in F^0$ and let

$$K=P(\mathbf{x}^0,t)$$

The quantity $K>v$ since

$$P(\mathbf{x}^0,t)=f(\mathbf{x}^0)-t\sum_{i=1}^{m}\frac{1}{g_i(\mathbf{x}^0)}>f(\mathbf{x}^0)\ge v$$

Define the sets

$$S_0=\{\mathbf{x} \mid f(\mathbf{x}) \le K \text{ and } g_i(\mathbf{x}) \le 0 \text{ for } i=1,\dots,m\}$$

$$S_i=\{\mathbf{x} \mid g_i(\mathbf{x}) \le \frac{-t}{K-v}\} \qquad i=1,\dots,m$$

We use these sets to define the new set

$$S=\bigcap_{i=0}^{m} S_i$$

The set S is compact because S_0 is compact by assumption and the other S_i are closed. Moreover, $S \subseteq F^0$ and it is nonempty because it contains the point $\mathbf{x}^0$. To see this, suppose the contrary; namely, suppose $\mathbf{x}^0 \notin S_j$ for some $j \ge 1$ ($\mathbf{x}^0 \in S_0$ by definition). Then

$$0>g_j(\mathbf{x}^0)>\frac{-t}{K-v}$$

or after suitable manipulation,

$$K+\frac{t}{g_j(\mathbf{x}^0)}<v$$

But this implies by the definition of K that

$$v<f(\mathbf{x}^0)-t\sum_{i\ne j}\frac{1}{g_i(x^0)}<v$$

which is impossible.

Since $P(\mathbf{x},t)$ is a differentiable function of x, the proof of the theorem is completed by showing that $P(\mathbf{x},t)\geq P(\mathbf{x}^0,t)$ for all $\mathbf{x}\in F^0-S$. For $\mathbf{x}\in F^0-S$ and $\mathbf{x}\notin S_0$, $P(\mathbf{x},t)\geq f(\mathbf{x})>K=P(\mathbf{x}^0,t)$ where the first inequality follows from the definition of P and the second from $\mathbf{x}\in S^0$. For $\mathbf{x}\in F^0-S$ and $\mathbf{x}\notin S_j$ for some j, we have

$$P(\mathbf{x},t)\geq f(\mathbf{x})-\frac{t}{g_j(\mathbf{x})}$$

$$\geq v+K-v=K=P(\mathbf{x}^0,t)$$

which establishes the theorem. ■

We conclude our discussion of SUMT by considering the special case when the functions f and the g_i in problem (7.11) are convex. In this case it can be shown that P is convex (see Exercise 7.8) and the solution of $\nabla P(\mathbf{x},t)=0$ as indicated by Theorem 7.11 is a sufficient as well as necessary condition for minimization of P over F^0. Moreover, the following theorem shows that there are implied dual variables each time $\nabla P(\mathbf{x},t)=0$ is solved.

THEOREM 7.13. Suppose f and the g_i in (7.11) are convex. Let $\bar{\mathbf{x}}\in F^0$ denote any point satisfying

$$\nabla P(\bar{\mathbf{x}},\bar{t})=0$$

and define the positive vector $\bar{\mathbf{u}}\in R^m$ by

$$\bar{u}_i=\frac{\bar{t}}{\left[g_i(\bar{\mathbf{x}})\right]^2}\qquad i=1,\ldots,m \tag{7.21}$$

Then

$$L(\bar{\mathbf{u}})=f(\bar{\mathbf{x}})+\sum_{i=1}^{m}\frac{\bar{t}}{g_i(\bar{\mathbf{x}})}\leq v\leq f(\bar{\mathbf{x}})$$

PROOF. By Theorem 7.12, the point $\bar{\mathbf{x}}$ satisfies

$$\nabla P(\bar{\mathbf{x}},\bar{t})=0$$

or

$$0=\nabla f(\bar{\mathbf{x}})+\bar{t}\sum_{i=1}^{m}\frac{1}{\left[g_i(\bar{\mathbf{x}})\right]^2}\nabla g_i(\bar{\mathbf{x}})$$

$$=\nabla f(\bar{\mathbf{x}})+\sum_{i=1}^{m}\bar{u}_i\nabla g_i(\bar{\mathbf{x}})=\nabla L(\bar{\mathbf{x}},\bar{\mathbf{u}})$$

This $\bar{\mathbf{x}}$ is optimal in the convex function $f(\mathbf{x})+\bar{\mathbf{u}}g(\mathbf{x})$ and the result follows from the Weak Duality Theorem 5.2. ∎

7.5 QUADRATIC AND LINEAR COMPLEMENTARY PROGRAMMING

The quadratic programming problem is

$$\begin{aligned} v = \min\ & \mathbf{cx} + \tfrac{1}{2}\mathbf{xDx} \\ \text{s.t.}\ & \mathbf{Ax} \geq \mathbf{b} \\ & \mathbf{x} \geq \mathbf{0} \end{aligned} \tag{7.22}$$

where for the moment we make no assumptions about the matrix $\mathbf{D}$ except the trivial one that it is symmetric. If $\mathbf{D}$ is not symmetric, then we can replace it by the symmetric matrix $\tilde{\mathbf{D}}=\frac{1}{2}(\mathbf{D}+\mathbf{D}^T)$ with the property that $\mathbf{x}\tilde{\mathbf{D}}\mathbf{x}=\mathbf{xDx}$. We will present a pivoting algorithm closely resembling the simplex method that is applicable to several classes of problems including problem (7.22). As we shall see, the algorithm's convergence to a globally optimal solution to (7.22) will require additional assumptions about the matrix $\mathbf{D}$, such as positive semidefiniteness. In the following section, we derive and analyze a similar pivoting algorithm with wide application to the computation of fixed points, economic equilibria, and convex programming problems.

As was the case for unconstrained optimization, the quadratic programming problem (7.22) can be used to approximate a linearly constrained, nonlinear programming problem. The quadratic programming problem also has applications in its own right as shown by the following example.

EXAMPLE 7.2. (Portfolio Selection). An investor wishes to invest his assets in a stock market portfolio. He considers n different stocks and must decide on the proportion x_j to invest in stock j, where $\sum_{j=1}^{n} x_j \leq 1$ and $x_j \geq 0$. From historical data, it is estimated that the expected return of a portfolio $\mathbf{x}$ is $\boldsymbol{\mu}\mathbf{x}=\sum_{j=1}^{n}\mu_j x_j$, and the variance is $\mathbf{xDx}=\sum_{i=1}^{n}\sum_{j=1}^{n}\sigma_{ij}x_i x_j$. The covariance matrix $\mathbf{D}$ is positive definite due to the statistical properties of the random stock returns. The investor is torn between wishing to maximize his expected return by investing in the stock j with the highest expected return μ_j, and wishing to reduce his risk as measured by the variance of the portfolio. For example, the risk can be reduced by investing in negatively correlated stocks. Because of his conflict, the investor decides to generate the set of all *efficient* solutions where a portfolio $\mathbf{x}$ is efficient if there does not exist a portfolio $\mathbf{y}$ satisfying $\boldsymbol{\mu}\mathbf{y} \geq \boldsymbol{\mu}\mathbf{x}$ and $\mathbf{yDy} \leq \mathbf{xDx}$, with at least one strict inequality. The entire set of efficient solutions can be generated by varying nonnegative values of θ in the parametric family of quadratic programming problems

$$\begin{aligned} v(\theta) = \max\ & \theta \boldsymbol{\mu}\mathbf{x} - \mathbf{xDx} \\ \text{s.t.}\ & \sum_{j=1}^{n} x_j \leq 1 \\ & \mathbf{x} \geq \mathbf{0} \end{aligned} \tag{7.23}$$

In this case, the objective function for fixed θ is concave. Note that the investor is given the option not to invest if the risk outweighs the expected return. ▲

Solution of the quadratic programming problem (7.22) is a special case of the *linear complementarity problem*: Given a p-vector $\mathbf{q}$ and a $p \times p$ matrix $\mathbf{M}$, find vectors $\mathbf{w}$ and $\mathbf{z}$ that satisfy the conditions

$$\begin{aligned} & \mathbf{w} = \mathbf{q} + \mathbf{Mz} \\ & \mathbf{zw} = 0 \\ & \mathbf{z} \geq \mathbf{0} \qquad \mathbf{w} \geq \mathbf{0} \end{aligned} \tag{7.24}$$

The conversion of the optimization problem (7.22) to the existence problem (7.24) is again effected by means of the Kuhn-Tucker conditions. Letting $\mathbf{u}$ denote the dual variables on the constraints, the Kuhn-Tucker conditions for (7.22) are

$$\begin{aligned} & \mathbf{y} = \mathbf{Dx} - \mathbf{A}^T\mathbf{u} + \mathbf{c} \\ & \mathbf{s} = \mathbf{Ax} - \mathbf{b} \\ & \mathbf{us} + \mathbf{yx} = 0 \\ & \mathbf{x} \geq \mathbf{0} \qquad \mathbf{u} \geq \mathbf{0} \qquad \mathbf{y} \geq \mathbf{0} \qquad \mathbf{s} \geq \mathbf{0} \end{aligned}$$

The correspondence with (7.22) is to let $\mathbf{w} = (\mathbf{y}, \mathbf{s})$, $\mathbf{z} = (\mathbf{x}, \mathbf{u})$, $\mathbf{q} = (\mathbf{c}, -\mathbf{b})$ and the matrix

$$\mathbf{M} = \begin{pmatrix} \mathbf{D} & -\mathbf{A}^T \\ \mathbf{A} & \mathbf{0} \end{pmatrix}$$

With no assumptions about $\mathbf{D}$, the Kuhn-Tucker conditions are necessary, but not sufficient, implying the possibility of multiple solutions to (7.24), including some that are not optimal. On the other hand, the matrix $\mathbf{M}$ is positive semidefinite if the matrix $\mathbf{D}$ is positive semidefinite because $\mathbf{zMz} = \mathbf{xDx}$ implying $\mathbf{zMz} \geq 0$ for all $\mathbf{z} = (\mathbf{x}, \mathbf{u})$ since $\mathbf{xDx} \geq \mathbf{0}$ for all $\mathbf{x}$. In this case, the Kuhn-Tucker conditions are sufficient as well as necessary. You are asked in Exercise 7.3 at the end of the chapter to show that a bimatrix game can be cast as a linear complementarity problem.

For $i = i, \ldots, p$, the variables z_i and w_i in (7.24) are called *complementary* and each is the *complement* of the other. A *complementary solution* is a $\mathbf{z}, \mathbf{w}$ pair satisfying $z_i w_i = 0$ for $1 = i, \ldots, p$. As in linear programming, we will be working with basic feasible solutions to the linear system $\mathbf{w}=\mathbf{q}+\mathbf{Mz}$. We will make throughout the usual *nondegeneracy assumption* that the values of all basic variables are positive in all basic feasible solutions. As we saw in Chapter 1, this can be ensured by lexicographic arguments for resolving degeneracies. A *complementary basic feasible* solution is one in which the complement of each basic variable is nonbasic.

The *linear complementary algorithm* for solving (7.24) begins by adding an artificial variable z_0 to the linear system with corresponding activity $\mathbf{0}_p$ equal to a p-vector of 1's. This gives us

$$\mathbf{w}=\mathbf{q}+\mathbf{0}_p z_0+\mathbf{Mz} \tag{7.25}$$

which always has nonnegative solutions since z_0 can be taken to be a sufficiently large number to ensure that $\mathbf{w}=\mathbf{q}+\mathbf{0}_p z_0\geq 0$. A solution to (7.25) is called *almost complementary* if $z_i w_i=0$ for $i=1,\ldots,p$, and is *complementary* if, in addition, $z_0=0$. Let

$$Z_0=\{(z_0,\mathbf{z})|\mathbf{w}=\mathbf{q}+\mathbf{0}_p z_0+\mathbf{Mz}\geq\mathbf{0},\ z_0\geq 0,\ \mathbf{z}\geq\mathbf{0}\}$$

Our assumption of nondegeneracy is equivalent to saying that each extreme point of Z_0 corresponds to exactly one $p\times p$ basis made up of p columns taken from the matrix $(\mathbf{I},-\mathbf{0}_p,-\mathbf{M})$. An extreme point is called almost complementary or complementary if the corresponding unique basis has those properties.

We will also be concerned with rays of Z_0 which are nonzero solutions $(\mathbf{w}^h;z_0^h,\mathbf{z}^h)$ satisfying

$$\mathbf{w}^h=0_p z_0^h+\mathbf{Mz}^h \tag{7.26}$$

and

$$\mathbf{w}^h\geq\mathbf{0},\ z_0^h\geq 0,\ \mathbf{z}^h\geq\mathbf{0}$$

A ray $(\mathbf{w}^h;z_0^h,\mathbf{z}^h)$ emanating from the extreme point $(\mathbf{w}^*;z_0^*,\mathbf{z}^*)$ generates the feasible solutions

$$\left(\mathbf{w}^*+\lambda\mathbf{w}^h;z_0^*+\lambda z_0^h,\mathbf{z}^*+\lambda\mathbf{z}^h\right)\qquad\text{for all }\lambda\geq 0$$

If the extreme point is almost complementary, then we say the ray is *almost complementary* if

$$\left(z_i^*+\lambda z_i^h\right)\left(w_i^*+\lambda w_i^h\right)=0\qquad i=1,\ldots,p$$

which in turn implies that

$$z_i^*w_i^*=z_i^*w_i^h=z_i^h w_i^*=z_i^h w_i^h=0 \tag{7.27}$$

The algorithm begins by calculating the smallest value of z_0, say $z_0=\tilde{z}_0$, such that $\mathbf{w}=\mathbf{q}+\mathbf{0}_p z_0\geq\mathbf{0}$. We assume that $\tilde{z}_0>0$ because otherwise $\mathbf{q}\geq\mathbf{0}$ and the solution $\mathbf{w}=\mathbf{q}$, $\mathbf{z}=\mathbf{0}$ is trivially a solution to (7.24). At $z_0=\tilde{z}_0$ there is exactly one variable equal to zero, say w_k. This variable is replaced by z_0 in the basis, giving us a nondegenerate almost complementary basic feasible solution to the linear system $\mathbf{Iw}-\mathbf{0}_p z_0-\mathbf{Mz}=\mathbf{q}$. Attached to the extreme point corresponding to this basis is the almost complementary ray $(\mathbf{w}^h; z_0^h, \mathbf{z})=(\mathbf{0}_p \bar{z}_0^h; \bar{z}_0^h, 0)$, where $\bar{z}_0^h$ is any positive number. We can envision that we arrived at our initial almost complementary basic feasible solution by coming down the ray as z_0 is reduced to its basic value $\tilde{z}_0$.

The algorithm proceeds by bringing into the basis the unique variable z_k that is the complement to the variable w_k just driven out of the basis. The change of basis is accomplished by the usual pivot operation [see Definition 1.1] that we used in the simplex method. The variable z_k is increased from zero to that positive value (assuming nondegeneracy), which drives to zero a unique blocking basic variable z_0 or w_l. Since the former basic variable w_k is now zero, the new basic solution with z_k positive is almost complementary in the case when z_0 remains in the basis, and complementary in the case when z_k replaces z_0. Note that z_k is the unique nonbasic variable that will maintain almost complementarity when brought into the basis. In the latter case, of course, the algorithm stops with a solution to the linear complementarity problem (7.24). A third possibility is that there is no blocking variable to z_k. The algorithm terminates when this happens, and we investigate below the significance of this occurrence.

At an arbitrary iteration of the algorithm, there is some variable, say z_l (or w_l), that has just been removed from the basis. We choose the complement to this variable, w_l (or z_l), and attempt to pivot it into the basis by determining the blocking variable. If the attempt fails because there is no blocking variable, we say the algorithm terminates in an almost complementary ray. On the other hand, if there is a blocking variable, then w_l (or z_l) is effectively substituted for it in the basis by a pivot. If the blocking variable is z_0, the algorithm terminates. Again the complement variable w_l (or z_l) is the unique nonbasic variable that maintains the almost complementary condition of the solution when brought into the basis. If the blocking variable is z_ν (or w_ν), $\nu \neq 0$, the procedure is repeated; that is, at the next iteration, we attempt to pivot the complement w_ν (or z_ν) into the basis. Note that the process of attempting to pivot into the basis the complement of the variable that just left preserves at each iteration the almost complementarity of the solutions.

In Section 1.4 we saw that the change of basis operation from one nondegenerate basic solution to another corresponds to movement along an edge from one extreme point of Z_0 to an adjacent one. The movement is reversible in that if we were to bring back into the basis the variable which just left, the variable that just entered in its place would be driven out. Thus, we can view the linear complementary algorithm as generating a path starting from the first almost complementary extreme point and visiting a sequence of almost complementary extreme points, each one connected to its predecessor by an edge of Z_0. Termination of the algorithm is a consequence of the following theorem.

THEOREM 7.14. Along an almost complementary path of extreme points generated by the linear complementary algorithm, the only almost complementary basic feasible solution which can reoccur is the initial one.

PROOF. We suppose the contrary and establish a contradiction. Suppose the situation is as shown in Figure 7.5 where the nodes represent extreme points of Z_0, and the arrows represent the direction of change in moving from one extreme point to an adjacent one along an edge. The edges are called almost complementary. By our nondegeneracy assumption, the extreme points will all be distinct. If extreme point P_k is visited twice by the path, the implication is that there are two distinct almost complementary edges out of P_k to the distinct extreme points P_{k+1} and P_l. These two edges correspond to the introduction of two distinct entering variables to the basis at P_k. But there is at each iteration of the algorithm a unique entering variable that maintains almost complementarity of the solution. It is the complement of the unique variable which just left the basis. But according to the contrary assumption, there are two distinct variables that could enter the basis, which is impossible. ■

COROLLARY 7.3. The linear complementary algorithm terminates after a finite number of iterations in a complementary basic feasible solution to the linear complementary problem (7.24), or in an almost complementary ray different from the starting almost complementary ray.

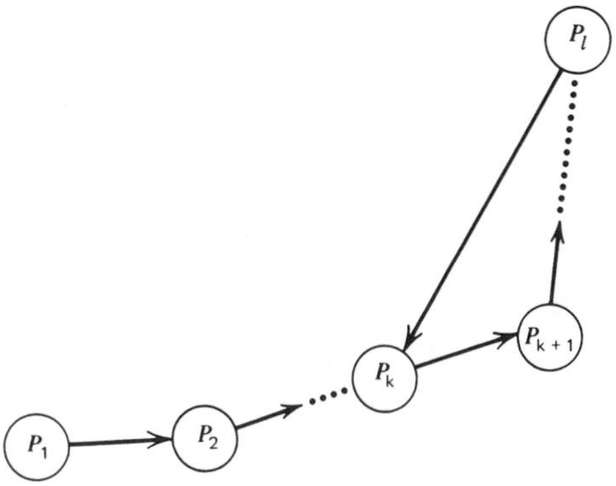

Figure 7.5. Repeating an almost complementary basic feasible solution along an almost complementary path.

PROOF. By Theorem 7.14 a basic extreme point solution to Z_0 can never be repeated by the algorithm except possibly the first. But the first is at the end of a ray so that if it were repeated, then there would be two distinct nonbasic variables that could be introduced to produce a new almost complementary basic solution. This is impossible since there is a unique entering nonbasic variable for the first basic feasible solution. Thus, we are forced to conclude that the algorithm must terminate finitely since otherwise some almost complementary basis would have to be repeated. ■

The linear complementarity problem (7.24) and the algorithm for solving it have been presented without assumptions on the matrix **M**. The algorithm could be applied, for example, to an arbitrary quadratic programming problem (7.22) posed in the form (7.24). If it terminated with a complementary basic feasible solution, then the Kuhn-Tucker conditions are satisfied but the corresponding **x** solution need not be a global or even local minimum in (7.22). Moreover, the linear complementarity algorithm may terminate in an almost complementary ray, and such an occurrence is difficult to interpret for an arbitrary problem (7.24). We show below that if **M** belongs to a class of matrices including those that are positive semidefinite, and if termination in an almost complementary ray occurs, then the linear complementary problem (7.24) has no solution. Thus, in the well-behaved case when (7.22) is a convex programming problem because **D** is positive semidefinite, then **M** in (7.24) is positive semidefinite and the linear complementary algorithm either produces a globally optimal solution or indicates that the problem has no solution because $\mathbf{Ax} \geq \mathbf{b}$ has no nonnegative solution or because the objective function in (7.22) is unbounded.

First, we need the following result characterizing further in the general case termination in an almost complementary ray.

THEOREM 7.15. Suppose the linear complementary algorithm terminates in the almost complementary ray $(\mathbf{w}^h; z_0^h, \mathbf{z}^h)$. Then $\mathbf{z}^h$ is nonzero and

$$z_i^h(\mathbf{M}\mathbf{z}^h)_i \leq 0 \qquad i=1,\ldots,p$$

PROOF. Let $(\mathbf{w}^*; z_0^*, \mathbf{z})$ denote the extreme point of Z_0 where the almost complementary ray was discovered. For every $\lambda \geq 0$, we have

$$(\mathbf{w}^* + \lambda\mathbf{w}^h) = \mathbf{q} + \mathbf{0}_p(z_0^* + \lambda z_0^h) + \mathbf{M}(\mathbf{z}^* + \lambda\mathbf{z}^h)$$

and because the ray is almost complementary,

$$(w_i^* + \lambda w_i^h)(z_i^* + \lambda z_i^h) = 0 \qquad i=1,\ldots,p \tag{7.28}$$

We can rule out the case that $\mathbf{z}^h = 0$ because otherwise we must have $z_0^h > 0$ and thus $\mathbf{w}^h$ has all positive components by (7.26). From (7.28) we conclude that $\mathbf{z}^* + \lambda\mathbf{z}^h = \mathbf{z}^* = \mathbf{0}$, which implies that the ray is the original one and this is impossible by Corollary 7.3.

Moreover, the individual equations of (7.26) are of the form

$$w_i^h = z_0^h + (\mathbf{Mz}^h)_i \qquad i = 1, \ldots, p$$

and therefore from (7.28), we have

$$z_i^h w_i^h = 0 = z_i^h z_0^h + z_i^h (\mathbf{Mz}^h)_i \qquad i = 1, \ldots, p$$

implying

$$z_i^h (\mathbf{Mz}^h)_i \leq 0 \qquad i = 1, \ldots, p$$

because

$$z_i^h \geq 0 \qquad \text{and} \qquad z_0^h \geq 0 \quad \blacksquare$$

We are interested in the class of matrices, called *copositive plus*, which satisfy the two conditions for all $\mathbf{y} \geq 0$

$$\mathbf{yMy} \geq 0 \tag{7.29}$$

and

$$(\mathbf{M} + \mathbf{M}^T)\mathbf{y} = 0 \quad \text{if } \mathbf{yMy} = 0 \tag{7.30}$$

The following lemma establishes that this class of matrices includes the relevant ones for convex quadratic programming.

LEMMA 7.2. If the matrix $\mathbf{M}$ is positive semidefinite, then $\mathbf{M}$ is copositive plus.

PROOF. Condition (7.29) is trivially true since $\mathbf{yMy} \geq 0$ for all $\mathbf{y}$ if $\mathbf{M}$ is positive semidefinite. To show condition (7.30), we use the property of a symmetric, positive semidefinite quadratic form $\mathbf{Q}$ that

$$\mathbf{yQy} = 0 \qquad \text{implies } \mathbf{Qy} = \mathbf{0}$$

To see this for $\mathbf{y} \neq \mathbf{0}$, we use the fact that $\mathbf{Q}$ can be diagonalized by the change of basis $\mathbf{y} = \mathbf{Rz}$, where $\mathbf{R}$ is nonsingular, $\mathbf{yQy} = \mathbf{zR}^T\mathbf{QRz} = \mathbf{zDz}$, and $\mathbf{D}$ is a diagonal matrix with nonnegative elements. Suppose now that $\mathbf{yQy} = 0$ and $\mathbf{y} \neq 0$. Then $\mathbf{z} \neq 0$ because $\mathbf{R}$ is nonsingular and $\mathbf{zDz} = 0$ by the transformation implying $\mathbf{Dz} = 0$ since $\mathbf{D}$ is diagonal. Therefore, $\mathbf{R}^T\mathbf{QRR}^{-1}\mathbf{y} = 0$ or $\mathbf{Qy} = 0$ since $\mathbf{R}^T$ is nonsingular.

Returning to (7.30), we note that $\mathbf{yMy} = \frac{1}{2}\mathbf{y}(\mathbf{M} + \mathbf{M}^T)\mathbf{y}$ implying $\mathbf{M} + \mathbf{M}^T$ is symmetric and positive semidefinite. Thus, if $\mathbf{yMy} = 0$ and therefore $\mathbf{y}(\mathbf{M} + \mathbf{M}^T)\mathbf{y} = 0$, we can conclude that $(\mathbf{M} + \mathbf{M}^T)\mathbf{y} = \mathbf{0}$. $\blacksquare$

LEMMA 7.3. The linear complementarity problem (7.24) has no solution if there exists a vector $\mathbf{r}$ such that

$$\mathbf{rM} \leq \mathbf{0} \qquad \mathbf{rq} < 0 \qquad \mathbf{r} \geq \mathbf{0}$$

PROOF. Suppose the contrary; that is, suppose there exist $\mathbf{w}, \mathbf{z}$ satisfying $\mathbf{w} = \mathbf{q} + \mathbf{Mz}$ for $\mathbf{w} \geq \mathbf{0}$ and $\mathbf{z} \geq \mathbf{0}$. Then $0 \leq \mathbf{rw} = \mathbf{rq} + \mathbf{rMz} \leq \mathbf{rq} < 0$, which is impossible. ■

THEOREM 7.16. If the matrix $\mathbf{M}$ in the linear complementarity problem (7.24) is copositive plus, and if the linear complementary algorithm terminates in an almost complementary ray, then (7.24) has no solution.

PROOF. Let $(\mathbf{w}^h; z_0^h, \mathbf{z}^h)$ denote the nonzero almost complementary ray emanating from the basic feasible solution $(\mathbf{w}^*; z_0^*, \mathbf{z}^*)$ to (7.25). This ray satisfies

$$0 = \mathbf{z}^h \mathbf{w}^h = \mathbf{z}^h \mathbf{0}_p z_0^h + \mathbf{z}^h \mathbf{M} \mathbf{z}^h$$

Since $\mathbf{M}$ is copositive plus and $\mathbf{z}^h \geq \mathbf{0}$, both terms on the right side of this equation are nonnegative and must therefore be zero. The scalar $z_0^h = 0$ because $\mathbf{z}^h \neq \mathbf{0}$ by Theorem 7.15 implying $\mathbf{z}^h \mathbf{0}_p > 0$. Again since $\mathbf{M}$ is copositive plus, $\mathbf{z}^h \mathbf{M} \mathbf{z}^h = 0$ and $\mathbf{z}^h \geq 0$ implies

$$\mathbf{M}\mathbf{z}^h + \mathbf{M}^T \mathbf{z}^h = \mathbf{0} \tag{7.31}$$

But since $(\mathbf{w}^h; z_0^h, \mathbf{z}^h)$ is a ray, we have $\mathbf{w}^h = \mathbf{M}\mathbf{z}^h \geq \mathbf{0}$ implying from (7.31) that $\mathbf{M}^T \mathbf{z}^h \leq \mathbf{0}$, or equivalently, $\mathbf{z}^h \mathbf{M} \leq \mathbf{0}$. Also, from (7.27), we have

$$0 = \mathbf{z}^* \mathbf{w}^h = \mathbf{z}^* \mathbf{M} \mathbf{z}^h = \mathbf{z}^* (-\mathbf{M}^T \mathbf{z}^h) = \mathbf{z}^h \mathbf{M} \mathbf{z}^*$$

and

$$\begin{aligned} 0 = \mathbf{z}^h \mathbf{w}^* &= \mathbf{z}^h \mathbf{q} + \mathbf{z}^h \mathbf{0}_p z_0^* + \mathbf{z}^h \mathbf{M} \mathbf{z}^* \\ &= \mathbf{z}^h \mathbf{q} + \mathbf{z}^h \mathbf{0}_p z_0^* \end{aligned}$$

Thus, $\mathbf{z}^h \mathbf{q} < 0$ because $\mathbf{z}^h \mathbf{0}_p z_0^* > 0$. The linear complementary problem is therefore inconsistent because $\mathbf{r} = \mathbf{z}^h$ satisfies the conditions of Lemma 7.3. ■

EXAMPLE 7.3. (Portfolio Selection Example 7.2 Continued). Analysis of historical data yields the following mean and variance data for three stocks.

TABLE 7.1. DATA FOR PORTFOLIO SELECTION PROBLEM

	Expected Return/\$	Standard Deviation
Stock 1	$0.12 = \mu_1$	$0.30 = \sigma_1$
Stock 2	$0.18 = \mu_2$	$0.50 = \sigma_2$
Stock 3	$0.08 = \mu_3$	$0.40 = \sigma_3$

The correlation coefficients for the three stocks are $\rho_{12}=\sigma_{12}/\sigma_1\sigma_2=0.12$, $\rho_{23}=\sigma_{23}/\sigma_2\sigma_3=0.20$, $\rho_{13}=\sigma_{13}/\sigma_1\sigma_3=-0.10$. Thus, the covariance matrix of the three stocks is

$$D=\begin{bmatrix} 0.09 & 0.018 & -0.012 \\ 0.018 & 0.25 & 0.04 \\ -0.012 & 0.04 & 0.16 \end{bmatrix}$$

Let x_1, x_2, x_3 denote the proportion of the portfolio invested in stocks 1, 2, 3, respectively. The linear complementarity problem (7.24) derived from (7.23) with this data using $\theta=1$ as the mixture of return and risk is

$$\begin{pmatrix} y_1 \\ y_2 \\ y_3 \\ s \end{pmatrix}=\begin{pmatrix} -0.120 \\ -0.180 \\ -0.080 \\ 1 \end{pmatrix}+\begin{pmatrix} 0.18 & 0.036 & -0.024 & 1 \\ 0.036 & 0.50 & 0.08 & 1 \\ -0.024 & 0.08 & 0.32 & 1 \\ -1 & -1 & -1 & 0 \end{pmatrix}\begin{pmatrix} x_1 \\ x_2 \\ x_3 \\ u \end{pmatrix}$$

$$x_1y_1+x_2y_2+x_3y_3+us=0 \qquad \text{all variables nonnegative}$$

We convert this problem to the form (7.25) by adding to it the variable z_0 with associated column 0_4. The result is given in Table 7.2; note that we have expressed the linear system in (7.25) as $-\mathbf{0}_p z_0+\mathbf{I}\mathbf{w}-\mathbf{M}\mathbf{z}=\mathbf{q}$.

TABLE 7.2 LINEAR COMPLEMENTARY ALGORITHM APPLIED TO PORTFOLIO SELECTION PROBLEM

Basic Variables	Current Values	z_0	y_1	y_2	y_3	s	x_1	x_2	x_3	u
y_1	-0.12	-1	1				-0.18	-0.036	0.024	-1
y_2	-0.18	-1		1			-0.036	-0.5	-0.08	-1
y_3	-0.08	-1			1		0.024	-0.08	-0.32	-1
s	1	-1				1	1	1	1	

We obtain our starting basic feasible solution by setting $z_0=0.18$, or equivalently, pivoting z_0 into the basis for the variable y_2. The result is shown in Table 7.3. Since y_2 was the variable which just left the basis, we choose the complementary variable x_2 as the entering variable. The blocking variable is y_1 and we pivot on the element 0.464.

Continuing with the linear complementary algorithm, we arrive at the basic feasible solution shown in Table 7.4. Since the variable which left the basis in constructing this solution was s, the new entering variable is u and the blocking variable is z_0. Thus, z_0 leaves the basis and we obtain the optimal portfolio solution $x_1=0.557$, $x_2=0.262$, $x_3=0.180$ and $u=0.015$ is

TABLE 7.3 LINEAR COMPLEMENTARY ALGORITHM APPLIED TO PORTFOLIO SELECTION PROBLEM

Basic Variables	Current Values	z_0	y_1	y_2	y_3	s	x_1	x_2	x_3	u
y_1	0.06		1	-1			-0.144	0.464	0.104	
z_0	0.18	1		-1			0.036	0.5	0.08	1
y_3	0.10			-1	1		0.06	0.42	-0.24	
s	1.18			-1		1	1.036	1.5	1.08	1

TABLE 7.4. LINEAR COMPLEMENTARY ALGORITHM APPLIED TO PORTFOLIO SELECTION PROBLEM

Basic Variables	Current Values	z_0	y_1	y_2	y_3	s	x_1	x_2	x_3	u
x_2	0.263		1.047	-1.836	0.88	0.095		1		0.095
z_0	0.0135	1	0.52	-0.274	-0.297	0.09				0.91
x_1	0.565		-2.73	2.09	1.16	0.519	1			0.519
x_3	0.184		1.155	1.472	-2.33	0.295			1	0.295

the optimal value of the dual variable on the constraint $x_1+x_2+x_3=1$. Note that the variance of the portfolio is given sufficient weight that it is optimal to invest in stock 3 despite its low expected return/\$ because it is negatively correlated with stock 1. ▲

7.6 ECONOMIC EQUILIBRIUM PROBLEMS

The existence and computation of economic equilibria is a subject that has received widespread attention from mathematical economists. For the most part, it has been studied without recourse to mathematical programming methods despite the close relationship that exists between the two subjects. On the one hand, mathematical programming methods can often be used to compute economic equilibria for a given numerical problem or establish that there are none. Conversely, constructive fixed point methods for computing economic equilibria can also be used to solve mathematical programming problems and other problems that are not obviously fixed point problems. It is logical for us to study constructive fixed point methods in this section because they have much in common with the linear complementary algorithm presented in the previous section.

We begin with a discussion of the classical economic model of exchange. This model consists of m individuals and n commodities. Individual i

initially holds a nonnegative quantity w_{ij} of each commodity j, and he buys and sells each commodity with his preferences dictated by his utility function and the market prices. Let x_{ij} denote the variable quantity of commodity j held by individual i, and let $\mathbf{x}_i = (x_{i1}, \ldots, x_{in})$. Let $U_i(\mathbf{x}_i)$ denote the utility to individual i of holding the vector $\mathbf{x}_i$ of the n commodities; the usual assumption is that U_i is a concave function.

Let Π_j denote the nonnegative market price for commodity j and let $\mathbf{\Pi} = (\Pi_1, \ldots, \Pi_n)$. Individual i selects his vector $\mathbf{x}_i$ of the commodities according to the mathematical programming problem

$$\begin{aligned} &\max U_i(\mathbf{x}_i) \\ &\text{s.t.} \sum_{j=1}^{n} \Pi_j x_{ij} = \sum_{j=1}^{n} \Pi_j w_{ij} \\ &x_{ij} \geq 0 \qquad j = 1, \ldots, n \end{aligned} \tag{7.32}$$

In words, individual i maximizes his utility subject to the constraint that his total capital for buying and selling commodities equals the value of his initial holdings according to the market prices. Let $x_{ij}(\mathbf{\Pi})$ denote the desired holdings of commodity j by individual i when the market prices are $\mathbf{\Pi}$. We define the *excess demand* for commodity j by individual i as $g_{ij}(\mathbf{\Pi}) = x_{ij}(\mathbf{\Pi}) - w_{ij}$, and the *market excess demand* for commodity j denoted by $g_j(\mathbf{\Pi})$ is defined as

$$g_j(\mathbf{\Pi}) = \sum_{i=1}^{m} g_{ij}(\mathbf{\Pi}) \tag{7.33}$$

According to the budget constraint in (7.32), the g_{ij} satisfy for all Π

$$\Pi_1 g_{i1}(\mathbf{\Pi}) + \cdots + \Pi_n g_{in}(\mathbf{\Pi}) = 0$$

and therefore (7.34)

$$\Pi_1 g_1(\mathbf{\Pi}) + \cdots + \Pi_n g_n(\mathbf{\Pi}) = 0$$

A market price vector is an *equilibrium price vector* if the market excess demand for each commodity at these prices is nonpositive, and actually equals zero if the price is positive. Mathematically, the price vector $\mathbf{\Pi}$ is an equilibrium price vector if

$$g_j(\mathbf{\Pi}) \leq 0 \qquad j = 1, \ldots, n$$

and (7.35)

$$g_j(\mathbf{\Pi}) = 0 \qquad \text{if } \Pi_j > 0$$

The existence of an equilibrium price vector requires some assumptions about the excess demand functions g_{ij}, thereby implying some assumptions about the utility functions U_i. The usual assumptions are

1. Each $g_{ij}(\mathbf{\Pi})$ is homogeneous of degree zero; that is, for all $\lambda>0$, $g_{ij}(\lambda\mathbf{\Pi})=g_{ij}(\mathbf{\Pi})$. This permits us to restrict our attention to prices on the simplex

$$S=\left\{\mathbf{\Pi}\,\middle|\,\sum_{j=1}^{n}\Pi_j=1,\ \mathbf{\Pi}\geq\mathbf{0}\right\} \tag{7.36}$$

2. Each g_{ij} is continuous on S.

These assumptions are satisfied, for example, if the utility functions are given by

$$U_i(\mathbf{x}_i)=\left(\sum_{j=1}^{n}(a_{ij})^{1-a_i}x_{ij}^{a_i}\right)^{1/a_i} \tag{7.37}$$

where $0<a_{ij}$ and $0<a_i<1$. In this case it can be shown (see Exercise 7.13) that the excess demand functions are given by

$$g_{ij}(\mathbf{\Pi})=\frac{a_{ij}\sum_{j=1}^{n}w_{ij}\Pi_j}{\Pi_j^{b_i}\sum_{j=1}^{n}a_{ij}\Pi_j^{1-b_i}}-w_{ij} \tag{7.38}$$

where $b_i=1/(1-a_i)$, and these functions satisfy properties 1 and 2.

With assumptions 1 and 2, the existence of an equilibrium price vector can be established by the following theorem, which we state without proof. References are given at the end of the chapter to proofs of the theorem. The constructive procedure to be given below does not rely on the theorem and moreover, it can be used to prove it.

THEOREM 7.17. (Brouwer's Fixed Point Theorem). Let f be a continuous mapping of the simplex S into itself. Then there exists a $\bar{\mathbf{\Pi}}\in S$ such that $f(\bar{\mathbf{\Pi}})=\bar{\mathbf{\Pi}}$.

We apply this theorem by constructing a mapping f of S into S that is continuous and with the property that any fixed point is an equilibrium price vector. The image of $\mathbf{\Pi}$ under f is denoted by $[f_1(\mathbf{\Pi}),f_2(\mathbf{\Pi}),\ldots,f_n(\mathbf{\Pi})]$ and given by

$$f_j(\mathbf{\Pi})=\frac{\Pi_j+\theta\max[0,g_j(\mathbf{\Pi})]}{1+\theta\sum_{k=1}^{n}\max[0,g_k(\mathbf{\Pi})]} \tag{7.39}$$

where $\theta>0$. It is easy to verify that f is continuous since each f_j is the ratio of continuous mappings. Moreover, $f_j(\mathbf{\Pi}) \geq 0$ and

$$\sum_{j=1}^{n} f_j(\mathbf{\Pi}) = \frac{\sum_{j=1}^{n} \Pi_j + \theta \sum_{j=1}^{n} \max\left[0, g_j(\mathbf{\Pi})\right]}{1 + \theta \sum_{k=1}^{n} \max\left[0, g_k(\mathbf{\Pi})\right]} = 1$$

and therefore f maps S into itself. Now suppose that $\overline{\mathbf{\Pi}}$ is a fixed point. We can argue that $\sum_{j=1}^{n} \max[0, g_j(\overline{\mathbf{\Pi}})] = 0$. For, otherwise $\sum_{j=1}^{n} \max[0, g_j(\overline{\mathbf{\Pi}})] > 0$ implying $\overline{\Pi}_j + \theta \max[0, g_j(\overline{\mathbf{\Pi}})] = K\overline{\Pi}_j$ with $K > 1$. Thus, $g_j(\overline{\mathbf{\Pi}}) > 0$ for all j such that $\overline{\Pi}_j > 0$. This violates the property (7.34) of excess demand functions that $\overline{\Pi}_1 g_1(\overline{\mathbf{\Pi}}) + \cdots + \overline{\Pi}_n g_n(\overline{\mathbf{\Pi}}) = 0$. Therefore $g_j(\overline{\mathbf{\Pi}}) \leq 0$ for all j and again appealing to (7.34), we must have $g_j(\overline{\mathbf{\Pi}}) = 0$ if $\overline{\Pi}_j > 0$ demonstrating that any fixed point $\overline{\mathbf{\Pi}}$ is an equilibrium price vector.

The classical economic model of exchange just presented assumes that we know the utility functions of the individuals buying and selling commodities and that the buying and selling is unconstrained by production or other limitations. A variety of other equilibrium models have been proposed and studied, including some that are more specific in their application, involve data, and address the issues of utility estimation and constraints on the interaction between economic individuals. One such model is the traffic equilibrium model that we discuss briefly and then illustrate with an example.

The traffic equilibrium model is defined over a network $G = [\mathcal{N}, \mathcal{A}]$ corresponding to a road system. There are some distinguished pairs of nodes, called Origin-Destination (OD) pairs, between which travel is desired. Let d_k denote the demand (number of trips) desired between OD pair k. This demand is variable and depends on the time it takes to make the trip; demand is monotonically decreasing in travel time. There are a number of chains (at least two) in the network connecting each origin to each destination. Let f_{kl} denote the nonnegative flow along chain l connecting OD pair k. The total flow F_a in each arc $a \in \mathcal{A}$ is given by

$$F_a = \sum_{k,l} \delta_{k,l,a} f_{kl}$$

where

$$\delta_{k,l,a} = \begin{cases} 1 & \text{if arc } a \text{ is included in chain } l \\ & \text{connecting OD pair } k \\ 0 & \text{otherwise} \end{cases}$$

The time to traverse the arc a is given by the volume delay function $\tau_a(F_a)$

which is monotonically increasing in the flow F_a. These times are used as "distances" on the arcs in the network G and the time to travel from origin to destination for each OD pair k is the shortest route time using these distances. Thus, we have three sets of interdependent variables: demands, flows, and travel times. These variables are said to be in *user equilibrium* if there is consistency between the variables, and no traveler between any OD pair can have his travel time reduced without increasing the travel time of travelers on another OD pair. In other words, if demand is given, then for each OD pair, the only chains with positive flows are the shortest route chains where the time on each arc in the shortest route calculation is computed by the volume delay function of the flows. In addition, user equilibrium assumes that the given demand is at the level implied by the shortest route travel times.

There are two remarks to be made about the traffic equilibrium model just presented. First, the travelers' demand for travel between each OD pair as a function of the travel time can be empirically estimated by statistical methods. Moreover, it implies the travelers' utility for making the trip in the following sense. Since demand $d_k(t_k)$ is a monotonically increasing function of travel time t_k, we can invert it to obtain the function d_k^{-1} that gives the travel time corresponding to a given demand for travel. It can be shown that under proper assumptions on the integratibility of d_k^{-1}, the conditions of a user equilibrium are the global optimality conditions on the problem of minimizing $\sum_k \int_0^{t_k} d_k^{-1}(\xi)d\xi$, subject to constraints that flows are along chains connecting OD pairs to meet demands implied by the travel times t_k. In other words, the demand functions d_k imply the objective functions $\int_0^{t_k} d_k^{-1}(\xi)d\xi$. The second remark about the traffic equilibrium model is that an alternative optimality criterion to user equilibrium is *system equilibrium*, which is simply to minimize the total travel times of all travelers in the system.

EXAMPLE 7.4. We illustrate the concept of traffic equilibrium by the following simple example shown in Figure 7.6. Travelers wish to go from nodes 1 and 2 to the city. Demands are fixed and the times are shown on the arcs as variables and functions of the flow f in each of the arcs. Solution of this problem will be given below after we devise methods for solving equilibrium problems. ▲

Our interest at this point is the construction, analysis, and use of the so-called *simplicial approximation algorithm* for the computation of approximate fixed points of a continuous mapping of the simplex S into itself. This algorithm begins in a corner of S and systematically searches over a subset of it until an approximate fixed point is found. The approximation is in the following sense: An exact fixed point $\overline{\Pi}$ satisfies

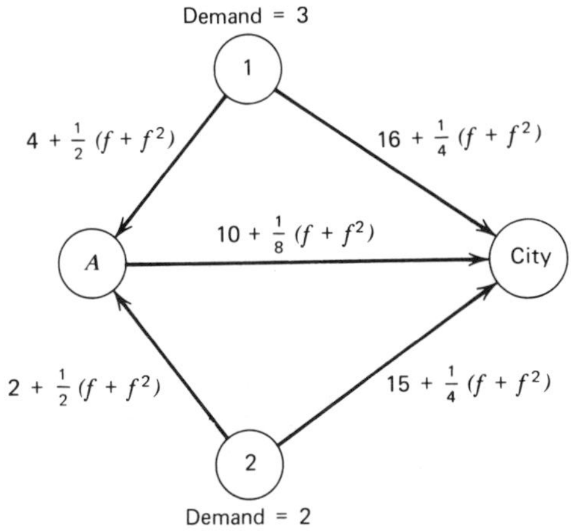

Figure 7.6. Traffic equilibrium problem.

$f_j(\overline{\mathbf{\Pi}}) \geq \overline{\Pi}_j$, since $\Sigma_{j=1}^n f_j(\overline{\mathbf{\Pi}}) = \Sigma_{j=1}^n \overline{\Pi}_j = 1$, and $f_j(\overline{\mathbf{\Pi}})$ and $\overline{\Pi}_j$ are nonnegative. The algorithm finds instead a collection of closely neighboring points $\mathbf{\Pi}^1, \ldots, \mathbf{\Pi}^n \in S$ such that $f_j(\mathbf{\Pi}^j) \geq \Pi_j^j$ for $j = 1, \ldots, n$. Since f is continuous, we would expect that $\mathbf{\Pi}^1, \ldots, \mathbf{\Pi}^n$ lie in a neighborhood of a fixed point and any one of them approximates a fixed point. The nearness cannot be quantified, however, and the approximation is only insured in the limiting sense that the error $\max_{i=1,\ldots,n} |f_i(\mathbf{\Pi}^j) - \Pi_i^j|$ for $\mathbf{\Pi}^j$ approximating a fixed point goes to zero as the maximal distance between the points of S used in the search goes to zero. Note, moreover, that $f_j(\mathbf{\Pi}^j) \geq \Pi_j^j$ for $j = 1, \ldots, n$, does *not* imply that there is a fixed point lying within the subsimplex

$$\left\{ \mathbf{\Pi} | \mathbf{\Pi} = \sum_{j=1}^{n} \lambda_j \Pi^j, \ \sum_{j=1}^{n} \lambda_j = 1, \lambda_j \geq 0, j = 1, \ldots, n \right\} \subseteq S$$

The simplex S is an $(n-1)$ dimensional affine subspace of R^n and it is the convex hull of the n linearly independent points

$$(1,0,\ldots,0), (0,1,0,\ldots,0), \ldots, (0,\ldots,0,1)$$

The simplicial approximation algorithm searches over certain subsets of S called subsimplices each of which is described by

$$\begin{aligned} \Pi_j &\geq a_j \qquad j = 1, \ldots, n \\ \sum_{j=1}^{n} \Pi_j &= 1 \\ \mathbf{\Pi} &\geq \mathbf{0} \end{aligned} \tag{7.40}$$

where $a_j \geq 0$ and $\Sigma_{j=1}^n a_j < 1$. Figure 7.7 shows two such subsimplices

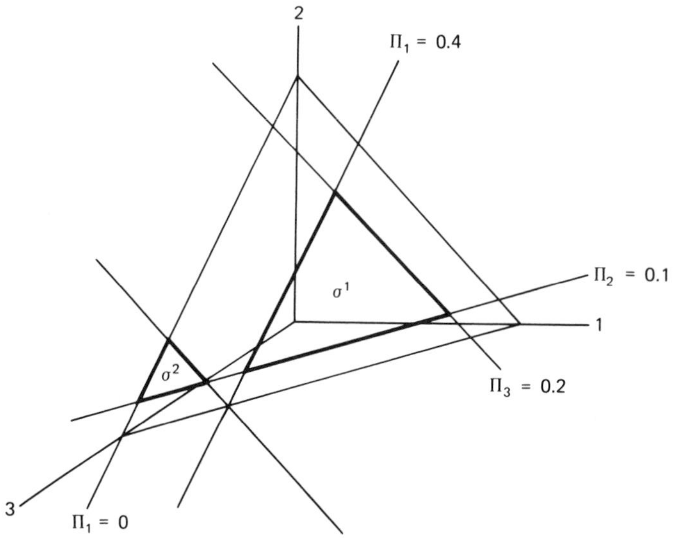

Figure 7.7. Subsimplices on a simplex.

corresponding to $\boldsymbol{\sigma}^1 = (a_1, a_2, a_3) = (0.4, 0.1, 0.2)$ and $\boldsymbol{\sigma}^2 = (a_1, a_2, a_3) = (0, 0.1, 0.8)$. These subsimplices could also be described as the convex hull of n linearly independent points on S, but the representation (7.40) is more convenient for our purposes.

Specifically, the subsimplices of the form (7.40) with which we will work are iteratively derived from a finite set P of n-vectors $\mathbf{\Pi}^1, \ldots, \mathbf{\Pi}^n$, $\mathbf{\Pi}^{n+1}, \ldots, \mathbf{\Pi}^K$. The vectors $\mathbf{\Pi}^{n+1}, \ldots, \mathbf{\Pi}^K$ are selected arbitrarily from the interior of S except for the nondegeneracy assumption that $\Pi_j^{k_1} = \Pi_j^{k_2}$ for any j and any $k_1 \neq k_2$. The nondegeneracy assumption is made to avoid ambiguity in the simplicial approximation algorithm, and for ease of exposition. A regular grid of points on S is theoretically permissible, and computationally more efficient, but we can avoid a significant amount of technical detail by presenting the algorithm with this assumption. The first n vectors $\mathbf{\Pi}^j$ for $j = 1, \ldots, n$, in the finite set P correspond to the edges of S with $\Pi_j = 0$ and are given as

$$
\begin{aligned}
\mathbf{\Pi}^1 &= (0, M_1, M_1, \ldots, M_1) \\
\mathbf{\Pi}^2 &= (M_2, 0, M_2, \ldots, M_2) \\
&\vdots \\
\mathbf{\Pi}^n &= (M_n, M_n, \ldots, M_n, 0)
\end{aligned}
\tag{7.41}
$$

where $M_1 > M_2 > \ldots M_n > 1$.

An arbitrary set of n vectors $\mathbf{\Pi}^{k_1},\ldots,\mathbf{\Pi}^{k_n}$ is used to define a subsimplex (7.40) by selecting

$$a_j=\min\left(\Pi_j^{k_1},\ldots,\Pi_j^{k_n}\right) \qquad j=1,\ldots,n \tag{7.42}$$

A set of n vectors is called a *primitive set* if there is no $\mathbf{\Pi}^k$ in P satisfying $\Pi_j^k>a_j$ for the a_j defined in (7.42). By the nondegeneracy assumption it follows immediately that there will be exactly one Π^{k_j} achieving the minimum in each of these inequalities. In Figure 7.8, the vectors $\mathbf{\Pi}^7,\mathbf{\Pi}^8,\mathbf{\Pi}^{10}$ form a primitive set; similarly, $\mathbf{\Pi}^1,\mathbf{\Pi}^5,\mathbf{\Pi}^6$ form a primitive set (recall that $\mathbf{\Pi}^1$ corresponds to the boundary of S for which $\Pi_1=0$).

The simplicial approximation algorithm works with a sequence of primitive sets that it evaluates by a labeling scheme. The label associated with a vector $\mathbf{\Pi}^k$ in P for $k>n$ is any integer j such that $f_j(\mathbf{\Pi}^k)\geq\Pi_j^k$. The unique label for $\mathbf{\Pi}^j$, $j\leq n$, is j. This is consistent with the labeling of the other Π^k since $\Pi_j^j=0$ implies $f_j(\mathbf{\Pi}^j)\geq\Pi_j^j$. The goal of the algorithm is to find a primitive subsimplex $\mathbf{\Pi}^{k_1},\ldots,\mathbf{\Pi}^{k_n}$ with a complete set of labels. As we have seen, any of the Π^{k_j} can be taken as an approximate fixed point in this case. Later on we will present a generalization of the algorithm in which a vector will be associated with each of the vectors $\mathbf{\Pi}^j$ in P. These vectors will be called *vector labels* to distinguish them from the *integer labels* we are now using.

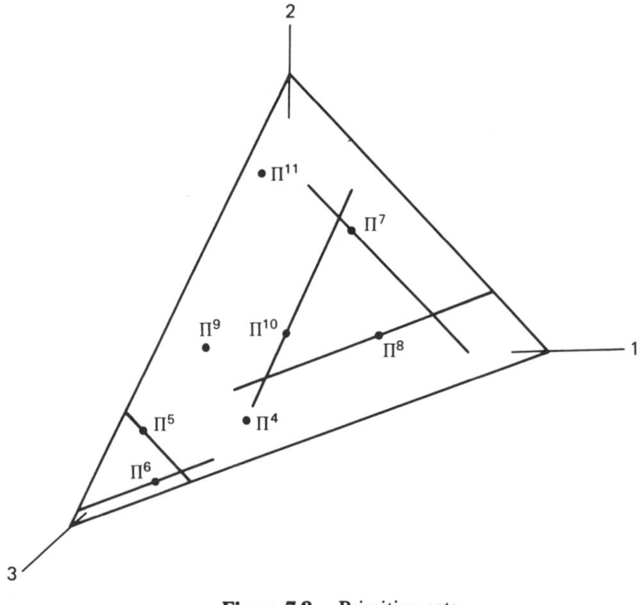

Figure 7.8. Primitive sets.

The following lemma is the main tool in the simplicial approximation algorithm. It says, in effect, that we can make a unique exchange of vectors from P in going from one primitive set to the next in much the same way the simplex algorithm in Chapter 1 changes bases or the linear complementarity algorithm discussed in the previous section made a unique change of basis at each pivot.

LEMMA 7.4. Let $\mathbf{\Pi}^{k_1},\ldots,\mathbf{\Pi}^{k_n}$ be a primitive set, and let $\mathbf{\Pi}^{k_\alpha}$ be a specific one of the set. Then, aside from one exceptional case, there is a unique vector $\mathbf{\Pi}^k \in P$, different from $\mathbf{\Pi}^{k_\alpha}$, such that $\mathbf{\Pi}^{k_1},\ldots,\mathbf{\Pi}^{k_{\alpha-1}},\mathbf{\Pi}^k,\mathbf{\Pi}^{k_{\alpha+1}},\ldots,\mathbf{\Pi}^{k_n}$ form a primitive set. The exceptional case is when the $n-1$ vectors $\mathbf{\Pi}^{k_j}$, $j\neq\alpha$, are all selected from the first n vectors of P, and in this case no replacement is possible.

PROOF. Without loss of generality, assume

$$\Pi_j^{k_j}=\min\left(\Pi_j^{k_1},\ldots,\Pi_j^{k_n}\right) \qquad j=1,\ldots,n$$

and furthermore, assume $\mathbf{\Pi}^{k_1}$ is the vector being removed from the primitive set. We treat the usual (nonexceptional) case first. Let $\mathbf{\Pi}^{k_\beta}$ be the unique vector in the primitive set with the second smallest value of its first coordinate. Consider all $\mathbf{\Pi}^k$ in P satisfying

$$\Pi_1^k > \Pi_1^{k_\beta}$$

$$\Pi_j^k > \Pi_j^{k_j}, \qquad j\neq 1,\beta$$

and among those with $\Pi_\beta^k < \Pi_\beta^{k_\beta}$, select the unique one $\mathbf{\Pi}^{k^*}$ with the maximal value of the j_βth coordinate. Note that $k_\beta > n$ since we are treating the non-exceptional case, and therefore, there is at least one $\mathbf{\Pi}^k$ satisfying these restrictions on $\mathbf{\Pi}^k$, namely, $\mathbf{\Pi}^\beta$ with $\Pi_\beta^\beta=0$.

It can be immediately verified that $\mathbf{\Pi}^{k^*},\mathbf{\Pi}^{k_2},\ldots,\mathbf{\Pi}^{k_n}$ is a primitive set, and moreover, that

$$\Pi_j^{k_j}=\min\left(\Pi_j^{k^*},\Pi_j^{k_2},\ldots,\Pi_j^{k_n}\right) \qquad j\neq 1,\beta$$

$$\Pi_1^{k_\beta}=\min\left(\Pi_1^{k^*},\Pi_1^{k_2},\ldots,\Pi_1^{k_n}\right)$$

$$\Pi_\beta^{k^*}=\min\left(\Pi_\beta^{k^*},\Pi_\beta^{k_2},\ldots,\Pi_\beta^{k_n}\right)$$

Thus, the vector $\mathbf{\Pi}^{k_1}$ in the original primitive set has been replaced by the unique vector $\mathbf{\Pi}^{k^*}$ in P and the first part of the lemma is established.

Consider the exceptional case and suppose without loss of generality that $\mathbf{\Pi}^{k_j}$ for $j\neq\alpha$, are the $n-1$ vectors selected from the first n vectors. In this case, there is only one choice for the vector $\mathbf{\Pi}^{k_\alpha}$ permitting the set $\mathbf{\Pi}^{k_1},\ldots,\mathbf{\Pi}^{k_n}$ to be primitive; namely, $\mathbf{\Pi}^{k_\alpha}$ must be the unique vector among the $\mathbf{\Pi}^j$ for $j>n$ with the maximal value of coordinate α. Thus, it is not possible in this case to replace Π^{j_α} by another vector and achieve a primitive set. ■

This lemma permits us to construct the simplicial approximation algorithm. The algorithm systematically searches through a sequence of primitive sets with the following properties:

1. The label 1 is not associated with any vector in the primitive set.
2. The vectors in the primitive set will be labeled differently, except for one pair of vectors with the same label.

The algorithm proceeds by taking one of the two vectors with the same label and removing it from the primitive set following the rules of Lemma 7.4, either obtaining another primitive set with the same properties or else finding one with a complete set of labels in which case it terminates. For each primitive set generated after the first one, one of the two vectors with the duplicate label has just entered the primitive set, and it is the other vector with the duplicate label which is removed. The first primitive set considered by the algorithm is the one considered as an exceptional case in Lemma 7.4; that is, it consists of $\Pi^2,\ldots,\Pi^n$ and Π^{k^*} for $k^*>n$ with the property that $\Pi_1^{k^*}$ is the maximal first coordinate among the Π^k in P for $k>n$. If the label of Π^{k^*} is not 1, say it is α, then it is the vector Π^α which is removed.

For completeness, we give the following step by step description of the algorithm.

Simplicial Approximation Algorithm with Integer Labels

Step 1. (Initialization). Let

$$\Pi^1=(0,M_1,M_1,\ldots,M_1)$$
$$\Pi^2=(M_2,0,M_2,\ldots,M_2)$$
$$\vdots$$
$$\Pi^n=(M_n,M_n,\ldots,M_n,0)$$

where $M_1>M_2>\cdots>M_n>1$, and select the vectors $\Pi^{n+1},\ldots,\Pi^K$ so that $\Pi_j^{k_1}\neq\Pi_j^{k_2}$ for all j and all $k_1\neq k_2$. Find the unique vector Π^{k^*} with the property that $\Pi_1^{k^*}$ has the maximal value of the first component among the Π^k for $k\geq n+1$. Then $(\Pi^{k^*},\Pi^2,\ldots,\Pi^n)$ is a primitive set. Go to Step 2, with this primitive set and $\Pi^k=\Pi^{k^*}$.

Step 2. If Π^k is one of the first n vectors, then assign the integer label $l(\Pi^k)=k$. Otherwise, compute $f(\Pi^k)=[f_1(\Pi^k),\ldots,f_n(\Pi^k)]$ and assign an integer label $l(\Pi^k)$ that is any index l such that $f_l(\Pi^k)\geq\Pi_l^k$. If $l(\Pi^k)=1$, the algorithm terminates and any one of the vectors in the primitive set approximates a fixed point. Otherwise, go to Step 2 indicating that the exiting vector from the primitive set is $\mathbf{\Pi}^{k_\alpha}$ for which $l(\mathbf{\Pi}^{k}) = l(\mathbf{\Pi}^{k_\alpha}) \geq 2$.

Step 3. Replace $\mathbf{\Pi}^{k_\alpha}$ in the primitive set $\mathbf{\Pi}^{k_1},\ldots,\mathbf{\Pi}^{k_{\alpha-1}}$, $\mathbf{\Pi}^{k_\alpha},\mathbf{\Pi}^{k_{\alpha+1}},\ldots,\mathbf{\Pi}^{k_n}$ by the uniquely determined $\mathbf{\Pi}^{k}$ in P as follows: For $j=1,\ldots,n$, let $p(j)$ be the index such that

$$\Pi^{k_j}_{p(j)} = \min\left(\Pi^{k_1}_j,\ldots,\Pi^{k_n}_j\right) \qquad j=1,\ldots,n$$

Let $\mathbf{\Pi}^{k_\beta}$ denote the unique vector in the primitive set with the second smallest value of coordinate $p(k_\alpha)$. Choose $\mathbf{\Pi}^{k}$ so that it satisfies

$$\Pi^{k}_j > \Pi^{k_j}_{p(j)} \qquad j \neq \alpha, \beta$$

$$\Pi^{k}_{p(k_\alpha)} \geq \Pi^{k_\beta}_{p(k_\alpha)}$$

and moreover, so that it has the maximal value of the coordinate $p(k_\beta)$ less than $\Pi^{k_\beta}_{p(k_\beta)}$.

Go to step two with the new primitive set and the indicated vector $\mathbf{\Pi}^{k}$.

THEOREM 7.18. The simplicial approximation algorithm with integer labels terminates after a finite number of iterations (i.e., after generating a finite number of primitive sets) with a completely labeled primitive set.

PROOF. The proof is basically the same as the proof of Theorem 7.14 and Corollary 7.3 in which we proved that the linear complementary algorithm converged finitely. Since the algorithm can terminate only if it finds a completely labeled primitive set, and since there are only a finite number of primitive sets, if it did not terminate then it would have to generate some primitive set more than once. Now each primitive set after the first primitive set is uniquely determined by its predecessor, and its successor is also uniquely determined, according to Lemma 7.4. Consider the first primitive set that is generated twice by the algorithm. It has two distinct predecessors and one successor, which is impossible. The first primitive set used by the algorithm has no predecessor and a unique successor. If it were the first primitive set repeated by the algorithm, then it would have two successors by inverting the exchange of vectors which produced it from its predecessor. This is also impossible and therefore we can conclude that the simplicial approximation algorithm terminates finitely, and by necessity, with a completely labeled subsimplex. ■

To analyze the behavior of the simplicial approximation algorithm as the approximation becomes more exact, we need the following definition. We say the sequence of sets $P^l = \mathbf{\Pi}^1,\ldots,\mathbf{\Pi}^n,\mathbf{\Pi}^{n+1},\ldots,\mathbf{\Pi}^{k_l}$ for $l=1,2,\ldots$ is *dense* on S if the distance between distinct points $\mathbf{\Pi}^k, \mathbf{\Pi}^t \in P^l$, say

$$\min_{k \neq t} \max_{j=1,\ldots,n} |\Pi_j^k - \Pi_j^t|$$

goes to zero as l goes to $+\infty$.

PROOF OF BROWER'S FIXED POINT THEOREM 7.17. Consider a sequence of sets P^l of n vectors $\mathbf{\Pi}^1,\ldots,\mathbf{\Pi}^n,\mathbf{\Pi}^{n+1},\ldots,\mathbf{\Pi}^{k_l}$ such that $\mathbf{\Pi}^1,\ldots,\mathbf{\Pi}^n$ are as defined in (7.41), the vectors $\mathbf{\Pi}^{n+1},\ldots,\mathbf{\Pi}^{k_l}$ are on the simplex S and also such that the vectors $\mathbf{\Pi}^{n+1},\ldots,\mathbf{\Pi}^{k_l}$ become dense on S. Let $\mathbf{\Pi}^{k_1},\ldots,\mathbf{\Pi}^{k_{n,l}}$ denote the primitive set with a complete set of integer labels, say $f(\Pi_j^{k_{j,l}}) \geq \Pi_j^{k_{j,l}}$ for $j=1,\ldots,n$. Since S is compact, and since the vectors became dense on S, there must be a subsequence of these primitive sets converging to a single point $\mathbf{\Pi}^*$. By the continuity of f, $\mathbf{\Pi}^*$ must satisfy $f_j(\mathbf{\Pi}^*) \geq \Pi_j^*$. ■

EXAMPLE 7.5. The traffic equilibrium problem of Example 7.4 can be formulated as a fixed point problem and solved by the simplicial approximation algorithm with integer labels. Let x denote the flow from node 1 to the city along the route $(1,A),(A,\text{city})$; then $3-x$ is the flow from node 1 to the city along the route $(1,\text{city})$. Similarly, let y denote the flow from node 2 to the city along route $(2,A),(A,\text{city})$; then $2-y$ is the flow from node 2 to the city along the route $(2,\text{city})$. We know *a priori* that $x+y \leq 5$. This is converted to an equality constraint and we work with the set F consisting of all x,y,z satisfying

$$x+y+z=5$$

$$x \geq 0 \qquad y \geq 0 \qquad z \geq 0$$

The next step is to give a continuous mapping of F into itself, say $(x,y,z) \in F$ is mapped into $[f(x,y,z),g(x,y,z),h(x,y,z)] \in F$, such that a fixed point is an equilibrium. Given $x,y,z \in F$, the mapping is constructed by considering the relative times to travel the alternate routes from nodes 1 and 2 to the city. Specifically, we look at the two differences

$$\begin{aligned}\Delta_1(x,y) &= 16+\tfrac{1}{4}\left[(3-x)+(3-x)^2\right] \\ &\quad -\left\{4+\tfrac{1}{2}(x+x^2)+10+\tfrac{1}{8}\left[x+y+(x+y)^2\right]\right\} \\ &= \frac{40-19x-3x^2-y-y^2-2xy}{8}\end{aligned}$$

and

$$\begin{aligned}\Delta_2(x,y) &= 15+\tfrac{1}{4}\left[(2-y)+(2-y)^2\right] \\ &\quad -\left\{2+\tfrac{1}{2}(y+y^2)+10+\tfrac{1}{8}\left[x+y+(x+y)^2\right]\right\} \\ &= \frac{36-x-x^2-15y-3y^2-2xy}{8}\end{aligned}$$

If $\Delta_1(x,y)>0$, then flow from node 1 to the city along the route (1, city) should be diverted to the route $(1,A),(A,\text{city})$, whereas the opposite is true if $\Delta_1(x,y)<0$. Similar statements are true for $\Delta_2(x,y)$. However, if $\Delta_1(x,y)=\Delta_2(x,y)=0$, then x and y have equalized the time to travel the two routes to the city from each of the nodes 1 and 2 and we are in equilibrium. We use Δ_1 and Δ_2 to construct the following mapping of F into itself.

$$f(x,y,z)=x+\begin{cases}\dfrac{(3-x)\Delta_1(x,y)}{|\Delta_1(x,y)|+|\Delta_2(x,y)|+0.001} & \text{if } \Delta_1(x,y)>0\\ 0 & \text{if } \Delta_1(x,y)=0\\ \dfrac{x\Delta_1(x,y)}{|\Delta_1(x,y)|+|\Delta_2(x,y)|+0.001} & \text{if } \Delta_2(x,y)<0\end{cases}$$

$$g(x,y,z)=y+\begin{cases}\dfrac{(2-y)\Delta_2(x,y)}{|\Delta_1(x,y)|+|\Delta_2(x,y)|+0.001} & \text{if } \Delta_2(x,y)>0\\ 0 & \text{if } \Delta_2(x,y)=0\\ \dfrac{y\Delta_2(x,y)}{|\Delta_2(x,y)|+|\Delta_2(x,y)|+0.001} & \text{if } \Delta_2(x,y)<0\end{cases}$$

and

$$h(x,y,z)=5-f(x)-g(y)$$

The reader can verify that any point satisfying $(\bar{x},\bar{y},\bar{z})=[f(\bar{x},\bar{y},\bar{z}),g(\bar{x},\bar{y},\bar{z}),h(\bar{x},\bar{y},\bar{z})]$ is a traffic equilibrium. In this regard, note for example, that we can have $\bar{x}=3=f(\bar{x},\bar{y},\bar{z})$ and $\Delta_1(\bar{x},\bar{y})>0$ indicating that the route $(1,A),(A,\text{city})$ carrying *all* the demand from node 1 is preferred for incremental demand to the route (1, city) carrying *no* demand from node 1.

Table 7.5 gives the points in F that we will use in applying the simplicial approximation algorithm with integer labels to this problem.

TABLE 7.5. TRIAL EQUILIBRIUM POINTS FOR SIMPLICIAL APPROXIMATION ALGORITHM

	Π^4	Π^5	Π^6	Π^7	Π^8	Π^9	Π^{10}
x	0.23	0.45	0.79	1.04	0.73	1.36	0.57
y	0.64	0.98	0.81	0.37	1.19	1.24	0.36
z	4.13	3.57	3.40	3.59	3.08	2.40	4.07

	Π^{11}	Π^{12}	Π^{13}	Π^{14}	Π^{15}	Π^{16}	Π^{17}	Π^{18}
x	1.75	1.50	2.03	1.97	2.50	3.42	1.26	1.08
y	0.72	0.83	0.49	1.12	1.20	0.86	1.71	1.53
z	2.53	2.67	2.48	1.91	1.30	0.72	2.03	2.39

TABLE 7.6. SIMPLICIAL APPROXIMATION ALGORITHM APPLIED TO TRAFFIC EQUILIBRIUM PROBLEM

Primitive Set	Labels	Δ_1	Δ_2	$\begin{cases} x \\ f \end{cases}$	$\begin{cases} y \\ g \end{cases}$	$\begin{cases} z \\ h \end{cases}$	Integer Label
$\Pi^1, \Pi^2, \mathbf{\Pi}^4$	1, 2, 1	3.91	3.08	$\begin{cases} 0.23 \\ 1.78 \end{cases}$	$\begin{cases} 0.64 \\ 1.24 \end{cases}$	$\begin{cases} 4.13 \\ 1.98 \end{cases}$	1
$\mathbf{\Pi}^{10}, \Pi^2, \Pi^4$	1, 2, 1	3.41	3.61	$\begin{cases} 0.57 \\ 1.75 \end{cases}$	$\begin{cases} 0.36 \\ 1.20 \end{cases}$	$\begin{cases} 4.07 \\ 2.05 \end{cases}$	1
$\Pi^{10}, \Pi^2, \mathbf{\Pi}^7$	1, 2, 2	1.97	3.39	$\begin{cases} 1.04 \\ 1.79 \end{cases}$	$\begin{cases} 0.37 \\ 1.44 \end{cases}$	$\begin{cases} 3.59 \\ 1.77 \end{cases}$	2
$\Pi^{10}, \mathbf{\Pi}^4, \Pi^7$	1, 1, 2	3.91	3.08	$\begin{cases} 0.23 \\ 1.78 \end{cases}$	$\begin{cases} 0.64 \\ 1.24 \end{cases}$	$\begin{cases} 4.13 \\ 1.98 \end{cases}$	1
$\mathbf{\Pi}^5, \Pi^4, \Pi^7$	1, 1, 2	3.50	2.11	$\begin{cases} 0.45 \\ 2.04 \end{cases}$	$\begin{cases} 0.98 \\ 1.36 \end{cases}$	$\begin{cases} 3.57 \\ 1.60 \end{cases}$	1
$\Pi^5, \mathbf{\Pi}^6, \Pi^7$	1, 1, 2	2.55	2.40	$\begin{cases} 0.79 \\ 1.93 \end{cases}$	$\begin{cases} 0.81 \\ 1.39 \end{cases}$	$\begin{cases} 3.40 \\ 1.68 \end{cases}$	1
$\mathbf{\Pi}^{12}, \Pi^6, \Pi^7$	2, 1, 2	0.09	1.91	$\begin{cases} 1.50 \\ 1.57 \end{cases}$	$\begin{cases} 0.83 \\ 1.94 \end{cases}$	$\begin{cases} 2.67 \\ 1.49 \end{cases}$	2
$\Pi^{12}, \Pi^6, \mathbf{\Pi}^8$	2, 1, 1	2.70	1.36	$\begin{cases} 0.73 \\ 2.24 \end{cases}$	$\begin{cases} 1.19 \\ 1.46 \end{cases}$	$\begin{cases} 3.08 \\ 1.30 \end{cases}$	1
$\Pi^{12}, \mathbf{\Pi}^9, \Pi^8$	2, 2, 1	0.30	0.77	$\begin{cases} 1.36 \\ 1.82 \end{cases}$	$\begin{cases} 1.24 \\ 1.79 \end{cases}$	$\begin{cases} 2.40 \\ 1.40 \end{cases}$	2
$\mathbf{\Pi}^1, \Pi^9, \Pi^8$	1, 2, 1	—	—	—	—	—	1
$\Pi^1, \Pi^9, \mathbf{\Pi}^{18}$	1, 2, 1	1.10	0.06	$\begin{cases} 1.08 \\ 2.90 \end{cases}$	$\begin{cases} 1.53 \\ 1.55 \end{cases}$	$\begin{cases} 2.39 \\ 0.55 \end{cases}$	1
$\mathbf{\Pi}^{17}, \Pi^9, \Pi^{18}$	3, 2, 1	0.30	−0.70	$\begin{cases} 1.26 \\ 1.78 \end{cases}$	$\begin{cases} 1.71 \\ 0.51 \end{cases}$	$\begin{cases} 2.03 \\ 2.71 \end{cases}$	3

The results of running the algorithm are shown in Table 7.6. The initial primitive set consists of the vectors Π^1, Π^2, Π^4, where Π^4 has the maximal *third* component among the Π^k, $k \geq 4$. For each primitive set shown, the boldface vector is the one that just entered the set and the calculations shown for that primitive set are the necessary ones to determine an appropriate label for the boldface vector. The exiting vector from a primitive set is always the other vector with the same integer label as the integer label of the boldface vector. The algorithm terminates when the primitive set $\Pi^{17}, \Pi^9, \Pi^{18}$ is reached with a complete set of labels. If we weight each of these vectors equally, we obtain as an approximate fixed point or traffic equilibrium

$$\Pi^* = \begin{bmatrix} 1.23 \\ 1.49 \\ 2.28 \end{bmatrix} \quad \blacktriangle$$

As we mentioned above, an important extension of the simplicial approximation algorithm is to associate a vector label $\mathbf{a}_k$ with each vector $\mathbf{\Pi}^k$

in the set P. The vector $\mathbf{a}_k$ has the same dimension as $\mathbf{\Pi}^k$, and it may be generated as needed during the running of the algorithm. The reason for using vector rather than integer labels for each $\mathbf{\Pi}^k$ is that vector labels contain more information about the problem to be solved. This problem may be a fixed point problem or it may be some other numerical problem that can be solved by the algorithm but is difficult to cast as a fixed point algorithm. In the latter category is the dual problem (5.4), which we will solve by using the simplicial approximation algorithm with vector labels. As we shall see, there is a close connection between the generalized primal-dual algorithm of Chapter 6 and this algorithm.

Consider the same set of points P used before by the integer label version of the algorithm. We associate with each $\mathbf{\Pi}^k$ for $k>n$ the n-vector $\mathbf{a}_k$, and we associate with each $\mathbf{\Pi}^j$ for $j \le n$ the unit n-vector $\mathbf{e}_j$. Let $\mathbf{A}$ denote the $n \times (K-n)$ matrix with columns $\mathbf{a}_k$, and let $\mathbf{b}$ denote an n-vector of right-hand sides with positive components. The linear system associated with the vectors in P is therefore

$$\mathbf{Is}+\mathbf{Ax}=\mathbf{b}$$

We wish to associate nonnegative basic solutions to this system, otherwise called basic feasible solutions. For technical reasons, we must make the additional nondegeneracy assumption that all basic feasible solutions to the linear system are nondegenerate. As with the previous nondegeneracy assumption on the $\mathbf{\Pi}^k$ in P, this assumption is made for expositional simplicity and can be dropped by using methods for resolving degeneracies, such as the lexicographic procedure given in Chapter 1.

The goal of the simplicial approximation algorithm with vector labels is to find a primitive subsimplex $\mathbf{\Pi}^{k_1},\ldots,\mathbf{\Pi}^{k_n}$ of vectors from P with vector labels $\mathbf{a}_{k_1},\ldots,\mathbf{a}_{k_n}$ such that these vectors are a basic feasible solution to the linear system $\mathbf{Is}+\mathbf{Ax}=\mathbf{b}$ (some of the $\mathbf{a}_{k_j}$ can be unit vectors). The algorithm begins with the same initial set of primitive vectors $\mathbf{\Pi}^{k^*},\mathbf{\Pi}^{k_2},\ldots,\mathbf{\Pi}^{k_n}$ where $\mathbf{\Pi}^{k^*}$ has the maximal value of the first coordinate among the $\mathbf{\Pi}^k$ in P for $k>n$. The label for $\mathbf{\Pi}^{k^*}$ is the vector $\mathbf{a}_{k^*}$, while the labels for the other vectors Π^j are the unit vectors $\mathbf{e}_j$. The matrix $\mathbf{I}$ with variables $\mathbf{s}$ is a feasible basis for the linear system because $\mathbf{b}$ is a positive n-vector. These vector labels $\mathbf{e}_1,\ldots,\mathbf{e}_n$ do not correspond completely to the initial primitive set because of the presence of $\mathbf{\Pi}^{k^*}$ in the primitive set, and the absence of $\mathbf{\Pi}^1$.

We attempt to achieve a correspondence between the primitive set and a feasible basis by pivoting the vector $\mathbf{a}_{k^*}$ into the basis $\mathbf{I}$. If the uniquely determined vector exiting from the basis is $\mathbf{e}_1$, then we have achieved our goal and the algorithm is terminated. If the exiting vector is $\mathbf{e}_j$ for $j \ge 2$, then there is a lack of correspondence between the primitive set and the vectors in the feasible basis in that $\mathbf{\Pi}^j$ is in the primitive set but $\mathbf{e}_j$ is not in the basis. Moreover, $\mathbf{e}_1$ is in the basis and $\mathbf{\Pi}^1$ is not. Again, we try to remedy the situation by taking $\mathbf{\Pi}^j$ out of the primitive set and uniquely replacing it by $\mathbf{\Pi}^k$ in P. If $\mathbf{\Pi}^k = \mathbf{\Pi}^1$ then we have also achieved our goal and the algorithm is terminated. If $\mathbf{\Pi}^k \neq \mathbf{\Pi}^1$, then we still lack correspondence between the primitive set and the vector labels constituting the feasible basis, and the basis pivoting operation is repeated.

Thus, at an intermediate point during computation with the simplicial approximation algorithm with vector labels, we have a primitive set $\mathbf{\Pi}^{k_1}, \mathbf{\Pi}^{k_2}, \ldots, \mathbf{\Pi}^{k_n}$ and the vectors $\mathbf{e}_1, \mathbf{a}_{k_2}, \ldots, \mathbf{a}_{k_n}$ constituting a feasible basis for the linear system. The vectors $\mathbf{a}_{k_j}$ can be unit vectors as well as columns of $\mathbf{A}$. First, we pivot the vector $\mathbf{a}_{k_1}$ into the basis. If $\mathbf{e}_1$ is the unique vector to leave the basis, the algorithm terminates. If some other vector $\mathbf{a}_{k_j}$ leaves the basis, then we replace $\mathbf{\Pi}^{k_j}$ in the primitive set by the uniquely determined $\mathbf{\Pi}^k$ in P. If $\mathbf{\Pi}^k = \mathbf{\Pi}^1$, the algorithm terminates; otherwise, the process is repeated.

Simplicial Approximation Algorithm with Vector Labels

Step 1. (Initialization). This step is the same as Step 1 of the algorithm with integer labels with the addition that the initial basic feasible solution to $\mathbf{Is} + \mathbf{Ax} = \mathbf{b}$ is $\mathbf{e}_1, \mathbf{a}_{k_2}, \ldots, \mathbf{a}_{k_n} = \mathbf{e}_1, \mathbf{e}_2, \ldots, \mathbf{e}_n$.

Step 2. If $\mathbf{\Pi}^k = \mathbf{\Pi}^1$, the algorithm terminates with the primitive set

$$\mathbf{\Pi}^1, \mathbf{\Pi}^{k_2}, \ldots, \mathbf{\Pi}^{k_n}$$

and the associated vector labels

$$\mathbf{e}_1, \mathbf{a}_{k_2}, \ldots, \mathbf{a}_{k_n}$$

Otherwise, compute the vector label $\mathbf{a}_k$ and pivot it into the feasible basis $\mathbf{e}_1, \mathbf{a}_k, \ldots, \mathbf{a}_{k_n}$. If $\mathbf{e}_1$ is the unique vector to leave the basis, the algorithm terminates with the primitive set

$$\mathbf{\Pi}^k, \mathbf{\Pi}^{k_2}, \ldots, \mathbf{\Pi}^{k_n}$$

and the associated vector labels

$$\mathbf{a}_k, \mathbf{a}_{k_2}, \ldots, \mathbf{a}_{k_n}$$

If $\mathbf{a}_{k_\alpha} \neq \mathbf{e}_1$ is the unique vector to leave the basis, go to Step 3 indicating that the exiting vector from the primitive set is $\mathbf{\Pi}^{k_\alpha}$.

Step 3. This step is the same as Step 3 of the algorithm with integer labels.

THEOREM 7.19. The simplicial approximation algorithm with vector labels terminates after a finite number of iterations (i.e., after generating a finite number of primitive sets) with a completely vector labeled primitive set.

PROOF. The proof is essentially the same as that of Theorem 7.18. Each time Steps 2 and 3 of the algorithm are taken, a unique succeeding pair of basis and primitive sets is determined. If the algorithm did not terminate finitely, a basis and primitive set combination would ultimately be repeated, which is impossible because this would indicate that the first such repeated basis and primitive set had two successors. ■

We conclude our discussion of constructive fixed point methods by showing how the simplicial approximation algorithm with vector labels can be used to solve the dual problem (5.4) discussed extensively in Chapters 5 and 6. The relevance of solving dual problems in optimizing the primal problems from which they are derived will not be repeated here.

We rewrite the dual problem as

$$\begin{aligned} d = \max\ & L(\mathbf{u}) \\ \text{s.t. } & \mathbf{u} \geq \mathbf{0} \end{aligned} \tag{7.43}$$

where the Lagrangean function is defined as

$$L(\mathbf{u}) = \min_{\mathbf{x} \in X} \{ f(\mathbf{x}) + \mathbf{u}g(\mathbf{x}) \}$$

We assume f is a continuous function from R^n to R^1, g is a continuous function from R^n to R^m with components g_i, and X is a nonempty, compact subset of R^n. As in Chapter 5, we make these assumptions for expositional expedience, and it is possible to have weaker conditions and the simplicial approximation algorithm would still find an approximately optimal solution to (7.43). We make the additional assumption that there exists an $\mathbf{x}^0 \in X$ satisfying $g_i(\mathbf{x}^0) < 0$ for $i = 1, \ldots, m$ (see Lemma 5.4 for a further discussion of this condition). The implication of the last assumption is that there exists an $M > 0$ such that all optimal solutions to the dual problem satisfy $\Sigma_{i=1}^m u_i \leq M - 1$. In other words, the dual problem (7.43) has an optimal solution and we can add a nonbinding constraint to it

$$\begin{aligned} d = \max\ & L(\mathbf{u}) \\ \text{s.t. } & \sum_{i=1}^{m} u_i \leq M \\ & \mathbf{u} \geq \mathbf{0} \end{aligned} \tag{7.44}$$

Procedurally, we will work with the simplex U in R^{m+1} given by

$$U = \left\{ \mathbf{u} \,\middle|\, \sum_{i=0}^{m} u_i = M, \mathbf{u} \geq \mathbf{0} \right\}$$

where u_0 is the slack variable on the previous inequality.

In Chapter 6 (see Section 6.5) we saw that the generalized primal-dual algorithm for solving (7.43), or equivalently (7.44), was based on the following necessary and sufficient condition for optimality. The point $\mathbf{u} \geq \mathbf{0}$ is optimal if and only if there exist points $\mathbf{x}^1, \ldots, \mathbf{x}^K$ in X and weights $\lambda_1, \ldots, \lambda_K$ such that

$$L(\mathbf{u}) = f(\mathbf{x}^k) + \mathbf{u}g(\mathbf{x}^k) \qquad k = 1, \ldots, K$$

and

$$\sum_{k=1}^{K} g(\mathbf{x}^k)\lambda_k \begin{cases} \leq 0 & \text{if } u_i = 0 \\ = 0 & \text{if } u_i > 0 \end{cases} \tag{7.45}$$

where

$$\sum_{k=1}^{K} \lambda_k = 1$$

and

$$\lambda_k \geq 0$$

The vectors $g(\mathbf{x}^k)$ are subgradients of the Lagrangean at $\mathbf{u}$, as are all convex combinations of them, and the condition says, in effect, that there exists a subgradient γ of L at $\mathbf{u}$ with the properties: $\gamma_i \leq 0$ for all i with equality if $u_i > 0$.

The simplicial approximation algorithm with vector labels effectively replaces the necessary and sufficient optimality condition (7.45) on subgradients computed at a single dual vector $\mathbf{u}$ by the same condition on subgradients computed at a collection of dual vectors that are close together in a primitive set. The algorithm works with the points $\mathbf{u}^0, \mathbf{u}^1, \ldots, \mathbf{u}^m, \mathbf{u}^{m+1}, \ldots, \mathbf{u}^K$ where

$$\mathbf{u}^0 = (0, M_0, M_0, \ldots, M_0)$$
$$\mathbf{u}^1 = (M_1, 0, M_1, \ldots, M_1)$$
$$\mathbf{u}^m = (M_m, M_m, \ldots, M_m, 0)$$

$M_0 > M_1 > M_2 > \cdots > M_m > M$, and the remaining $\mathbf{u}^{m+1}, \ldots, \mathbf{u}^K$ are chosen arbitrarily on the simplex U except that the usual nondegeneracy assumption that $u_i^{k_1} \neq u_i^{k_2}$ for any i and any $k_1 \neq k_2$. The situation is depicted schematically by

$$\begin{array}{lcccccccc}
\text{dual vectors} & \mathbf{u}^0 & \mathbf{u}^1 & \cdots & \mathbf{u}^m & \mathbf{u}^{m+1} & \cdots & \mathbf{u}^K \\
 & \left[\begin{array}{c} 1 \\ 0 \\ 0 \\ \vdots \\ 0 \\ 0 \end{array}\right. & \begin{array}{c} 0 \\ 1 \\ 0 \\ \vdots \\ 0 \\ 0 \end{array} & & \begin{array}{c} 0 \\ 0 \\ 0 \\ \vdots \\ 0 \\ 1 \end{array} & \begin{array}{c} 1 \\ \gamma_1^{m+1}+1 \\ \cdot \\ \vdots \\ \\ \gamma_m^{m+1}+1 \end{array} & & \left.\begin{array}{c} 1 \\ \gamma_1^K+1 \\ \cdot \\ \vdots \\ \\ \gamma_m^K+1 \end{array}\right] \\
\text{vector labels} & & & & & & & \\
\text{variables} & s_0 & s_1 & \cdots & s_m & x_{m+1} & \cdots & x_K
\end{array}$$

We refer to the vector labels of $\mathbf{u}^i$ for $i=0,1,\ldots,m$, as $\mathbf{e}_i$ for $i=0,1,\ldots,m$. The vector γ^k that is the vector label of $\mathbf{u}^k$ is any Lagrangean subgradient of L at $\mathbf{u}^k$. The right-hand side in the linear system $\mathbf{Is}+\mathbf{Ax}=\mathbf{b}$ is the $(m+1)$-vector of 1's. We have added $+1$ to the components γ_i^k to ensure nondegeneracy of basic feasible solutions including the initial basic solution of the algorithm $\mathbf{s}=\mathbf{b}$, $\mathbf{x}=\mathbf{0}$.

Let us consider the consequences of termination of the simplicial approximation algorithm with the primitive set $\mathbf{u}^{k_0},\mathbf{u}^{k_1},\ldots,\mathbf{u}^{k_m}$ and the corresponding feasible basis made up of $m+1$ columns from (7.46). Let $\tilde{s}_i,\tilde{x}_k$ denote the values of the variables in the basic feasible solution. Let $I=\{i|s_i$ is in the basis$\}$ and without loss of generality, we can assume that $\mathbf{u}^{k_i}=\mathbf{u}^i$ for all $i\in I$. Let I^c denote the complementary set $\{0,1,\ldots,m\}-I$ corresponding to the x_{k_i} variables for $k_i>m$; I^c is nonempty because the set $\mathbf{u}^0,\mathbf{u}^1,\ldots,\mathbf{u}^m$ is not a primitive set. There are two possible terminations of the algorithm to distinguish depending upon whether or not $\mathbf{u}^0$ is in the primitive set. The usual case is when $\mathbf{u}^0$ is not in the primitive set or in other words, when the algorithm terminates a sufficient distance from the boundary of U with the slack $u_0=0$. In this case, we have from the top row of the matrix (7.46) that

$$\sum_{i\in I^c}\tilde{x}_{k_i}=1$$

For any other row i, we have

$$\tilde{s}_i+\sum_{i\in I^c}\gamma_i^{k_i}\tilde{x}_{k_i}=1-\sum_{i\in I^c}\tilde{x}_{k_i}=0$$

For all $i\in I^c$, that is, those rows where s_i is not in the basis and thus $\tilde{s}_i=0$, we have

$$\sum_{i\in I^c}\gamma_i^{k_i}\tilde{x}_{k_i}=0$$

This is consistent with the optimality condition (7.45) since $\mathbf{u}^i$ is not in the primitive set indicating that the optimal solution to (7.44) found by the algorithm has $u_i>0$. For $i\in I$, then s_i is basic and $\tilde{s}_i>0$ by the nondegeneracy assumption, and we have

$$\sum_{i\in I^c}\gamma_i^{k_i}\tilde{x}_{k_i}=-\tilde{s}_i<0$$

This condition is also consistent with the optimality condition (7.45) because $\mathbf{u}^i$ is in the primitive set indicating that the approximately optimal solution to (7.43) found by the algorithm has $u_i=0$, or at worst, $u_i\approx 0$. To be specific, we can take any one of the $\mathbf{u}^{k_i}$ for $i\in I$ as the approximately optimal solution to (7.43), or perhaps a better choice is

$$\bar{\mathbf{u}}^P = \sum_{i \in I^c} \mathbf{u}^{k_i} \tilde{x}_{k_i} \tag{7.47}$$

The unusual and unexpected termination of the simplicial approximation algorithm applied to the dual problem (7.43) is when $\mathbf{u}^0$ is in the terminal primitive set $\mathbf{u}^0, \mathbf{u}^{k_1}, \ldots, \mathbf{u}^{k_m}$. This is the case when the algorithm has terminated near the boundary $\sum_{i=1}^m u_i = M$ and we do not expect it to occur because the constraint $\sum_{i=1}^m u_i \le M$ on (7.43) is nonbinding. Nevertheless, due to the approximation inherent in the approach, there is no guarantee that this case will not occur. However, it cannot occur infinitely often as the points $\mathbf{u}^{m+1}, \ldots, \mathbf{u}^K$ are made dense on the set U. Note that since s_0 is a basic variable in this case with value $\tilde{s}_0 > 0$, we have from the top row in (7.46) that

$$0 < \sum_{i \in I^c} \tilde{x}_{k_i} = 1 - \tilde{s}_0 < 1$$

This implies for the other rows i that

$$\tilde{s}_i + \sum_{i \in I^c} \gamma_i^{k_i} \tilde{x}_{k_i} = 1 - \sum_{i \in I} \tilde{x}_{k_i}$$

or

$$\tilde{s}_i + \sum_{i \in I^c} \gamma_i^{k_i} \tilde{x}_{k_i} - \tilde{s}_0 = 0$$

for $\tilde{s}_0 > 0$. This positive quantity represents a dual variable or shadow price on the binding constraint $\sum_{i=1}^m u_i = M$ implied by the presence of $\mathbf{u}^0$ in the primitive set.

THEOREM 7.20. Suppose the simplicial approximation algorithm with vector labels is applied to the dual problem (7.43) an infinite number of times with points drawn from the sets P^l for $l = 1, 2, \ldots$, with the property that the maximal distance between distinct points $\mathbf{u}^k$ in P^l for $k > m$ goes to zero in the limit as l goes to infinity. Let $\bar{\mathbf{u}}^{P^l}$ denote the approximately optimal dual solution calculated in (7.47) found by the algorithm. Then any limit point of the sequence $\{\bar{\mathbf{u}}^{P^l}\}_{l=1}^{\infty}$ is an optimal solution to the dual problem.

PROOF. The proof is left as an exercise.

7.7 EXERCISES

7.1 Show that the optimal gradient descent algorithm (7.2) has the property that successive gradients are orthogonal; namely, $\nabla f(\mathbf{x}^{k+1}) \cdot \nabla f(\mathbf{x}^k) = 0$.

7.2 Consider the quadratic programming problem

$$d = \max \mathbf{bu} - \tfrac{1}{2}\mathbf{xDx}$$
$$\text{s.t. } -\mathbf{Dx} + \mathbf{A}^T\mathbf{u} \le \mathbf{c}$$
$$\mathbf{u} \ge \mathbf{0}$$

(1) Convert this problem to a linear complementarity problem.

(2) Show that this problem is dual to (7.22) in the sense that $v = d$ in the case when either v or d is finite.

7.3 (Lemke, 1965) A *bimatrix game*, $\Gamma(\mathbf{A},\mathbf{B})$, is given by a pair of $m \times n$ matrices **A** and **B**. The *row player* has m pure strategies corresponding to the rows of **A**. The *column player* has n pure strategies corresponding to the columns of **B**. If the row player uses strategy i and the column player uses strategy j, then their respective losses are a_{ij} and b_{ij}. Using *mixed strategies*,

$$\mathbf{x} = (x_1, \ldots, x_m) \ge \mathbf{0}, \ \sum_{i=1}^{m} x_i = 1$$

$$\mathbf{y} = (y_1, \ldots, y_n) \ge \mathbf{0}, \ \sum_{j=1}^{n} y_j = 1$$

their expected losses are **xAy** and **xBy**, respectively. A pair $(\mathbf{x}^0, \mathbf{y}^0)$ of mixed strategies is a (Nash) equilibrium point of $(\mathbf{A},\mathbf{B})$ if

$$\mathbf{x}^0\mathbf{A}\mathbf{y}^0 \le \mathbf{x}\mathbf{A}\mathbf{y}^0 \qquad \text{for all mixed integers } \mathbf{x}$$
$$\mathbf{x}^0\mathbf{B}\mathbf{y}^0 \le \mathbf{x}^0\mathbf{B}\mathbf{y} \qquad \text{for all mixed strategies } \mathbf{y}$$

(1) Prove that if $(\mathbf{x}^0, \mathbf{y}^0)$ is an equilibrium point of $\Gamma(\mathbf{A},\mathbf{B})$, then it is also an equilibrium point for $\Gamma(\mathbf{A}^1,\mathbf{B}^1)$ where

$$\mathbf{A}^1 = (a_{ij} + K) \qquad \mathbf{B}^1 = (b_{ij} + L)$$

where K and L are arbitrary scalars. Thus, we assume in the remainder of this exercise that all components of **A** and **B** are positive.

(2) Prove that the solution $(\mathbf{x}^0, \mathbf{y}^0)$ is an equilibrium point of $\Gamma(\mathbf{A},\mathbf{B})$ if and only if

$$(\mathbf{x}^0\mathbf{A}\mathbf{y}^0)\mathbf{0}_m \le \mathbf{A}\mathbf{y}^0$$
$$(\mathbf{x}^0\mathbf{B}\mathbf{y}^0)\mathbf{0}_n \le \mathbf{B}^T\mathbf{x}^0$$

where **A** and **B** have positive components, and $\mathbf{0}_m$ and $\mathbf{0}_n$ are vectors of 1's in R^m and R^n, respectively.

(3) Prove that for **A** and **B** with positive components, if $\mathbf{u}^*, \mathbf{v}^*, \mathbf{x}^*, \mathbf{y}^*$ is a

solution of the system

$$\mathbf{u}=\mathbf{A}\mathbf{y}-\mathbf{0}_m \qquad \mathbf{u}\geq\mathbf{0} \qquad \mathbf{y}\geq\mathbf{0}$$

$$\mathbf{v}=\mathbf{B}^T\mathbf{x}-\mathbf{0}_n \qquad \mathbf{v}\geq\mathbf{0} \qquad \mathbf{x}\geq\mathbf{0}$$

$$\mathbf{x}\mathbf{u}+\mathbf{y}\mathbf{v}=0$$

then

$$(\mathbf{x}^0,\mathbf{y}^0)=\left(\frac{\mathbf{x}^*}{\mathbf{x}^*\mathbf{0}_m},\frac{\mathbf{y}^*}{\mathbf{y}^*\mathbf{0}_n}\right)$$

is an equilibrium point of $\Gamma(\mathbf{A},\mathbf{B})$. Conversely, if $(\mathbf{x}^0,\mathbf{y}^0)$ is an equilibrium point of $\Gamma(\mathbf{A},\mathbf{B})$, then

$$(\mathbf{x}^*,\mathbf{y}^*)=\left(\frac{\mathbf{x}^0}{\mathbf{x}^0\mathbf{B}\mathbf{y}^0},\frac{\mathbf{y}^0}{\mathbf{x}^0\mathbf{A}\mathbf{y}^0}\right)$$

is a solution of the above system.

(4) Show that the problem of finding an equilibrium point can be cast as a linear complementarity problem.

(5) Prove that a bimatrix game has an equilibrium point.

HINT. Adapt the linear complementary algorithm for the problem of point (4). Do not add the artificial column $\mathbf{0}_p$ to the problem with variable z_0. Instead, initiate the algorithm by choosing the smallest positive value of x_1, say x_1^0, such that

$$\mathbf{v}=-\mathbf{0}_n+\mathbf{B}_1^T x_1^0\geq\mathbf{0}$$

where $\mathbf{B}_1^T$ is the first column of $\mathbf{B}^T$. Let $\mathbf{v}^0$ denote this value of $\mathbf{v}$; it will have (assuming nondegeneracy) one zero component, say $v_r^0=0$. Take the complement y_r of v_r and decrease it towards zero from large values. The ray so generated is complementary except possibly x_1u_1 might not equal zero. The initial extreme point is obtained for some positive value of y_r. Show that the linear complementary algorithm cannot terminate in an almost complementary ray and therefore it must find an equilibrium point.

7.4 Show how the set of all efficient portfolios in Examples 7.2 and 7.3 can be parametrically generated by the linear complementary algorithm.

7.5 Prove Theorem 7.9 using a second-order Taylor series expansion of L and the continuity of the second partials of f and the g_i.

7.6 Prove Theorem 7.20.

7.7 Consider the unconstrained quadratic problem

$$\min \mathbf{cx} + \tfrac{1}{2}\mathbf{xDx}$$

where $\mathbf{D}$ is an $n \times n$ symmetric positive definite matrix. Two vectors $\mathbf{q}^1, \mathbf{q}^2$ are said to be *conjugate* if $\mathbf{q}^1\mathbf{D}\mathbf{q}^2 = 0$. A finite set of nonzero vectors $\mathbf{q}^0, \mathbf{q}^1, \ldots, \mathbf{q}^k$ is said to be a *conjugate* set if $\mathbf{q}^i\mathbf{D}\mathbf{q}^j = 0$ for all $i \neq j$.

(1) Prove that the vectors in any conjugate set are linearly independent.

(2) Let $\{\mathbf{q}^i\}_{i=0}^{n-1}$ be a conjugate set. For any $\mathbf{x}^0 \in R^n$, consider the sequence

$$\mathbf{x}^{k+1} = \mathbf{x}^k + \alpha_k \mathbf{q}^k \qquad k = 0, 1, 2$$

with

$$\alpha_k = \frac{-\mathbf{q}^k \mathbf{q}^k}{\mathbf{q}^k \mathbf{D} \mathbf{q}^k}$$

and

$$\mathbf{q}^k = \mathbf{c} + \mathbf{D}\mathbf{x}^k$$

Prove that $\mathbf{x}^n$ is minimal in the quadratic unconstrained minimization problem.

7.8 Prove that the penalty function $P(\mathbf{x}, t)$ given in (7.20) is convex if f and the g_i are convex.
HINT. Prove and use Theorems A.11 and A.12 in Appendix A.

7.9 (Scarf, 1967) Let $C_1, \ldots, C_n$ be closed sets on the simplex S defined in (7.36) with the following properties:

(1) Every vector in S is a member of at least one C_i

(2) Every vector on the side $x_i = 0$ of S is contained in C_i for $i = 1, \ldots, n$.

Prove that the set

$$\bigcap_{i=1}^{n} C_i$$

is not empty.

HINT. Use the simplicial approximation algorithm with integer labels and let the grid become dense on S.

7.10 (Nakayama et al., 1975) Consider the generalized Lagrangean function given in (7.12) derived from

$$\mathcal{L}(\mathbf{x}, \boldsymbol{\sigma}, t) = f(\mathbf{x}) + \sum_{i=1}^{m} \begin{cases} t g_i^2(\mathbf{x}) + \sigma_i g_i(\mathbf{x}) & \text{if } g_i(\mathbf{x}) \geq 0 \\ \dfrac{\sigma_i^2 g_i(\mathbf{x})}{\sigma_i - t g_i(\mathbf{x})} & \text{if } g_i(\mathbf{x}) < 0 \end{cases}$$

Assume for all $\boldsymbol{\sigma} \geq \mathbf{0}$ that the vector $\mathbf{x}(\boldsymbol{\sigma})$ satisfying $\mathcal{L}(\mathbf{x}\boldsymbol{\sigma}, t) = \min \mathcal{L}(\mathbf{x}, \boldsymbol{\sigma}, t)$ is unique, and moreover, that $\mathcal{L}(\boldsymbol{\sigma}, t)$ is strictly concave in $\boldsymbol{\sigma}$. Finally, assume that the optimal solution $\boldsymbol{\sigma}^*$ of the dual problem

$$\begin{aligned} &\max \mathcal{L}(\boldsymbol{\sigma}, t) \\ &\text{s.t. } \boldsymbol{\sigma} \geq \mathbf{0} \end{aligned}$$

satisfies $\sigma_i^* > 0$ for $i = 1, \ldots, m$, or in other words, $\nabla \mathcal{L}(\boldsymbol{\sigma}^*, t) = 0$.

(1) Show that the multiplier function used in $\mathcal{L}(\mathbf{x}, \boldsymbol{\sigma}, t)$ satisfies definition 7.1.

(2) Define the sequence $\{\boldsymbol{\sigma}^k\}_{k=1}^{\infty}$ by the rule

$$\sigma_i^{k+1} = \begin{cases} \sigma_i^k + 2tg_i(\mathbf{x}^k) & \text{if } g_i(x^k) \geq 0 \\ \dfrac{(\sigma_i^k)^3}{[\sigma_i^k - tg_i(\mathbf{x}^k)]^2} & \text{if } g_i(x^k) < 0 \end{cases}$$

where $\boldsymbol{\sigma}^1$ is an arbitrary m-vector with positive components. Show that the mapping F given by $F(\boldsymbol{\sigma}^k) = \boldsymbol{\sigma}^{k+1}$ is a contraction mapping and therefore it has a unique fixed point.

(3) Show that the unique fixed point of F is the optimal dual solution $\boldsymbol{\sigma}^*$.

7.11 Consider the nonlinear programming problem

$$\begin{aligned} &\min \mathbf{cx} \\ &\text{s.t. } g_i(\mathbf{x}) \leq 0 \qquad i = 1, \ldots, m \\ &\mathbf{x} \in R^n \end{aligned}$$

where the g_i are convex and have continuous first partials. Assume the set

$$F = \{\mathbf{x} \mid g_i(\mathbf{x}) \leq 0, i = 1, \ldots, m\}$$

is bounded. Assume further that we have a (bounded) polyhedron $P^0 \supseteq F$. We generate a sequence $\{\mathbf{x}^k\}_{k=0}^{\infty}$ by the following algorithm called the *convex cutting plane algorithm.*

Step 0. Select an initial polyhedron $P^0 \supseteq F$ and go to Step 1 with $k = 0$.

Step 1. Solve the linear programming problem

$$\begin{aligned} &\min \mathbf{cx} \\ &\text{s.t. } \mathbf{x} \in P^k \end{aligned}$$

Let $\mathbf{x}^k$ denote the optimal solution found by the simplex method. If $g(\mathbf{x}^k) \leq 0$, stop; the solution $\mathbf{x}^k$ is optimal in the given nonlinear programming problem. Otherwise, go to Step 2.

Step 2. Let i be an index for which $g_i(\mathbf{x}^k)$ is maximal and define

$$P^{k+1} = P^k \cap \{\mathbf{x} | g_i(\mathbf{x}^k) + \nabla g_i(\mathbf{x}^k)(\mathbf{x} - \mathbf{x}^k) \leq 0\}$$

Go to Step 1 with the polyhedron P^{k+1}.

(1) Show that $\nabla g_i(\mathbf{x}^k) \neq 0$ for i defined in Step 2 and morover, that the hyperplane $H = \{x | g_i(\mathbf{x}^k) + \nabla g_i(\mathbf{x}^k)(\mathbf{x} - \mathbf{x}^k) = 0\}$ strictly separates F and the point $\mathbf{x}^k$.

(2) Prove that any limit point of the sequence $\{\mathbf{x}^k\}_{k=0}^{\infty}$ given by the algorithm is optimal in the given nonlinear programming problem.

7.12 Suppose the assumptions of Theorem 7.5 are satisfied. Show that for some K (K is unknown), the feasible directions algorithm generates feasible $\mathbf{x}^k$ for $k \leq K$ with the property

$$\frac{f(\mathbf{x}^k) - v}{f(\mathbf{x}^{k-1}) - v} \leq \frac{1}{2}$$

7.13 Use the Kuhn-Tucker condition to derive the excess demand functions (7.38) for the individual utility maximization problems (7.32) with the utility function given in (7.37).

7.14 (1) Show that an optimal solution to the problem

$$\begin{aligned} &\min f(\mathbf{x}) \\ &\text{s.t.} \quad \mathbf{Ax} = \mathbf{b} \\ &\qquad \mathbf{x} \geq \mathbf{0} \end{aligned}$$

where f is concave and the feasible region is bounded, lies at an extreme point of the feasible region.

(2) Suppose f is strictly concave. Show that each extreme point of the feasible region is a strict local minimum.

7.15 Let f be a continuously differentiable concave function, $\mathbf{c}$ an n vector, and $\mathbf{B}$ an $r \times n$ matrix $r \leq n$ with linearly independent rows. Suppose we must find the optimal solution to the problem

$$\begin{aligned} &\max f(\mathbf{x}) \\ &\text{s.t.} \ \mathbf{Bx} = \mathbf{c} \end{aligned}$$

(1) Show that the Kuhn-Tucker conditions for $\mathbf{x}^*$ to be optimal for this problem are

$$\begin{aligned} \mathbf{Bx}^* &= \mathbf{c} \\ \nabla f(\mathbf{x}^*) + \mathbf{B}^T\mathbf{u} &= \mathbf{0} \end{aligned}$$

where $\mathbf{u}$ is an unconstrained r-vector.

Define the $n \times n$ matrix

$$\mathbf{I} - \mathbf{B}^T(\mathbf{B}\mathbf{B}^T)^{-1}\mathbf{B}$$

where $(\mathbf{B}\mathbf{B}^T)^{-1}$ is the inverse of an $r \times r$ matrix $(\mathbf{B}\mathbf{B}^T)$.

(2) Let $\mathbf{x}$ be a point feasible for the problem, and suppose

$$\left[\mathbf{I} - \mathbf{B}^T(\mathbf{B}\mathbf{B}^T)^{-1}\mathbf{B}\right]\nabla f(\mathbf{x}) = \mathbf{0}$$

Prove that $\mathbf{x}$ is optimal for the problem.

(3) Prove

$$\left[\mathbf{I} - \mathbf{B}^T(\mathbf{B}\mathbf{B}^T)^{-1}\mathbf{B}\right]^T\left[\mathbf{I} - \mathbf{B}^T(\mathbf{B}\mathbf{B}^T)^{-1}\mathbf{B}\right] = \left[\mathbf{I} - \mathbf{B}^T(\mathbf{B}\mathbf{B}^T)^{-1}\mathbf{B}\right]$$

HINT.

$$\left[\mathbf{B}^T(\mathbf{B}\mathbf{B}^T)^{-1}\mathbf{B}\right]^T = \left[\mathbf{B}^T(\mathbf{B}\mathbf{B}^T)^{-1}\mathbf{B}\right]$$

Now define

$$\mathbf{d} = \left[\mathbf{I} - \mathbf{B}^T(\mathbf{B}\mathbf{B}^T)^{-1}\mathbf{B}\right]\nabla f(\mathbf{x})$$

(4) Prove

$$\nabla f(\mathbf{x})^T\mathbf{d} = \mathbf{d}^T\mathbf{d}$$

(5) If $\mathbf{x}$ is not optimal for the problem, then show that

$$\nabla f(\mathbf{x})^T\mathbf{d} > 0$$

(6) Let $\mathbf{x}$ be feasible for the problem. Prove $\mathbf{x} + \tau\mathbf{d}$ is also feasible for all τ.

7.16 (Duffin, 1962)

(1) Let $g(\mathbf{x})$ be a differentiable convex function defined on an open convex subset $X \subseteq R^n$. Define the function

$$W(\mathbf{x}) = g(\mathbf{x}) - \nabla g(\mathbf{x})\mathbf{x}$$

Prove that $W(\mathbf{x}^1) \geq W(\mathbf{x}^2) + [\nabla g(\mathbf{x}^2) - \nabla g(\mathbf{x}^1)]\mathbf{x}^1$ for any $\mathbf{x}^1, \mathbf{x}^2 \in X$.

(2) Show that $W(\mathbf{x}^1) = W(\mathbf{x}^2)$ if $\nabla g(\mathbf{x}^1) = \nabla g(\mathbf{x}^2)$.

(3) Define the set

$$T = \{\mathbf{y} \mid \mathbf{y} = \nabla g(\mathbf{x}) \text{ for some } \mathbf{x} \in X\}$$

and define the function $w(\mathbf{y})$ on T by

$$w(\mathbf{y}) = W(\mathbf{x})$$

where $\mathbf{y} = \nabla g(\mathbf{x})$. By part (2), the function w is well defined; it is called the *Legendre transform* of $g(\mathbf{x})$. In the special case when g is strictly convex and possesses continuous second partials and a nonzero

Hessian for all $\mathbf{x} \in X$, show that $\mathbf{y} = g(\mathbf{x})$ is a one-to-one mapping of X onto T. In addition, prove that $w(\mathbf{y})$ has continuous first partials and

$$\frac{\partial w(\mathbf{y})}{\partial y_j} = -x_j$$

where $\mathbf{y} = \nabla g(\mathbf{x})$.

(4) Suppose P and Q are orthogonal complementary subspaces of R^n with the properties that $P \cap S$ and $Q \cap T$ are not empty. Show that

$$\inf_{\mathbf{x} \in P \cap S} g(\mathbf{x}) \geq \sup_{\mathbf{y} \in D \cap T} w(\mathbf{y})$$

In addition, prove that equality is attained if $g(\mathbf{x})$ attains its infimum in $P \cap S$.

(5) Consider the nonlinear programming problem

$$\min g(\mathbf{t}) = \sum_{j=1}^{n} c_j t_1^{a_{j1}} \cdots t_m^{a_{jm}}$$

$$\text{s.t. } t_i > 0 \qquad i = 1, \ldots, m$$

The parameters $c_j > 0$, $j = 1, \ldots, n$. The function $g(\mathbf{t})$ is called a *posynomial* and this problem is an example of a *geometric programming problem*. Use the Legendre transform to transform the problem to

$$\max \prod_{j=1}^{n} \left(\frac{c_j}{\delta_j} \right)^{\delta_j}$$

$$\text{s.t. } \sum_{j=1}^{n} \delta_j = 1$$

$$\sum_{j=1}^{n} a_{ji} \delta_j = 0 \qquad i = 1, \ldots, m$$

Use part (4) to show that solution of the transformed problem solves the original problem.

HINT. Verify and use the following facts. Let $g(\mathbf{x}) = \sum_{j=1}^{n} c_j e^{x_j}$ where the x_j are given by $x_j = \sum_{i=1}^{m} a_{ji} \log t_i$ for $t_i > 0$, $i = 1, \ldots, m$. Choose P to be the subspace of R^n spanned by the columns of the matrix (a_{ji}). The set T is the interior of the positive orthant of R^n. The subspace D is the solution space of the linear homogeneous system

$$\sum_{j=1}^{n} a_{ji} y_j = 0 \qquad \text{for } i = 1, \ldots, m$$

7.8 NOTES

SECTION 7.1. There are a large number of books devoted entirely to nonlinear programming. Avriel (1976) covers many topics in the area and contains an extensive list of references.

SECTION 7.2. The Fibonacci search method and its properties was discovered by Kiefer (1953). Avriel (1976) and Luenberger (1973) give fairly extensive treatments of unconstrained optimization techniques.

SECTION 7.3. The feasible directions algorithm is due to Frank and Wolfe (1956).

SECTION 7.4. Our development of generalized duality was motivated by that of Nakayama et al. (1975) and Evans and Gould (1970). Fiacco and McCormick (1968) is a basic reference for the sequential unconstrained minimization technique as well as other topics in nonlinear programming.

SECTION 7.5. Our treatment of linear complementarity problems follows that of Cottle and Dantzig (1968). The algorithm is due to Lemke and Howson (1965). See also Gould and Tolle (1974) for a number of extensions including nonlinear complementarity problems arising from more general mathematical programming problems.

SECTION 7.6. The simplicial approximation algorithm for the calculation of Brouwer's fixed points first appeared in Scarf (1967). We based our development on that paper. More recent developments in the approximation of fixed points can be found in Scarf and Hansen (1973), Eaves (1972), and Todd (1976). Stoer and Witzgall (1970) have a classical existential treatment of fixed point theorems that complements the constructive methods. Arrow and Hahn (1971) give a thorough treatment of the theory of economic equilibria from a mathematical economics viewpoint. Florian and Nguyen (1974) give an algorithm based on Benders' method for solving traffic equilibrium problems. The particular applications of simplicial approximation methods to solving dual mathematical programming problems is found in Fisher et al. (1975). Shapiro (1978) discusses the relationship between mathematical programming and economic equilibria arising in energy planning models.

8
INTEGER PROGRAMMING AND COMBINATORIAL OPTIMIZATION

8.1 INTRODUCTION

This final chapter is concerned with the structure of integer programming and combinatorial optimization problems and methods for their solution. We regard integer programming problems as a special class of combinatorial optimization problems involving the minimization of a linear objective function subject to linear constraints where the variables are also constrained to take on nonnegative integer values. Mixed integer programming problems are linear optimization problems where some of the variables are constrained to be nonnegative integer and the remainder can take on continuous nonnegative values. Combinatorial optimization problems are the more general class of problems involving decision variables that must take on values from a discrete set and with structure that may be represented by means other than linear systems, for example, by networks. The first part of this chapter is devoted to integer programming problems, whereas the second part is devoted to specific combinatorial optimization problems including the traveling salesman and network synthesis problems.

Integer programming and combinatorial optimization problems often arise as generalizations or syntheses of the network, graph, and dynamic programming problems discussed in Chapters 3 and 4. The special structures can be exploited by the price and resource directive decomposition methods discussed in Chapter 6. In the case of price or dual decomposition, the Lagrangean function to be maximized in the implied dual problem [see problem (5.4)] is not differentiable. This is in contrast to the nonlinear programming problems of Chapter 7 where we relied heavily on the differentiability of the objective and constraint functions. The discrete nature of the Lagrangean calculations also produces dual problems that can be represented as large-scale linear programming problems. Thus, we are implicitly using large-scale linear programming approximations to analyze the integer programming and combinatorial optimization problems. Moreover, the approximations are relaxations of the given problems, and often the relaxations are not exact. In other words, there can be a duality gap between a primal integer programming or combinatorial optimization problem and its dual implying an optimal solution to the dual can fail to provide a solution to the primal.

For this reason, the constructive dual methods presented in this chapter need to be used in a more flexible fashion than they were used in the solution of convex nonlinear programming problems. As we shall see, if a dual problem fails to solve a given primal problem, then a stronger dual problem, or equivalently, a tighter relaxation approximation, can be constructed. Alternatively, the primal problem can be perturbed thereby producing a different dual problem for which there may be no duality gap. This is one of the main purposes of branch and bound methods that we present for the zero-one integer programming problem.

The taxonomy of integer programming and combinatorial optimization problems is so rich and complex that it is misleading to imply that all such problems are well-defined generalizations and syntheses of the simpler problems of Chapters 3 and 4. It sometimes takes considerable insight into a given combinatorial optimization problem, such as the traveling salesman problem discussed in Section 8.7, to identify the proper special structures to be exploited. The ambiguity is compounded by the fact that a large fraction of combinatorial optimization problems have integer programming formulations, but these formulations are not necessarily the most efficient, because it can be awkward to use linear equations or inequalities to represent combinatorial structures. Moreover, the representation of integer programming problems is generally not unique and some representations are easier to solve than others.

The analysis and solution of integer programming problems can be facilitated by the application of number theory or group theoretic constructs. The result is an induced special structure that can be exploited by dual methods in the same way as special structures arising naturally in other combinatorial optimization problems. The relevant number and group theory is reviewed in Appendix C.

8.2 INTEGER PROGRAMMING MODELING

The restriction of mathematical programming variables to integer values, often values of zero or one, arises naturally in many applications. For example, the number of airline crews flying a particular sequence of flights in a crew scheduling problem is either zero or one. Similarly, we may require an integer number of heavy-duty dump trucks on a construction site to meet earth moving requirements. More generally, the essence of integer programming is the restriction of some decision variables to small integer values. On the other hand, if a variable x_j corresponds to a quantity that must take on integer values, but these can be large, say 20 or more, then there is usually no practical need to explicitly require x_j to be integer. For example, if $x_j = 21.68$ dump trucks is the optimal value when x_j is permitted to vary continuously rather than in integer amounts, then taking $x_j = 22$ should be satisfactory for practical purposes since any error should be small relative to errors in the data.

As we shall see in Section 8.4, the magnitude of the coefficients in an integer programming problem are an important factor in determining its difficulty of solution. There are a number of ways in which the coefficients can be reduced, either exactly or approximately. As a simple example, consider the inequality

$$93x_1 + 102x_2 + 118x_3 \leq 146$$

where x_1, x_2, and x_3 are required to be integer. Clearly, an equivalent constraint is

$$1x_1 + 1x_2 + 1x_3 \leq 1$$

In general, such an extreme simplification is not possible. Other procedures based on the idea are given in the exercises at the end of the chapter (see Exercises 8.7 and 8.9).

Integer variables can also be used in a variety of ways to model constraints on decision problems that cannot be effectively represented by any of the mathematical programming models discussed in previous chapters. For example, suppose we wish to add to a linear programming problem the restriction that at most one of the variables x_1 and x_2 can be positive. Let δ be a zero-one variable where $\delta = 0$ corresponds to permitting x_1 to be positive but not x_2, and $\delta = 1$ corresponds to permitting x_2 to be positive but not x_1. The desired restriction is effected by the addition of the constraints

$$x_1 \leq M_1(1-\delta)$$
$$x_2 \leq M_2\delta$$

where M_1 and M_2 are upper bounds on the values of x_1 and x_2, respectively. A variety of other logical conditions can be imposed in a similar fashion on subsets of decision variables in a linear or nonlinear programming problem.

An important use of integer variables is to represent nonconvex nonlinear separable cost and constraint functions (see Section 1.5 for a discussion of linear programming representations of convex separable functions). As a specific example, suppose the function $f(x)$ shown in Figure 8.1 is a component in an objective function to be minimized in a mathematical programming problem. We have approximated f by a piecewise linear function consisting of three pieces, and we can expand x by the substitution

$$x = y_1 + y_2 + y_3$$

$$0 \le y_1 \le M_1 \qquad 0 \le y_2 \le M_2 \qquad 0 \le y_3 \le M_3$$

The function approximation is

$$f(x) \approx \Delta_1 y_1 + \Delta_2 y_2 + \Delta_3 y_3$$

where we also require that

$$y_2 > 0 \qquad \text{implies} \qquad y_1 = M_1$$

and

$$y_3 > 0 \qquad \text{implies} \qquad y_2 = M_2$$

The two additional conditions are required because the slopes $\Delta_1, \Delta_2, \Delta_3$ of the piecewise linear approximation are not increasing; without the additional conditions, the optimizing algorithm would prefer to increase y_3

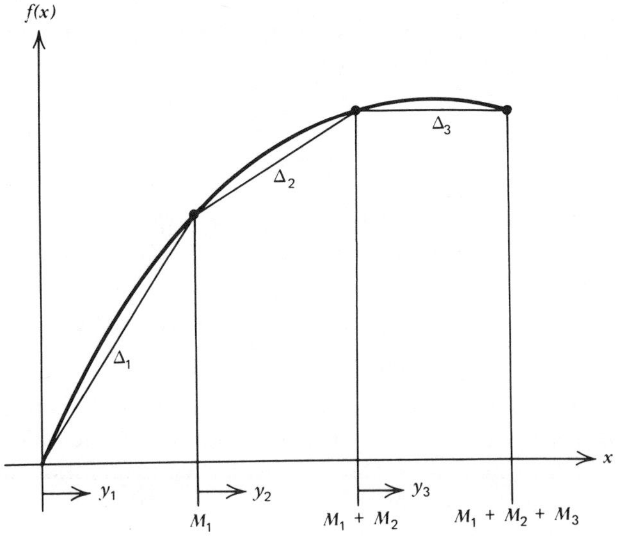

Figure 8.1. Piecewise linear approximation to concave separable minimization problem.

from zero before increasing y_1 or y_2 from zero because $\Delta_3 < \Delta_2$, and to increase y_2 from zero before y_1 because $\Delta_2 < \Delta_1$. The conditions are converted to a mathematical programming statement by the addition of two zero-one variables δ_1 and δ_2, and the additional constraints

$$M_1\delta_1 - y_1 \leq 0 \quad \text{and} \quad y_2 - M_2\delta_1 \leq 0$$
$$M_2\delta_2 - y_2 \leq 0 \quad \text{and} \quad y_3 - M_3\delta_2 \leq 0$$

Graph coloring problems provide another illustration of the flexibility of integer programming modeling techniques. Let $G=[\mathfrak{N}, \mathcal{E}]$ denote an arbitrary graph with m nodes and suppose we wish to use four different colors to color the nodes so that no pair of adjacent nodes (nodes connected by an edge) have the same color. In the special case when G is a planar graph (that is, a graph that can be drawn in the plane without edges crossing one another), this is the famous four color problem for which it has recently been proven that a solution always exists. Let $0, 1, 2, 3$ correspond to the colors and let $x_i \in \{0, 1, 2, 3\}$ be the variable which selects the color for node i. We require $x_i \neq x_j$ if $\langle i,j \rangle \in \mathcal{E}$, or in other words, $x_i - x_j \geq 1$ or $x_j - x_i \geq 1$. This can be accomplished by the introduction of a zero-one variable δ_{ij} and the constraints

$$x_i - x_j \geq 1 - \delta_{ij}$$

and

$$x_j - x_i \geq -3 + 4\delta_{ij}$$

Thus, the graph G can be colored with four colors and a four coloring solution has been found if $v=0$ is the optimal value in the integer programming problem

$$v = \min \sum_{\langle i,j \rangle \in \mathcal{E}} \{s_{ij} + t_{ij}\}$$

$$\begin{aligned} \text{s.t. } x_i - x_j + s_{ij} &\geq 1 - \delta_{ij} && \text{for all } \langle i,j \rangle \in \mathcal{E} \\ x_j - x_i + t_{ij} &\geq -3 + 4\delta_{ij} \\ x_i &= 0, 1, 2, 3 && \text{for } i = 1, \ldots, m \\ \delta_{ij} &= 0 \text{ or } 1 && \text{for all } \langle i,j \rangle \in \mathcal{E} \\ s_{ij}, t_{ij} &= 0, 1, 2, 3, 4, \ldots \end{aligned}$$

The following model illustrates how an integer programming formulation can be inefficient because it does not permit the exploitation of special structures. It also illustrates how a complex combinatorial optimization problem can arise as a synthesis of the simple dynamic programming problems of Chapter 4.

Consider a manufacturing system consisting of I items for which production is to be scheduled at minimum cost over T time periods. The demand for item i in period t is the nonnegative integer r_{it}; this demand must be met by stock from inventory or by production during the period. Let the variable x_{it} denote the production of item i in period t. The inventory of item i at the end of period t is

$$y_{it}=y_{i,t-1}+x_{it}-r_{it} \qquad t=1,\ldots,T$$

where we assume $y_{i,0}=0$, or equivalently, initial inventory has been netted out of the r_{it}. Associated with x_{it} is a direct unit cost of production c_{it}. Similarly, associated with y_{it} is a direct unit cost of holding inventory h_{it}. The problem is complicated by the fact that positive production of item i in period t uses up a quantity $a_i+b_ix_{it}$ of a scarce resource q_t to be shared among the I items. The parameters a_i and b_i are assumed to be nonnegative.

This problem can be written as the mixed integer programming problem

$$v=\min \sum_{i=1}^{I}\sum_{t=1}^{T}(c_{it}x_{it}+h_{it}y_{it}) \tag{8.1a}$$

$$\text{s.t.} \sum_{i=1}^{I}(a_i\delta_{it}+b_ix_{it})\le q_t \qquad t=1,\ldots,T \tag{8.1b}$$

and for $i=1,\ldots,I$

$$\sum_{t=1}^{s}x_{it}-y_{is}=\sum_{t=1}^{s}r_{it} \qquad s=1,\ldots,T \tag{8.1c}$$

$$x_{it}\le M_{it}\delta_{it} \qquad t=1,\ldots,T \tag{8.1d}$$

$$x_{it}\ge 0 \qquad y_{it}\ge 0$$

$$\delta_{it}=0 \text{ or } 1 \qquad t=1,\ldots,T \tag{8.1e}$$

where $M_{it}=\Sigma_{s=t}^{T}r_{is}$ is an upper bound on the amount we would want to produce of i in period t. The constraints (8.1b) state that shared resource usage cannot exceed q_t. The constraints (8.1c) relate accumulated production and demand through period t to ending inventory in period t, and the nonnegativity of the y_{it} implies that demand must be met and not delayed (backlogged). The constraints (8.1d) ensure that $\delta_{it}=1$ and therefore the fixed charge resource usage a_i is incurred, if production x_{it} is positive in period t. Problem (8.1) is a mixed integer programming problem with IT zero-one variables, $2IT$ continuous variables and $T+2IT$ constraints. For example, if $I=200$ and $T=10$, these figures are 2000 zero-one variables, 4000 continuous variables, and 4010 constraints, which is a mixed integer programming problem of immense size.

For future reference, define the set

$$N_i = \{(\delta_{it}, x_{it}, y_{it}) \text{ for } t = 1, \ldots, T \mid \delta_{it}, x_{it}, y_{it} \text{ satisfies (8.1c), (8.1d), (8.1e)}\}.$$

This set describes a feasible production schedule for item i ignoring the joint constraints (8.1b). The integer programming formulation (8.1) is not effective because it fails to exploit the special network structure of the sets N_i. This can be accomplished by dual decomposition techniques as follows. Assign dual variables $u_t \geq 0$ to the scarce resources q_t and place the constraints (8.1b) in the objective function to form the Lagrangean

$$L(u) = -\sum_{t=1}^{T} u_t q_t + \min_{(\delta_{it}, x_{it}, y_{it}) \in N_i} \sum_{i=1}^{I} \sum_{t=1}^{T} \{(c_{it} + u_t b_i) x_{it} + u_t a_i \delta_{it} + h_{it} y_{it}\}$$

Letting

$$L_i(u) = \min_{(\delta_{it}, x_{it}, y_{it}) \in N_i} \sum_{t=1}^{T} \{(c_{it} + u_t b_i) x_{it} + u_t a_i \delta_{it} + h_{it} y_{it}\} \tag{8.2}$$

the Lagrangean function clearly separates to become

$$L(u) = -\sum_{t=1}^{T} u_t q_t + \sum_{i=1}^{I} L_i(u)$$

Each of the problems (8.2) is a simple dynamic programming shortest-route calculation for scheduling item i exactly of the type studied in Example 4.1, in which the dual variables on shared resources adjust the costs as shown. Notice that it is easy to add additional constraints on the problem of scheduling item i that can be accommodated by the network representation; for example, permitting production in period t only if inventory falls below a preassigned level.

Unfortunately, we must give up something in using dual decomposition techniques on the mixed integer programming problem (8.1) to exploit the special structure of the sets N_i. In the context of this application, the global optimality conditions we seek but may not achieve involve dual variables that permit each of the I items to be separately scheduled by the dynamic programming calculation L_i while achieving a global minimum. This can be approximately accomplished if the number of joint constraints (8.1b) is small relative to I. In general, a global minimum can be achieved only by imbedding the dual decomposition in a branch and bound scheme as described in Section 8.6. These points are discussed in more detail in Exercise 8.11.

8.3 GROUP THEORY AND INTEGER SOLUTIONS TO LINEAR SYSTEMS

A fundamental concern in the study of integer programming problems is a constructive characterization of the solutions to

$$\begin{aligned} \mathbf{Ax} &= \mathbf{b} \\ \mathbf{x} &\text{ integer} \end{aligned} \tag{8.3}$$

where $\mathbf{A}$ is $m \times n$ with columns denoted by $\mathbf{a}_j$, $\mathbf{b}$ is $m \times 1$ and we assume the coefficients of $\mathbf{A}$ and $\mathbf{b}$ are integer. Furthermore, we assume for convenience that the rank of $\mathbf{A}$ is m. The assumption that all the coefficients in $\mathbf{A}$ and $\mathbf{b}$ are integer may seem restrictive, but if the coefficients were rational, they could be made into integers by multiplying the entire system by a sufficiently large integer. Pathologies in the theory to be presented can occur if the coefficients are irrational, but this is only of mathematical interest rather than practical concern since a computer must be given rational data. The integer coefficients can be very large in magnitude, however, which as we said in the previous section, can cause numerical difficulties. This difficulty, and methods for circumventing it, are discussed in exercises at the end of the chapter.

Constructive methods for characterizing the integer solutions to linear systems will be utilized in the following sections, in conjunction with other procedures, to account for nonnegativity constraints on the variables and to select an optimal solution from among the set of feasible solutions. This approach is very much in the spirit of the simplex method for linear programming which, as we discussed in Chapter 1, begins with constructive methods for characterizing continuous solutions to linear systems. The characterization of integer solutions begins where the continuous theory leaves off.

The new element we will use in this section is elementary number theory; Appendix C contains a review of the relevant constructs. Considerable insight as well as expositional convenience is gained by using the formalism of abelian group theory to summarize and work with the number theory methods. Appendix C also contains a review of the relevant constructs of group theory.

The procedures to be developed in this section will be summarized by homomorphisms ϕ from Z^m, the abelian group of integer m-vectors under ordinary addition, into various finite abelian groups $\mathcal{G}$. A homomorphism of this type will usually be given by specifying the images $\varepsilon_i = \phi(\mathbf{e}_i)$ for the unit vectors $\mathbf{e}_i \in Z^m$. Thus, for any $\mathbf{a} \in Z^m$, we have

$$\mathbf{a} = a_1\mathbf{e}_1 + \ldots + a_m\mathbf{e}_m$$

implying

$$\phi(\mathbf{a}) = a_1\phi(\mathbf{e}_1) + \dots + a_m\phi(\mathbf{e}_m)$$

since ϕ is a homorphism. The first observation about any such homomorphism is that it produces an aggregation of the linear system.

LEMMA 8.1.

$$\{\mathbf{x}|\mathbf{Ax}=\mathbf{b}, \mathbf{x} \text{ integer}\} \subseteq \left\{\mathbf{x}\Big| \sum_{j=1}^{n} \phi(\mathbf{a}_j)x_j = \phi(\mathbf{b}), \mathbf{x} \text{ integer}\right\}$$

PROOF. Suppose $\tilde{\mathbf{x}}$ is integer and satisfies $\mathbf{A}\tilde{\mathbf{x}}=\mathbf{b}$. Note that $\mathbf{A}\tilde{\mathbf{x}}$ and $\mathbf{b}$ are in Z^m. We apply ϕ to both sides of this equation with the result

$$\phi(\mathbf{A}\tilde{\mathbf{x}}) = \sum_{j=1}^{n} \phi(\mathbf{a}_j)\tilde{x}_j = \phi(\mathbf{b})$$

where the first equality follows because ϕ is a homomorphism. ■

EXAMPLE 8.1. Consider the system

$$\begin{aligned} 2x_1 - x_2 + x_3 + 3x_4 - 2x_5 + 2x_6 - 2x_7 &= 1 \\ -4x_1 - 3x_2 + 2x_3 + x_4 + x_5 + x_6 - 2x_7 &= -4 \\ x_j \text{ integer} \end{aligned}$$

We define a homomorphism from Z^2 into $\mathcal{G} = Z_5$ by specifying the images of the unit vectors $\mathbf{e}_1$ and $\mathbf{e}_2$. Specifically, we take $\phi(\mathbf{e}_1) = \varepsilon_1 = 1$ and $\phi(\mathbf{e}_2) = \varepsilon_2 = 1$ implying, for example, that

$$\begin{aligned} \phi\begin{pmatrix} 2 \\ -4 \end{pmatrix} &= \phi(2\mathbf{e}_1 - 4\mathbf{e}_2) \\ &= 2\phi(\mathbf{e}_1) - 4\phi(\mathbf{e}_2) = -2 \equiv 3 \pmod 5 \end{aligned}$$

The image of the entire linear system is

$$\begin{aligned} 3x_1 + 1x_2 + 3x_3 + 4x_4 + 4x_5 + 3x_6 + 1x_7 &\equiv 2 \pmod 5 \\ x_j \text{ integer} \end{aligned}$$

The solution $x_2 = x_7 = 1$ and the other variables zero satisfies the congruence but not the linear system. The solution $x_1 = x_3 = x_7 = 1$ and the other variables zero satisfies both the congruence and the linear system. ▲

It is easier to optimize over the group equation

$$\sum_{j=1}^{n} \phi(\mathbf{a}_j)x_j = \phi(\mathbf{b})$$

than $\mathbf{Ax}=\mathbf{b}$ because the group equation is defined over the finite group $\mathcal{G}$

rather than the infinite group Z^m. For the aggregation to be effective in characterizing integer solutions to $\mathbf{Ax}=\mathbf{b}$, we need to consider specific homomorphisms induced by $m\times m$ nonsingular matrices.

THEOREM 8.1. Let $\mathbf{B}$ be any $m\times m$ nonsingular matrix of integers made up of m columns of $\mathbf{A}$, and without loss of generality, suppose $\mathbf{A}=(\mathbf{N},\mathbf{B})$ and $\mathbf{x}=(\mathbf{x_N},\mathbf{x_B})$. Then $\mathbf{x}=(\mathbf{x_N},\mathbf{x_B})$ is an integer solution to $\mathbf{Ax}=\mathbf{b}$ if and only if

$$\mathbf{x_B}=\mathbf{B}^{-1}\mathbf{b}-\mathbf{B}^{-1}\mathbf{N}\mathbf{x_N}$$

and

$$\phi_{\mathbf{B}}(\mathbf{N}\mathbf{x_N})=\sum_{j=1}^{n-m}\phi_{\mathbf{B}}(\mathbf{a}_j)x_j=\phi_{\mathbf{B}}(\mathbf{b})$$

$$x_j \text{ integer}, j=1,\ldots,n-m$$

where $\phi_{\mathbf{B}}$ is a homomorphism from Z^m onto a finite abelian group

$$\mathcal{G}_{\mathbf{B}}=Z_{q_1}\oplus Z_{q_2}\oplus\ldots\oplus Z_{q_r}$$

where for all i, q_i divides q_{i+1}, $q_i\geq 2$, $\prod_{i=1}^{r}q_i=|\det\mathbf{B}|$, and the q_i are uniquely determined by $\mathbf{B}$. Moreover, $\phi_{\mathbf{B}}(\mathbf{a}_j)=0$ for $\mathbf{a}_j$ basic.

The Smith reduction procedure for diagonalizing an integer matrix is the main tool used in proving Theorem 8.1. We need to establish this procedure before we can prove the theorem.

Definition 8.1. An *elementary row operation* on an integer matrix $\mathbf{B}$ is the permutation of two rows, the addition or subtraction of an integer multiple of one row to another, or the multiplication of a row by -1. An *elementary column operation* is similarly defined.

LEMMA 8.2. An elementary row operation on a matrix $\mathbf{B}$ corresponds to multiplication of $\mathbf{B}$ on the left by a matrix of integers $\mathbf{R}$ with $\det\mathbf{R}=\pm 1$. An elementary column operation corresponds to multiplication on the right by a matrix of integers $\mathbf{C}$ with $\det\mathbf{C}=\pm 1$.

PROOF. An elementary row operation on $\mathbf{B}$ is achieved by performing the same operation on an identity matrix and multiplying $\mathbf{B}$ on the left by this transformed identity matrix. For example, to interchange row k and row l of $\mathbf{B}$, interchange row k and row l of the identity matrix. The interchange has the effect of changing the determinant of the matrix from $+1$ to -1. As another example, if λ times row k is to be added to row l, $k\neq l$, the element on row l, column k of the identity matrix is changed from 0 to λ. This change clearly leaves the determinant of the transformed matrix equal to 1. ■

LEMMA 8.3. Let $\mathbf{B}$ be an $m \times m$ nonsingular matrix of integers. A diagonal matrix

$$\boldsymbol{\Delta} = \begin{bmatrix} p_1 & & 0 \\ & \ddots & \\ 0 & & p_m \end{bmatrix}$$

can be constructed from $\mathbf{B}$ by a finite number of elementary row and column operations, where $\boldsymbol{\Delta} = \mathbf{RBC}$, $\mathbf{R}$ and $\mathbf{C}$ are integer matrices, $\det \mathbf{R} = \pm 1$, $\det \mathbf{C} = \pm 1$, $p_i \geq 1$ and integer, p_i divides p_{i+1} for $i = 1, \ldots, m-1$, and $\prod_{i=1}^{m} p_i = |\det \mathbf{B}|$.

PROOF. Recall that

$$\mathbf{B} = \begin{pmatrix} a_{1,n-m+1} & \cdots & a_{1,n} \\ a_{m,n-m+1} & \cdots & a_{m,n} \end{pmatrix}$$

Let a_{kl} be an element of $\mathbf{B}$ which is smallest in magnitude among the nonzero elements. Perform the following elementary column operations on $\mathbf{B}$:

if $a_{kl} > 0$, subtract $[a_{kj}/a_{kl}]$ times column k from each column j, $j \neq l$, $j = 1, \ldots, m$,

if $a_{kl} < 0$, add $[a_{kj}/-a_{kl}]$ times column k to each column j, $j \neq l$, $i = 1, \ldots, m$,

where $[t]$ denotes the integer part of t; that is, $[t]$ is the largest integer not exceeding t. Perform similar elementary row operations on rows i, $i \neq k$ for $i = 1, \ldots, m$.

Let v_{ij} denote the elements of the matrix that result after these elementary operations are performed. It is easy to see by our choice of row and column operations that every v_{kj} and v_{il} is a nonnegative integer less than $|a_{kl}|$. The procedure of selecting a minimal element among the nonzero elements and then performing the indicated row and column operations is repeated. Since the magnitude of the minimal nonzero element v_{kl} strictly decreases in integer amounts at each step, the procedure will, in a finite number of steps, produce such an element with the property that $v_{kj} = 0$, $j \neq l, j = 1, \ldots, m$, and $v_{il} = 0$, $i \neq l$, $i = 1, \ldots, m$. A sufficient but not necessary condition for this to occur is for $v_{kl} = +1$ or -1 implying the stated elementary row and column operations will make the other elements on row k and column l equal to zero.

When the kth row and the lth column become zero except for v_{kl}, the next step is to permute the rows and columns until row k becomes row 1 and column l becomes column 1. The procedure is then repeated on the submatrix with row 1 and column 1 excluded. It is clear that **B** will become diagonalized in a finite number of steps by a sequence of elementary row and column operations. In matrix form, we have

$$\mathbf{\Delta} = \mathbf{R}_Q \ldots \mathbf{R}_2 \mathbf{R}_1 \mathbf{B} \mathbf{C}_1 \ldots \mathbf{C}_S$$

where $\mathbf{R}_1, \ldots, \mathbf{R}_Q$ correspond to the successive row operations, and $\mathbf{C}_1, \ldots, \mathbf{C}_S$ to the successive column operations. Since the determinants of the **R** and **C** matrices are all $+1$ or -1, we have $\det \mathbf{\Delta} = \prod_{i=1}^{m} p_i = |\det \mathbf{B}|$, and without loss of generality, we can assume $p_i \geq 1$.

The only point left to prove is that $\mathbf{\Delta}$ can be constructed so that $p_i | p_{i+1}$, $i = 1, \ldots, m$. To see that this is so, suppose that $\mathbf{\Delta}$ is diagonal and that the positive diagonal elements $p_i', \ldots, p_m'$ satisfy $p_1' \leq p_{i+1}'$ for $i = 1, \ldots, m$, but for some i, p_i' does not divide p_{i+1}'. In this case, we can perform the column operation of adding column i to column $i+1$, and then perform the row operation of subtracting row i multiplied by $[p_{i+1}'/p_i']$ from row $i+1$. The remainder $v_{i+1,i+1}$ is positive and equals

$$p_{i+1}' - \left[\frac{p_{i+1}'}{p_i'} \right] p_i' < p_i$$

Thus, if we were to continue to perform elementary row and column operations on the submatrix consisting of the last $m-i+1$ rows and columns of $\mathbf{\Delta}$, we would obtain a diagonal submatrix where the minimal element is less than p_i'. Continuing in this way, we can guarantee the condition $p_i | p_{i+1}$ for $i = 1, \ldots, m$, because the reduction process just described will continue until this condition holds or all the p_i equal 1, except $p_m = |\det \mathbf{B}|$. ■

PROOF OF THEOREM 8.1. The continuous solutions to $\mathbf{Ax} = \mathbf{b}$ are characterized by choosing any values for the independent variables $\mathbf{x_N}$ and setting the values of the dependent variables by the equation

$$\mathbf{x_B} = \mathbf{B}^{-1}\mathbf{b} - \mathbf{B}^{-1}\mathbf{N}\mathbf{x_N} \tag{8.4}$$

We are concerned here with the choices of $\mathbf{x_N}$ as an integer vector that produces integer vectors $\mathbf{x_B}$. The theorem says, in effect, that not any choice of $\mathbf{x_N}$ integer will make $\mathbf{x_B}$ integer, but that the independent variables must satisfy a group equation, or equivalently, a system of congruences.

Let $\mathbf{\Delta} = \mathbf{RBC}$ be the diagonal matrix of Lemma 8.3. Thus $\mathbf{B}^{-1} = \mathbf{C}\mathbf{\Delta}^{-1}\mathbf{R}$, and substituting in the linear system $\mathbf{x_B} = \mathbf{B}^{-1}\mathbf{b} - \mathbf{B}^{-1}\mathbf{N}\mathbf{x_N}$, we have

$$\mathbf{x_B} = \mathbf{C}\mathbf{\Delta}^{-1}\mathbf{R}\mathbf{b} - \mathbf{C}\mathbf{\Delta}^{-1}\mathbf{R}\mathbf{N}\mathbf{x_N}$$

The next step is to prove the intermediate result that $\mathbf{x_B}$ is integer if and only if

$$\mathbf{C}^{-1}\mathbf{x_B}=\boldsymbol{\Delta}^{-1}\mathbf{Rb}-\boldsymbol{\Delta}^{-1}\mathbf{RNx_N}$$

is integer. If $\mathbf{C}^{-1}\mathbf{x_B}$ is an integer vector, then premultiplying by $\mathbf{C}$ gives an integer vector because $\mathbf{C}$ consists of integer coefficients. On the other hand, if $\mathbf{x_B}$ is integer, then $\mathbf{C}^{-1}\mathbf{x_B}$ is integer because $\mathbf{C}$ is an integer matrix and $\det \mathbf{C}=\pm 1$.

Thus, the requirement that the dependent variables $\mathbf{x_B}$ be integer is equivalent to the requirement that the independent variables $\mathbf{x_N}$ satisfy

$$\boldsymbol{\Delta}^{-1}\mathbf{RNx_N}\equiv\boldsymbol{\Delta}^{-1}\mathbf{Rb} \pmod 1$$

where vector congruence modulo 1 means that each element of $\boldsymbol{\Delta}^{-1}\mathbf{RNx_N}$ must differ from $\boldsymbol{\Delta}^{-1}\mathbf{Rb}$ by an integer. Theorem 8.1 is established by considering each row in the system. To do this, let $\mathbf{RN}=(\tilde{a}_{ij})$, $\mathbf{Rb}=(\tilde{b}_i)$, and recalling that $\boldsymbol{\Delta}$ is diagonal, we have

$$\boldsymbol{\Delta}^{-1}=\begin{bmatrix} 1/p_1 & & 0 \\ & \ddots & \\ 0 & & 1/p_m \end{bmatrix}$$

We rewrite row i as

$$\frac{1}{p_i}\sum_{j=1}^{n-m}\tilde{a}_{ij}x_j=\frac{1}{p_i}b_i \pmod 1$$

or

$$\sum_{j=1}^{n-m}\tilde{a}_{ij}x_j=\tilde{b}_i \pmod{p_i} \tag{8.5}$$

Since the coefficients $\tilde{a}_{ij}$ and $\tilde{b}_i$ are integer, if $p_i=1$ the ith congruence (8.3) is satisfied by any integer $\mathbf{x_N}$. Thus, the only congruences that constrain the independent variables are the congruences such that $p_i\geq 2$, say $p_{m-r+1},\ldots,p_m$. As a notational convenience, let $q_i=p_{m-r+i}$ for $i=1,\ldots,r$.

Without loss of generality, the elements $\tilde{a}_{ij}$ and $\tilde{b}_i$ on rows $m-r+i$ for $i=1,\ldots,r$ can be reduced mod q_i; that is, $\tilde{a}_{m-r+i,j}\rightarrow\alpha_{ij}\in Z_{q_i}$ where $\alpha_{ij}\equiv\tilde{a}_{m-r+i,j} \pmod{q_i}$, and $\tilde{b}_{m-r+i}\rightarrow\beta_i\in Z_{q_i}$ where $\beta_i\equiv b_{m-r+i} \pmod{q_i}$. Thus, $\mathbf{x_B}=\mathbf{B}^{-1}\mathbf{b}-\mathbf{B}^{-1}\mathbf{Nx_N}$ is an integer vector if and only if the independent integer variables $\mathbf{x_N}$ satisfy the system of congruences

$$\sum_{j=1}^{n-m}\alpha_{ij}x_j\equiv\beta_i \pmod{q_i} \qquad i=1,\ldots,r \tag{8.6}$$

The system of congruences (8.6) in group theoretic terms is an equation over the finite abelian group $\mathcal{G}_{\mathbf{B}}=Z_{q_1}\oplus\ldots\oplus Z_{q_r}$ with the required properties on the q_i. This group consists of $D=\Pi_{i=1}^{r}q_i$ elements, where each element is an r-tuple $\omega=(\omega_1,\ldots,\omega_r)$ and the $\omega_i \in Z_{q_i}$. Exercise 8.3 contains the demonstration that the q_i are uniquely determined by $\mathbf{B}$. What remains to be discussed are the properties of the homomorphism $\phi_{\mathbf{B}}$ mapping Z^m into $\mathcal{G}_{\mathbf{B}}$.

First, note that each nonbasic activity $\mathbf{a}_j$ in the linear system (8.4) is mapped into an r-tuple $\alpha_j=(\alpha_{1j},\ldots,\alpha_{rj})$ in the system of congruences (8.6) and $\mathbf{b}$ is mapped into the r-tuple $\beta=(\beta_1,\ldots,\beta_r)$. Specifically, the matrix $\mathbf{R}$ in the equation $\boldsymbol{\Delta}=\mathbf{RBC}$ is used to map the integer m-vector $\mathbf{a}_j$ into $\mathbf{Ra}_j=\tilde{\mathbf{a}}_j$ which in turn is mapped into the r-tuple α_j by reduction modulo q_i of the coefficients $\tilde{a}_{m-r+i,j}$ for $i=1,\ldots,r$. More generally, any integer m-vector $\mathbf{w}$ with coefficients w_l for $l=1,\ldots,m$, is mapped into the r-tuple $\phi_{\mathbf{B}}(\mathbf{w})=\omega$ by the rule

$$\omega_i \equiv \sum_{l=1}^{m} w_l \varepsilon_{il} \pmod{q_i} \qquad i=1,\ldots,r \tag{8.7}$$

where $\varepsilon_l=(\varepsilon_{1l},\ldots,\varepsilon_{rl})$ is the image of $\phi_{\mathbf{B}}(\mathbf{e}_l)$ of the unit vector $\mathbf{e}_l \in R^m$; that is, $\mathbf{e}_l$ is mapped into $\mathbf{Re}_l=\tilde{\mathbf{e}}_l$ and the coefficients $\tilde{e}_{m-r+i,l}$ are reduced modulo q_i. The validity of (8.7) can be seen by noting that $\mathbf{w}=w_1\mathbf{e}_1+\ldots+w_m\mathbf{e}_m$ implying that $\mathbf{Rw}=w_1\mathbf{Re}_1+\ldots+w_m\mathbf{Re}_m$.

Similarly, we can easily see that each basic activity $\mathbf{a}_{n-m+l}$ for $l=1,\ldots,m$, is mapped into $\phi_{\mathbf{B}}(\mathbf{a}_{n-m+l})=\alpha_{n-m+l}=(0,\ldots,0)$. This follows because $\boldsymbol{\Delta}=\mathbf{RBC}$ implies $\mathbf{Ra}_{n-m+l}=\boldsymbol{\Delta}\mathbf{h}$ where $\mathbf{h}$ is column l of $\mathbf{C}^{-1}$; $\mathbf{h}$ is integer since $\mathbf{C}$ is integer and $\det\mathbf{C}=\pm 1$. Thus, for $i=1,\ldots,r$, we have $\alpha_{i,n-m+l}\equiv p_{m-r+i}h_{m-r+i,l} \pmod{q_i} \equiv 0 \pmod{q_i}$ since $p_{m-r+i}=q_i$. In other words, $\phi_{\mathbf{B}}(\mathbf{a}_{n-m+l})=0$.

To see that $\phi_{\mathbf{B}}$ maps Z^m *onto* $\mathcal{G}_{\mathbf{B}}$, consider any r-tuple $\omega \in \mathcal{G}_{\mathbf{B}}$ and extend it to an m-vector by $\bar{\omega}^T=(0,\ldots,0,\omega)^T$. The vector $\bar{\mathbf{w}}=\mathbf{R}^{-1}\bar{\omega}$ is in Z^m because $\mathbf{R}$ is integer and unimodular; moreover, the last r rows of $\mathbf{R}\bar{\mathbf{w}}=\bar{\omega}$ give the group image of $\bar{\mathbf{w}}$ and this is simply ω. Thus, for every $\omega \in \mathcal{G}_{\mathbf{B}}$ there is a $\mathbf{w} \in Z^m$ such that $\phi_{\mathbf{B}}(\mathbf{w})=\omega$ and the mapping is onto $\mathcal{G}_{\mathbf{B}}$.

In summary, the diagonalization of $\mathbf{B}$ produces the system of congruences (8.6) that characterizes the integer values of the independent variables $\mathbf{x}_{\mathbf{N}}$ producing integer values for the dependent variables $\mathbf{x}_{\mathbf{B}}=\mathbf{B}^{-1}\mathbf{b}-\mathbf{BN}\mathbf{x}_{\mathbf{N}}$. In group theoretic terms, the characterization is that the nonbasics satisfy $\sum_{j=1}^{n-m}\phi_{\mathbf{B}}(\mathbf{a}_j)x_j=\phi_{\mathbf{B}}(\mathbf{b})$ and x_j integer. ■

EXAMPLE 8.2. We consider the linear system in integer variables of Example 8.1. Let

$$\mathbf{B}=\begin{pmatrix} 3 & -2 \\ 1 & 1 \end{pmatrix}$$

be a basis consisting of the fourth and fifth columns of the system. This matrix is diagonalized by four elementary operations:

$$\begin{pmatrix} 3 & -2 \\ 1 & 1 \end{pmatrix} \xrightarrow{\mathbf{R}_1=\begin{pmatrix} 1 & -3 \\ 0 & 1 \end{pmatrix}} \begin{pmatrix} 0 & -5 \\ 1 & 1 \end{pmatrix} \xrightarrow{\mathbf{C}_1=\begin{pmatrix} 1 & -1 \\ 0 & 1 \end{pmatrix}} \begin{pmatrix} 0 & -5 \\ 1 & 0 \end{pmatrix}$$

$$\begin{pmatrix} 0 & -5 \\ 1 & 0 \end{pmatrix} \xrightarrow{\mathbf{R}_2=\begin{pmatrix} 0 & 1 \\ 1 & 0 \end{pmatrix}} \begin{pmatrix} 1 & 0 \\ 0 & -5 \end{pmatrix} \xrightarrow{\mathbf{R}_3=\begin{pmatrix} 1 & 0 \\ 0 & -1 \end{pmatrix}} \boldsymbol{\Delta}=\begin{pmatrix} 1 & 0 \\ 0 & 5 \end{pmatrix}$$

In matrix form, we have $\boldsymbol{\Delta}=\mathbf{R}_3\mathbf{R}_2\mathbf{R}_1\mathbf{B}\mathbf{C}_1$, or $\boldsymbol{\Delta}=\mathbf{RBC}$ where $\mathbf{R}=\mathbf{R}_3\mathbf{R}_2\mathbf{R}_1$ and $\mathbf{C}=\mathbf{C}_1$. According to the proof of Theorem 8.1, the second row of

$$\mathbf{R}=\begin{pmatrix} 0 & 1 \\ -1 & 3 \end{pmatrix}$$

gives the images of the unit vectors $\mathbf{e}_1, \mathbf{e}_2 \in Z^2$, which in turn describes the homomorphism $\phi_{\mathbf{B}}$. Specifically, we have

$$\varepsilon_1=\phi_{\mathbf{B}}(\mathbf{e}_i)=-1\equiv 4 \pmod 5$$

and

$$\varepsilon_2=\phi_{\mathbf{B}}(\mathbf{e}_2)=3\equiv 3 \pmod 5$$

Note that the group images of the basis vectors $\phi_{\mathbf{B}}\begin{pmatrix} 3 \\ 1 \end{pmatrix}=3\varepsilon_1+1\varepsilon_2\equiv 0$ (mod 5) and $\phi_{\mathbf{B}}\begin{pmatrix} -2 \\ 1 \end{pmatrix}=-2\varepsilon_1+1\varepsilon_2\equiv 0$ (mod 5) ensuring that the computation was done correctly.

We use this homomorphism to map the linear system into the congruence or group equation

$$1x_1+2x_2+0x_3+0x_4+0x_5+1x_6+1x_7\equiv 2 \pmod 5$$
$$x_j \text{ integer}$$

If we consider only the nonbasics, then the set of all integer solutions to the linear system is characterized by

$$x_4=-\tfrac{7}{5}-\left(-\tfrac{6}{5}\right)x_1-\left(-\tfrac{7}{5}\right)x_2-\left(\tfrac{5}{5}\right)x_3-\left(\tfrac{4}{5}\right)x_6-\left(-\tfrac{6}{5}\right)x_7$$
$$x_5=-\tfrac{13}{5}-\left(-\tfrac{14}{5}\right)x_1-\left(-\tfrac{8}{5}\right)x_2-\left(\tfrac{5}{5}\right)x_3-\left(\tfrac{1}{5}\right)x_6-\left(-\tfrac{4}{5}\right)x_7$$

where

$$1x_1+2x_2+0x_3+1x_6+1x_7\equiv 2 \pmod 5$$
$$x_1, x_2, x_3, x_6, x_7 \text{ integer}$$

For example, $x_1=3$, $x_2=1$, $x_3=0$, $x_6=1$, $x_7=1$ satisfies the congruence and gives us $x_4=4$ and $x_5=8$. ▲

There is an interesting geometrical interpretation of the group induced by integer bases. The following theorem provides the central mathematical insight required to establish the geometry.

THEOREM 8.2. Let $\mathbf{w}^1, \mathbf{w}^2$ be any integer m-vectors and let $\phi_\mathbf{B}$ be the homomorphism induced by the basis $\mathbf{B}$ in Theorem 8.1. Then $\phi_\mathbf{B}(\mathbf{w}^1)=\phi_\mathbf{B}(\mathbf{w}^2)$ if and only if $\mathbf{w}^1=\mathbf{w}^2+\mathbf{B}\mathbf{x}_\mathbf{B}$ for some integer.

PROOF. (1) $\mathbf{w}^1=\mathbf{w}^2+\mathbf{B}\mathbf{x}_\mathbf{B}$ for $\mathbf{x}_\mathbf{B}$ integer$\Rightarrow\phi_\mathbf{B}(\mathbf{w})^1=\phi_\mathbf{B}(\mathbf{w}^2)$. This result follows from the result of Theorem 8.1 that $\phi_\mathbf{B}(\mathbf{a}_{n-m+i})=0$ for $\mathbf{a}_{n-m+i}$ a column of $\mathbf{B}$. Thus, $\phi_\mathbf{B}(\mathbf{w}^1)=\phi_\mathbf{B}(\mathbf{w}^2)+\sum_{i=1}^{m}\phi_\mathbf{B}(\mathbf{a}_{n-m+i})x_{n+i}=\phi_\mathbf{B}(\mathbf{w}^2)$.

(2) $\phi_\mathbf{B}(\mathbf{w}^1)=\phi_\mathbf{B}(\mathbf{w}^2)\Rightarrow\mathbf{w}^1=\mathbf{w}^2+\mathbf{B}\mathbf{x}_\mathbf{B}$ for some $\mathbf{x}_\mathbf{B}$ integer. Consider the two linear systems $\mathbf{A}\mathbf{x}=\mathbf{w}^1$, $\mathbf{x}$ integer, and $\mathbf{A}\mathbf{x}=\mathbf{w}^2$, $\mathbf{x}$ integer. By Theorem 8.1, the integer solutions are characterized by

$$\mathbf{x}_\mathbf{B}=\mathbf{B}^{-1}\mathbf{w}^1-\mathbf{B}^{-1}\mathbf{N}\mathbf{x}_\mathbf{N}$$

$$\mathbf{x}_\mathbf{B}=\mathbf{B}^{-1}\mathbf{w}^2-\mathbf{B}^{-1}\mathbf{N}\mathbf{x}_\mathbf{N}$$

where $\mathbf{x}_\mathbf{N}$ satisfies

$$\sum_{j=1}^{n-m}\phi_\mathbf{B}(\mathbf{a}_j)x_j=\phi_\mathbf{B}(\mathbf{w}^1)=\phi_\mathbf{B}(\mathbf{w}^2)$$

$$x_j \text{ integer}, j=1,\ldots,n-m$$

Let $\bar{\mathbf{x}}_\mathbf{N}$ be an integer $(n-m)$-vector satisfying the group equation. Then

$$\mathbf{x}_\mathbf{B}^1=\mathbf{B}^{-1}\mathbf{w}^1-\mathbf{B}^{-1}\mathbf{N}\bar{\mathbf{x}}_\mathbf{N}$$

and

$$\mathbf{x}_\mathbf{B}^2=\mathbf{B}^{-1}\mathbf{w}^2-\mathbf{B}^{-1}\mathbf{N}\bar{\mathbf{x}}_\mathbf{N}$$

arc integer by Theorem 8.1. This implies directly that

$$\mathbf{B}^{-1}\mathbf{w}^1=\mathbf{B}^{-1}\mathbf{w}^2+\mathbf{x}_\mathbf{B}^1-\mathbf{x}_\mathbf{B}^2$$

or

$$\mathbf{w}^1=\mathbf{w}^2+\mathbf{B}(\mathbf{x}_\mathbf{B}^1-\mathbf{x}_\mathbf{B}^2)$$

which is what we wanted to show. ■

COROLLARY 8.1. The group $\mathcal{G}_\mathbf{B}$ induced by the integer matrix $\mathbf{B}$ in Theorem 8.1 is isomorphic to the quotient group $Z^m/Z^m(\mathbf{B})$ where $Z^m(\mathbf{B})$ is the subgroup of Z^m consisting of integer m-vectors of the form $\mathbf{B}\mathbf{x}_\mathbf{B}$ for $\mathbf{x}_\mathbf{B}$ integer.

PROOF. Two groups are isomorphic if there is a one-to-one and onto correspondence or mapping from one group to the other. Each element of the group $Z^m/Z^m(\mathbf{B})$ is a coset of the form $\mathbf{w}+\mathbf{B}\mathbf{x}_\mathbf{B}$ for all $\mathbf{x}_\mathbf{B}$ integer. The isomorphic image of this coset in $\mathcal{G}_\mathbf{B}$ is $\phi_\mathbf{B}(\mathbf{w})$. By Theorem 8.2, any $\mathbf{w}$ in a given coset of $Z^m/Z^m(\mathbf{B})$ is mapped into the same element $\phi_\mathbf{B}(\mathbf{w})$ of $\mathcal{G}_\mathbf{B}$, and any $\mathbf{w}^1,\mathbf{w}^2 \in Z^m$, which are mapped into the same element of $\mathcal{G}_\mathbf{B}$, are in the same coset. Thus, there is a one-to-one correspondence between elements (cosets) of $Z^m/Z^m(\mathbf{B})$ and $\mathcal{G}_\mathbf{B}$. The correspondence is onto because, as we established in Theorem 8.1, for every $\omega \in \mathcal{G}_\mathbf{B}$ there is a $\mathbf{w} \in Z^m$, and therefore an element (coset) $\{\mathbf{w}+\mathbf{B}\mathbf{x}_\mathbf{B} | \mathbf{x}_\mathbf{B} \text{ integer}\}$ in $Z^m/Z^m(\mathbf{B})$, such that $\phi_\mathbf{B}(\mathbf{w})=\omega$. ■

EXAMPLE 8.3. We illustrate the geometry implied by Corollary 8.1 by considering the system in Example 8.2 and the basis we used there. The situation is depicted in Figure 8.2 where we have shown the lattice of points that can be spanned by integer combinations of the vectors (3, 1) and (−2, 1). The circled vectors are the representatives we have chosen of the five cosets constituting the factor group $Z^m/Z^m(\mathbf{B})$; namely, (0,0), (1,2), (1,1), (1,0), (1, −1). Every $\mathbf{w} \in Z^m$ can be spanned by adding $\mathbf{B}\mathbf{x}_\mathbf{B}$ for suitable $\mathbf{x}_\mathbf{B}$ to one of these vectors. ▲

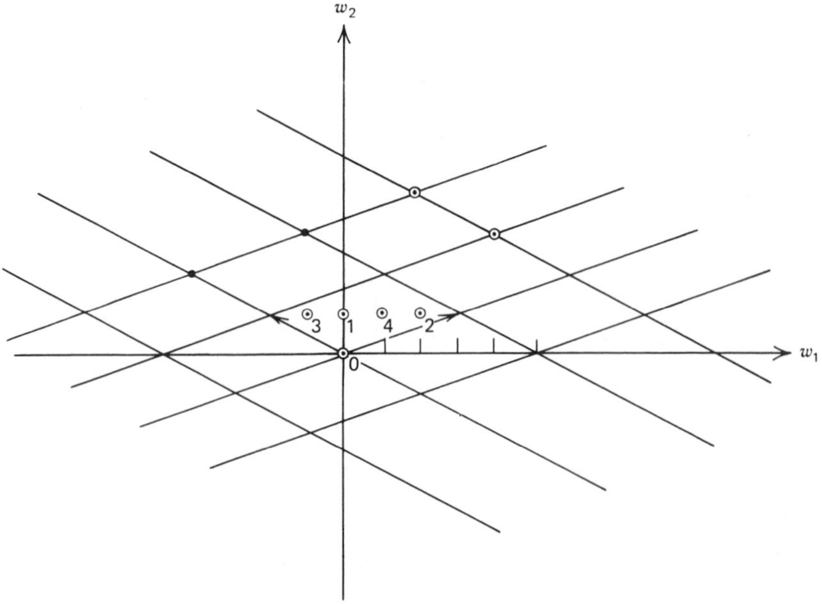

Figure 8.2. Factor group defined by a basis.

EXAMPLE 8.4. As a preview to the application of group theoretic methods to integer programming, consider problem (2.30) for which we illustrated the cutting plane method. The feasible region can be expressed in terms of the nonbasic variables x_1 and x_2 by the constraints

$$\tfrac{4}{7}x_1 - \tfrac{8}{7}x_2 \le \tfrac{30}{7}$$

$$-\tfrac{6}{7}x_1 + \tfrac{33}{7}x_2 \le \tfrac{60}{7}$$

$$1x_1 + 5x_2 \equiv 4 \pmod{7}$$

$$x_1, x_2 \text{ nonnegative integer}$$

The inequality constraints ensure that the basic variables x_3 and x_4, which are the slacks on these constraints, are nonnegative. The congruence is derived from the optimal basis as suggested by Theorem 8.1 and it ensures that the dependent basic variables will be integer when we solve for them; that is, $x_3 = (\frac{30}{7}) - (\frac{4}{7})x_1 + (\frac{8}{7})x_2$ and $x_4 = (\frac{60}{7}) + (\frac{6}{7})x_1 - (\frac{33}{7})x_2$ will be integer if x_1 and x_2 are integer and satisfy $1x_1 + 5x_2 \equiv 4 \pmod{7}$.

The situation is depicted in Figure 8.3 where the shaded region indicates the values of the variables x_1 and x_2 satisfying the inequality constraints

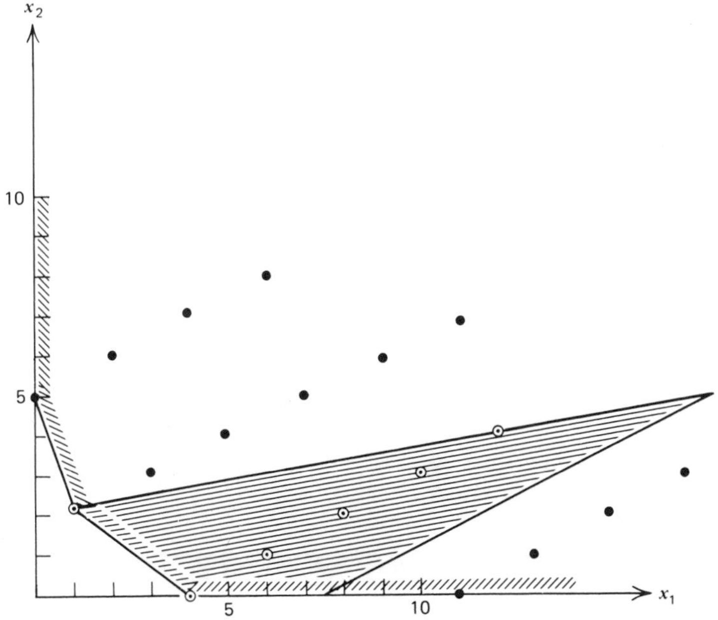

Figure 8.3. Representation of an integer programming problem in terms of its optimal basis and induced group.

lying inside the convex hull of the solutions satisfying the congruence. As we shall see in Section 8.5, the integer programming dual problem derived from this reformulation of the problem is equivalent to the linear programming problem of minimizing the original objective function over the intersection of these two polyhedra. The dots indicate the values of x_1 and x_2 satisfying the congruence. The circled dots are the six feasible solutions satisfying both types of constraints. For problem (2.30), the minimum over the intersection of the polyhedra occurs at the point $x_1=1$, $x_2=2$ which gives us by substitution the basic values $x_3=6$, $x_4=0$. Note that two of the four extreme points of the intersection polyhedron give feasible integer solutions while two extreme points do not. ▲

8.4 GROUP OPTIMIZATION PROBLEMS

In this section, we formulate and solve group optimization problems suggested by and derived from the constructions in the previous section. In the following section, we show how the group optimization problems are used to solve integer programming problems. As we shall see, the problems resemble some of the shortest route problems discussed in Chapter 4, particularly the knapsack problem discussed in Section 4.3.

The *unconstrained group optimization* problem is

$$
\begin{aligned}
v(\beta) = \min \sum_{j=1}^{n} c_j x_j \\
\text{s.t.} \sum_{j=1}^{n} \alpha_j x_j = \beta \\
\mathbf{x} \geq 0 \text{ and integer}
\end{aligned}
\tag{8.8}
$$

where the α_j and β are elements of a finite abelian group $\mathcal{G} = Z_{q_1} \oplus \ldots \oplus Z_{q_r}$, the q_i are positive integers greater than one, and we assume for convenience that q_i divides q_{i+1} for all i. The group has $D = \prod_{i=1}^{r} q_i$ elements. For convenience, we assume the c_j are integers. The following lemma demonstrates, in effect, that problem (8.8) can be represented as a shortest route problem.

LEMMA 8.4. Let $\mathbf{x}$ be an optimal solution to the unconstrained group optimization problem, and let $\mathbf{y}$ be any nonnegative integer vector satisfying $\mathbf{y} \leq \mathbf{x}$. Then $\mathbf{y}$ is optimal in (8.8) with the group right-hand-side element β replaced by $\sum_{j=1}^{n} \alpha_j y_j$.

PROOF. Suppose to the contrary that there exists a nonnegative integer vector $\mathbf{z}$ such that $\sum_{j=1}^{n} \alpha_j z_j = \sum_{j=1}^{n} \alpha_j y_j$ and $\mathbf{cz} < \mathbf{cy}$. Then the n-vector $\mathbf{z}+(\mathbf{x}-\mathbf{y})$ is nonnegative integer and satisfies

$$\sum_{j=1}^{n} \alpha_j(z_j + x_j - y_j) = \sum_{j=1}^{n} \alpha_j x_j + \sum_{j=1}^{n} \alpha_j z_j - \sum_{j=1}^{n} \alpha_j y_j = \Sigma \alpha_j x_j = \beta$$

Thus $\mathbf{z}+(\mathbf{x}-\mathbf{y})$ is feasible in (8.8), and moreover, $\mathbf{c}[\mathbf{z}+(\mathbf{x}-\mathbf{y})] = \mathbf{cx}+\mathbf{cz}-\mathbf{cy} < \mathbf{cx}$ contradicting the optimality of $\mathbf{x}$. ■

THEOREM 8.3. Let $v(\omega)$ be the cost of an optimal solution to the unconstrained group optimization problem (8.8) with group right-hand side ω. Then

$$\begin{aligned} v(\omega) &= \min_{j=1,\ldots,n} \{c_j + v(\omega - \alpha_j)\} \qquad \text{for } \omega \neq 0 \\ v(0) &= 0 \end{aligned} \tag{8.9}$$

PROOF. Suppose first that (8.8) has an optimal solution for all group right-hand sides ω. Clearly, $\mathbf{x}=0$ is optimal when $\omega=0$ in (8.8). Let $\mathbf{x}$ be an optimal solution for a given $\omega \neq 0$. We must have $\mathbf{x} \neq 0$ which implies for some j, say $j=j^*$, that $x_{j^*} > 0$. Let $\mathbf{y} = \mathbf{x} - \mathbf{e}_{j^*}$; by Lemma 8.4, $\mathbf{y}$ is optimal in (8.8) with group right-hand side $\omega - \alpha_{j^*}$. In other words, $v(\omega - \alpha_{j^*}) = \mathbf{cy} = \mathbf{cx} - c_{j^*} = v(\omega) - c_{j^*}$, or $v(\omega) = c_{j^*} + v(\omega - \alpha_{j^*})$ which implies directly that $v(\omega) \geq \min_{j=1,\ldots,n} \{c_j + v(\omega - \alpha_j)\}$. To show the reverse inequality, let $\mathbf{y}^j$ be an optimal solution in (8.8) with group right-hand side $\omega - \alpha_j$. Then $v(\omega - \alpha_j) = \mathbf{cy}^j$ and moreover $\mathbf{y}^j + \mathbf{e}_j$ with cost $v(\omega - \alpha_j) + c_j$ is feasible in (8.8) with group right-hand side ω. This implies for all j that $v(\omega) \leq c_j + v(\omega - \alpha_j)$, or $v(\omega) \leq \min_{j=1,\ldots,n}\{c_j + v(\omega - \alpha_j)\}$. Thus, the recursion is established in the special case when (8.8) has an optimal solution for all group right-hand sides ω.

Suppose now that (8.8) has no feasible solution for group right-hand side ω. In other words, $v(\omega) = +\infty$. Then for all j, (8.8) has no feasible solution for group right-hand side $\omega - \alpha_j$ because otherwise we could take such a feasible solution, add e_j to it, and thereby obtain a feasible solution for (8.8) with group right-hand side ω. Thus, $v(\omega) = +\infty$ implies for all j that $v(\omega - \alpha_j) = +\infty$ and the recursion is still valid.

Finally, suppose $v(\omega) = -\infty$ for some $\omega \in G$. This means there is a sequence of nonzero, nonnegative integer vectors $\{\mathbf{x}^k\}_{k=1}^{\infty}$ which are feasible in (8.8) with group right-hand side ω and such that $\lim_{k\to\infty} \mathbf{cx}^k = -\infty = v(\omega)$. There must be some component j, say $j=j^*$, such that $x_{j^*}^k > 0$ for k occurring infinitely often. This implies $x_{j^*}^k - e_{j^*}$ is nonnegative integer for k occurring infinitely often, and since these solutions are feasible in (8.8) with group right-hand side $\omega-\alpha_{j^*}$, we can conclude that $v(\omega-\alpha_{j^*})=-\infty$. Thus,

$$v(\omega)=c_{j^*}+v(\omega-\alpha_{j^*})=\min_{j=1,\ldots,n}\{c_j+v(\omega-\alpha_j)\}$$

and the theorem is proven. ■

The recursion (8.9) is similar to the ones seen in Chapter 4 for shortest route problems, particularly recursion (4.9) for the knapsack problem. In fact, it is useful to view (8.8) as a shortest route problem in the *group network* which consists of D nodes, one for each element $\omega \in G$, and arcs $(\omega, \omega+\alpha_j)$ with length c_j for $j=1,\ldots,n$. Problem (8.8) is solved by using the recursion (8.9) to find a shortest route chain from the origin node 0 to the node β in the group network. The corresponding solution $\mathbf{x}$ to the unconstrained group optimization problem is derived from a chain in the network by letting x_j equal the number of times an arc of the form $(\omega, \omega+\alpha_j)$ is used in the chain. Unlike the knapsack problem, the cyclical nature of the group optimization problem implies that it is difficult to rank *a priori* the variables x_j by increasing value of c_j per "length". For example, if $\mathcal{G}$ is a cyclic group, it is less likely to expect $x_1>x_2$ in an optimal solution to (8.8) if $c_1/\alpha_1<c_2/\alpha_2$ than a knapsack problem of similar size.

There is an ambiguity, however, to be resolved in the correspondence between chains in the group network and nonnegative integer n-vectors $\mathbf{x}$ in the unconstrained group optimization problem. Each $\mathbf{x}$ corresponds to a number of chains in the group network because the arcs in the chain can be taken in any order. For example, the chains $(0,\alpha_1)$, $(\alpha_1,\alpha_1+\alpha_2)$ and $(0,\alpha_2)$, $(\alpha_2,\alpha_2+\alpha_1)$ both correspond to $\mathbf{x}=(1,1,0,\ldots,0)$. In general, the different chains are the various permutations of the arcs given by the combination of arcs in $\mathbf{x}$. We resolve the ambiguity by associating with each nonnegative integer $\mathbf{x}$ the unique chain in the group network that begins at the origin with x_1 arcs of the form $(\omega,\omega+\alpha_1)$, followed by x_2 arcs of the form $(\omega,\omega+\alpha_2)$, etc., and finishes up with x_n arcs of the form $(\omega,\omega+\alpha_n)$. We call this the rule of *chain augmentation to the right*.

The following theorem specializes for the group optimization problem the shortest route result that there exists a circuitless shortest route chain if there is a shortest route chain and all the arc lengths are nonnegative.

THEOREM 8.4. Suppose there is a feasible solution to the unconstrained group optimization problem (8.8). If all of the c_j are nonnegative, then there exists an optimal solution to (8.8) satisfying $\sum_{j=1}^{n} x_j \le D-1$. If $c_k<0$ for some k, then (8.8) does not have an optimal solution and $v(\beta)=-\infty$.

PROOF. If all of the $c_j \ge 0$, then every chain in the group network has nonnegative length and problem (8.8) is well defined; that is, an optimal solution exists. Let $\mathbf{x}$ be any optimal solution to (8.8) and suppose $\sum_{j=1}^{n} x_j$

$=K$. There are K arcs in any chain in the group network corresponding to $\mathbf{x}$ and the chain visits $K+1$ nodes. If $K \geq D$, then the chain visits $D+1$ nodes and therefore it must visit some node more than once. In other words, the chain contains a circuit that can be deleted without loss of optimality because it has nonnegative length (actually, the circuit must have zero length). The result is a new shortest route chain with a strictly smaller number of arcs, and the corresponding solution to (8.8) has a strictly smaller sum. This reduction process can be continued as long as the sum exceeds $D-1$ and we have proved the first part of the theorem.

If $c_k < 0$ for some k, then we can construct a sequence of chains whose lengths converge to $-\infty$. We simply take any feasible solution $\tilde{\mathbf{x}}$ to (8.8) and consider $\tilde{\mathbf{x}} + tD\mathbf{e}_k$ for $t = 1, 2, \ldots$. We have for all t that

$$\sum_{j=1}^{n} \alpha_j \tilde{x}_j + tD\alpha_k = \sum_{j=1}^{n} \alpha_j \tilde{x}_j = \beta$$

because $D\alpha_k = 0$ in $\mathcal{G}$, and the objective function terms go to $-\infty$. ■

EXAMPLE 8.5. Consider the group optimization problem

$$\begin{aligned} &\min 3x_1 + 8x_2 \\ &\text{s.t. } 1x_1 + 0x_2 \equiv 0 \pmod 2 \\ &\quad\;\; 1x_1 + 3x_2 \equiv 1 \pmod 4 \\ &\quad\;\; x_1, x_2 \text{ nonnegative integer} \end{aligned}$$

The representation of this problem as the group shortest-route problem is shown in Figure 8.4. Note that we have omitted arcs drawn to the origin node since it never pays to return to that node. A shortest route path with

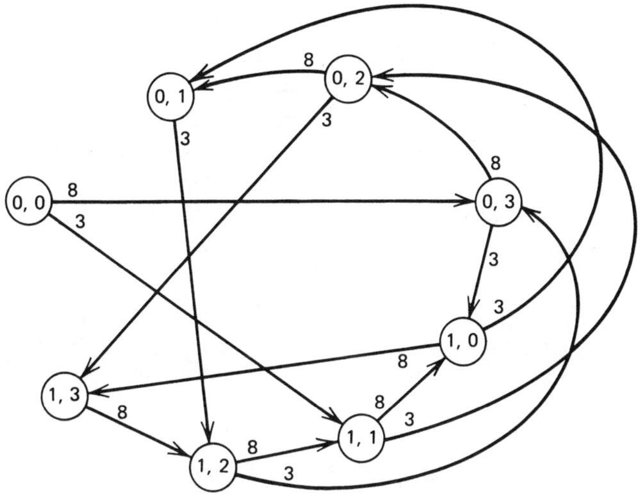

Figure 8.4. Group shortest route problem.

length 17 is [(0,0),(1,1)], [(1,1),(0,2)], [(0,2),(0,1)] corresponding to the solution $x_1=2$, $x_2=1$. We are forced to use x_2 although its cost is great relative to x_1 because $\alpha_1=(1,1)$ alone generates a subgroup of order four of $Z_2 \oplus Z_4$ that does not include $\beta=(0,1)$. ▲

The special structure of the group network and the recursion (8.9) for computing shortest routes can be exploited in a special algorithm for solving the unconstrained group optimization problem (8.8). The algorithm generates a sequence $\{\mathbf{x}^k\}$ of nonnegative integer solutions satisfying $\mathbf{cx}^k \leq \mathbf{cx}^{k+1}$ and the first $\mathbf{x}^k$ satisfying $\Sigma_{j=1}^n \alpha_j x_j^k = \beta$ is optimal. The algorithm can be easily extended to find optimal solutions for all group right-hand sides ω in (8.8).

The algorithm assumes that all the c_j are positive integers. We have already assumed they are integer, and they must be nonnegative for (8.8) to have an optimal solution according to Theorem 8.4. The c_j's can be perturbed as follows if any of them are zero. First, multiply all the c_j's by D, the number of elements in the group defining (8.8). Then increase the c_j's equal to zero to a cost of one. A nonoptimal solution to (8.8) cannot become optimal as a result of the perturbation. For suppose $\mathbf{x}^1$ and $\mathbf{x}^2$ satisfy $\mathbf{cx}^1 < \mathbf{cx}^2$, say $\mathbf{cx}^1 + 1 \leq \mathbf{cx}^2$ implying

$$D\mathbf{cx}^1 + D \leq D\mathbf{cx}^2$$

Let $\tilde{\mathbf{c}}$ denote the perturbed objective function. The cost of $\mathbf{x}^1$ satisfies

$$\tilde{\mathbf{c}}\mathbf{x}^1 \leq D\mathbf{cx}^1 + \sum_{j=1}^{n} x_j^1 \leq D\mathbf{cx}^1 + (D-1)$$

where the last inequality follows from Theorem 8.4. Comparing these two inequalities, we have

$$\tilde{\mathbf{c}}\mathbf{x}^1 < D\mathbf{cx}^2 \leq \tilde{\mathbf{c}}\mathbf{x}^2$$

which is what we require for the perturbation.

The $\mathbf{x}^k$ are generated by the algorithm in a unique fashion by incrementing the first component x_1^k times, then incrementing the second component x_2^k times, etc. This manner of generating nonnegative integer solutions expresses the rule of chain augmentation to the right. For example, suppose the algorithm generates the solution $x=(1,1,2)$ at some iteration. Then this solution was preceded by the generated solutions in the order (0,0,0), (1,0,0), (1,1,0), and (1,1,1). At iteration k, the algorithm considers a list of previously generated solutions $\mathbf{x}^{k_j}$ for $j=1,\ldots,n$, called *candidates*. The new solution $\mathbf{x}^k$ is generated from the candidates by the rule

$$\mathbf{x}^k = \mathbf{x}^{k_{j_k}} + \mathbf{e}_{j_k}$$

where the index j_k is the minimal one satisfying

$$\mathbf{cx}^{k_{j_k}} + c_{j_k} = \min_{j=1,\ldots,n} \{\mathbf{cx}^{k_j} + c_j\} \tag{8.10}$$

By construction, the $\mathbf{x}^{k_j}$ have the property that $x_l^{k_j}=0$ for $l>j$ so that the method of generating $\mathbf{x}^k$ is consistent with the rule of chain augmentation to the right. The solution $\mathbf{x}^k$ is optimal in (8.8) if $\omega^k=\sum_{j=1}^n \alpha_j x_j^k=\beta$ and the algorithm stops. Otherwise, the algorithm continues and tests $\mathbf{x}^k$ to see if $\mathbf{cx}^k$ is greater than the cost of previously generated solutions for (8.8) with group right-hand side $\sum_{j=1}^n \alpha_j x_j^k$. If so, then $\mathbf{x}^k$ is *dominated* and it is ruled out as a candidate. Otherwise, $\mathbf{x}^k$ is called *undominated* and it is optimal for the group right-hand side it spans.

Strictly speaking, if a solution $\mathbf{x}^g$ has already been generated for (8.8) with right-hand side $\sum_{j=1}^n \alpha_j x_j^k$, then $\mathbf{x}^k$ could be considered dominated even if $\mathbf{cx}^k=\mathbf{cx}^g$ (we cannot have $\mathbf{cx}^k<\mathbf{cx}^g$). The proof that the algorithm works in this case is more difficult, however, and thus we include $\mathbf{x}^k$ in our list of undominated solutions if $\mathbf{cx}^k=\mathbf{cx}^g$. Alternatively, we could further perturb the cost coefficients c_j so that ties in cost cannot occur.

Once the index j_k has been determined by (8.10) the algorithm updates for future reference the candidate $\mathbf{x}^{k_{j_k}}$ for index j_k. This is accomplished by the previously generated undominated solution $\mathbf{x}^l$ with minimal index l satisfying $l>k_{j_k}$ and $x_j^l=0$ for $j=j_k+1,\ldots,n$. If there is no such $\mathbf{x}^l$, the implication is that the algorithm will make no further use of arc type $(\omega,\omega+\alpha_{j_k})$ in extending chains in the group network toward the node corresponding to the group right-hand side β.

Unconstrained Group Optimization Algorithm

Step 0. (Initialization). Perturb the c_j if necessary to make them all positive. Set $\hat{v}(\omega)=+\infty$ for all $\omega\neq 0$. Also set

$$k_j=0 \qquad j=1,\ldots,n$$

$$f_j=c_j \qquad j=1,\ldots,n$$

Go to Step 1 with $k=1$.

Step 1. Find the minimal index j_k such that

$$f_{j_k}=\min_{j=1,\ldots,n}\{f_j\}$$

Stop if $f_{j_k}=+\infty$; in this case, there is no feasible solution to (8.8). Otherwise, set

$$\mathbf{x}^k=\mathbf{x}^{k_{j_k}}+\mathbf{e}_{j_k}$$

$$\omega^k=\omega^{k_{j_k}}+\alpha_{j_k}$$

Stop if $\omega^k=\beta$; then $\mathbf{x}^k$ is optimal in the unconstrained group optimization problem (8.8); continue otherwise. The solution $\mathbf{x}^k$ is dominated if

$$\mathbf{cx}^k>\hat{v}(\omega^k)$$

Otherwise, it is undominated. If $\mathbf{c}\mathbf{x}^k < \hat{v}(\omega^k)$, set $\hat{v}(\omega^k) = v(\omega^k) = \mathbf{c}\mathbf{x}^k$.

Step 2. Update the index k_{j_k} by

$$k_{j_k} = \min\{\, l \mid x^l \text{ undominated, } l > k_{j_k}, \text{ and } x_j^l = 0 \text{ for } j = j_k + 1, \ldots, n \}$$

and use the new index to update f_{j_k} by

$$f_{j_k} = \mathbf{c}\mathbf{x}^{k_{j_k}} + c_{j_k}$$

If there is no such undominated $\mathbf{x}^l$, then take $k_{j_k} = 0$ and $f_{j_k} = +\infty$. Index k to $k+1$ and return to Step 1.

THEOREM 8.5. The unconstrained group optimization algorithm finds an optimal solution to the unconstrained group optimization problem (8.8), or establishes that there is none.

PROOF. Since the c_j are all positive, the algorithm cannot generate an infinite sequence of nonnegative integer vectors. Specifically, we know from Theorem 8.4 that no undominated solutions will be generated with cost equal to or greater than $D \max c_j$ since such a solution cannot be optimal for problem (8.8) with the group right-hand side it spans. If (8.8) does not have a feasible solution, then the algorithm will terminate with this information when $f_j = +\infty$ for all j in Step 1.

Suppose now that $\tilde{\mathbf{x}}$ is an optimal solution for (8.8) with group right-hand side β. By Lemma 8.4, all nonnegative integer $\mathbf{y}^l$ satisfying $\mathbf{y}^l \le \mathbf{x}$ are optimal in (8.8) with group right-hand side $\Sigma_{j=1}^n \alpha_j y_j^l$. Suppose the $\mathbf{y}^l$ are ordered by the rule of chain augmentation to the right. All of these $\mathbf{y}^l$ are undominated, including $\tilde{\mathbf{x}}$, and will be generated in order by the algorithm unless an alternative optimal solution is found first. ■

EXAMPLE 8.6. Consider the unconstrained group optimization problem

$$\begin{aligned} \min\ & 4x_1 + 7x_2 + 8x_3 + 11x_4 + 16x_5 \\ \text{s.t. } & 10x_1 + 1x_2 + 3x_3 + 7x_4 + 9x_5 \equiv 5 (\bmod 13) \\ & x_j \text{ nonnegative integer} \end{aligned} \tag{8.11}$$

The application of the unconstrained group optimization problem to this problem is given in Table 8.1. The row labeled c gives the cost of each generated solution $\mathbf{x}$, the row α gives the right-hand side $\Sigma_{j=1}^n \alpha_j x_j$, the row r gives the greatest index such that $x_r > 0$ and the row p gives the index of the predecessor of $\mathbf{x}$. Only the undominated $\mathbf{x}$'s are numbered at the top of the table. The rows labeled $k_j(f_j)$ give the index of the candidates $\mathbf{x}^{k_j}$ and the cost of their extension, $\mathbf{c}\mathbf{x}^{k_{j_k}} + c_{j_k}$.

As an illustration of the calculations in Table 8.1, consider the generation of the eighth undominated solution. This solution is $(x_1^8, x_2^8, x_3^8, x_4^8, x_5^8)$

TABLE 8.1(a) UNCONSTRAINED GROUP OPTIMIZATION ALGORITHM

	0	1	2	3	4	5		6
c	0	4	7	8	8	11	11	12
α	0	10	1	7	3	11	7	4
r		1	2	1	3	2	4	1
p		0	0	1	0	1	0	3
x_1	0	1	0	2	0	1	0	3
x_2	0	0	1	0	0	1	0	0
x_3	0	0	0	0	1	0	0	0
x_4	0	0	0	0	0	0	1	0
x_5	0	0	0	0	0	0	0	0
$k_1(f_1)$	0(4)	1(8)	1(8)	3(12)	3(12)	3(12)	3(12)	6(16)
$k_2(f_2)$	0(7)	0(7)	1(11)	1(11)	1(11)	2(14)	2(14)	2(14)
$k_3(f_3)$	0(8)	0(8)	0(8)	0(8)	1(12)	1(12)	1(12)	1(12)
$k_4(f_4)$	0(11)	0(11)	0(11)	0(11)	0(11)	0(11)	1(15)	1(15)
$k_5(f_5)$	0(16)	0(16)	0(16)	0(16)	0(16)	0(16)	0(16)	0(16)
		$v(10)=4$	$v(1)=7$	$v(7)=8$	$v(3)=8$	$v(11)=11$		$v(4)=12$

TABLE 8.1(b) UNCONSTRAINED GROUP OPTIMIZATION ALGORITHM

		7	8				
c	12	14	15	15	15	16	16
α	0	2	8	4	4	1	10
r	3	2	2	3	4	1	3
p	1	2	3	2	1	5	3
x_1	1	0	2	0	1	4	2
x_2	0	2	1	1	0	0	0
x_3	1	0	0	1	0	0	1
x_4	0	0	0	0	1	0	0
x_5	0	0	0	0	0	0	0
$k_1(f_1)$	6(16)	6(16)	6(16)	6(16)	6(16)	—	—
$k_2(f_2)$	2(14)	3(15)	5(18)	5(18)	5(18)	5(18)	5(18)
$k_3(f_3)$	2(15)	2(15)	2(15)	3(16)	3(16)	3(16)	4(16)
$k_4(f_4)$	1(15)	1(15)	1(15)	1(15)	2(18)	2(18)	2(18)
$k_5(f_5)$	0(16)	0(16)	0(16)	0(16)	0(16)	0(16)	0(16)
		$v(2)=14$	$v(8)=15$				

TABLE 8.1(c) UNCONSTRAINED GROUP OPTIMIZATION ALGORITHM

	9	10	11		12
c	16	16	18	18	19
α	6	9	12	8	5
r	3	5	2	4	2
p	4	0	5	2	6
x_1	0	0	1	0	3
x_2	0	0	2	1	1
x_3	2	0	0	0	0
x_4	0	0	0	1	0
x_5	0	1	0	0	0
$k_1(f_1)$	—	—	—	—	—
$k_2(f_2)$	5(18)	5(18)	6(19)	6(19)	
$k_3(f_3)$	5(19)	5(19)	5(19)	5(19)	
$k_4(f_4)$	2(18)	2(18)	2(18)	3(19)	
$k_5(f_5)$	0(16)	1(20)	1(20)	1(20)	
	$v(6)=16$	$v(9)=16$	$v(12)=18$		$v(5)=19$

$=(2,1,0,0,0)$ and it was selected in Step 1 of the algorithm because $f_2=15$ was minimal among the f_j and $j=2$ was the minimal index among the f_j with value 15. Since $k_2=3$, we know $x^8=x^3+e_2=(2,0,0,0,0)+(0,1,0,0,0)$. Since the group right-hand side 8 had not been previously spanned, we have $v(8)=15$. The index $k_2(f_2)$ is updated in Step 2 by considering in order the undominated solutions x^4, x^5, x^6, x^7 to see if they can be augmented to the right by e_2. The solution $x^4=(0,0,1,0,0)$ with $r=3$ cannot be, but the solution $x^5=(1,1,0,0,0)$ with $r=2$ can be so augmented. Thus, k_2 becomes 5 and $f_2=cx^5+c_2=11+7=18$.

The algorithm finds the optimal solution $(x_1,x_2,x_3,x_4,x_5)=(3,1,0,0,0)$ to the stated problem with a cost (or length in the group network) equal to 19. Note that in the process of discovering this solution, the algorithm also found an optimal solution to the problem with all group right-hand sides in Z_{13}. This is not usually the case, but due to our construction of a numerical example designed to illustrate all aspects of the algorithm. ▲

The unconstrained group optimization algorithm can be modified easily to solve the *zero-one group optimization problem*

$$z(\beta)=\min \sum_{j=1}^{n} c_j x_j$$

$$\text{s.t.} \sum_{j=1}^{n} \alpha_j x_j=\beta \tag{8.12}$$

$$x_j=0 \text{ or } 1$$

Zero-One Group Optimization Algorithm

Step 0. (Initialization). Determine the set $J^{-}=\{j|c_j<0\}$ and make upper bound substitutions $x_j=1-x_j$ for $j\in J^{-}$. The cost of x_j becomes $-c_j>0$ and it has the group representation $-\alpha_j$. The group right-hand side to be spanned becomes $\sigma=\beta-\Sigma_{j\in J^{-}}\alpha_j$. Perturb the nonnegative c_j if necessary to make them all positive.

For the transformed problem with all c_j positive, set $\mathbf{x}^0=0$ and $\omega^0=0$. Set $\hat{z}(\omega)=+\infty$ for $\omega\neq 0$. Also set

$$k_j=0 \qquad j=1,\dots,n$$

$$f_j=c_j \qquad j=1,\dots,n$$

Go to Step 1 with $k=1$.

Step 1. Find the minimal index j_k such that

$$f_{j_k}=\min_{j=1,\dots,n}\{f_j\}$$

Stop if $f_{j_k}=+\infty$; in this case, there is no feasible solution to (8.12). Otherwise, set

$$\mathbf{x}^k=\mathbf{x}^{k_{j_k}}+\mathbf{e}_{j_k}$$

$$\omega^k=\omega^{k_{j_k}}+\alpha_{jk}$$

Stop if $\omega^k=\beta$; then $\mathbf{x}$ given by $x_j=1-x_j^k$ for $j\in J^{-}$ and $x_j=x_j^k$ for $j\in(J^{-})^c$, is optimal in the zero-one group optimization problem (8.12). Otherwise, continue. The solution $\mathbf{x}^k$ is dominated if

$$\mathbf{c}\mathbf{x}^k\geq\hat{z}(\omega^k)$$

and

$$\min\{j|x_j^k=1\}\geq\min\{j|x_j^l=1\}$$

where $\mathbf{x}^l$ is the previously generated solution such that $\mathbf{c}\mathbf{x}^l=\hat{z}(\omega^k)$. Otherwise, $\mathbf{x}^k$ is undominated. If $\mathbf{c}\mathbf{x}^k<\hat{z}(\omega^k)$, set $\hat{z}(\omega^k)=\mathbf{c}\mathbf{x}^k$.

Step 2. Update the index k_{j_k} by

$$k_{j_k}=\min\{l|x^l\text{undominated, } l>k_{j_k}\text{, and } x_j^l=0 \text{ for } j=j_k,j_k+1,\dots,n\}$$

and use the new index to update f_{j_k} by

$$f_{j_k}=\mathbf{c}\mathbf{x}^{k_{j_k}}+c_{j_k}$$

If there is no such undominated $\mathbf{x}^l$, then take $k_{j_k}=0$ and $f_{j_k}=+\infty$. Index k to $k+1$ and return to Step 1.

There are three types of changes to the unconstrained algorithm in constructing the zero-one algorithm. First, the zero-one problem (8.12) always has an optimal solution if it has a feasible solution, even if some c_j are negative, because the x_j are bounded. Thus, the zero-one algorithm begins in the initialization step by making upper bound substitutions on the x_j with $c_j < 0$. These substitutions are then accounted for in Step 1 when an optimal solution to (8.12) is found. Second, a zero-one solution $\mathbf{x}^k$ in the transformed problem with all c_j positive is dominated in Step 1 if (1) $\mathbf{c}\mathbf{x}^k$ is at least as great as the cost $\hat{z}(\omega^k)$ of the best known solution with right-hand side ω^k, and (2) it does not have more flexibility with respect to augmentations to the right than the best known solution. The final change is in Step 2 where the index k_{j_k} is selected for the candidate corresponding to j_k. In looking over previously generated undominated solutions to augment to the right, we consider as a new candidate $x^{k_{j_k}}$ only those $\mathbf{x}^l$ with component j_k equal to zero. This permits us to extend $\mathbf{x}^l$ by the unit vector $\mathbf{e}_{j_k}$ without violating the zero-one constraints.

EXAMPLE 8.7. We solve problem (8.11) of Example 8.6 with x_j constrained to be zero-one instead of nonnegative integer. The results are shown in Table 8.2. The optimal solution found by the algorithm is $(x_1, x_2, x_3, x_4, x_5) = (1, 1, 0, 1, 0)$ with a cost equal to 22. As we should expect, this cost is higher than the cost of an optimal solution with the x_j's permitted to take on any nonnegative integer values. ▲

TABLE 8.2(a) ZERO-ONE GROUP OPTIMIZATION ALGORITHM

	0	1	2	3	4	5		6
c	0	4	7	8	11	11	12	15
α	0	10	1	3	11	7	0	4
r		1	2	3	2	4	3	3
p		0	0	0	1	0	1	2
x_1	0	1	0	0	1	0	1	0
x_2	0	0	1	0	1	0	0	1
x_3	0	0	0	1	0	0	1	1
x_4	0	0	0	0	0	1	0	0
x_5	0	0	0	0	0	0	0	0
$k_1(f_1)$	0(4)	—	—	—	—	—	—	—
$k_2(f_2)$	0(7)	0(7)	1(11)	1(11)	—	—	—	—
$k_3(f_3)$	0(8)	0(8)	0(8)	1(12)	1(12)	1(12)	2(15)	4(19)
$k_4(f_4)$	0(11)	0(11)	0(11)	0(11)	0(11)	1(15)	1(15)	1(15)
$k_5(f_5)$	0(16)	0(16)	0(16)	0(16)	0(16)	0(16)	0(16)	0(16)
		$z(10)=4$	$z(1)=7$	$z(3)=8$	$z(11)=11$	$z(7)=11$		$z(4)=15$

TABLE 8.2(b) ZERO-ONE GROUP OPTIMIZATION ALGORITHM

		7	8			9	
c	15	16	18	19	19	20	22
α	4	9	8	1	10	6	5
r	4	5	4	3	4	5	4
p	1	0	2	4	3	1	4
x_1	1	0	0	1	0	1	1
x_2	0	0	1	1	0	0	1
x_3	0	0	0	1	1	0	0
x_4	1	0	1	0	1	0	1
x_5	0	1	0	0	0	1	0
$k_1(f_1)$	—	—	—	—	—	—	
$k_2(f_2)$	—	—	—	—	—	—	
$k_3(f_3)$	4(19)	4(19)	4(19)	—	—	—	
$k_4(f_4)$	2(18)	2(18)	3(19)	3(19)	4(22)	6(24)	
$k_5(f_5)$	0(16)	1(20)	1(20)	1(20)	2(23)	2(23)	
		$z(9)=16$	$z(8)=18$			$z(6)=20$	$z(5)=22$

8.5 INTEGER PROGRAMMING DUALITY

In this section we apply the group theoretic methods of the previous two sections in conjunction with the constructs of mathematical programming duality and nondifferentiable optimization discussed in Chapters 5 and 6 to the integer programming problem

$$\begin{aligned} v = \min\ & \mathbf{cx} \\ \text{s.t. } & \mathbf{Ax} = \mathbf{b} \\ & x_j = 0 \text{ or } 1 \qquad \text{for all } j \end{aligned} \tag{8.13}$$

where $\mathbf{A}$ is $m \times n$ and $\mathbf{A}$, $\mathbf{b}$, and $\mathbf{c}$ have integer coefficients. We have chosen a zero-one form for the problem because the essence of integer programming is the selection of small integer values for the variables. Note, also, that if x_j can take on any nonnegative integer value between 0 and M, then it can be expanded in zero-one variables as

$$x_j = \sum_{k=0}^{K} 2^k y_{jk}$$
$$y_{jk} = 0 \text{ or } 1$$

where K is the largest integer such that $2^K \leq M$. Substitution of the y_{jk} for x_j in (8.13) produces a zero-one problem from a bounded one. In practice, however, it is unnecessary and sometimes inefficient to make these substitutions and the methods to be discussed can deal with upper bounds different from one in exactly the same way that we will deal with the upper bounds of one.

The convexification equivalence to dualization that we saw in Section 5.3 provides important insights about integer programming because of convex structures inherent in problem (8.13). Let

$$F=\{\mathbf{x}|\mathbf{Ax}=\mathbf{b}, x_j=0 \text{ or } 1 \text{ for all } j\} \tag{8.14}$$

and consider the linear programming problem

$$\begin{aligned} &\min \mathbf{cx} \\ &\text{s.t. } \mathbf{x}\in[F]^c \end{aligned} \tag{8.15}$$

Recall that $[F]^c$ denotes the convex hull of F. The extreme points of the linear programming problem (8.15) are all zero-one, and therefore, if we used the simplex method to solve it, we would naturally find such an extreme point solution that is optimal in the integer programming problem (8.13) as well. The difficulty with this method for solving (8.13) is that the convex hull $[F]^c$ is not explicitly given to us and it is generally too complex to derive. The dual approach to be discussed in this section analyzes (8.13), in effect, by large-scale linear programming approximations. The approximations can be iteratively refined until one is obtained that approximates $[F]^c$ in a neighborhood of an optimal extreme point of $[F]^c$; that is, an optimal zero-one solution to (8.13).

The integer programming dual construction begins with a homomorphism ϕ from Z^m onto a finite abelian group $\mathcal{G}$. For the moment, we need not concern ourselves with the specification of ϕ because this will follow naturally from our development of how it is used to analyze the integer programming problem (8.13) and how it can be improved if an optimal solution is not found. By Lemma 8.1, all feasible zero-one solutions $\mathbf{x}$ to (8.13) also satisfy $\Sigma_{j=1}^{n}\phi(\mathbf{a}_j)x_j=\phi(\mathbf{b})$ and therefore we can add this group equation to (8.13) without altering the problem. The result is

$$v=\min \mathbf{cx} \tag{8.16a}$$

$$\text{s.t. } \mathbf{Ax}=\mathbf{b} \tag{8.16b}$$

$$\sum_{j=1}^{n}\phi(\mathbf{a}_j)x_j=\phi(\mathbf{b}) \tag{8.16c}$$

$$x_j=0 \text{ or } 1 \text{ for all } j \tag{8.16d}$$

and for future reference, we define

$$X=\{\mathbf{x}|\mathbf{x} \text{ satisfies (8.16c) and (8.16d)}\}$$

This set is finite and we denote $X=\{\mathbf{x}^t\}_{t=1}^{T}$. Note also that $F\subseteq X$.

An integer programming dual problem is constructed by dualizing on the constraints (8.16b) and letting the remaining group equations (8.16c) represent them in an aggregate or approximate sense. Specifically, for any $\mathbf{u} \in R^m$, we define the Lagrangean by

$$Z(\mathbf{u}) = -\mathbf{ub} + \min_{\mathbf{x} \in X} (\mathbf{c} + \mathbf{uA})\mathbf{x}$$

The computation of $Z(\mathbf{u})$ is performed by the zero-one group optimization problem given in the previous section. An integer programming dual problem is therefore

$$\begin{aligned} z = \max\, & Z(\mathbf{u}) \\ \text{s.t. } & \mathbf{u} \in R^m \end{aligned} \tag{8.17}$$

where we have the usual duality properties including Z concave, continuous, and nondifferentiable, and $Z(\mathbf{u}) \leq v$ for all $\mathbf{u} \in R^m$ implying $z \leq v$. The lower bounds provided by Z are important to the branch and bound approach to integer programming discussed in the following section. In this section we concentrate on the construction of dual problems that provide an optimal solution to the integer programming problem.

The relationship between the integer programming primal and dual problems (8.13) and (8.17) that we seek, but may not find, is expressed by the global optimality conditions (see Definition 5.1) for this particular pair of problems.

Global Optimality Conditions for Integer Programming

The pair $(\bar{\mathbf{x}}, \bar{\mathbf{u}})$ with $\bar{\mathbf{x}} \in X$ and $\bar{\mathbf{u}} \in R^m$ is said to satisfy the optimality conditions if

(1) $Z(\bar{\mathbf{u}}) = -\bar{\mathbf{u}}\mathbf{b} + (\mathbf{c} + \bar{\mathbf{u}}\mathbf{A})\bar{\mathbf{x}}$

(2) $\mathbf{Ax} = \mathbf{b}$

We have demonstrated in Chapter 5 that if $(\bar{\mathbf{x}}, \bar{\mathbf{u}})$ satisfies the global optimality conditions, then $\bar{\mathbf{x}}$ is optimal in the integer programming problem (8.13) and $\bar{\mathbf{u}}$ is optimal in the integer programming dual problem (8.17). The strategy indicated by the global optimality conditions is to find a $\bar{\mathbf{u}}$ that is optimal in (8.17) and then try to find a complementary $\mathbf{x} \in X$ satisfying conditions (1) and (2). The optimization of (8.17), if it fails to produce an optimal solution to (8.13), provides the necessary insight to change the group and the homomorphism defining X and thereby strengthen $Z(\mathbf{u})$ and (8.17).

The set X used in defining $Z(\mathbf{u})$ consists of the finite collection of solutions $\{\mathbf{x}^t\}_{t=1}^{T}$. This implies that the integer programming dual problem (8.17) is equivalent to the large-scale linear programming problem

$$\begin{aligned} z = \max\, & v \\ v \leq -\mathbf{ub} + (\mathbf{c} + \mathbf{uA})\mathbf{x}^t \qquad & t = 1, \ldots, T \end{aligned}$$

because for any $\mathbf{u} \in R^m, \nu(\mathbf{u})$ will take on its maximal value, which equals $-\mathbf{ub} + \min_{t=1,\ldots,T}(\mathbf{c} + \mathbf{uA})\mathbf{x}^t = Z(\mathbf{u})$. The linear programming dual of this problem is

$$\begin{aligned} z = \min \sum_{t=1}^{T} (\mathbf{cx}^t)\lambda_t& \\ \text{s.t.} \sum_{t=1}^{T} (\mathbf{Ax}^t)\lambda_t = \mathbf{b}& \\ \sum_{t=1}^{T} \lambda_t = 1& \\ \lambda_t \geq 0& \end{aligned} \tag{8.18}$$

For future reference we call (8.18) the *integer programming dual relaxation problem*. It solves the integer programming dual problem (8.17) in the sense that any vector of optimal dual variables for (8.18) is optimal in (8.17). This result is precisely what we should have expected from Section 5.4 where we showed how generalized linear programming can be used to solve dual problems using master linear programming problems of the form of (8.18). The enormous number of columns to be expected in (8.18) can be generated as needed by the generalized linear programming algorithm, or the generalized primal-dual algorithm in Section 6.5. Alternatively, the dual problem could be solved approximately by the subgradient optimization algorithm given in Section 6.4.

Note that the integer programming dual relaxation problem can be made equivalent to the integer programming problem (8.13) from which it is derived by imposing integrality constraints on the variables λ_t in (8.18). If this were done, then optimization of the problem would produce a single zero-one solution $\mathbf{x}^t \in X$ such that $\mathbf{Ax}^t = \mathbf{b}$ and $\mathbf{cx}^t$ is minimal; that is, $\mathbf{x}^t$ would be optimal in (8.13). This observation is important to our iterative construction of integer programming dual problems.

Our concern is to investigate the conditions under which a given dual problem solves the integer programming problem (8.13), and what to do if it does not solve it.

THEOREM 8.6. Suppose the integer programming dual relaxation problem (8.18) is solved by the simplex method and let $\bar{\lambda}_t, t = 1, \ldots, T$, denote the optimal solution it finds. Without loss of generality, suppose further that $\bar{\lambda}_t > 0$ for $t = 1, \ldots, K$, and $\bar{\lambda}_t = 0$ for $t = K+1, \ldots, T$. If $K = 1$, that is, if $\bar{\lambda}_1 = 1$, then $\mathbf{x}^1$ is an optimal solution to the integer programming problem (8.13). On the other hand, if $K \geq 2$, then $\mathbf{x}^t$ is infeasible in (8.13) for $t = 1, \ldots, K$.

PROOF. Case 1: $K=1$. Let $\bar{\theta}\in R^1$ denote the optimal linear programming dual variable or shadow price on the convexity constraint $\Sigma_{t=1}^{T}\lambda_t=1$ and let $-\bar{\mathbf{u}}\in R^m$ denote the optimal shadow prices on the other rows. Since $\bar{\lambda}_1>0$, we have

$$(\mathbf{c}+\bar{\mathbf{u}}\mathbf{A})\mathbf{x}^1-\bar{\theta}=0$$

and since $\bar{\mathbf{u}},\bar{\theta}$ are optimal, we have for all t that

$$(\mathbf{c}+\bar{\mathbf{u}}\mathbf{A})\mathbf{x}^t-\bar{\theta}\geq 0$$

This implies

$$-\bar{\mathbf{u}}\mathbf{b}+\bar{\theta}=Z(\bar{\mathbf{u}})=-\bar{\mathbf{u}}\mathbf{b}+(\mathbf{c}+\bar{\mathbf{u}}\mathbf{A})\mathbf{x}^1$$

Since $\bar{\lambda}_1=1$, we have from (8.18) that $\mathbf{A}\mathbf{x}^1=\mathbf{b}$. Thus, conditions 1 and 2 of the global optimality conditions hold and x^1 is optimal in the integer programming problem (8.13).

Case 2: $K\geq 2$. Since $\lambda_t>0$, the columns $\begin{pmatrix}\mathbf{A}\mathbf{x}^t\\ 1\end{pmatrix}$ for $t=1,\ldots,K$, are basis columns in an optimal basis for the dual problem (8.18). This implies they are linearly independent. Suppose to the contrary that one of the $\mathbf{x}^t$, $t=1,\ldots,K$, is feasible in (8.13); without loss of generality, say it is $\mathbf{x}^1$. We show a contradiction to the linear independence. The optimal solution to (8.18) found by the simplex method is

$$\begin{pmatrix}\mathbf{b}\\ 1\end{pmatrix}\bar{\lambda}_1+\sum_{t=2}^{K}\begin{pmatrix}\mathbf{A}\mathbf{x}^t\\ 1\end{pmatrix}\bar{\lambda}_t=\begin{pmatrix}\mathbf{b}\\ 1\end{pmatrix}$$

where we have used the fact that $\begin{pmatrix}\mathbf{A}\mathbf{x}^1\\ 1\end{pmatrix}=\begin{pmatrix}\mathbf{b}\\ 1\end{pmatrix}$. Rearranging, we have

$$\sum_{t=2}^{K}\begin{pmatrix}\mathbf{A}\mathbf{x}^t\\ 1\end{pmatrix}\bar{\lambda}_t=\begin{pmatrix}\mathbf{b}\\ 1\end{pmatrix}(1-\bar{\lambda}_1)$$

and since $0<\bar{\lambda}_1<1$, we have

$$\sum_{t=2}^{K}\begin{pmatrix}\mathbf{A}\mathbf{x}^t\\ 1\end{pmatrix}\frac{\bar{\lambda}_t}{1-\bar{\lambda}_1}=\begin{pmatrix}\mathbf{b}\\ 1\end{pmatrix}=\begin{pmatrix}\mathbf{A}\mathbf{x}^1\\ 1\end{pmatrix}$$

In other words, the column $\begin{pmatrix}\mathbf{A}\mathbf{x}^1\\ 1\end{pmatrix}$ is linearly dependent on the other columns in the basis. ■

When the integer programming dual relaxation problem (8.18) fails to solve the integer programming problem (8.13) from which it is derived, Theorem 8.6 tells us it is because there are two or more zero-one solutions that are infeasible in (8.13) but that, when taken in convex combination, is feasible in (8.13). Moreover, the convex combination achieves a cost z that is no greater than the cost v of an optimal solution in (8.13). The dual analysis can continue, however, by strengthening the dual problem by recalling that we would like to impose integrality restrictions on the variables in (8.18).

THEOREM 8.7. Suppose the integer programming dual relaxation problem (8.18) fails to solve the integer programming problem (8.13) because, according to Theorem 8.6, the optimal solution to (8.18) found by the simplex method has more than one positive variable; say $\bar{\lambda}_t > 0$ for $t = 1, \ldots, K$, $\bar{\lambda}_t = 0$ for $t = K+1, \ldots, T$, where $K \geq 2$. Then the columns $\begin{pmatrix} \mathbf{Ax} \\ 1 \end{pmatrix}, \ldots, \begin{pmatrix} \mathbf{Ax}^K \\ 1 \end{pmatrix}$ can be used to construct a homomorphism σ from Z^m into a finite abelian group $\mathcal{H}$ such that

$$\sum_{j=1}^{n} \sigma(\mathbf{a}_j) x_j^t \neq \sigma(\mathbf{b}) \qquad \text{for } t = 1, \ldots, K$$

PROOF. The columns $\begin{pmatrix} \mathbf{Ax}^1 \\ 1 \end{pmatrix}, \ldots, \begin{pmatrix} \mathbf{Ax}^K \\ 1 \end{pmatrix}$ form part of an optimal basis for the linear system in (8.18). By Theorem 8.1, this basis determines a finite abelian group $\mathcal{H}$ and a homomorphism σ from Z^{m+1} onto $\mathcal{H}$ with the property that

$$\sigma\begin{pmatrix} \mathbf{Ax}^t \\ 1 \end{pmatrix} = 0 \qquad \text{for } t = 1, \ldots, K$$

Since the basic feasible solution corresponding to this basis is not integer, we also have

$$\sigma\begin{pmatrix} \mathbf{b} \\ 1 \end{pmatrix} \neq 0$$

We have for any $t = 1, \ldots, K$,

$$\sigma\begin{pmatrix} \mathbf{Ax}^t \\ 1 \end{pmatrix} = \sum_{i=1}^{m} \sigma(\mathbf{e}_i) \sum_{j=1}^{n} a_{ij} x_j^t + \sigma(\mathbf{e}_{m+1}) = 0$$

and

$$\sigma\begin{pmatrix} \mathbf{b} \\ 1 \end{pmatrix} = \sum_{i=1}^{m} \sigma(\mathbf{e}_i) b_i + \sigma(e_{m+1}) \neq 0$$

Rearranging the orders of summation, there results

$$\sum_{j=1}^{n} \left[\sum_{i=1}^{m} \sigma(\mathbf{e}_i) a_{ij} \right] x_j^t = -\sigma(e_{m+1})$$

$$\neq \sum_{i=1}^{m} \sigma(\mathbf{e}_i) b_i$$

Thus, if we define the homomorphism σ from Z^m into $\mathcal{H}$ by $\sigma(w)=\sum_{i=1}^{m}\sigma(\mathbf{e}_i)w_i$, we obtain immediately that for any $t=1,\ldots,K$

$$\sum_{j=1}^{n}\sigma(\mathbf{a}_j)x_j^t\neq\sigma(\mathbf{b}) \quad \blacksquare$$

The implication of Theorem 8.7 to strengthening the integer programming dual problem is clear. We restrict the set over which the Lagrangean is optimized to the set

$$X'=\left\{\mathbf{x}\,|\,\sum_{j=1}^{n}\phi(\mathbf{a}_j)x_j=\phi(\mathbf{b}),\ \sum_{j=1}^{n}\sigma(\mathbf{a}_j)x_j=\sigma(\mathbf{b}),\text{ and }x_j=0\text{ or }1\text{ for all }j\right\}$$

The new Lagrangean Z' and the new integer programming dual problem are stronger because, by construction, we have eliminated all the $\mathbf{x}^t\in X$ satisfying

1. $Z(\mathbf{u})=-\mathbf{ub}+(\mathbf{c}+\mathbf{uA})\mathbf{x}^t$ for $\mathbf{u}$ optimal in the integer programming dual problem, and
2. $\lambda_t>0$ for the column derived from the infeasible solution $\mathbf{x}^t$ in the integer programing dual relaxation problem.

In strengthening the dual, the size of the group over which the Lagrangean is minimized has increased from $|\mathcal{G}|$ to $|\mathcal{G}|\cdot|\mathcal{H}|$ which, in general, is the price we have to pay. Sometimes, however, the new group equation $\sum_{j=1}^{n}\sigma(\mathbf{a}_j)x_j=\sigma(\mathbf{b})$ implies the old group equation $\sum_{j=1}^{n}\phi(\mathbf{a}_j)x_j=\phi(\mathbf{b})$ and the latter can be dropped.

We formalize the construction by the following iterative method.

Iterative Integer Programming Dual Method

Step 0. Let $\mathbf{A},\mathbf{b},\mathbf{c}$ be the data for the integer programming problem (8.13) where $\mathbf{A}$ is $m\times n$ and $\mathbf{A}$, $\mathbf{b}$, and $\mathbf{c}$ have integer coefficients. Let $\mathcal{G}^0$ be an arbitrary finite abelian group and ϕ^0 a homomorphism from Z^m onto $\mathcal{G}^0$. Go to Step 1 with $l=0$.

Step 1. Define the set
The Lagrangean

$$Z^l(\mathbf{u})=-\mathbf{ub}+\min_{\mathbf{x}\in X^l}(\mathbf{c}+\mathbf{uA})\mathbf{x}$$

where

$$X^l=\left\{\mathbf{x}\,|\,\sum_{j=1}^{n}\phi^l(\mathbf{a}_j)x_j=\phi^l(\mathbf{b}),\ x_j=0\text{ or }1\text{ for all }j\right\}=\{\mathbf{x}^t\}_{t=1}^{T^l}$$

and the integer programming dual problem

$$z^l=\max Z^l(\mathbf{u})$$
$$\text{s.t. }\mathbf{u}\in R^m$$

which we effectively solve by using the simplex method to solve the integer programming dual relaxation problem

$$\begin{aligned} z^l = \max \sum_{t=1}^{T^l} (\mathbf{c}\mathbf{x}^t)\lambda_t \\ \text{s.t.} \sum_{t=1}^{T^l} (\mathbf{A}\mathbf{x}^t)\lambda_t = \mathbf{b} \\ \sum_{t=1}^{T^l} \lambda_t = 1 \\ \lambda_t \geq 0 \end{aligned} \tag{8.19}$$

If $z^l = +\infty$, then the integer programming problem (8.13) is infeasible and the method terminates. Otherwise, if only one λ_t is positive in the optimal solution to the integer programming dual relaxation problem found by the simplex method, then the corresponding zero-one $\mathbf{x}^t$ is optimal in (8.13). If more than one λ_t is positive in the optimal solution, then the corresponding zero-one $\mathbf{x}^t$ is optimal in (8.13). If more than one λ_t is positive in the optimal solution, then the corresponding zero-one $\mathbf{x}^t$ are infeasible in (8.13). In this case, go to Step 2.

Step 2. Construct a group $\mathcal{H}^l$ and a homomorphism σ^l from Z^m onto $\mathcal{H}^l$, which excludes the infeasible solutions found in Step 1. Add the group equation

$$\sum_{j=1}^{n} \sigma^l(\mathbf{a}_j)x_j = \sigma^l(\mathbf{b})$$

to the group equations defining X^l. The result is the new set $X^{l+1} \subsetneq X^l$, a new group $\mathcal{G}^{l+1} = \mathcal{G}^l \oplus \mathcal{H}^l$, and a homomorphism ϕ^{l+1} from Z^m onto $\mathcal{G}^{l+1}$. Return to Step 1.

THEOREM 8.8. The iterative integer programming dual method converges finitely to an optimal solution to the integer programming problem (8.13), or proves it is infeasible.

PROOF. The solution of each integer programming dual problem by solving (8.19) is finite because X^l is finite. If the optimal solution to this integer programming dual relaxation problem does not yield an optimal solution to (8.13), then the new finite set X^{l+1} defining the new integer programming dual problem has at least two fewer vectors than X^l. This follows from the construction of Theorem 8.7. Since X^0 is finite, the reduction process must terminate finitely either with an optimal solution to (8.13) or prove that (8.13) is infeasible because we find $z = +\infty$. ■

THEOREM 8.9. The integer programming dual relaxation problem (8.18) is equivalent to the linear programming problem

$$d = \min \mathbf{cx}$$
$$\text{s.t. } \mathbf{x} \in \{\mathbf{x} | \mathbf{Ax} = \mathbf{b},\ 0 \le x_j \le 1 \text{ for all } j\} \cap [X]^c \tag{8.20}$$

where $[X]^c$ denotes the convex hull of X.

PROOF. The two problems are equivalent in the sense that

$$\{(\mathbf{cx}, \mathbf{x}) | \mathbf{x} \text{ feasible in (8.20)}\} = \left\{ (\mathbf{cy}, \mathbf{y}) | \mathbf{y} = \sum_{t=1}^{T} x^t \lambda_t, \right.$$
$$\left. \lambda_t \text{ feasible in (8.18)} \right\} \tag{8.21}$$

Suppose $\tilde{\mathbf{x}} \in \{\mathbf{x} | \mathbf{Ax} = \mathbf{b},\ 0 \le x_j \le 1\} \cap [X]^c$. The fact that $\tilde{\mathbf{x}} \in [X]^c$ implies that there exists a convex combination λ_t for $t = 1, \ldots, T$, such that $\tilde{\mathbf{x}} = \sum_{t=1}^{T} \mathbf{x}^t \tilde{\lambda}_t$. Since $\mathbf{A}\tilde{\mathbf{x}} = \mathbf{b}$, we have $\sum_{t=1}^{T} (\mathbf{Ax}^t)\tilde{\lambda}_t = \mathbf{b}$ and therefore the convex combination is feasible in (8.18). Thus, the left-hand set in (8.21) is included in the right-hand set. Conversely, if $\tilde{\mathbf{y}} = \sum_{t=1}^{T} \mathbf{x}^t \tilde{\lambda}_t$ for $\tilde{\lambda}_t$ feasible in (8.18), then $\tilde{\mathbf{y}}$ is in $[X]^c$ by definition. Moreover, $\mathbf{A}\tilde{\mathbf{y}} = \mathbf{b}$ and $0 \le \tilde{y}_j \le 1$. This establishes that the right-hand set in (8.21) is included in the left-hand set. ■

Theorem 8.9 tells us that the integer programming dual problem is effectively the linear programming problem that results if we intersect the feasible region of the ordinary linear programming relaxation of (8.13) with the convex polyhedron $[X]^c$. The extreme points of $[X]^c$ are zero-one vectors, but we may not find one that is optimal in (8.20) and therefore (8.13) because of the intersection.

Problem (8.20) can be shown to be the ordinary linear programming relaxation if the group $\mathcal{G}$ used in the construction of X is the trivial one consisting of the singleton $Z_1 = \{0\}$. In this case, the group equation is $\sum_{j=1}^{n} 0 \cdot x_j = 0$ and X is simply the set of all 2^n zero-one vectors x. Thus, $[X]^c = \{\mathbf{x} | 0 \le x_j \le 1\}$ and (8.20) reduces to the ordinary linear programming relaxation. The choice $\mathcal{G}^0 = Z_1$ in the iterative integer programming dual method is an attractive one from an implementation standpoint because then we can begin by solving the ordinary linear programming relaxation using the simplex method directly rather than by column generation via generalized linear programming. Even if the group $\mathcal{G}$ used in defining X is nontrivial but arbitrarily chosen, it may be that $\{\mathbf{x} | \mathbf{Ax} = \mathbf{b},\ 0 \le x_j \le 1\} \subseteq [X]^c$ in which case we do not cut off any of the feasible region of the ordinary linear programming relaxation by intersecting it with $[X]^c$. It is for this reason that we need to use the group theoretic procedures of Section 8.3 in conjunction with integer programming dual relaxation problem to select the first nontrivial group in the iterative method.

The polyhedron $[X]^c$ has been studied in its own right because its faces can be used as cuts in the cutting plane method discussed in Section 2.7. These cuts are related to the dual constructs in the following way. For any $\mathbf{u} \in R^m$, the inequality

$$(\mathbf{c}+\mathbf{uA})\mathbf{x} \geq Z(\mathbf{u})+\mathbf{ub} \tag{8.22}$$

is valid for any $\mathbf{x} \in X$ because of the definition of the Lagrangean $Z(\mathbf{u})$. Since X contains all the feasible solutions to (8.13), the inequality can be added to any linear programming relaxation of (8.13). For example, for any $\mathbf{u} \in R^m$, consider the linear programming relaxation

$$\begin{aligned} l = \min\ & \mathbf{cx} \\ \text{s.t. } & \mathbf{Ax} = \mathbf{b} \\ & (\mathbf{c}+\mathbf{uA})\mathbf{x} \geq Z(\mathbf{u})+\mathbf{ub} \\ & 0 \leq x_j \leq 1 \text{ for all } j \end{aligned} \tag{8.23}$$

It is easy to see that $l \geq Z(\mathbf{u})$ because for any feasible $\tilde{\mathbf{x}}$, we have

$$(\mathbf{c}+\mathbf{uA})\tilde{\mathbf{x}} \geq Z(\mathbf{u})+\mathbf{ub}$$

implying

$$\mathbf{c}\tilde{\mathbf{x}} \geq Z(\mathbf{u})+\mathbf{u}(\mathbf{b}-\mathbf{A}\tilde{\mathbf{x}}) = Z(\mathbf{u})$$

Since $\tilde{\mathbf{x}}$ could be optimal, we can conclude that $l \geq Z(\mathbf{u})$. Thus, the strongest cut (8.22) to add in terms of producing the highest value of the objective function in (8.23) is one written with respect to $\mathbf{u}$ that is optimal in the integer programming dual problem (8.17). To derive the strongest cut, we need to solve the dual problem which to a large extent makes the use of the cut superfluous since in so doing, we already know $Z(\mathbf{u})$ and any optimal solution to the integer programming dual relaxation problem (8.18) is optimal in (8.23).

The inequality (8.22) is a supporting hyperplane of the polyhedron $[X]^c$, but it is not a face of $[X]^c$ when there are an insufficient number of $\mathbf{x}^t \in X$ satisfying $Z(\mathbf{u}) = -\mathbf{ub}+(\mathbf{c}+\mathbf{uA})\mathbf{x}^t$. Formally, a supporting hyperplane $\Pi\mathbf{x} \geq \Pi_0$ of $[X]^c$ is a face if there are q linearly independent zero-one n-vectors $\mathbf{x}^t$ such that $\Pi\mathbf{x}^t = \Pi_0$ where $q \leq n$ is the dimension of $[X]^c$. Given these $\mathbf{x}^t$, a face can be derived by setting $\Pi_0 = 1$ and then setting the n-vector $\Pi = \mathbf{P}^T(\mathbf{PP}^T)^{-1}\mathbf{0}_q$, where $\mathbf{P}$ is the $q \times n$ matrix with columns $\mathbf{x}^t$, and $\mathbf{0}_q$ is the q-vector of ones. Thus, if we have $k < q$ linearly independent points satisfying $\Pi(\mathbf{u})\mathbf{x}^t = 1$, where

$$\Pi(\mathbf{u}) = \frac{1}{Z(\mathbf{u})+\mathbf{ub}}(\mathbf{c}+\mathbf{uA})$$

from the inequality or cut (8.22), we can try to extend it to a number of cuts by perturbing the Π_j's to discover $q-k$ more linearly independent $\mathbf{x}^t$. This process is not unique and therefore a number of faces will in general be produced from a single cut of the form (8.22).

EXAMPLE 8.8. We illustrate the principles of integer programming duality applied to the integer programming problem

$$\begin{aligned} v = \min\ & -4x_1 - 7x_2 + 2x_3 + 4x_4 - x_5 + 8x_6 + 2x_7 \\ \text{s.t.}\ & 2x_1 - x_2 + x_3 + 3x_4 - 2x_5 + 2x_6 - 2x_7 = 1 \\ & -4x_1 - 3x_2 + 2x_3 + x_4 + x_5 + x_6 - 2x_7 = -4 \\ & x_j = 0 \text{ or } 1 \text{ for all } j \end{aligned} \tag{8.24}$$

The zeroth integer programming dual problem is derived from this problem by taking $\mathcal{G}^0 = Z_1 = \{0\}$ and ϕ^0 is the trivial mapping of integer vectors in Z^2 into $0 \in \mathcal{G}^0$. The integer programming dual problem induced by $\mathcal{G}^0$ and ϕ^0 is the ordinary linear programming relaxation of (8.24) obtained by replacing $x_j = 0$ or 1 by $0 \le x_j \le 1$. The integer programming dual relaxation problem (8.18), which we use to solve the zeroth dual problem has, in principle, 2^7 columns but far fewer will be generated, the exact number depending on the columns with which we start.

An optimal solution to the zeroth dual problem is obtained by taking a convex combination of the two solutions $x_1^1 = x_2^1 = x_3^1 = x_4^1 = 1$ and $x_1^2 = x_2^2 = x_3^2 = x_5^2 = 1$ (we will not mention explicitly the variables equal to zero). The corresponding columns for the integer programming dual relaxation problem (8.18) are therefore

$$\mathbf{c}\mathbf{x}^1 = -5 \qquad\qquad \mathbf{c}\mathbf{x}^2 = -10$$

$$\begin{pmatrix} \mathbf{A}\mathbf{x}^1 \\ 1 \end{pmatrix} = \begin{bmatrix} 5 \\ -4 \\ 1 \end{bmatrix} \qquad \begin{pmatrix} \mathbf{A}\mathbf{x}^2 \\ 1 \end{pmatrix} = \begin{bmatrix} 0 \\ -4 \\ 1 \end{bmatrix}$$

Thus, the zeroth dual is solved by solving

$$\begin{aligned} z_2^0 = \min & -5\lambda_1 - 10\lambda_2 \\ & 5\lambda_1 + 0\lambda_2 = 1 \\ & -4\lambda_1 - 4\lambda_2 = -4 \\ & \lambda_1 + \lambda_2 = 1 \\ & \lambda_1 \ge 0, \lambda_2 \ge 0 \end{aligned} \tag{8.25}$$

This linear programming problem has optimal objective function value $z_2^0 = -9$, $\bar{\lambda}_1 = \frac{1}{5}$, $\bar{\lambda}_2 = \frac{4}{5}$ and optimal dual variables $(\bar{u}_1, \bar{u}_2) = (-1, -1)$. The dual variable on the convexity row is $\bar{\theta} = 6$. The fact that (8.25) solves the zeroth dual problem is verified by pricing out the seven activities in (8.25) and setting $x_j = 0$ or 1 depending on whether the reduced cost $c_j + u_1 a_{1j} + ua_{2j}$ is positive or negative. The reader can verify that no new columns for (8.25) are indicated at this value of $\bar{\mathbf{u}}$.

We use the Smith reduction scheme given in Section 8.3 to diagonalize the optimal basis for (8.26),

$$\begin{bmatrix} 0 & 5 & 0 \\ 1 & -4 & -4 \\ 0 & 1 & 1 \end{bmatrix}$$

where the first column is an artificial column required to make up a basis. This basis induces the group Z_5 and the homomorphism σ^1 from Z^3 to Z_5 given by $\sigma^1(\mathbf{e}_1) = 1$, $\sigma^1(\mathbf{e}_2) = \sigma^1(\mathbf{e}_3) = 0$. We use this homomorphism to construct a zero-one group optimization problem restricting the set of zero-one solutions to a nontrivial congruence. Specifically, the Lagrangean calculation is

$$\begin{aligned} Z^1(u_1, u_2) = & -u_1 + 4u_2 \\ & + \min(-4 + 2u_1 - 4u_2)x_1 + (-7 - u_1 - 3u_2)x_2 + (2 + u_1 + 2u_2)x_3 \\ & \quad + (4 + 3u_1 + u_2)x_4 + (-1 - 2u_1 + u_2)x_5 + (8 + 2u_1 + u_2)x_6 \\ & \quad + (2 - 2u_1 - 2u_2)x_7 \end{aligned}$$

$$\text{s.t. } 2x_1 + 4x_2 + x_3 + 3x_4 + 3x_5 + 2x_6 + 3x_7 \equiv 1 \pmod 5 \tag{8.26}$$
$$x_j = 0 \text{ or } 1 \text{ for all } j$$

The value $Z^1(-1, -1) = -8$ corresponding to the optimal solution in (8.26) given by $x_1 = 1$, $x_2 = 1$; this is higher than the optimal solution value in the zeroth dual $Z^0(-1, -1) = -9$.

We maximize $Z^1(u_1, u_2)$ by solving the integer programming dual relaxation problem for this first dual problem. The optimal solution is obtained by taking a convex combination of the two solutions $x_2^3 = x_3^3 = x_4^3 = x_5^3 = 1$ and $x_1^4 = x_2^4 = 1$. The corresponding columns are

$$\mathbf{cx}^3 = -2 \qquad \mathbf{cx}^4 = -11$$

$$\begin{pmatrix} \mathbf{Ax}^3 \\ 1 \end{pmatrix} = \begin{bmatrix} 1 \\ 1 \\ 1 \end{bmatrix} \qquad \begin{pmatrix} \mathbf{Ax}^4 \\ 1 \end{pmatrix} = \begin{bmatrix} 1 \\ -7 \\ 1 \end{bmatrix}$$

and the optimal basis is

$$\begin{bmatrix} 1 & 1 & 1 \\ 0 & 1 & -7 \\ 0 & 1 & 1 \end{bmatrix}$$

where the first column is again an artificial added to make up a basis. The corresponding optimal dual vector is $(\bar{u}_1^1, \bar{u}_2^1) = (0, -\frac{9}{8})$ and the maximal dual lower bound is $z^1 = -7\frac{5}{8}$. The cut (8.23) written with respect to this dual problem and dual solution is

$$\tfrac{4}{8}x_1 - \tfrac{29}{8}x_2 - \tfrac{2}{8}x_3 + \tfrac{23}{8}x_4 - \tfrac{17}{8}x_5 + \tfrac{55}{8}x_6 + \tfrac{34}{8}x_7 \geq \tfrac{-25}{8}$$

The basis induces the group Z_8 and the homomorphism σ^2 from Z^3 to Z_8 given by $\sigma^2(\mathbf{e}_1)=0$, $\sigma^2(\mathbf{e}_2)=1$, $\sigma^2(\mathbf{e}_3)=-1$. We add the congruence

$$4x_1 + 5x_2 + 2x_3 + x_4 + x_5 + x_6 + 6x_7 \equiv 4 \pmod 8$$

to the previous one. In other words, the new Lagrangean Z^2 is defined over the abelian group $Z_5 \oplus Z_8$. The second dual problem has an optimal solution $(\bar{u}_1^2, \bar{u}_2^2) = (0,0)$ with the corresponding solution $x_1^5 = x_3^5 = x_7^5 = 1$ optimal in the Lagrangean. This solution is optimal in (8.24) because the global optimality conditions are satisfied. ▲

8.6 BRANCH AND BOUND FOR INTEGER PROGRAMMING

Branch and bound is a method guaranteed to find an optimal solution to the integer programming problem (8.13), or establish that it is infeasible, by an implicitly exhaustive and nonredundant search of the 2^n zero-one solutions. Branch and bound is complementary to the integer programming dual methods discussed in the previous section because it uses the dual methods to limit and direct the search. Conversely, branch and bound can be viewed as a method for perturbing an integer programming problem and any one of its duals if the dual fails to yield an optimal solution to the primal by establishing the global optimality conditions for some primal-dual pair.

Traditional branch and bound methods for integer programming use the ordinary linear programming relaxation of (8.13) which, as we have seen, is just the first in a series of increasingly strong linear programming relaxations. Considerable implementation expertise is required to choose the correct balance between the extent of a branch and bound search and the extent of the dual analysis of each integer programming subproblem that arises during the search. The correct balance is largely a function of the specific problem structure that may only become apparent while the problem is being optimized.

There are many types of branch and bound schemes that have been proposed for integer programming under a number of different names. Our purpose here is simply to illustrate the idea with one realization of the approach. Branch and bound has also found use in a wide variety of other combinatorial optimization problems such as the traveling salesman problem which is discussed in the following section. Branch and bound can be used to solve the mixed integer programming problem (6.55) by straightforward modifications to the scheme presented in this section. Benders' algorithm for mixed integer programming discussed in Section 6.6 can be integrated with branch and bound (see Exercise 8.21).

Let P^0 denote the set consisting of the 2^n zero-one solutions to the integer programming problem (8.13). At any stage of computation, the least cost known solution $\hat{\mathbf{x}} \in P^0$ satisfying $\mathbf{A}\hat{\mathbf{x}} = \mathbf{b}$ is called the *incumbent* with *incumbent cost* $\hat{v} = \mathbf{c}\hat{\mathbf{x}}$. Branch and bound generates a sequence of subproblems

$$\begin{aligned} v(P^k) = \min\, & \mathbf{cx} \\ \text{s.t. } & \mathbf{Ax} = \mathbf{b} \\ & \mathbf{x} \in P^k \end{aligned} \tag{8.27}$$

where $P^k \subseteq P^0$. For example, P^k may be of the form $\{\mathbf{x} | x_{j_1} = 0, x_j = 0 \text{ or } 1, j \neq j_1\}$. A variable constrained to be zero or one in P^k is called a *fixed* variable; the other variables are called *free* variables. Another type of set used to define subproblems is $P^k = \{\mathbf{x} | x_{j_1} + x_{j_2} \leq 1, x_j = 0 \text{ or } 1\}$. In this case, we would incorporate the constraint $x_{j_1} + x_{j_2} \leq 1$ in any linear programming relaxation of (8.27). The set P^k is chosen in general to preserve the integer programming structure of the original problem (8.13).

The role of branch and bound as a perturbation scheme for integer programming dual problems can be seen from the results of Theorem 8.6. Specifically, suppose the integer programming dual relaxation problem (8.18) fails to solve the primal problem (8.13) because for $K \geq 2$, $\bar{\lambda}_t > 0$ for $t = 1, \ldots, K$ in (8.18). The corresponding solutions x^t are infeasible according to Theorem 8.6, and their columns in (8.18) must be linearly independent. This implies that there is at least one variable, say x_j, such that $x_j^{t_1} = 0$ and $x_j^{t_2} = 1$ for $\bar{\lambda}_{t_1} > 0$ and $\bar{\lambda}_{t_2} > 0$ in (8.18). If we were to separate (8.13) by creating two subproblems, one with $x_j = 0$ and one with $x_j = 1$, then the integer programming dual for each subproblem would no longer be optimized because for each dual relaxation problem there would be eliminated at least one column with $\bar{\lambda}_t > 0$.

If we can find an optimal solution to (8.27), then we have implicitly optimized all subproblems of the form (8.27) with P^k replaced by $P^l \subseteq P^k$ and such subproblems do not have to be explicitly enumerated. The same conclusion holds if we can ascertain that $v(P^k) \geq \hat{v}$ without actually discovering the precise value of $v(P^k)$. If either of these two cases obtain, then we say that the subproblem (8.30) has been *fathomed*. The latter case is called *fathoming by bound*. If the subproblem is not fathomed, then we *separate* (8.27) into new subproblems of the same form with P^k replaced by

P^l for $l=1,\ldots,L$, and

$$\bigcup_{l=1}^{L} P^l = P^k, \qquad P^{l_1} \cap P^{l_2} = \phi, \qquad l_1 \neq l_2$$

The new subproblems are called *descendants* of (8.27), and (8.27) is called the *ancestor* of the new subproblems.

The strategy in analyzing a given subproblem (8.27) is to try to fathom it by bound by solution of a dual problem

$$\begin{aligned} z(P^k) = \max\ & Z(\mathbf{u}; P^k) \\ \text{s.t. } & \mathbf{u} \in R^m \end{aligned}$$

where

$$\begin{aligned} Z(\mathbf{u}; P^k) = -\mathbf{ub} + \min\{&(\mathbf{c} + \mathbf{uA})\mathbf{x}\} \\ \text{s.t. } \sum_{j=1}^{n} &\phi(\mathbf{a}_j) x_j = \phi(\mathbf{b}) \\ &\mathbf{x} \in P^k \end{aligned}$$

and ϕ is a homomorphism from Z^m onto a finite abelian group. We assume the simple constraints defining P^k can be easily incorporated into the group optimization algorithm. Figure 8.5 gives a typical branch and bound scheme which we discuss step by step.

Step 1. Often the initial subproblem list consists of only one subproblem corresponding to the given integer programming problem (8.13).

Step 2. The selection of the next subproblem to be analyzed can depend upon a variety of criteria. For example, each subproblem usually has associated with it a lower bound that it inherited from its immediate ancestor. The selected subproblem could be the one with the least lower bound on the assumption that it is the most attractive of the subproblems from an objective function point of view. Conversely, the subproblem with the greatest lower bound might be a better choice on the assumption that the lower bounds must be increased for an optimal or good feasible integer programming solution to be found. Another possibility is to choose the subproblem according to some user specified preference about the variables; for example, choose subproblem k before subproblem l if $x_j = 1$ in subproblem k and $x_j = 0$ in subproblem l.

Step 3. Construction of a dual problem is a trivial task if we wish to use the ordinary linear programming relaxation to try to fathom the subproblem. If a stronger dual problem is desired, for example, as the result of

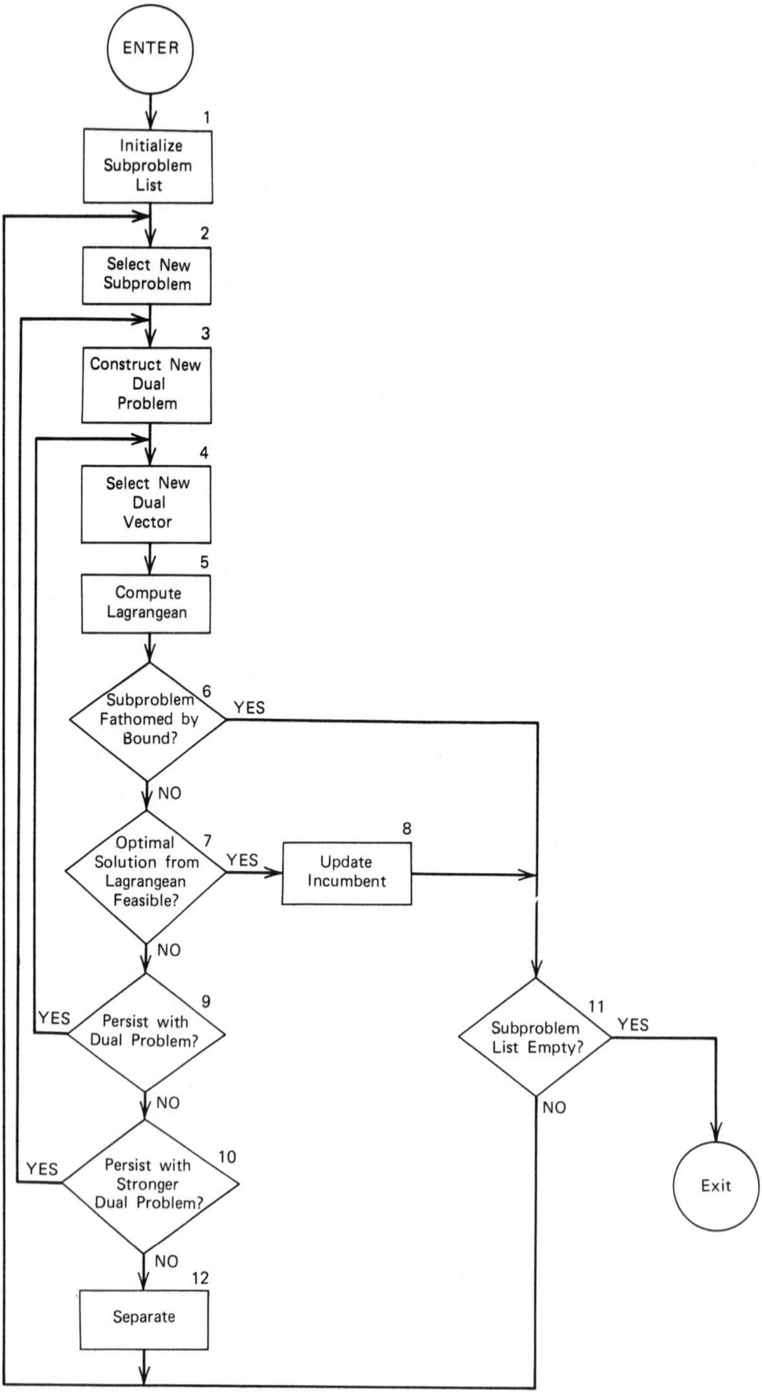

Figure 8.5. Branch and bound scheme.

previously unsuccessful dual analysis, then we use the group theoretic procedures such as the Smith reduction in this step. Included in this step are various procedures for controlling the size of the groups used in the dual construction.

Step 4. Each time this step is reached there are a number of options for selecting the new dual vector. One method is to solve a generalized linear programming master problem (see Section 5.4) using columns previously generated for the given subproblem and columns generated from solutions to other subproblems that are admitted in the given subproblem. The solution of the master can be a great deal of work just to select a new dual vector and an attractive alternative is to use the subgradient optimization method discussed in Section 6.4. A hybrid approach is subgradient optimization followed by or mixed with the generalized primal-dual algorithm discussed in Section 6.5.

Step 5. The Lagrangean can be computed using the group optimization algorithms given in Section 8.4.

Step 6. One result of Step 5 is the lower bound $Z(\mathbf{u}; P^k)$ on the minimal objective function value on the subproblem $v(P^k)$. It should be clear that the subproblem is fathomed if $Z(\mathbf{u}; P^k) \geq \hat{v}$. Fathoming a subproblem by bound is usually a more frequent event than fathoming it by finding an optimal solution to it. For this reason the main thrust of the dual analysis is to obtain quickly good lower bounds and also to solve the dual by a method that provides monotonically increasing lower bounds. The subgradient optimization method is attractive along both these dimensions, although it can sometimes behave erratically. Note that we can use the incumbent cost $\hat{v}$ as a target value in the formula (6.13) for selecting the step length in subgradient optimization. If this value is fairly easy to attain by $Z(\mathbf{u}; P^k)$ for some $\bar{\mathbf{u}}$, then the subgradient optimization method has a good chance to find such a $\bar{\mathbf{u}}$. Conversely, if the target $\hat{v}$ is too high, say $\hat{v} > z(P^k)$, then the subgradient optimization method should oscillate, indicating the need either to construct a stronger dual problem for the given subproblem or to separate it.

Steps 7 and 8. Suppose for some $\bar{\mathbf{u}}$ that $Z(\bar{\mathbf{u}}; P^k) = -\bar{\mathbf{u}}\mathbf{b} + (\mathbf{c} + \bar{\mathbf{u}}\mathbf{A})\bar{\mathbf{x}}$ for $\bar{\mathbf{x}} \in P^k$ and $\bar{\mathbf{x}}$ satisfying $\mathbf{A}\bar{\mathbf{x}} = \mathbf{b}$. Then by the global optimality conditions, $\bar{\mathbf{x}}$ is optimal in the subproblem and it is fathomed. Since the subproblem was not fathomed by bound, we must have $Z(\bar{\mathbf{u}}; P^k) = \mathbf{c}\bar{\mathbf{x}} < \hat{v}$ and therefore the incumbent $\hat{\mathbf{x}}$ should be replaced by $\bar{\mathbf{x}}$ with new incumbent cost $\mathbf{c}\bar{\mathbf{x}}$. If the optimal zero-one solution $\bar{\mathbf{x}}$ in the Lagrangean is infeasible, but not seriously so, then it may be appropriate to try some heuristic adjustment of it to try to achieve a good feasible solution that is better than the incumbent.

Step 9. The decision to persist with the dual problem is definitely negative if we discover that the dual problem has been optimized. More generally, it may be that the dual analysis is proving ineffectual, in which case it should be abandoned prior to optimizing it. The latter case is manifested by the absence of significant increase in the dual lower bounds during several iterations through Step 5.

Step 10. The decision to replace a given dual problem by a stronger one depends on empirical evidence such as the general effectiveness of dual analysis on the various subproblems tested thus far and the size of the group optimization problems that are likely to result.

Step 11. The subproblem list might have contained some subproblems before the last Lagrangean calculation that can now be eliminated because a new incumbent has been found. In addition, the lower bounds computed for the last subproblem might also be valid for some of the subproblems on the list which then become fathomed by bound. Sometimes it is worthwhile exiting although the subproblem list is not empty because the associated lower bounds on the unfathomed subproblems are sufficiently high to indicate that no great improvement over the incumbent will be found.

Step 12. The dual analysis should be used to help select the separation variable or variables in this step. For example, it is possible to compute lower bounds on the new dual bounds as the result of fixing a single free variable x_j to zero or one. These are called penalties (see Exercise 2.19). The separation could then be effected, for example, by creating two new subproblems with the previously free variable $x_k = 0$ and $x_k = 1$, which has the highest penalty. The greatest dual lower bound found for the subproblem just analyzed is valid for all of its descendants at the separation step.

EXAMPLE 8.9. We apply branch and bound to the numerical example (8.24) of Example 8.8 using the first dual with Lagrangean (8.26) to try to fathom subproblems. The results are given in Table 8.3. Throughout the table we use "x^*" to denote fixed values for the variables. Subgradient optimization is used to try to ascend in the dual problems; the relaxation parameter λ in formula (6.13) for picking the step length is fixed at the constant value 0.2. The initial incumbent cost is taken to be 10 without knowledge of any feasible solutions. This is a safe value for a fathoming bound that does not eliminate any feasible solutions. Five steps with the subgradient optimization algorithm are taken for each subproblem. If the subproblem is not fathomed as a result of the analysis, but the lower

TABLE 8.3 BRANCH AND BOUND EXAMPLE

u_1	u_2	$Z^1(u_1,u_2)$	x_1	x_2	x_3	x_4	x_5	x_6	x_7	γ_1	γ_2
Subproblem 0											
0	−1.13	−8	1	1	0	0	0	0	0	0	−3
0	−1.13	−8	0	1	1	1	1	0	0	0	5
Subproblem 1											
created from subproblem 0; lower bound = −8; incumbent cost = 10											
0	−1.13	−8	1*	1	0	0	0	0	0	0	−3
0	−2.33	−10.00	1*	0	1	0	1	0	0	0	3
0	−1.00	−8	1*	1	0	0	0	0	0	0	−3
0	−2.20	−9.6	1*	0	1	0	1	0	0	0	3
0	−0.90	−8.3	1*	1	0	0	0	0	0	0	−3
separated to create subproblems 3, 4											
Subproblem 2											
created from subproblem 0; lower bound = −8; incumbent cost = 10											
0	−1.13	−8	0*	1	1	1	1	0	0	0	5
0	−0.40	−4.8	0*	1	1	0	1	0	1	−5	2
−0.50	−0.20	−3.0	0*	1	1	1	1	0	0	0	5
−0.50	0.30	1.1	0*	1	1	0	1	0	1	−5	2
−0.86	0.42	−0.1	0*	1	1	1	1	0	0	0	5
separated to create subproblems 5, 6											
Subproblem 3											
created from subproblem 1; lower bound = −8; incumbent cost = 10											
0	−1.00	−6.00	1*	0	1*	0	1	0	0	0	3
0	0.06	−4.06	1*	0	1*	0	1	0	0	0	3
0	1.00	−5.0	1*	1	1*	1	1	0	1	0	1
0	4.0	−8.0	1*	1	1*	1	1	0	1	0	1
0	7.6	−11.6	1*	1	1*	1	1	0	1	0	1
separated to create subproblems 7, 8											
Subproblem 4											
created from subproblem 1; lower bound = −8; incumbent cost = 10											
0	−2.20	−4.4	1*	1	0*	0	0	0	0	0	−3
0	−3.16	−1.52	1*	1	0*	0	0	0	0	0	−3
0	−3.93	0.79	1*	1	0*	0	0	0	0	0	−3
0	−4.65	1.00	1*	0	0*	1	1	0	1	0	0
fathomed because optimal solution and new incumbent discovered; new incumbent cost = 1											
Subproblem 5											
created from subproblem 2; lower bound = 1.1; incumbent cost = 1											
			0*	0*							
fathomed by bound											

TABLE 8.3 (Continued)

u_1	u_2	$Z^1(u_1,u_2)$	x_1	x_2	x_3	x_4	x_5	x_6	x_7	γ_1	γ_2
Subproblem 6											
created from subproblem 2; lower bound = 1.1; incumbent cost = 1											
			0*	1*							
fathomed by bound											
Subproblem 7											
created from subproblem 3; lower bound = −4.06; incumbent cost = 1.											
0	.06	−2.82	1*	0*	1*	0	1	0	0	0	3
0	.31	−2.07	1*	0*	1*	0	1	0	0	0	3
0	.51	−1.47	1*	0*	1*	0	1	0	0	0	3
0	.67	−0.99	1*	0*	1*	0	1	0	0	0	3
0	.80	−0.60	1*	0*	1*	0	1	0	0	0	3
0	.92	−0.24	1*	0*	1*	0	1	0	0	0	3
0	1.00	0	1*	0*	1*	0	0	0	1	0	0
fathomed because optimal solution and new incumbent discovered; new incumbent cost = 0											
Subproblem 8											
created from subproblem 3; lower bound = −4.06; incumbent cost = 0.											
0	−1.0	−3	1*	1*	1*	1	1	0	1	0	−1
fathomed by bound because $Z^1(u_1,u_2) = -4 - u_2$ for all u_1, u_2.											

bounds are still increasing, then subgradient optimization is continued. Otherwise, the subproblem is separated to create two new subproblems each with an additional variable fixed at zero or one.

Subproblem 0 refers to the entire problem (8.24) and we have listed the two zero-one solutions, which, when taken in convex combination, are optimal in the dual relaxation problem for the first dual problem. Here we choose to branch on the variable x_1 rather than construct a stronger dual problem. The result is subproblems 1 and 2 where we have starred the fixed values of x_1. The initial lower bound associated with these two subproblems is -8 and we take $\mathbf{u} = (0, -1.13)$ as our initial starting dual solution. The ascent from this point to the new dual solution $(0, -2.33)$ is computed by the formula (6.12) using the incumbent cost as the target value; that is,

$$(0, -2.33) = (0, -1.13) + \frac{0.2[10-(-8)]}{9}(0, -3)$$

The value $Z^1(0, -2.33) = -10$ is computed by using the zero-one group optimization algorithm to solve (8.26) with x_1 fixed at a value of one. Note that subgradient optimization for both subproblems 1 and 2 oscillates between two distinct solutions. Neither subproblem is fathomed and the new subproblems 3, 4 and 5, 6 are created by fixing x_3 and x_2, respectively.

Subproblem 4 yields the first incumbent because $Z^1(0, -4.65)$ produces a feasible and therefore optimal solution in the subproblem. The new incumbent cost = 1 permits subproblems 5 and 6 to be fathomed by bound without further analysis. Subproblem 7 produces a better incumbent. Subproblem 8 is fathomed because there is only one solution to the zero-one group optimization problem (8.26) with x_1, x_2, x_3 fixed at one. This implies $Z^1(u_1, u_2) = -4 - u_2$ for all u_1, u_2. Thus, $Z^1(u_1, u_2)$ goes to $+\infty$ as u_2 goes to $-\infty$ implying there is no feasible solution to (8.24) with x_1, x_2, x_3 fixed at one. ▲

8.7 THE TRAVELING SALESMAN PROBLEM

In this section we consider a combinatorial optimization problem that can be formulated as an integer programming problem of immense size but can be solved in an efficient manner by exploiting some imbedded graph structure. The exploitation of this structure can be achieved by the same dual methods used in studying the integer programming problem.

A traveling salesman starts out at his home city and must visit $m-1$ other cities exactly once before returning home. His objective is to minimize the total distance traveled in making his *tour*. In graph theoretic terms, the traveling salesman problem is to find a minimal length simple cycle of m edges in a complete graph G with m nodes and lengths c_{ij} associated with the edges $\langle i,j \rangle$. The simple cycles in this graph are the tours. Note that there is an implied symmetry on the intercity distances; namely, $c_{ij} = c_{ji}$ for all i,j. In addition, we assume that the triangle inequality holds; namely, $c_{ik} \le c_{ij} + c_{jk}$ for all i,j,k. The requirement that the salesman visit each city exactly once is consistent with the goal of minimizing total distance traveled only if the triangle inequality holds. An arbitrary incomplete graph of m nodes, with associated edge lengths c_{ij}, can be extended to a complete graph with edge lengths obeying the triangle inequality by finding the shortest route path from each node i to every other node. The new edge length between i and j is the shortest route distance between i and j (see Chapter 4 for a discussion of and algorithms for this shortest route problem).

The traveling salesman problem is a combinatorial optimization problem studied by many authors. It arises as a substructure in a variety of mathematical programming models including vehicle routing and machine scheduling. For these problems the traveling salesman problem may be solved a number of times as a subproblem in a larger mathematical programming problem. In a vehicle routing problem from a central depot, for example, a traveling salesman problem needs to be solved each time the customers or stops assigned to each vehicle are specified. This assignment needs to be varied in a branch and bound fashion to discover an optimal assignment of customers to vehicles.

There are good heuristic algorithms for solving the traveling salesman problem, due in large part to the fact that feasible solutions are easy to generate and evaluate. The practical implication is that the convergent methods to be discussed below may not be required for a specific application such as the one discussed above. Nevertheless, the structures we identify and use in the convergent methods are constructs fundamental to the theory of combinatorial optimization and merit study in their own right.

It is clearly possible to solve the traveling salesman problem by a branch and bound scheme that implicitly enumerates and tests for optimality all the tours in a complete graph. The term "Branch and Bound" was in fact first used to describe such an approach to the problem using a simple assignment problem relaxation of the traveling salesman problem. Our development will focus rather on a formulation of the traveling salesman problem that permits a dual problem to be constructed that captures most of the primal problem structure.

The basic graph structure of the traveling salesman problem is closely related to trees.

Definition 8.2. A *1-tree* defined on a node set $1,\ldots,m$ is a graph consisting of a tree on the nodes $2,\ldots,m$, together with two edges connecting node 1 to the tree.

Every traveling salesman tour of the complete graph G is a 1-tree; but there are many 1-trees that are not tours. A tour is a 1-tree with the additional property that the degree at each node is 2. We use these properties to write a mathematical programming formulation of the problem.

Any subgraph of G can be defined by a zero-one vector $\mathbf{x}\in R^{\frac{m(m-1)}{2}}$ where $x_{ij}=1$ if $\langle i,j\rangle$ is included in the subgraph and $x_{ij}=0$ if $\langle i,j\rangle$ is not included. For any such $\mathbf{x}$, define the objective function

$$f(\mathbf{x})=\sum_{i=1}^{m-1}\sum_{j=i+1}^{m} c_{ij}x_{ij}$$

and for $i=1,\ldots,m$, define the degree function

$$g_i(\mathbf{x})=\sum_{j<i} x_{ji}+\sum_{i<j} x_{ij}$$

Let

$$T = \{\mathbf{x} | \mathbf{x} \text{ is a 1-tree}\}$$

Then the traveling salesman problem can be written as

$$\begin{aligned} v = \min & f(\mathbf{x}) \\ \text{s.t. } & g_i(\mathbf{x}) = 2 \qquad i = 1, \ldots, m \\ & \mathbf{x} \in T \end{aligned} \tag{8.28}$$

The traveling salesman dual problem is constructed by dualizing with respect to the complicating constraints $g_i(\mathbf{x}) = 2$. Specifically, for $\mathbf{u} \in R^m$, we define the Lagrangean

$$L(\mathbf{u}) = -2 \sum_{i=1}^{m} u_i + \min_{\mathbf{x} \in T} \left\{ f(\mathbf{x}) + \sum_{i=1}^{m} u_i g_i(\mathbf{x}) \right\}$$

and the traveling salesman dual problem is

$$\begin{aligned} d = \max & L(\mathbf{u}) \\ \text{s.t. } & \mathbf{u} \in R^m \end{aligned} \tag{8.29}$$

Note that the minimization in the Lagrangean is the problem of finding the minimum spanning 1-tree in the complete graph G where the edge lengths are $c_{ij} + u_i + u_j$. If an edge $\langle i,j \rangle$ is included in a 1-tree, then the degree at nodes i and j are increased by one implying the edge's contribution to total length is $c_{ij} + u_i + u_j$. A minimum spanning 1-tree can be found by a simple modification of the minimum spanning tree algorithm given in Section 3.8. Specifically, we compute a minimum spanning tree on the nodes $2, \ldots, m$ using the lengths $c_{ij} + u_i + u_j$ and connect node 1 to this tree by the two minimal length edges out of node 1.

Problem (8.29) is another example of a nondifferentiable concave optimization problem that can be solved by the methods of Chapter 6. Solving the traveling salesman dual (8.29) provides an optimal solution to the primal (8.28) if for any $\mathbf{u} \in R^m$, the minimum spanning 1-tree Lagrangean calculation produces a tour. In this case, the tour is optimal in (8.28), $\mathbf{u}$ is optimal in (8.29) and the objective functions are equal; namely, $d = v$. Once again there is no guarantee that $d = v$ and the traveling salesman dual will solve the primal. A global strategy for solving the primal is to use branch and bound with the dual problem providing bounds for fathoming in exactly the same manner as described in Section 8.6 for the integer programming problem. The traveling salesman dual also provides information useful in separating a subproblem, and this is discussed in an exercise at the end of the chapter.

An important final point about the traveling salesman problem is that it can be formulated as an integer programming problem, although not usefully solved that way, at least not directly. The implicit constraints

$\mathbf{x} \in T$ in the formulation (8.28) can be written as inequalities in integer variables. The result is

$$\min \sum_{i=1}^{m-1} \sum_{j=i+1}^{m} c_{ij} x_{ij} \tag{8.30a}$$

$$\text{s.t.} \sum_{j<i} x_{ji} + \sum_{i<j} x_{ij} = 2 \qquad i = 1, \ldots, m \tag{8.30b}$$

$$\sum_{\substack{i \in S \\ j \in S \\ i<j}} x_{ij} \leq |S| - 1 \text{ for any proper subset } S \subset \{2, 3, \ldots, m\} \tag{8.30c}$$

$$x_{ij} = 0 \text{ or } 1 \qquad \text{for all } \langle i,j \rangle \tag{8.30d}$$

The constraints (8.30c) are satisfied by the zero-one solutions $\mathbf{x}$ representing the 1-trees of the graph. The difficulty with this formulation is that there are on the order of 2^m constraints in (8.30c). Thus, a 50-city traveling salesman problem requires an astronomically large integer programming formulation. It should be clear that our previous formulation and the dual analysis offer great advantages because the 1-tree structure is dealt with directly in the Lagrangean optimization rather than as the very large system of inequalities (8.30c).

EXAMPLE 8.10. Table 8.4 gives the edge lengths for a 6-city traveling salesman problem.

We apply the subgradient optimization algorithm of Section 6.3 starting at the point $\mathbf{u}^1 = (0,0,0,0,0,0)$. The minimum spanning 1-tree with edge weights c_{ij} calculated by the greedy algorithm of Section 3.8 is shown in Figure 8.6. The 1-tree provides the lower bound $L(\mathbf{u}^1) = 29$ on the length of a minimal traveling salesman tour. This 1-tree is not a tour and we use the subgradient vector $\boldsymbol{\gamma}^1$ with components γ_i^1 = degree of 1-tree at node i minus 2, as the direction for subgradient optimization. For the 1-tree of Figure 8.6, $\boldsymbol{\gamma}^1 = (0,0,1,0,-1,0)$. The new dual solution selected in this direction is $\mathbf{u}^2 = (0,0,2,0,-2,0)$ and the new 1-tree calculation with edge lengths $c_{ij} + u_i^2 + u_j^2$ produces a lower bound $L(\mathbf{u}^2) = 28$. Note that the lower

TABLE 8.4 DISTANCES BETWEEN CITIES FOR A TRAVELING SALESMAN EXAMPLE

i	2	3	4	5	6
1	8	11	2	9	6
2		4	5	12	13
3			14	3	9
4				6	11
5					10

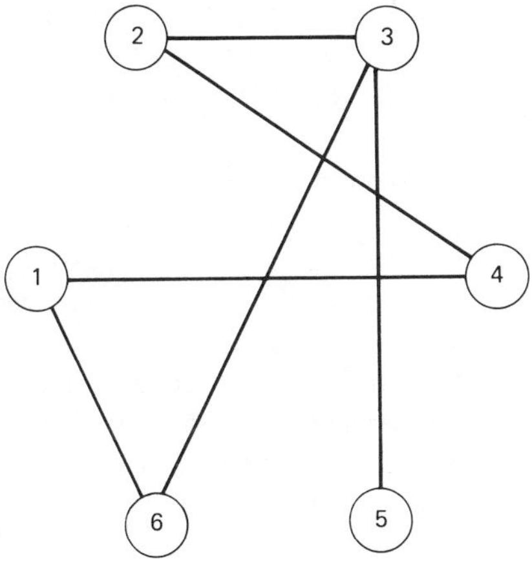

Figure 8.6. Traveling salesman example—minimum spanning 1-tree.

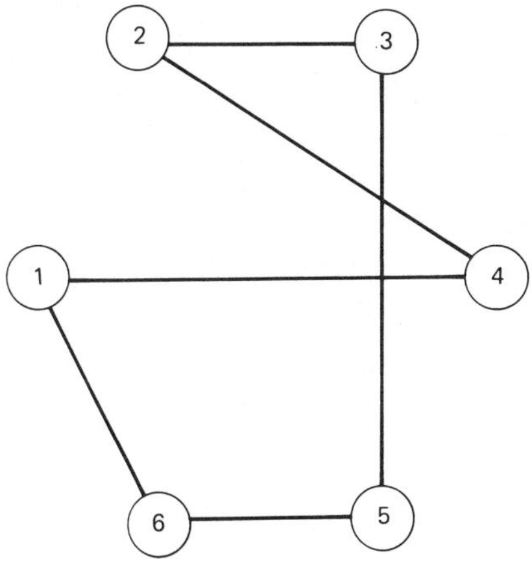

Figure 8.7. Traveling salesman example—minimal length tour.

bound did not increase above $L(\mathbf{u}^1)$. The new subgradient direction is $\gamma^2=(0,-1,-1,1,1,0)$ and the new dual solution $\mathbf{u}^3=(0,-1,1,1,-1,0)$. Calculation of $L(\mathbf{u}^3)$ produces the 1-tree shown in Figure 8.7. This 1-tree is a tour implying it is the minimal length tour with total length equal to 30 and the traveling salesman problem is solved. ▲

8.8 NETWORK SYNTHESIS

Generally speaking, a network synthesis problem is one in which we are given a set of nodes and required to connect the nodes by arcs with sufficient capacities to ensure that specified network flow requirements can be met. This is in contrast to the network optimization problems of Chapter 3 for which all the parameters of the problems were given and we were required simply to analyze the networks to find the minimal cost or maximal flow solution.

Specifically, suppose we are given m nodes representing m locations, and asked to provide connections between some of the nodes with sufficient capacity that the maximal flow between any pair of nodes k,l is at least r_{kl}. We assume the requirements are symmetric, namely $r_{kl}=r_{lk}$, and thus the connections between nodes are (undirected) edges. Associated with each edge $\langle i,j\rangle$ is a capacity q_{ij}, which may be zero, and a nonnegative unit cost c_{ij} for each unit of capacity installed. For convenience, we assume there is a complete set of edges on the m nodes, but edges with high unit cost c_{ij} will tend to have $q_{ij}=0$. We give a formulation of the problem which exploits the max-flow/min-cut Theorem 3.2.

Let $\mathbf{q}\in R^{\frac{m(m-1)}{2}}$ denote the vector of edge capacities to be determined. To ensure that the maximal flow possible from node k to node l is at least r_{kl}, the results of Section 3.5 tell us that we must require the cut capacity of every cut separating k and l to be at least r_{kl}. Let $\sigma=(R,Q)$ be any such cut of the node set $\mathfrak{N}=\{1,\dots,m\}$; that is, $R\cup Q=\mathfrak{N}$, $R\cap Q=\phi$, and $k\in R$, $l\in Q$. Let $\mathbf{p}_\sigma^{kl}\in R^{\frac{m(m-1)}{2}}$ be a zero-one vector with $p_{ij\sigma}^{kl}=1$ if $\langle i,j\rangle$ satisfies $i\in R$, $j\in Q$ or $j\in R$, $i\in Q$, and $p_{ij\sigma}^{kl}=0$ otherwise. Then the cut capacity of σ is $\mathbf{p}_\sigma^{kl}\cdot\mathbf{q}$. The set of all cuts separating k and l is denoted by $\rho(k,l)$. With this background, we can write the network synthesis problem as

$$\min \sum_{i=1}^{m-1}\sum_{j=i+1}^{m} c_{ij}q_{ij}$$

$$\begin{aligned}\text{s.t.}\quad \sum_{i=1}^{m-1}\sum_{j=i+1}^{m} p_{ij\sigma}^{kl}q_{ij}\ge r_{kl}\qquad &\text{for all } \sigma\in\rho(k,l)\\ k=1,\dots,m-1&\\ l=k+1,\dots,m&\\ q_{ij}\ge 0&\end{aligned}\tag{8.31}$$

The size of (8.31) can be reduced by considering an imbedded tree structure as shown by the following two theorems.

THEOREM 8.10. A necessary and sufficient condition that the nonnegative numbers F_{ij}, $i=1,\dots,m-1$, $j=i+1,\dots,m$, represent the maximal flows in the edges of a graph defined over m nodes is that

$$F_{ik}\ge\min(F_{ij},F_{jk})\qquad \text{for all } i,j,k \tag{8.32}$$

PROOF. 1. (Necessity). Suppose the F_{ik} are maximal flows and suppose to the contrary that $F_{ik} < \min(F_{ij}, F_{jk})$ for some i,j,k. By the max-flow/min-cut theorem, F_{ik} is equal to the minimal cut capacity of all cuts separating i and k. Let (R,Q) be such a minimal cut. If $j \in R$, then this cut also separates j and k implying the maximal flow between j and k is no greater than F_{ik} which in turn is less than the maximal flow F_{jk}, a contradiction. If the node $j \in Q$, then the same contradiction obtains on the maximal flow between i and j. Thus, if the F_{ik} are maximal flows, the inequalities (8.32) must hold.

2. (Sufficiency). Suppose the inequalities (8.32) hold. It is easy to show by induction that this implies

$$F_{ik} \geq \min(F_{ij}, F_{jl}, \ldots, F_{tk}) \tag{8.33}$$

for any path $\langle i,j,l,\ldots,t,k \rangle$ connecting node i to node k. To prove sufficiency, we must exhibit a graph with the maximal flows F_{ik}. Specifically, we will show that a maximal spanning tree of the m nodes, where the F_{ik} are taken as lengths, is the graph we seek. The capacity of the edge $\langle i,j \rangle$ in the tree equals F_{ij}. A maximal spanning tree can be computed by the greedy algorithm of Chapter 3 where the algorithm begins with the longest edge first and works down through the list of edges ordered by decreasing length. Let $\langle i,j,l,\ldots,t,k \rangle$ be any path in the maximal spanning tree found by the algorithm. We must have

$$F_{ik} \leq \min(F_{ij}, F_{jl}, \ldots, F_{tk}) \tag{8.34}$$

since otherwise, say $F_{ik} > F_{yz}$, we could replace the edge $\langle y,z \rangle$ in the tree by the edge $\langle i,k \rangle$ and have a spanning tree of greater length. The replacement produces a tree because the new graph consists of $m-1$ edges connecting m nodes. Comparing (8.33) and (8.34), we have for any path in the maximal spanning tree that

$$F_{ik} = \min(F_{ij}, F_{jl}, \ldots, F_{tk}) \tag{8.35}$$

Consider now the graph with $q_{ij} = F_{ij}$ for $\langle i,j \rangle$ in the maximal spanning tree, and $q_{ij} = 0$ otherwise. The maximal flow between any pair of nodes i and k is clearly determined by (8.35) since the path in the maximal spanning tree connecting i to k is unique. Thus, we have exhibited a graph with the maximal flows and sufficiency is established. ■

The maximal spanning tree used in the proof of Theorem 8.10 is the central construct to be used in reducing the size of the network synthesis problem (8.31).

THEOREM 8.11. Let r_{kl}, $k=1,\ldots,m-1$, $l=k+1,\ldots,m$, be given non-negative maximal flow requirements, and let τ denote a maximal spanning tree of the m nodes using the r_{kl} as edge lengths. Then the maximal flow requirements are satisfied for all node pairs if they are satisfied for the $m-1$ node pairs corresponding to the edges of the maximal spanning tree.

PROOF. Consider any node pair i,j such that $\langle i,j\rangle \in \tau$, and consider the unique path $\langle i,k,l,\ldots,t,j\rangle$ connecting i to j in τ. The maximal flow F_{ij} attainable between i and j satisfies

$$F_{ij} \geq \min(F_{ik}, F_{kl}, \ldots, F_{tj})$$

which in turn implies

$$F_{ij} \geq \min(r_{ik}, r_{kl}, \ldots, r_{tj}) \tag{8.36}$$

since the maximal flows are satisfied for edges in τ. On the other hand, since τ is a maximal spanning tree when the edge lengths are the flow requirements, we have

$$r_{ij} \leq \min(r_{ik}, r_{kl}, \ldots, r_{tj}) \tag{8.37}$$

by the same reasoning as in the previous theorem. The theorem is established by comparing (8.36) and (8.37). ■

Thus, the network synthesis problem (8.31) can be reduced by retaining only the constraints corresponding to node pairs which are edges of the tree τ. This gives us

$$\begin{aligned} \min \quad & \sum_{i=1}^{m-1} \sum_{j=i+1}^{m} c_{ij} q_{ij} \\ \text{s.t.} \quad & \sum_{i=1}^{m-1} \sum_{j=i+1}^{m} p_{ij\sigma}^{kl} q_{ij} \geq r_{kl} \qquad \text{for all } \sigma \in \rho(k,l) \\ & \qquad\qquad\qquad\qquad\qquad\quad \text{for all } \langle k,l\rangle \in \tau \\ & q_{ij} \geq 0 \end{aligned} \tag{8.38}$$

There is a remaining difficulty with this problem due to the very large number of rows. It is more convenient to solve the dual of (8.38) which is

$$\begin{aligned} \max \quad & \sum_{\langle k,l\rangle \in \tau} \sum_{\sigma \in \rho(k,l)} r_{kl} u_{kl\sigma} \\ \text{s.t.} \quad & \sum_{\langle k,l\rangle \in \tau} \sum_{\sigma \in \rho(k,l)} \mathbf{p}_{\sigma}^{kl} u_{kl\sigma} \leq \mathbf{c} \\ & u_{kl\sigma} \geq 0 \end{aligned} \tag{8.39}$$

The $m(m-1)/2$-vector $\mathbf{c}$ has components equal to the costs c_{ij}. Note that (8.39) has a feasible solution, and (8.38) a (bounded) optimal solution, only if $c_{ij} \geq 0$ for all i,j.

The formulation (8.39) has its number of rows equal to the number of edges and a large number of columns, one for each cut. These columns can be generated as needed by the generalized linear programming algorithm given in Section 5.4. The master linear programming problem at each iteration is (8.39) with a restricted number of columns $\mathbf{p}_\sigma^{kl}$. Let v_{ij} be the optimal dual variables on the constraints; these v_{ij} represent trial capacities on the edges. To test the optimality of the v_{ij}, or equivalently, to see if additional columns could be profitably added to (8.39), we solve for all $\langle k,l\rangle \in \tau$ the subproblem

$$\max_{\sigma \in \rho(k,l)} \left\{ r_{kl} - \sum_{i=1}^{m-1} \sum_{j=i+1}^{m} v_{ij} p_{ij\sigma}^{kl} \right\}$$

Ignoring the constant r_{kl} term, this reduces to

$$\min_{\sigma \in \rho(k,l)} \left\{ \sum_{i=1}^{m-1} \sum_{j=i+1}^{m} v_{ij} p_{ij\sigma}^{kl} \right\} \tag{8.40}$$

which is the problem of finding the minimal cut capacity between k and l when the edge capacities are v_{ij}. The minimal cut capacity equals the maximal flow, and the maximal flow algorithms of Section 3.5 also yield a minimal cut. If, for any $\langle k,l\rangle \in \tau$, the solution of (8.40) produces a cut $\sigma \in \rho(k,l)$ such that

$$\sum_{i=1}^{m-1} \sum_{j=i+1}^{m} v_{ij} p_{ij\sigma}^{kl} > r_{kl}$$

then the column $\mathbf{p}_\sigma^{kl}$ is added to the master linear programming problem and it is resolved. The implication when this occurs is that the v_{ij} are too large permitting the maximal flow between k and l to exceed r_{kl} and the cost must be too high. On the other hand, if

$$\sum_{i=1}^{m-1} \sum_{j=i+1}^{m} v_{ij} p_{ij\sigma}^{kl} \leq r_{kl}$$

for all $\sigma \in \rho(k,l)$ and all $\langle k,l\rangle \in \tau$, then the v_{ij} are the optimal capacities.

The network synthesis model just discussed has the important weakness that it does not consider simultaneous demands on the system. The maximal flow requirements are met for each node pair k,l on the assumption that the edge capacities will not be used by other node pairs. We wish to consider briefly the network synthesis problem that results when the requirements are to be met simultaneously. In formulating this problem, it is necessary to account for the direction of the flows. Thus, we consider a network defined on m nodes with a complete set of arcs (i,j) with capacities q_{ij} to be determined, which may not be symmetric, and a cost c_{ij}

per unit of capacity installed. Each node pair k,l is viewed as having a separate commodity to be transported through the network from k to l. The total requirement is r_{kl}, and we need not assume $r_{kl}=r_{lk}$. Let x_{ij}^{kl} denote the flow in the arc (i,j) of the commodity k,l.

The mathematical statement of this multi commodity flow problem is

$$\min \sum_{\substack{i=1 \\ i \neq j}}^{m} \sum_{j=1}^{m} c_{ij} q_{ij} \tag{8.41a}$$

$$\text{s.t.} \quad \left.\begin{aligned} &\sum_{\substack{j=1 \\ j \neq k}}^{m} x_{kj}^{kl} = r_{kl} \\ &\sum_{\substack{i=1 \\ i \neq l}}^{m} x_{il}^{kl} = r_{kl} \end{aligned}\right\} \quad \text{for all } k,l \tag{8.41b}$$

$$\sum_{\substack{j=1 \\ j \neq i}}^{m} x_{ij}^{kl} - \sum_{\substack{j=1 \\ j \neq i}}^{m} x_{ji}^{kl} = 0 \qquad \begin{array}{l} \text{for all } i \neq k,l \\ \text{for all } k,l \end{array} \tag{8.41c}$$

$$\sum_{l=1}^{m} \sum_{\substack{k=1 \\ k \neq l}}^{m} x_{ij}^{kl} - q_{ij} \leq 0 \qquad \begin{array}{c} \text{for all } i,j \\ i \neq j \end{array} \tag{8.41d}$$

$$x_{ij}^{kl} \geq 0 \qquad q_{ij} \geq 0$$

The constraints (8.41b) state that the requirements r_{kl} must be met by flow out of k and into l. The constraints (8.41c) preserve conservation of flow for each commodity at all intermediate nodes. The constraints (8.41d) state that the capacity in each arc (i,j) must be at least as great as the flow.

An important generalization of the multicommodity model (8.41) is to include a fixed cost as well as a variable cost on the installation of capacity. Specifically, this is a nonnegative cost K_{ij} that must be incurred if $q_{ij}>0$. Problem (8.41) can be extended to incorporate these fixed costs by the addition of a zero-one variable δ_{ij} for each arc (i,j). The objective function has the additional term

$$\sum_{\substack{i=1 \\ i \neq j}}^{m} \sum_{j=1}^{m} K_{ij} \delta_{ij}$$

and there are the additional constraints $q_{ij} - Q_{ij}\delta_{ij} \leq 0$ for all i and j, $i \neq j$, where Q_{ij} is an upper bound on the capacity to be installed in the arc (i,j). This extension of problem (8.41) is an order of magnitude more difficult to solve. The use of Benders' decomposition method discussed in Chapter 6 is an attractive approach to solving the resulting mixed integer programming problem.

8.9 EXERCISES

8.1 Let S_{ij} denote positive numbers for $i=1,\ldots,m$; $j=1,\ldots,m$. We wish to calculate x_i, $i=1,\ldots,m$, which are optimal in

$$\max \sum_{i,j} |x_i - x_j|$$
$$\text{s.t. } |x_i - x_j| \le S_{ij} \qquad \text{for all } i,j$$

Formulate this problem as an integer programming problem.

8.2 A set of words (ace, bc, dab, dfg, fe) is to be transmitted as messages. We wish to investigate the possibility of representing each word by one of the letters *in the word* such that the words will be uniquely represented. If such a representation is possible, we can transmit a single letter instead of a complete word. Formulate an integer programming problem which determines if such a representation is possible.

8.3 (1) For any $m \times m$ matrix **A** with integer coefficients, we denote by $d_p[\mathbf{A}]$ the greatest common divisor of all the $p \times p$ minors of **A**, provided they are not all zero. For $p=0$, we set $d_0[\mathbf{A}]=1$. Prove that for any matrix **A** of integers, $d_{p-1}[\mathbf{A}]$ divides $d_p[\mathbf{A}]$ for $p=1,2,\ldots,r$, where r is the largest integer for which the $r \times r$ minors are not all zero.

(2) Let **A** be a matrix of integers with rank r. Let $d_p[\mathbf{A}]$, $p=0,1,\ldots,r$ be the integers defined above. We denote the quotient of $d_p[\mathbf{A}]$ by $d_{p-1}[\mathbf{A}]$ by $q_p[\mathbf{A}]$. That is,

$$d_p[\mathbf{A}] = q_p[A]\, d_{p-1}[\mathbf{A}]$$

The integer $q_p[\mathbf{A}]$ is called the pth torsion order of **A**. We define $q_p[\mathbf{A}]$ to be zero for $p>r$. Prove that the integers $d_p[\mathbf{A}]$ and therefore the torsion orders $q_p[\mathbf{A}]$ do not change when **A** is transformed with respect to elementary row and column operations (see definition 8.1).

(3) Use the results of parts (1) and (2) to prove that the q_i defining the group $\mathcal{G}_{\mathbf{B}}$ in Theorem 8.1 are unique.

8.4 Show that the set of ω for which $v(\omega)=+\infty$ in (8.9) is a subgroup of the group $\mathcal{G}$ used in defining the unconstrained group optimization problem (8.8). Similarly, show that the set of ω for which $v(\omega)=-\infty$ is also a subgroup of $\mathcal{G}$.

8.5 Consider the collection of cuts (2.29) that can be derived from a linear programming tableau for the integer programming problem (2.26). Show that this collection forms an abelian group where we define the group operation "+" by

$$\left(\sum_{j=1}^{n} f_j x_j \geq f_0\right) \text{"+"} \left(\sum_{j=1}^{n} g_j x_j \geq g_0\right) = \left(\sum_{j=1}^{n} h_j x_j \geq h_0\right)$$

where

$$h_j \equiv f_j + g_j (\text{mod } 1) \qquad j=0,1,\ldots,n$$

Moreover, show that this group is isomorphic to a subgroup of the group $\mathcal{G}_{\mathbf{B}}$ of Theorem 8.1 where **B** is the basis defining the tableau.

8.6 Let x^* denote an optimal solution to the unconstrained group optimization problem (8.8) and suppose $\Pi_{j=1}^{n}(1+x_j^*) \geq D+1$. Show that the optimal solution is not unique.

8.7 (Gorry et al., 1972) Consider an arbitrary constraint

$$\sum_{j=1}^{n} a_{ij} x_j \leq b_i$$

where the variables x_j are required to take on nonnegative values. Let θ_i be an arbitrary positive number and construct the new constraint

$$\sum_{j=1}^{n} \left[\frac{a_{ij}}{\theta_i}\right] x_j \leq \left[\frac{b_i}{\theta_i}\right]$$

where [ ] denotes "integer part of"; that is, $[t]$ = largest integer $\leq t$. Show that the new constraint is a relaxation of the original constraint. In other words show that any nonnegative integer variables x_j satisfying the original constraint also satisfy the new constraint.

8.8 (Garfinkel and Nemhauser, 1972) The integer programming problem

$$\begin{aligned} \min \sum_{j=1}^{n} c_j x_j & \\ \text{s.t.} \sum_{j=1}^{n} a_{ij} x_j \geq 1 & \qquad i=1,\ldots,m \\ x_j = 0 \text{ or } 1 & \end{aligned}$$

where the c_j are positive and the a_{ij} are equal to zero or one is called a *covering problem* and denoted by *CP*. If the constraints are equalities instead of inequalities, then the problem is called a *set partitioning problem* and denoted by *PP*. Applications of these problems as well as their properties are discussed in Garfinkel and Nemhauser (1972). Let $\mathbf{r}_i$ denote row i of such a problem, and $\mathbf{a}_j$ denote column j.

(1) Establish the validity of the following assertions for the indicated problems

(i) (CP, PP) If $\mathbf{r}_i = \mathbf{e}_k$ (the kth unit vector) for some i, k, then $x_k = 1$ in every feasible solution, and $\mathbf{a}_k$ may be deleted. In addition, every row t such that $a_{kt} = 1$ may be deleted.

(ii) (PP) In addition to the deletions in (i), every column $q \neq k$ such that $a_{tq} = a_{ik} = 1$ for some t must be deleted.

(iii) (CP, PP) If $\mathbf{r}_t \geq \mathbf{r}_p$ for some t and p, then $\mathbf{r}_t$ may be deleted.

(iv) (PP) In addition to deleting row t in (iii), every column k such that $a_{tk} = 1$ and $a_{pk} = 0$ must be deleted.

(v) (PP, CP) If for some set of columns S and some column k, $\Sigma_{j \in S} \mathbf{a}_j = \mathbf{a}_k$ and $\Sigma_{j \in S} c_j \leq c_k$, column k may be deleted.

(vi) (CP) If for some set of columns S and some column k, $\Sigma_{j \in S} \mathbf{a}_j \geq \mathbf{a}_k$ and $\Sigma_{j \in S} c_j \leq c_k$, column k may be deleted.

(2) Let the zero-one n-vector $\mathbf{x}^*$ denote any feasible solution to CP. This solution is said to be *redundant* if $\mathbf{x}^* - \mathbf{e}_k$ is a feasible solution to CP for any k such that $x_k^* = 1$. A cover that is not redundant is called *prime*. Prove that, if we let $\mathbf{x}^*$ be any prime solution to CP, then $\mathbf{x}^*$ is an extreme point of the linear programming relaxation of CP obtained by replacing $x_j = 0$ or 1 by $0 \leq x_j \leq 1$.

(3) Prove that any optimal solution to CP is an extreme point of the linear programming relaxation of CP.

8.9 (Bradley et al., 1974) We are concerned with characterizations of equivalent inequalities (inequalities with identical sets of feasible solutions) to a given inequality

$$\sum_{j=1}^{n} a_j x_j \leq a_0 \qquad x_j = 0 \text{ or } 1 \tag{*}$$

(1) Show that we lose no generality in seeking equivalent inequalities to consider inequalities

$$\sum_{j=1}^{n} b_j x_j \leq b_0 \qquad x_j = 0 \text{ or } 1 \tag{**}$$

satisfying the conditions

(i) $b_j \geq 0, j = 1, \ldots, n$

(ii) All infeasible zero-one vectors $\mathbf{x}$ satisfy $\Sigma_{j=1}^{n} b_j x_j \geq b_0 + 1$

(iii) $b_1 \geq b_2 \geq \ldots \geq b_n$

Thus, we can assume the original inequality (*) satisfies these conditions.

(2) A zero-one n-vector $\mathbf{x}^*$ is called a *ceiling solution* of the inequality (*) if

(i) $\sum_{j=1}^n a_j x_j^* \leq a_0$.
(ii) If $x_k^* = 0$, then $\sum_{j=1}^n a_j x_j^* + a_k \geq a_0 + 1$.
(iii) If $x_k^* = 0$ and $x_{k+1}^* = 1$, then $\sum_{j=1}^n a_j x_j^* + a_k - a_{k+1} \geq a_0 + 1$.

The set $\{k | x_k^* = 1\}$ is called a *ceiling* of (*). Prove that if every ceiling solution of (*) is a feasible solution for (**), then every feasible solution of (*) is a feasible solution of (**) [i.e., (**) is a relaxation of (*)].

(3) A zero-one n-vector $\mathbf{x}^*$ is called a *roof solution* of (*) if

(i) $\sum_{j=1}^n a_j x_j^* \geq a_0 + 1$.
(ii) If $x_k^* = 1$, then $\sum_{j=1}^n a_j x_j^* - a_k \leq a_0$.
(iii) If $x_k^* = 1$ and $x_{k+1}^* = 0$, then $\sum_{j=1}^n a_j x_j^* - a_k + a_{k+1} \leq a_0$.

The set $\{k | x_k^* = 1\}$ is called a roof of (*). Prove that if every roof solution of (*) is an infeasible solution for (**), then every infeasible solution of (*) is an infeasible solution of (**) [i.e., (**) is a restriction of (*)].

(4) Let $\mathcal{C}$ be the collection of ceilings of (*) and $\mathcal{R}$ the collection of roofs. Prove that an inequality (**) is equivalent to (*) if and only if

$$\sum_{j \in C} b_j \leq b_0 \qquad \text{for all } C \in \mathcal{C}$$

$$\sum_{j \in R} b_j \geq b_0 + 1 \qquad \text{for all } R \in \mathcal{R}$$

$$b_1 \geq b_2 \geq \ldots \geq b_n \geq 0$$

8.10 (Cornuejols et al., 1977) Consider the integer programming problem

$$v = \max \sum_{i=1}^{m} \sum_{j=1}^{n} c_{ij} x_{ij} - \sum_{j=1}^{n} d_j y_j \tag{i}$$

$$\text{s.t.} \sum_{j=1}^{n} x_{ij} = 1 \qquad i = 1, \ldots, m \tag{ii}$$

$$1 \leq \sum_{j=1}^{n} y_j \leq K \tag{iii}$$

$$0 \leq x_{ij} \leq y_j \leq 1 \text{ for all } i \text{ and } j \tag{iv}$$

$$x_{ij} \text{ and } y_j \text{ integral for all } i \text{ and } j \tag{v}$$

This problem arises, for example, when a company is contemplating opening up bank accounts in various locations j to pay clients i. The quantity d_j is the cost per period of maintaining an account in location j. The quantity c_{ij} is the money saved (the float) per period by paying client i from location j as a result of the number of days it takes to clear checks.

(1) For any $\mathbf{u} \in R^m$, we define the Lagrangean

$$Z(\mathbf{u}) = \sum_{i=1}^{m} u_i + \max \sum_{i=1}^{m} \left\{ \sum_{j=1}^{n} (c_{ij} - u_i)x_{ij} - d_j y_j \right\}$$
$$\text{s.t. (iii), (iv), (v)}$$

and the dual problem

$$z = \min Z(\mathbf{u})$$
$$\text{s.t. } \mathbf{u} \in R^m$$

As usual, we have $z \geq v$ with no guarantee that $z = v$. Define

$$p_j(\mathbf{u}) = \sum_{i=1}^{m} \{\max(0, c_{ij} - u_i)\} - d_j$$

Let $J^+(\mathbf{u}) = \{j | p_j(\mathbf{u}) > 0\}$ and set $J(\mathbf{u}) = J^+(\mathbf{u})$ if $1 \leq |J^+(\mathbf{u})| \leq K$. Otherwise, let $J(\mathbf{u})$ be an index set corresponding to the K largest $p_j(\mathbf{u})$ if $|J^+(\mathbf{u})| > K$ or the largest single $p_j(\mathbf{u})$ if $|J^+(\mathbf{u})| = 0$. Show that

$$Z(\mathbf{u}) = \sum_{i=1}^{m} u_i + \sum_{j \in J(u)} p_j(\mathbf{u})$$

(2) Consider the following greedy heuristic:

Step 1. Let $k = 1$, $J^* = \phi$, and $u_i^1 = \min_{j=1,\ldots,n} c_{ij}$ for $i = 1, \ldots, m$.

Step 2. Let $p_j(\mathbf{u}^k) = \sum_{i=1}^{m} \max(0, c_{ij} - u_i^k) - d_j$ for $j \in J^*$. Find $j_k \notin J^*$ such that $p_k = p_{j_k}(\mathbf{u}^k) = \max_{j \notin J^*} p_j(\mathbf{u}^k)$. If $p_k < 0$ and $|J^*| \geq 1$, set $k = k - 1$ and go to Step 4. Otherwise set $J^* = J^* \cup \{j_k\}$. If $|J^*| = K$, go to Step 4; otherwise, go to Step 3.

Step 3. Let $k = k + 1$. For $i = 1, \ldots, m$, set $u_i^k = \max_{j \in J^*} c_{ij} = u_i^{k-1} + \max(0, c_{ij_{k-1}} - u_i^{k-1})$. Go to Step 2.

Step 4. Stop; the greedy solution is given by $y_j = 1, j \in J^*$, and $y_j = 0$ otherwise.

Show that the value of the greedy solution is

$$v_g = \sum_{i=1}^{m} u_i^1 + \sum_{j=1}^{k} p_j$$

where $k = |J^*|$.

(3) Suppose we have the special case that the $c_{ij} \geq 0$, $\text{minimum}_{j=1,\ldots,n} c_{ij} = 0$ for $i = 1, \ldots, m$, and $d_j = 0$ for $j = 1, \ldots, n$. With this data, the greedy

heuristic will always select K locations. Show that

$$\frac{v_g}{v} > \frac{1}{2}$$

HINT. Show that $v_g/z > \frac{1}{2}$ and use $z \geq v$. You may find it useful to show and use the relationships

$$\sum_{i=1}^{m} u_i^K = \sum_{k=1}^{K-1} p_k$$

and

$$z \leq Z(\mathbf{u}^K) \leq \sum_{k=1}^{K-1} p_k$$

8.11 (1) Specialize the generalized linear programming method discussed in Chapters 5 and 6 to the solution of the dual problem to the multi-item scheduling problem (8.1). Show that the master problem can be interpreted as selecting production strategies for each of the I items, and that (8.1) will not in general be solved by the dual decomposition approach because for some items, the master will select a convex combination of production strategies rather than a unique or pure strategy. If $I > T$, argue that the master will select pure production strategies for at least $I - T$ items.

(2) Develop a branch and bound scheme similar to the one given in Section 8.6 to find an optimal solution to (8.1) using the dual problem with the Lagrangean (8.2) as a fathoming procedure.

8.12 (Rubin, 1970) The Fibonacci sequence $\{S_k\}_{k=1}^{\infty}$ is given by

$$S_1 = 1 \qquad S_2 = 1 \qquad S_k = S_{k-1} + S_{k-2} \qquad k = 3,4,\ldots$$

For $k = 1,2,\ldots,$ consider the family of linear programming problems with constraint set

$$S_{2k}x + S_{2k+1}y \leq S_{2k+1}^2 - 1 \qquad (LP_k)$$

$$x,y \geq 0$$

Let I_k denote the convex hull of the integer solutions contained in L_k, the feasible region of LP_k. Prove that the convex hull I_k of the integer solutions to (LP_k) has $k+3$ extreme points.

HINT. Extreme points can be characterized as basic solutions to the appropriate linear system (see Lemma 1.9 of Chapter 1). It may be useful to use some or all of the following results about the Fibonacci numbers. You must prove any result you use.

LEMMA 1.

$$S_{2k+1}^2 = S_{2k}S_{2k+2} + 1$$
$$S_{2k}^2 = S_{2k-1}S_{2k+1} - 1 \qquad \text{for all } k \geq 0$$

LEMMA 2. The sequence $\{-S_{2k}/S_{2k+1}\}_{k=0}^{\infty}$ is strictly decreasing. The sequence $\{-S_{2k-1}/S_{2k}\}_{k=1}^{\infty}$ is strictly increasing.

LEMMA 3. $S_{2k}S_{2j+1} - S_{2j}S_{2k+1} \geq 0$ for all $k \geq 0$ and $0 \leq j \leq k$.

LEMMA 4. $S_{2k+1}S_{2j-1} - S_{2k}S_{2j} \geq 1$ for all $k \geq 0$ and $0 \leq j \leq k+1$.

8.13 (Miller, 1970) Consider the problem: Find the infimum of $f: X \to R$ where X is a *discrete rectangle*,

$$X = \{\mathbf{x} | \mathbf{x} \in Z^n,\ a_i \leq x_i \leq b_i,\ i = 1, \ldots, n\}$$

where Z^n is the set of integer n vectors, a_i, b_i are integers or infinite, and $R = (-\infty, +\infty)$. Let $\| \ \|$ denote the max norm of points in R^n; namely, if $\mathbf{y} \in R^n$,

$$\|\mathbf{y}\| = \max_{i=1,\ldots,n} |y_i|$$

Definition. The *discrete neighborhood* of a point $\mathbf{y} \in R^n$ is the set $N(\mathbf{y}) = \{\mathbf{x} | \mathbf{x} \in Z^n,\ \|\mathbf{x} - \mathbf{y}\| < 1\}$

(1) Prove the following: Suppose $X \subset Z^n$ has the property that if $\mathbf{x}^1, \mathbf{x}^2 \in X$, then $N[\alpha\mathbf{x}^1 + (1-\alpha)\mathbf{x}^2] \subset X$ for any α, $0 \leq \alpha \leq 1$. Then $\mathbf{x}^1, \mathbf{x}^2 \in X$ implies
$$V = \left\{\mathbf{v} | \mathbf{v} \in Z^n, \min\left(x_i^1, x_i^2\right) \leq v_i \leq \max\left(x_i^1, x_i^2\right),\ i = 1, \ldots, n\right\} \subset X$$

(2) Use part (1) to prove that the set $X \subset Z^n$ has the property that $\mathbf{x}^1, \mathbf{x}^2 \in X$ implies $N[\alpha\mathbf{x}^1 + (1-\alpha)\mathbf{x}^2] \subset X$ for any α, $0 \leq \alpha \leq 1$, if and only if X is a discrete rectangle.

Definition. The function $f: X \to R$ is *discretely convex* if X is a discrete rectangle and given $\mathbf{x}^1, \mathbf{x}^2 \in X$, $0 \leq \alpha \leq 1$,

$$\min_{\mathbf{x} \in N[\alpha x^1 + (1-\alpha)x^2]} f(\mathbf{x}) \leq \alpha f(\mathbf{x}^1) + (1-\alpha) f(\mathbf{x}^2)$$

(3) Use part (2) to prove the theorem: Let the function $f: X \to R$ be discretely convex and $\mathbf{x}^0 \in X$. If $f(\mathbf{x}^0) \leq f(\mathbf{x})$ for all $\mathbf{x} \in X$ such that $\|\mathbf{x}^0 - \mathbf{x}\| = 1$, then $f(\mathbf{x}^0) \leq f(\mathbf{x})$ for all $\mathbf{x} \in X$, i.e., $\mathbf{x}^0$ is optimal.

8.14 The quadratic assignment problem is

$$\min \sum_{i=1}^{n} \sum_{j=1}^{n} \sum_{k=1}^{m} \sum_{l=1}^{m} c_{ijkl} x_{ik} x_{jl}$$

$$\text{s.t.} \sum_{i=1}^{n} x_{ik} = 1 \qquad k = 1, \ldots, m$$

$$\sum_{k=1}^{m} x_{ik} = 1 \qquad i = 1, \ldots, n$$

$$x_{ik} = 0 \text{ or } 1 \qquad i = 1, \ldots, n; \; k = 1, \ldots, m$$

(1) Show how to convert this problem to a zero-one integer programming problem (that is, one with linear objective and constraint functions).

(2) Show that the traveling salesman problem discussed in Section 8.7 can be expressed as a quadratic assignment problem.

8.15 Consider the following single-facility scheduling problem. Given n jobs with positive integer execution times t_i and associated cost functions $c_i(t)$, execute them consecutively, once each, beginning at time 0, so as to minimize $\Sigma_{i=1}^{n} c_i(d_i)$ where d_i is the termination time of job i. For convenience, assume job 1 has execution time $t_1 = 1$.

(1) A simpler related problem is as follows: execute jobs consecutively without gaps throughout the interval $[0, \Sigma_{i=1}^{n} t_i]$ so as to minimize the sum of the costs associated with all terminations; the number of executions of each job can be more than one. Formulate this simpler problem as a mathematical programming problem in integer variables. In addition, give a dynamic programming algorithm for solving it.

(2) Show that the single-facility scheduling problem can be formulated as the mathematical programming problem in integer variables of part (1) with additional constraints.

(3) Use the results of parts (1) and (2) to construct a dual problem to the single-facility scheduling problem where the Lagrangean minimization is the simpler problem of part (1).

HINT. Adding constants π_i to the cost functions $c_i(t)$ does not change the single-facility scheduling problem but it does change the problem of part (1).

8.16 The nonsymmetric traveling salesman problem is defined over a network $[\mathcal{N}, \mathcal{A}]$ with m nodes for which we associate a length c_{ij} with arc (i,j) and we permit $c_{ij} \neq c_{ji}$. The problem is to find a (directed) circuit of minimal length visiting every node exactly once. How would you extend the dual approach of Section 8.7 to this problem?

8.17 (Held and Karp, 1971) A branch and bound approach for the traveling salesman problem using the traveling salesman dual problem (8.29) uses the following branching or separation strategy (see Section 8.6 for the approximate terminology and ideas). A subproblem in the branch and bound search is characterized by two sets:

X = set of edges included in a possible minimal length tour.

Y = set of edges excluded in a possible minimal length tour.

If the dual analysis of the subproblem fails to produce a fathoming at a dual solution $\mathbf{u}$, then separation can be derived by ordering the edges which have not yet been included or excluded according to the amount by which the lower bound would increase if the edges were excluded. The result is that we have the ordering $e_1, e_2, \ldots, e_r$ satisfying

$$L_{X, Y \cup \{e_1\}}(\mathbf{u}) \geq L_{X, Y \cup \{e_2\}}(\mathbf{u}) \geq \ldots \geq L_{X, Y \cup \{e_r\}}(\mathbf{u})$$

where $L_{Q,P}(u)$ is the Lagrangean calculation of (8.29) with edges in Q included and edges in P excluded. The indicated separation of X, Y is

$$\begin{array}{ll} X_1 = X & Y_1 = Y \cup \{e_i\} \\ X_2 = X \cup \{e_1\} & Y_2 = Y \cup \{e_2\} \\ \vdots & \\ X_q = X \cup \{e_1, e_2, \ldots, e_{q-1}\} & Y_q = Y \cup R_i \end{array}$$

where q is the smallest integer for which there exists a node i such that X does not contain two edges incident to i, but X_q does, and R_i denotes the edges incident to i not in X_q. If there are two such nodes i and j, then we take $Y_q = Y \cup R_i \cup R_j$.

(1) Show how the greedy algorithm for computing minimum spanning 1-trees can be amended to give the indicated ordering .

(2) Show that the indicated separation of X, Y is exhaustive and nonredundant.

(3) Apply the separation strategy to the numerical example in Section 8.7 at the point $\mathbf{u}^1$.

8.18 Prove that if $c_{ij} = 1$ for all i and j, then the network synthesis problem (8.31) is solved by finding the maximal spanning tree T using the r_{ij} as edge lengths, and setting $q_{ij} = v_{ij}$ if $\langle i,j \rangle \in T$, $q_{ij} = 0$ if $\langle i,j \rangle \notin T$.

8.19 Show how the multicommodity flow problem (8.41) can be reformulated as an arc-chain flow problem similar to problem (4.4). Show how generalized linear programming can be used to solve the new formulation by generating chains connecting all node pairs.

8.20 (Wagner and Falkson, 1975) Consider the location/demand model

$$\begin{aligned} \max \ & \sum_{j=1}^{n} f_j(d_j) - \sum_{i=1}^{m}\sum_{j=1}^{n} c_{ij}s_{ij} - \sum_{i=1}^{m} f_i y_i \\ \text{s.t.} \ & \sum_{i=1}^{m} s_{ij} - d_j = 0 \qquad j=1,2,\ldots,n \\ & \sum_{j=1}^{n} s_{ij} - My_i \le 0 \qquad i=1,\ldots,m \\ & s_{ij} \ge 0 \qquad d_j \ge 0 \qquad y_i = 0 \text{ or } 1 \end{aligned}$$

where the f_i are positive for all i, $c_{ij} \ge 0$ for all i and j, M is a large positive number representing an upper bound on the size of any plant in any location i, and the $f_j(\cdot)$ are nonnegative increasing concave functions for all j.

Show that this problem is equivalent to the pure location problem

$$\begin{aligned} \max \ & \sum_{(i,j)\in P} \{f_j(d_{ij}^*) - c_{ij}^*\}x_{ij} - \sum_{i=1}^{m} f_i y_i \\ \text{s.t.} \ & \sum_{i=1}^{m} x_{ij} \le 1 \qquad j=1,2,\ldots,n \\ & \sum_{j=1}^{n} x_{ij} - ny_i \le 0 \qquad i=1,\ldots,m \\ & x_{ij} \ge 0 \qquad y_i = 0 \text{ or } 1 \end{aligned}$$

where

$$P = \left\{ (i,j) \,|\, \frac{df_j(0)}{dd_j} > c_{ij} \right\}$$

and the d_{ij}^* satisfy

$$\frac{df(d_{ij}^*)}{dd_j} = c_{ij}$$

Interpret the quantities d_{ij}^*.

8.21 Discuss the integration of Benders' algorithm for the mixed integer programming problem (6.55) given in Section 6.6 with a branch and bound approach similar to the one developed in Section 8.6. If branch and bound is used to solve the zero-one integer programming subproblem (6.58) each time a cut is added to it, then there is some duplication of effort. Your discussion should propose means to overcome the duplication. In addition, your discussion should address the use of Benders' algorithm as a fathoming procedure.

8.10 NOTES

SECTION 8.1. There are a number of books on integer programming and combinatorial optimization that treat the subject with different emphases. Included are the books of Garfinkel and Nemhauser (1972), Salkin (1975), Greenberg (1971), Hu (1969), and Murty (1976). Recent results in these areas are contained in the book edited by Hammer et al. (1977) and the special issue of Mathematical Programming edited by Balinski (1974). Scott (1971) gives an overview of a number of topics and applications of networks and integer and combinatorial optimization. Geoffrion and Marsten (1972) provide an extensive survey of computational methods for integer programming up to that date. Land and Powell (1977) have compiled a recent survey of computational codes and experience with them. More details about computational experience with the group theoretic and integer programming dual methods discussed in this chapter are given by Gorry et al. (1973) and Fisher et al. (1975). Balas and Padberg (1976) give a survey of the special class of problems called set monitoring problems.

SECTION 8.2. The multi-item production scheduling model was proposed by Dzielinski and Gomory (1965). This model has been implemented by Lasdon and Terjung (1971).

SECTION 8.3. The importance of group theory to integer programming was first discovered by Gomory (1965). The construction in this section was taken from Wolsey (1970).

SECTION 8.4. The group optimization algorithms given in this section are due to Glover (1969). Shapiro (1968) and Gorry and Shapiro (1971) give alternate algorithms for the unconstrained and zero-one group optimization problems. Burdet and Johnson (1974) give an algorithm for solving the unconstrained group optimization problem that does not require an explicit group representation or knowing the order of the group. Denardo and Fox (1977) give an overview of related problems and algorithms.

SECTION 8.5. The exposition on integer programming duality follows Bell and Shapiro (1977). Earlier papers treating this subject are by Shapiro (1971), Fisher and Shapiro (1974), and Bell (1976). Bell (1978) gives an alternate way of strengthening the integer programming dual problems using number theory. Burdet and Johnson (1977) have a related approach to solving integer programming problems using group theory and subadditive functions along with branch and bound. The integer polyhedra implicit in the integer programming dual approach has been studied extensively by Gomory (1968) and Gomory and Johnson (1972). Jeroslow (1977) gives some recent results on integer programming cutting planes.

SECTION 8.6. Land and Doig (1960) was the first paper proposing a branch and bound approach for integer programming problems. The term was coined by Little et al. (1963) who gave the first experimental demonstration of the method applied to the traveling salesman problem. Garfinkel and Nemhauser (1972) have an extensive treatment of the different realizations of the branch and bound approach. The scheme depicted in Figure 8.5 is similar to one given by Geoffrion and Marsten (1972).

SECTION 8.7. Our development follows that of Held and Karp (1970, 1971). Miliotis (1976) has had success recently with a cutting plane approach to the problem based on the integer programming formulation (8.30), which is generated as needed rather than stated in its entirety *a priori*.

SECTION 8.8. Our development in this section follows Gomory and Hu (1961, 1964).

Appendix A
CONVEX SETS AND CONVEX FUNCTIONS

We review briefly the concepts of convex analysis needed for the developments in the body of the text. There are a number of books that go into detail about various aspects of convex analysis including Rockafellar (1970), Stoer and Witzgall (1970), Berge (1963), Nikaido (1968), and Hadley (1961). We give abbreviated proofs of some of the theorems to be presented and the reader is referred to these books for more details. It is assumed that the reader is familiar with basic concepts of real analysis including open and closed sets, boundary and interior points of a set, continuity of functions, etc. We also assume throughout that the vector spaces of interest are finite dimensional and real. Many of the definitions and results remain valid for more general vector spaces, but the generalizations are not a concern of this book (see Luenberger, 1969). In this appendix, we do not include $+\infty$ or $-\infty$ in the set R of real numbers.

Definition A.1. A set $X \subseteq R^n$ is *convex* if $\lambda \mathbf{x}^1 + (1-\lambda)\mathbf{x}^2 \in X$ for any $\mathbf{x}^1, \mathbf{x}^2 \in X$ and any $\lambda \in [0, 1]$.

Definition A.2. For any $\mathbf{a} \in R^n, \mathbf{a} \neq \mathbf{0}$, and any scalar a_0, the set

$$H = \{\mathbf{x} | \mathbf{ax} = a_0\}$$

is called a *hyperplane*. The set

$$\{\mathbf{x} | \mathbf{ax} \geq a_0\}$$

is a *closed halfspace* and the set

$$\{\mathbf{x} | \mathbf{ax} > a_0\}$$

is an *open halfspace*.

Definition A.3. A hyperplane $H = \{\mathbf{x} | \mathbf{ax} = a_0\} \subseteq R^n$ *separates* the two convex sets $X^1, X^2 \subseteq R^n$ if $X^1 \subseteq \{\mathbf{x} | \mathbf{ax} \geq a_0\}$ and $X^2 \subseteq \{\mathbf{x} | \mathbf{ax} \leq a_0\}$. The hyperplane *strictly separates* X^1 and X^2 if $X^1 \subseteq \{\mathbf{x} | \mathbf{ax} > a_0\}$ and $X^2 \subseteq \{\mathbf{x} | \mathbf{ax} < a_0\}$.

Definition A.4. Let $\mathbf{y}$ be a point on the boundary of a closed convex set X. The hyperplane $H = \{\mathbf{x} | \mathbf{ax} = a_0\}$ *supports* X at $\mathbf{y}$ if $\mathbf{ay} = a_0$ and $X \subseteq \{\mathbf{x} | \mathbf{ax} \geq a_0\}$.

Definition A.5. The set C is a *cone* if $\lambda\mathbf{x} \in C$ for all $\mathbf{x} \in C$ and all $\lambda \geq 0$.

THEOREM A.1. Let $X^\alpha, \alpha \in I$, be a collection of convex sets. Then $\bigcap_{\alpha \in I} X^\alpha$ is a convex set.

PROOF. Straightforward. ■

The feasible region of a linear programming or nonlinear convex programming problem equals the intersection of a finite number of convex sets. Thus, the feasible region is convex.

Definition A.6. A *polytope* is a convex set formed by the intersection of a finite number of half spaces. A *polyhedron* is a bounded polytope.

Definition A.7. Let $X \subseteq R^n$ be a convex set. The point $\mathbf{x} \in X$ is *extreme* if there exist no $\mathbf{x}^1, \mathbf{x}^2 \in X, \mathbf{x}^1 \neq \mathbf{x}^2$, such that $\mathbf{x} = \lambda\mathbf{x}^1 + (1-\lambda)\mathbf{x}^2$ for $\lambda \in (0,1)$.

Consider the problem

$$\begin{aligned} &\min f(\mathbf{x}) \\ &\text{s.t. } \mathbf{x} \in X \subseteq R^n \end{aligned} \tag{A.1}$$

where f is differentiable and X is convex. Optimizing over a convex set is a well behaved problem as shown by the following result.

THEOREM A.2. Suppose f achieves its minimum at $\mathbf{x}^*$ in problem (A.1). Then the directional derivative in all feasible directions at $\mathbf{x}^*$ is nonnegative; that is,

$$\nabla f(\mathbf{x}^*; \mathbf{x} - \mathbf{x}^*) = \nabla f(\mathbf{x}^*)(\mathbf{x} - \mathbf{x}^*) \geq 0 \text{ for all } \mathbf{x} \in X$$

PROOF. Suppose to the contrary that there exists an $\mathbf{x} \in X$ such that $\nabla f(\mathbf{x}^*)(\mathbf{x} - \mathbf{x}^*) < 0$. Since X is convex, $\lambda\mathbf{x} + (1-\lambda)\mathbf{x}^* \in X$ for all $\lambda \in [0,1]$. Since f is differentiable, we have

$$\lim_{\lambda \to 0^+} \frac{f[\mathbf{x}^* + \lambda(\mathbf{x} - \mathbf{x}^*)] - f(\mathbf{x}^*)}{\lambda} = \nabla f(\mathbf{x}^*)(\mathbf{x} - \mathbf{x}^*) < 0.$$

Thus, for λ positive but sufficiently small,

$$\frac{f[\mathbf{x}^* + \lambda(\mathbf{x} - \mathbf{x}^*)] f(\mathbf{x}^*)}{\lambda} < 0$$

or $f[\mathbf{x}^* + \lambda(\mathbf{x} - \mathbf{x}^*)] < f(\mathbf{x}^*)$ contradicting the optimality of $\mathbf{x}^*$. ■

THEOREM A.3. Let X be a closed and convex set in R^n and suppose $\mathbf{y} \notin X$. Then there exists a hyperplane strictly separating $\mathbf{y}$ from X.

PROOF. Let $\mathbf{x}^*$ be an optimal solution to the problem

$$\begin{aligned}&\text{minimum}\|\mathbf{x}-\mathbf{y}\|^2=(\mathbf{x}^*-\mathbf{y})^T(\mathbf{x}^*-\mathbf{y})\\&\text{s.t. } x \in X\end{aligned}$$

where $\|\cdot\|$ denotes the Euclidean norm. Since the objective function is continuous and X is closed, an optimal solution exists. By Theorem A.2, $\mathbf{x}^*$ satisfies

$$(\mathbf{x}^*-\mathbf{y})^T(\mathbf{x}-\mathbf{x}^*) \geq 0 \qquad \text{for all } \mathbf{x} \in X.$$

Since $\mathbf{y} \notin X$, we also have

$$(\mathbf{x}^*-\mathbf{y})^T(\mathbf{x}^*-\mathbf{y}) > 0$$

Combining these last two inequalities gives us

$$(\mathbf{x}^*-\mathbf{y})^T\mathbf{x} > (\mathbf{x}^*-\mathbf{y})^T\mathbf{y} \qquad \text{for all } \mathbf{x} \in X$$

Let v be the minimum of $(\mathbf{x}^*-\mathbf{y})^T\mathbf{x}$ over X. Since X is closed, $v > (\mathbf{x}^*-\mathbf{y})^T\mathbf{y}$. Thus, the hyperplane

$$\left\{\mathbf{x} \mid (\mathbf{x}^*-\mathbf{y})^T\mathbf{x} = \tfrac{1}{2}\left[(\mathbf{x}^*-\mathbf{y})^T\mathbf{y}+v\right]\right\}$$

strictly separates $\mathbf{y}$ from X. ■

THEOREM A.4. Suppose X and Y are two closed, convex, and disjoint sets. Then there exists a hyperplane that strictly separates them.

PROOF. Define the set

$$X-Y=\{\mathbf{x}-\mathbf{y} \mid \mathbf{x} \in X, \mathbf{y} \in Y\}$$

This set is convex and does not contain the origin. The hyperplane that strictly separates the origin from $X-Y$ can be shown to separate X and Y. ■

THEOREM A.5. Let $\mathbf{y}$ be a point on the boundary of a closed convex set X. Then there exists a supporting hyperplane of X at Y.

PROOF. Choose a sequence $\{\mathbf{y}^k\}$ converging to $\mathbf{y}$ from outside of X. Let $H^k=\{\mathbf{x} \mid \mathbf{a}^k\mathbf{x}=a_0^k\}$ be hyperplanes which strictly separate $\mathbf{y}^k$ from $\mathbf{y}$ by Theorem A.3. Without loss of generality, we can assume $\|(\mathbf{a}^k, \mathbf{a}_0^k)\|=1$. This implies that there exists a subsequence $\{\mathbf{a}^{k_i}, \mathbf{a}_0^{k_i}\}$ converging to $(\mathbf{a}^*, \mathbf{a}_0^*)$. The hyperplane $H=\{\mathbf{x} \mid \mathbf{a}^*\mathbf{x}=a_0^*\}$ supports X at $\mathbf{y}$ because for all i,

$$\mathbf{a}^{k_i}\mathbf{x} > a_0^{k_i} \qquad \text{for all } \mathbf{x} \in X$$

and

$$\mathbf{a}^{k_i}\mathbf{y}^{k_i} < a_0^{k_i}$$

implying in the limit that

$$\mathbf{a}^*\mathbf{x} \geq a_0^* \qquad \text{for all } \mathbf{x} \in X$$

and

$$\mathbf{a}^*\mathbf{y} \leq a_0^* \quad \blacksquare$$

THEOREM A.6. A convex set is the intersection of the half spaces containing it.

PROOF. Straightforward. ■

Definition A.8. The point $\mathbf{x} \in R^n$ is a *convex combination* of the points $\mathbf{x}^1, \ldots, \mathbf{x}^r$ in R^n if there exist weights $\lambda_1, \ldots, \lambda_r$ satisfying $\Sigma_{k=1}^r \lambda_k = 1$ and $\lambda_k \geq 0$ for $k = 1, \ldots, r$, such that $\mathbf{x} = \Sigma_{k=1}^r \lambda_k \mathbf{x}^k$.

Definition A.9. Let X be an arbitrary set in R^n. Then the *convex hull* of X is

$$\{\mathbf{x} | \mathbf{x} \text{ is a convex combination of } \mathbf{x}^1, \ldots, \mathbf{x}^r \text{ in } X \text{ and } r \text{ is an arbitrary positive integer}\}$$

We let $[X]^c$ denote the convex hull.

THEOREM A.7. (Caratheodory). Suppose $\mathbf{y} \in [X]^c$ for $X \subseteq R^n$. Then $\mathbf{y}$ is a convex combination of at most $n+1$ points from X.

PROOF. Suppose to the contrary that at least $r > n+1$ points are required to express $\mathbf{y}$; that is, $\mathbf{y} = \Sigma_{k=1}^r \lambda_k \mathbf{x}^k$ for $\lambda_k > 0$ and $\Sigma_{k=1}^r \lambda_k = 1$. Since the dimension of the space is n, the vectors $\mathbf{x}^k - \mathbf{x}^r$ for $k = 1, 2, \ldots, r-1$, must be linearly dependent. This implies we can find a_k not all zero such that

$$\sum_{k=1}^{r-1} a_k(\mathbf{x}^k - \mathbf{x}^r) = 0$$

Define $a_r = -\Sigma_{k=1}^{r-1} a_k$; then $\Sigma_{k=1}^r a_k = \Sigma_{k=1}^r a_k \mathbf{x}^k = 0$. We can choose $\alpha > 0$ small enough such that $\lambda_k + \alpha a_k \geq 0$ for all k and $\lambda_j + \alpha a_j = 0$ for some j. Define $\theta_k = \lambda_k + \alpha a_k$; we have $\theta_k \geq 0, \Sigma_{k=1}^r \theta_k = \Sigma_{k=1}^r \lambda_k = 1$ implying $\Sigma_{k=1}^r \theta_k \mathbf{x}^k = \Sigma_{k=1}^r \lambda_k \mathbf{x}^k = 1$. But $\theta_j = 0$ contradicting the assumption that at least r points are required to express $\mathbf{y}$.

THEOREM A.8. The set $[X]^c$ equals the intersection of all convex sets containing X.

PROOF. The set $[X]^c$ is convex by a straightforward induction argument and Definition A.1 of a convex set. Thus $[X]^c$ contains the intersection. Since the intersection contains X and the intersection of convex sets is convex, any point $\Sigma_{k=1}^{r}\lambda_k\mathbf{x}^k$ for $\mathbf{x}^k \in X$ and $\Sigma_{k=1}^{r}\lambda_k = 1, \lambda_k \geq 0$, is in the intersection. Thus, $[X]^c$ is in the intersection. ■

Definition A.10. The n-vector $\mathbf{d}$ is a *ray* of the convex set $X \subseteq R^n$ if

$$\mathbf{x} + \mu\mathbf{d} \in X \text{ for all } \mathbf{x} \in X \text{ and all } \mu \geq 0$$

The set of all rays of the convex set X is a convex cone. A vector $\mathbf{d}$ is an *extreme ray* if there exist no $\mathbf{d}^1, \mathbf{d}^2 \in D, \mathbf{d}^1 \neq \mathbf{d}^2$, such that $\mathbf{d} = \lambda\mathbf{d}^1 + (1-\lambda)\mathbf{d}^2$ for some $\lambda \in (0, 1)$.

THEOREM A.9. If X is a closed convex set containing no hyperplanes, then

$$X = [E]^c + [D]^c$$

where E is the set of extreme points of X and D is the set of extreme rays.

PROOF. See Rockafellar (1970); sections 18 and 19. ■

Definition A.11. A function f from R^n to R is *convex* if

$$f[\lambda\mathbf{x}^1 + (1-\lambda)\mathbf{x}^2] \leq \lambda f(\mathbf{x}^1) + (1-\lambda)f(\mathbf{x}^2)$$

for all $\mathbf{x}^1$, $\mathbf{x}^2 \in R^n$ and any $\lambda \in [0, 1]$. The function is *strictly convex* if the inequality is strict. The function f is *concave* if $-f$ is convex.

Recall that we do not include $+\infty$ or $-\infty$ in R. The definition of a convex function would remain consistent if we permitted $f(x) = +\infty$ for some $\mathbf{x} \in R^n$. Note also that $f(\mathbf{x}) = -\infty$ for all $\mathbf{x} \in R^n$ is a (trivial) convex function. We have chosen f finite for all $\mathbf{x} \in R^n$ for expositional simplicity, but the following theorems could be extended to include the more general case.

THEOREM A.10. The following are equivalent:

(1) f is convex.

(2) Given any $\mathbf{y} \in R^n$, there exists a *subgradient* $\gamma_{\mathbf{y}} \in R^n$ such that

$$f(\mathbf{x}) \geq f(\mathbf{y}) + \gamma_{\mathbf{y}}(\mathbf{x} - \mathbf{y}) \qquad \text{for all } \mathbf{x} \in R^n$$

(3) If f is differentiable, then

$$f(\mathbf{x}) \geq f(\mathbf{y}) + \nabla f(\mathbf{y})(\mathbf{x} - \mathbf{y}) \qquad \text{for all } \mathbf{x}, \mathbf{y} \in R^n$$

(4) If f has second partials, then the Hessian Matrix

$$H_f(\mathbf{x}) \equiv \frac{\partial f(\mathbf{x})}{\partial x_i \partial x_j}$$

is positive semidefinite; that is,

$$\mathbf{y}^T H_f(\mathbf{x})\mathbf{y} \geq 0 \qquad \text{for all } \mathbf{x}, \mathbf{y} \in R^n$$

(5) The set $\{(w,\mathbf{x})\in R^{n+1}|\mathbf{x}\in R^n \text{ and } w\geq f(\mathbf{x})\}$ is convex.

PROOF. We prove only (2)⇒(1)⇒(5)⇒(2) and leave the other relationships as exercises for the reader.

(2)⇒(1): Consider any $\mathbf{x}^1,\mathbf{x}^2\in X$ and the point $\mathbf{z}=\lambda\mathbf{x}^1+(1-\lambda)\mathbf{x}^2$ for $\lambda\in[0,1]$. By assumption, there exists a subgradient $\gamma_\mathbf{z}$ such that

$$f(\mathbf{x}^1)\geq f(\mathbf{z})+\gamma_\mathbf{z}(\mathbf{x}^1-\mathbf{z})$$

and

$$f(\mathbf{x}^2)\geq f(\mathbf{z})+\gamma_\mathbf{z}(\mathbf{x}^2-\mathbf{z})$$

Weighting the first inequality by λ and the second by $(1-\lambda)$, we obtain

$$\begin{aligned}\lambda f(\mathbf{x}^1)+(1-\lambda)f(\mathbf{x}^2)&\geq f(\mathbf{z})+\gamma_\mathbf{z}\left[\lambda\mathbf{x}^1+(1-\lambda)\mathbf{x}^2-\mathbf{z}\right]\\&=f(\mathbf{z})=f\left[\lambda\mathbf{x}^1+(1-\lambda)\mathbf{x}^2\right]\end{aligned}$$

(1)⇒(5): Consider any $(w^1,\mathbf{x}^1)$, $(w^2,\mathbf{x}^2)$ satisfying $w^1\geq f(\mathbf{x}^1)$ and $w^2\geq f(\mathbf{x}^2)$. For any $\lambda\in[0,1]$, we have

$$\begin{aligned}\lambda w^1+(1-\lambda)w^2&\geq\lambda f(\mathbf{x}^1)+(1-\lambda)f(\mathbf{x}^2)\\&\geq f\left[\lambda\mathbf{x}^1+(1-\lambda)\mathbf{x}^2\right]\end{aligned}$$

which is what we needed to show.

(5)⇒(2): Consider the point $\mathbf{z}\in R^n$ and the value $w_\mathbf{z}=f(\mathbf{z})$. By assumption, the set $\{(w,\mathbf{x})\in R^{n+1}|\mathbf{x}\in R^n,\ w\geq f(\mathbf{x})\}$ is convex implying by Theorem A.5 that there is a supporting hyperplane of the set through the point $(w_z,\mathbf{z})$. Specifically, there are the coefficients a_0, $b\in R$ and nonzero $\mathbf{a}\in R^n$ such that

$$\{(w,\mathbf{x})|bw+\mathbf{ax}\geq a_0\}\supseteq\{(w,\mathbf{x})|w\geq f(\mathbf{x})\}$$

and

$$bf(\mathbf{z})+\mathbf{az}=a_0$$

It must be that $b\neq 0$ because otherwise we have $\mathbf{ax}\geq\mathbf{az}$ for all $\mathbf{x}\in R^n$. If we take $\mathbf{x}=-\mathbf{a}+\mathbf{z}$, there results $\|\mathbf{a}\|^2\leq 0$ which is impossible since $\mathbf{a}\neq 0$. Select any $\mathbf{x}\in R^n$ and consider the point $[f(\mathbf{x}),\ \mathbf{x}]$. It satisfies

$$bf(\mathbf{x})+\mathbf{ax}\geq a_0=bf(\mathbf{z})+\mathbf{az},$$

or

$$f(\mathbf{x})\geq f(\mathbf{z})+\left(\frac{-a}{b}\right)(\mathbf{x}-\mathbf{z})$$

which is what we needed to show. ■

THEOREM A.11. Let g be a function from R^n to R^m and let f be a function from R^m to R. If f is convex and nondecreasing and if g is convex, then the composite function $f[g(\mathbf{x})]$ is convex.

PROOF. An exercise for the reader. ■

THEOREM A.12. If g and h are convex functions from R^n to R, then so are $g+h$ and $-1/g$.

PROOF. Use Theorem A.11. ■

THEOREM A.13. A convex function f defined over $X \subseteq R^n$ with values in R is continuous on the interior of X.

PROOF. See Rockafellar (1970), p. 82. ■

Definition A.11. The set of all subgradients of a convex function f at a point $\mathbf{x}$ is called the *subdifferential* and it is denoted by $\partial f(\mathbf{x})$.

THEOREM A.14. The subdifferential of a convex function defined on R^n is a nonempty, closed, and bounded convex set.

PROOF. See Rockafellar (1970); Section 23. ■

By Theorem A.10, the subdifferential of a convex function is nonempty everywhere. If the subdifferential consists of the singleton $\boldsymbol{\gamma}$ at the point $\mathbf{x}$, then f is differentiable at $\mathbf{x}$ and $\boldsymbol{\gamma} = \nabla f(\mathbf{x})$. The next theorem states that the directional derivatives of f exist everywhere even if there are points where it is not differentiable because $\partial f(\mathbf{x})$ is larger than a singleton.

THEOREM A.15. If f is convex, then the directional derivative $\nabla f(\mathbf{x};\mathbf{d})$ of f exists in any direction $\mathbf{d}$. Moreover,

$$\nabla f(\mathbf{x};\mathbf{d}) = \underset{\boldsymbol{\gamma} \in \partial f(\mathbf{x})}{\text{maz}} \ \boldsymbol{\gamma}\mathbf{d}$$

PROOF. See Rockafellar (1970); Section 23. ■

THEOREM A.16. If f is convex and differentiable on R^n, then it is continuously differentiable on R^n.

PROOF. See Rockafellar (1970); Section 25. ■

Finally, convex functions have the desirable property that the necessary conditions for optimality of Theorem A.2 are also sufficient if f is convex.

THEOREM A.17. Suppose the function f in problem (A.1) is convex and the set X is convex. Then $\mathbf{x}^*$ is a minimum if and only if the directional derivative in all feasible directions at $\mathbf{x}^*$ is nonnegative.

PROOF. We show sufficiency since necessity was established in Theorem A.2. For any $\mathbf{y} \in X$ and any $\gamma \in \partial f(\mathbf{x}^*)$, we have

$$f(\mathbf{y}) - f(\mathbf{x}^*) \geq \gamma(\mathbf{y} - \mathbf{x}^*)$$

Since from Theorem A.15. we know $\nabla f(\mathbf{x}^*; \mathbf{y} - \mathbf{x}^*) = \max_{\gamma \in \partial f(\mathbf{x}^*)} \gamma(\mathbf{y} - \mathbf{x}^*) \geq 0$, we have for some $\gamma^* \in \partial f(\mathbf{x}^*)$ that $\gamma^*(\mathbf{y} - \mathbf{x}^*) \geq \mathbf{0}$ implying $f(\mathbf{y}) - f(\mathbf{x}^*) \geq 0$. ■

Appendix B
GRAPH THEORY

A *graph* $G=(\mathcal{N},\mathcal{E})$ consists of a finite node set $\mathcal{N}$ and edge set $\mathcal{E}$ containing unordered pairs of distinct nodes from $\mathcal{N}$. For convenience, we let $\mathcal{N}$ be denoted by the set $\{1,\ldots,m\}$, and a generic edge is denoted by $\langle i,j\rangle$ for $i,j\in\mathcal{N}, i\neq j$, and the edge is said to be *incident* to i and j. There is no direction associated with the edge $\langle i,j\rangle$. A graph is *complete* if there is an edge $\langle i,j\rangle$ for all $i,j\in\mathcal{N}$, $i\neq j$. Two nodes are adjacent if there is an edge connecting them. The *degree* of node i in the graph G, denoted by $d(i)$, is the number of edges drawn to node i.

A *path* in G is denoted by $\langle i_1,i_2,\ldots,i_k\rangle$ where $\langle i_j,i_{j+1}\rangle\in\mathcal{E}$ for all j. A path is a *cycle* if $i_1=i_k$. A path is a *simple path* if it does not contain any cycles. A path (cycle) is a *simple cycle* if all the nodes are distinct except for the first and last nodes. A *subgraph* $H=(\mathcal{N}',\mathcal{E}')$ of the graph G satisfies $\mathcal{N}'\subseteq\mathcal{N}$, $\mathcal{E}'\subseteq\mathcal{E}$. A subgraph is a *spanning subgraph* if $\mathcal{N}'=\mathcal{N}$.

A *network* $G=(\mathcal{N},\mathcal{A})$ consists of a finite node set $\mathcal{N}$ and arc set $\mathcal{A}$ containing ordered pairs of distinct nodes from $\mathcal{N}$. Letting $\mathcal{N}=\{1,\ldots,m\}$, a generic edge is denoted by (i,j) for $i,j\in\mathcal{N}$, $i\neq j$. There is a direction associated with the arc (i,j). A *chain* in G is denoted by $(i_1,i_2,\ldots,i_k)$ where $(i_j,i_{j+1})\in\mathcal{A}$ for all j. A chain is a *circuit* if $i_1=i_k$. A chain is a *simple chain* if it does not contain any circuits. A chain (circuit) is a *simple circuit* if all the nodes are distinct except for the first and last nodes.

A central construct in studying graphs is the *tree*. A *tree* is a graph characterized by the following equivalent properties:

1. It has m nodes, $m-1$ edges, and is connected,
2. It has m nodes, $m-1$ edges, and no cycles,
3. There is a unique path from each node to every other node,
4. It has no cycles but exactly one cycle is created by adding an edge,
5. It is connected but ceases to be connected if any edge is removed.

Berge (1962) demonstrates these equivalences and generally provides an overview of graph theory. An *end* of a tree is a node of degree one.

LEMMA B.1. A tree consisting of at least two nodes has at least two ends.

A *forest* is a graph consisting of one or more trees. Equivalently, a forest is a graph without cycles. Each connected component of a forest is a tree. A *spanning tree* T of a graph G is a spanning subgraph of G and a tree. A *spanning forest* F of a graph G is a spanning subgraph of G and a forest.

The *node-edge incidence matrix* of a graph with m nodes and n edges is an $m \times n$ matrix $\mathbf{A}$ defined as follows. Each column $\mathbf{a}_l$ of $\mathbf{A}$ corresponds to a different edge $r_l = \langle i_\ell, j_\ell \rangle$ where

$$a_{kl} = \begin{cases} 1 & \text{if } k = i_\ell \text{ or } j_\ell \\ 0 & \text{otherwise} \end{cases}$$

The *edge-path incidence matrix* of a graph with m nodes and n edges is a matrix with n rows and a column for each distinct simple path connecting node 1 to node m. The column corresponding to a particular simple path has 1's on the rows corresponding to edges in the path and zeros otherwise.

The *node-arc incidence matrix* $\mathbf{A}$ of a network with m nodes and n arcs has m rows and n columns. Column $\mathbf{a}_l$ corresponding to a different arc $r_l = (i_l, j_l)$ where

$$a_{kl} = \begin{cases} 1 & \text{if } k = i_l \\ -1 & \text{if } k = j_l \\ 0 & \text{otherwise} \end{cases}$$

The *arc-chain incidence matrix* of a network with m nodes and n arcs is a matrix with n rows and a column for each distinct simple chain connecting node 1 to node m. The column corresponding to a particular simple chain has 1's on the rows corresponding to arcs in the chain and zeroes otherwise.

Appendix C
NUMBER THEORY AND GROUP THEORY

A relation R defined on a set A is a subset of $A \times A$. We write xRy if $(x,y) \in R$. A relation R is said to be (1) *reflexive* if xRx for any $x \in A$; (2) *symmetric* if xRy implies yRx for any $x,y \in A$; (3) transitive if xRy and yRz implies xRz for any $x,y,z \in A$. A relation R that is reflexive, symmetric and transitive over a set A is called an *equivalence relation*. It can easily be shown that an equivalence relation partitions A into mutually exclusive and exhaustive subsets of the form $xR = \{y \in A | xRy\}$, called *equivalence classes*.

EXAMPLE C.1. Let $G = (\mathcal{N}, \mathcal{E})$ be a complete graph defined on m nodes. A 1-tree defined on G is a subgraph of G which is a tree on the nodes $2,\ldots,m$, and node 1 is connected to this tree by 2 edges. An equivalence relation R defined on the set τ of all 1-trees is given by TRS if $d_T(i) = d_S(i)$ for all nodes i; that is, if the degrees of the 1-trees T and S at all nodes are equal. The traveling salesman tours discussed in Section 8.7 constitute the equivalence class with $d_T(i) = 2$ for all i. ▲

EXAMPLE C.2. We say two integers *a and b are congruent modulo k*, written $a \equiv b (\text{mod } k)$, if $a = b + tk$ for some integer t. The congruence relation is an equivalence relation on the set of integers Z dividing it into the k equivalence classes $\{a | a \equiv j (\text{mod } k)\}$ for $j = 0, 1, 2, \ldots, k-1$. For future reference, we define $Z_k = \{0, 1, 2, \ldots, k-1\}$ and define a binary operation $*$ on Z_k by $j_1 * j_2 = j_2 * j_1 = j_3$ where j_3 is the unique element of Z_k satisfying $j_3 \equiv j_1 + j_2 (\text{mod } k)$. ▲

The last example introduces us to elementary number theory, which is primarily concerned with questions of divisibility of integers. We review quite briefly some of the main results from this theory, particularly as it relates to integer programming.

Suppose m and n are any two integers. Then there are unique integers q and r such that $n = qm + r$, where the remainder r is some integer in the set Z_q. The integer q is the integer part of n/m, denoted by $[n/m]$, which is the largest integer less than or equal to n/m.

Definition C.1. Let m and n be any two nonzero integers. An integer d is called the greatest common divisor (*gcd*) of m and n if it satisfies the following:

1. $d > 0$.
2. d divides m and n.
3. Any integer that divides both m and n divides d.

If the greatest common divisor of m and n is 1, then we say m and n are *relatively prime*. A classic problem in elementary number theory is the calculation of the greatest common divisor of two integers.

Euclidean Algorithm

The greatest common divisor d of two positive integers m and n can be calculated as follows. Let $a_0 = m$, $a_1 = n$ and express a_0 uniquely as $a_0 = q_1 a_1 + a_2$ where a_2 is the remainder ($a_2 \varepsilon Z_{q_1}$). If $a_2 = 0$, then a_1 is the *gcd* of m and n. If $a_2 \neq 0$, then express a_1 uniquely as $a_1 = q_2 a_2 + a_3$ where a_3 is the remainder. If $a_3 = 0$, then a_2 is the *gcd* of m and n. If $a_3 \neq 0$, then express a_2 uniquely as $a_2 = q_3 a_3 + a_4$ where a_4 is the remainder. Continue in this fashion until a zero remainder is obtained. The last nonzero remainder is the *gcd* of m and n.

We now give some of the basic results about congruences.

THEOREM C.1.

(1) If $a \equiv b \pmod{k}$ and $b \equiv c \pmod{k}$, then $a \equiv c \pmod{k}$.

(2) If $a \equiv a' \pmod{k}$ and $b \equiv b' \pmod{k}$, then $a + b \equiv a' + b' \pmod{k}$, $a - b \equiv a' - b' \pmod{k}$ and $ab \equiv a'b' \pmod{k}$.

(3) If $a \equiv b \pmod{k}$, then $a^t \equiv b^t \pmod{k}$ for any positive integer t.

(4) If $ac \equiv bc \pmod{k}$ and if c and k are relatively prime, then $a \equiv b \pmod{k}$.

This theorem shows that congruences can be treated as ordinary equalities except for cancellation; that is, except for cancelling factors. Part (4) of the theorem states that cancellation in a congruence is allowed provided the factor being cancelled is relatively prime to k.

Let k be any positive integer. Then any integer k is congruent to one and only one of the integers in $Z_k = \{0, 1, 2, \ldots, k-1\}$. This is the reason in

Example C.2 that taking congruences (modulo k) is an equivalence relation on the set Z of integers. The equivalence classes are called the residue classes of the integers (modulo k).

These constructs of elementary number theory need to be generalized to higher dimensional sets to be useful for integer programming. The theory of groups greatly facilitates such a generalization.

Definition C.2. A set $\mathcal{G}$ with a binary operation * is called a *group* if

(1) $a*b \in \mathcal{G}$ for all $a,b \in \mathcal{G}$;
(2) $a*(b*c)=(a*b)*c$ for all $a,b,c \in \mathcal{G}$;
(3) There exists an identity element, denoted by e, such that $a*e=e*a=a$ for all $a \in \mathcal{G}$;
(4) for all $a \in \mathcal{G}$, there exists an inverse, denoted by a^{-1}, satisfying $a*a^{-1}=a^{-1}*a=e$.

The number of elements in a group $\mathcal{G}$, denoted by $|\mathcal{G}|$, is called its *order*. A subset $\mathcal{H}$ of a group $\mathcal{G}$ is called a *subgroup* if it is closed under the binary operation; that is, if $a*b \in \mathcal{H}$ for all $a,b \in \mathcal{H}$.

The stated properties of a group are not axiomatic in the sense that fewer properties of a group can be assumed and the remainder derived. An example of a group is the set of all $m \times m$ nonsingular matrices with matrix mutliplication as the binary operation. A subgroup of this group are $m \times m$ matrices with determinants equal to $+1$ or -1. Another example of a group is the set $Z_q=\{0,1,\dots,q-1\}$ with the binary operation defined for all $j_1,j_2 \in Z_q$ by $j_1*j_2=j_3$ where $j_3 \equiv j_1+j_2$ (mod q). For this group, the binary operations is commutative; namely $j_1*j_2=j_2*j_1$ for all $j_1,j_2 \in G$. (Note that the group of nonsingular matrices with matrix multiplication as its operation is not commutative). Commutative groups are called *abelian groups*. The notational convention for abelian groups is to denote the operation "*" by "+", the identity element e by 0, and the inverse of a by $-a$. This is because all abelian groups behave very much like the groups Z_q and Z. We come back to this point shortly.

Since our group theoretic concern in this book is mainly with abelian groups, the remainder of this appendix addresses them exclusively. Most of the constructs and results to be presented hold in modified form for arbitrary groups.

A mapping h from an abelian group $\mathcal{G}$ into an abelian group $\mathcal{H}$ is called a *homomorphism* if $h(a+b)=h(a)+h(b)$ for all $a,b \in \mathcal{G}$. The set $K=\{a|a \in \mathcal{G}$ such that $h(a)=0\}$ is a subgroup of $\mathcal{G}$ called the *kernel* of the homomorphism. A homomorphism is an *isomorphism* if it is also one-to-one and onto. The direct sum $\mathcal{G}=\mathcal{G}^1 \oplus \mathcal{G}^2$ of two abelian groups is a group whose element are all two-tuples (a,x) with $a \in \mathcal{G}^1$ and $x \in \mathcal{G}^2$. The group operation consists of the operations of $\mathcal{G}^1$ and $\mathcal{G}^2$ in the respective components of the two tuples; that is, $(a,x)+(b,y)=(c,z)$ where $a+b=c$ and $x+y=z$. It can be shown that every finite abelian group $\mathcal{G}$ is isomorphic to $Z_{q_1}\oplus\ldots\oplus Z_{q_r}$ where the q_i satisfy $q_i\geq 2$, $q_i|q_{i+1}$, $i=1,\ldots,r-1$, $\Pi_{i=1}^r q_i=|\mathcal{G}|$. These q_i are uniquely determined by $\mathcal{G}$.

Any subgroup $\mathcal{H}$ of an abelian group $\mathcal{G}$ naturally induces a homomorphism in the following way. For any $a\in\mathcal{G}$, let $a+\mathcal{H}=\{b\in\mathcal{G}\mid$ there exists an $x\in\mathcal{H}$ such that $b=a+x\}$. The set $a+\mathcal{H}$ is called a *coset*. The relation $R_{\mathcal{H}}$ defined on $\mathcal{G}$ by $aR_{\mathcal{H}}b$ if a and b are in the same coset is an equivalence relation. The collection of equivalence classes $\{a+\mathcal{H}|a\in\mathcal{G}\}$ forms a group, denoted by $\mathcal{G}/\mathcal{H}$ and called a quotient group, with the group operation + given by $(a+\mathcal{H})+(b+\mathcal{H})=c+\mathcal{H}$ where $c=a+b$. The mapping $\mathcal{H}$ from $\mathcal{G}$ onto $\mathcal{G}/\mathcal{H}$ given by $\phi_{\mathcal{H}}(a)=a+\mathcal{H}$ is called the *canonical homomorphism* induced by the subgroup $\mathcal{H}$. Its kernel is $\mathcal{H}$.

EXAMPLE C.3. Consider the group Z^m consisting of the set of all integer m-vectors under ordinary addition. Let $\mathbf{B}$ be any any $m\times m$ nonsingular integer matrix. The set $K_{\mathbf{B}}=\{\mathbf{a}|\mathbf{a}=\mathbf{B}\mathbf{x},\ \mathbf{x}$ integer$\}$ is a subgroup of Z^m. It can be shown using the Smith reduction procedure of Chapter 8 that the quotient group $Z^m/K_{\mathbf{B}}$ is a finite group consisting of $D=|\det\mathbf{B}|$ elements. ▲

BIBLIOGRAPHY

Agmon, S. (1954), "The relaxation method for linear inequalities," *Can. J. Math.*, **6**, 382–392.

Arrow, K. J. and F. H. Hahn (1971), *General Competitive Analysis*, Holden-Day.

Avriel, M. (1976), *Nonlinear Programs: Analysis and Methods*, Prentice-Hall.

Balas, E. and M. W. Padberg (1976), "Set partitioning: A survey," *SIAM Rev.*, **18**, 710–760.

Balinski, M. L., ed. (1974), *Mathematical Programming Study 2: Approaches to Integer Programming*, North-Holland.

Balinski, M. L. and P. Wolfe, eds. (1975), *Mathematical Programming Study 3: Nondifferentiable Optimization*, North-Holland.

Barr, R. S., F. Glover, and D. Klingman (1974), "An improved version of the out-of-kilter method and a comparative study of computer codes," *Math. Program.*, **7**, 60–87.

Bartels, R. H., G. H. Golub, and M. A. Saunders (1970), "Numerical techniques in mathematical programming," in *Nonlinear Programming*, J. B. Rosen, O. L. Mangarsarian, and K. Ritter, eds., Academic Press.

Baumol, W. J. and T. Fabian (1964), "Decomposition, pricing for decentralization, and external economies," *Manage. Sci.*, **11**, 1–32.

Bazaraa, M. S. and J. J. Jarvis (1977), *Linear Programming and Network Flows*, Wiley.

Bell, D. E. (1976), "Constructive group relaxations for integer programs," *SIAM J. Appl. Math.*, **30**, 708–719.

Bell, D. E. (1978), "Efficient group cuts for integer programs," Harvard Business School working paper HBS 78–10.

Bell, D. E. and J. F. Shapiro (1977), "A convergent duality theory for integer programming," *Oper. Res.*, **25**, 419–434.

Bellman, R. (1957), *Dynamic Programming*, Princeton University Press.

Bellman, R. and S. E. Dreyfus (1962), *Applied Dynamic Programming*, Princeton University Press.

Benders, J. F. (1962), "Partitioning procedures for solving mixed-variables programming problems," *Numer. Math.*, **4**, 238–252.

Berge, C. (1962), *The Theory of Graphs and Its Applications*, Methuen. Translated by A. Doig from the original French edition, Dunod, 1958.

Berge, C. (1963), *Topological Spaces*, MacMillan. Translated by E. M. Patterson.

Bertsekas, B. P. and S. K. Mitter (1973), "A descent numerical method for optimization problems with nondifferentiable cost functionals," *SIAM J. Control,* **11**, 637–652.

Bertsekas, D. P. (1976), *Dynamic Programming and Stochastic Control*, Academic Press.

Blackwell, D. (1962), "Discrete dynamic programming," *Ann. Math. Statist.*, **33**, 719–726.

Blackwell, D. (1965), "Discounted dynamic programming," *Ann. Math. Statist.*, **36**, 226–235.

Bland, R. G. (1977), "New finite pivoting rules for the simplex method," *Math. Oper. Res.*, **2**, 103–107.

Boesch, F. J., ed. (1975), *Large-Scale Networks: Theory and Design*, IEEE Press.

Bradley, G. H., G. Brown, and G. Graves (1977), "Design and implementation of large scale primal transshipment algorithms," *Manage. Sci.*, **24**, 1–34.

Bradley, G. H., P. L. Hammer, and L. A. Wolsey (1974), "Coefficient reduction for inequalities in 0-1 variables," *Math. Program.*, **7**, 263–282.

Bradley, S. P., A. C. Hax, and T. L. Magnanti (1977), *Applied Mathematical Programming*, Addison-Wesley.

Burdet, C. A. and E. L. Johnson (1974), "A subadditive approach to the group problem of integer programming," in *Mathematical Programming Study 2: Approaches to Integer Programming*, M. L. Balinski, ed., North-Holland, pp. 51–71.

Burdet, C. A. and E. L. Johnson (1977), "A subadditive approach to solve linear integer programs," in *Studies in Integer Programming*, P. L. Hammer, E. L. Johnson, B. H. Korte, and G. L. Nemhauser, eds., North-Holland, pp. 117–144.

Busacker, R. G. and T. L. Saaty (1965), *Finite Graphs and Networks: An Introduction with Applications*, McGraw-Hill.

Charnes, A. (1952), "Optimality and degeneracy in linear programming," *Econometrica*, **20**, 166–170.

Christofides, N. (1975), *Graph Theory: An Algorithmic Approach*, Academic Press.

Cornuejols, G., M. L. Fisher, and G. L. Nemhauser (1977), "Location of bank accounts to optimize float: an analytic study of exact and approximate algorithms," *Manage. Sci.*, **23**, 789–810.

Cottle, R. W. and G. B. Dantzig (1968), "Complementary pivot theory of mathematical programming," *Lin. Alg. Appl.*, **1**, 103–125.

Cunningham, W. H. (1976), "A network simplex method," *Math. Program.*, **11**, 105–116.

Danskin, J. M. (1967), *Theory of Max-Min and Its Application to Weapons Allocation Problems*, Springer-Verlag.

Dantzig, G. B. (1963), *Linear Programming and Extensions*, Princeton University Press.

Dantzig, G. B., A. Orden, and P. Wolfe (1955), "Notes on linear programming: part I—the generalized simplex method for minimizing a linear form under linear inequality restraints," *Pac. J. Math.*, **5**, 183–195.

Dantzig, G. B. and R. M. VanSlyke (1967), "Generalized upper bounding techniques," *J. Comp. Sci.*, **1**, 213–226.

Dantzig, G. B. and R. M. VanSlyke (1970), "Generalized linear programming," Chapter 3 in *Optimization Methods for Large Scale Systems*, D. A. Wismer, ed., McGraw-Hill.

Dantzig, G. B. and A. Wolfe (1961), "The decomposition algorithm for linear programming," *Econometrica*, **29**, 767–778.

Demyanov, V. F. (1968), "Algorithms for some minimax problems," *J. Comp. Syst. Sci.*, **2**, 342–380.

Denardo, E. V. (1967), "Contraction mappings in the theory underlying dynamic programming," *SIAM Rev.*, **9**, 165–177.

Denardo, E. V. and B. Fox (1968), "Multichain Markov renewal programs," *SIAM J. Appl. Math.*, **16**, 468–487.

Denardo, E. V. and B. Fox (1977), "Shortest route methods: 1. Reaching, pruning and buckets," Dept. of Adm. Sci., Yale Univ.

Denardo, E. V. and B. L. Miller (1968), "An optimality condition for discrete dynamic programming with no discounting," *Ann. Math. Statist.*, **39**, 1220–1227.

D'Epenoux, F. (1960), "Sur un problème de production et de stockage dans l'aléatoire," *Rev. Franç. Rech. Opérationelle*, **14**, 3–16.

D'Epenoux, F. (1963), "A probabilistic production and inventory problem," *Manage. Sci.*, **10**, 98–108.

Dijkstra, E. W. (1959), "A note on two problems in connexion with graphs," *Numer. Math.*, **1**, 269–271.

Dreyfus, S. E. (1969), "An appraisal of some shortest-path algorithms," *Oper. Res.*, **17**, 395–412.

Duffin, R. J. (1962), "Dual programs and minimum cost," *SIAM J.*, **10**, 119–123.

Dzielinski, B. P. and R. E. Gomory (1965), "Optimal programming of lot sizes, inventory and labor allocations," *Manage. Sci.*, **11**, 874–890.

Eaves, B. C. (1972), "Homotopies for computation of fixed points," *Math. Program*, **3**, 1–22.

Edmonds, J. (1965a), "Paths, trees, and flowers," *Can. J. Math.*, **17**, 449–467.

Edmonds, J. (1965b), "Maximum matching and a polyhedron with 0, 1-vertices," *J. Res. Nat. Bur. Stds.*, **69B**, 125–130.

Edmonds, J. (1968), "Optimum branchings," in *Mathematics of the Decision Sciences*, G. B. Dantzig and A. F. Vernott, eds., American Mathematical Society.

Edmonds, J. and R. M. Karp (1972), "Theoretical improvements in algorithmic efficiency for network flow problems," *J. ACM*, **19**, 248–264.

Elmaghraby, S. (1970), *Some Network Models in Management*, Springer-Verlag.

Evans, J. P. and F. J. Gould (1970), "Stability and exponential penalty function techniques in nonlinear programming," Institute of Statistics Memo Series No. 723, Department of Statistics, Univ. of North Carolina at Chapel Hill.

Everett, H., III (1963), "Generalized LaGrange multiplier method for solving problems of optimum allocation of resources," *Oper. Res.*, **11**, 399–417.

Fiacco, A. V. and G. P. McCormick (1968), *Nonlinear Programming, Sequential Unconstrained Minimization Techniques*, Wiley.

Fisher, M. L., W. D. Northup, and J. F. Shapiro (1975), "Using duality to solve discrete optimization problems: theory and computational experience," in *Mathematical Programming Study 3: Nondifferentiable Optimization*, 56–94, M. L. Balinski and P. Wolfe, eds., North-Holland.

Fisher, M. L. and J. F. Shapiro (1974), "Constructive duality in integer programming," *SIAM J. Appl. Math*, **27**, 31–52.

Florian, M. and S. Nguyen (1974), "A method for computing network equilibrium with elastic demands," *Trans. Sci.*, **8**, 321–332.

Floyd, R. (1962), "Algorithm 97: Shortest path," *Commun. ACM*, **5**, 345.

Ford, L. R., Jr., and D. R. Fulkerson (1962), *Flows in Networks*, Princeton University Press.

Fox, B. (1966), "Markov renewal programming by linear fractional programming," *SIAM J. Appl. Math.*, **14**, 1418–1432.

Frank, M. and I. T. Frisch (1971), *Communication, Transmission, and Transportation Networks*, Addison-Wesley.

Frank, M. and P. Wolfe (1956), "An algorithm for quadratic programming," *Naval Res. Logistics Quart.*, **3**, 95–110.

Fulkerson, D. R. (1961), "An out-of-kilter method for minimal cost flow problems," *J. Soc. Ind. Appl. Math.*, **9**, 18–27.

Gale, D. (1960), *The Theory of Linear Economic Models*, McGraw-Hill.

Gale, D. (1967), "A geometric duality theorem with economic application," *Rev. Econ. Studies*, **34**, 19–24.

Gale, D. (1968), "Optimal assignments in an ordered set: an application of matroid theory," *J. Comb. Theory*, **4**, 176–180.

Garfinkel, R. S. and G. L. Nemhauser (1972), *Integer Programming*, Wiley.

Geoffrion, A. M. (1971), "Duality in nonlinear programming: a simplified applications-oriented development," *SIAM Rev.*, **13**, 1–37.

Geoffrion, A. M. (1972), "Generalized Benders decomposition," *J. Opt. Theory Appl.*, **10**, 237–260.

Geoffrion, A. M. (1974), "Lagrangean relaxations for integer programming," *Mathematical Programming Study 2: Approaches to Integer Programming*, 82–114, M. L. Balinski, ed., North-Holland.

Geoffrion, A. M. and G. W. Graves (1974), Multicommodity distribution system design by Benders decomposition," *Manage. Sci.*, **20**, 822–844.

Geoffrion, A. M. and R. E. Marsten (1972), "Integer programming: A framework and state-of-the-art survey," *Ibid.*, **18**, 465–491.

Gilmore, P. C. and R. E. Gomory (1963), "A linear programming approach to the cutting stock problem, part II," *Oper. Res.*, **11**, 863–888.

Glover, F. (1969), "Integer programming over a finite additive group," *SIAM J. Control*, **7**, 213–231.

Golden, B. (1976), "Shortest-path algorithms: a comparison," *Oper. Res.*, **24**, 1164–1168.

Gomory, R. E. (1958), "Outline of an algorithm for integer solutions to linear programs," *Bull. Amer. Math. Soc.*, **64**, 275–278.

Gomory, R. E. (1965), "On the relation between integer and non-integer solutions to linear programs," *Proc. Natl. Acad. Sci.*, **53**, 260–265.

Gomory, R. E. (1969), "Some polyhedra related to combinatorial problems," *Lin. Alg. Appl.*, **2**, 451–558.

Gomory, R. E. and T. C. Hu (1961), "Multi-terminal network flows," *J. Soc. Indust. Appl. Math.*, **9**, 551–570.

Gomory, R. E. and T. C. Hu (1964), "Synthesis of a communication network," *J. Soc. Indust. Appl. Math.*, **12**, 348–369.

Gomory, R. E., and E. L. Johnson (1972), "Some continuous functions related to corner polyhedra I," *Math. Program.*, **3**, 23–85.

Gorry, G. A., W. D. Northup, and J. F. Shapiro (1973), "Computational experience with a group theoretic integer programming algorithm," *Math. Program.*, **4**, 171–192.

Gorry, G. A. and J. F. Shapiro (1971), "An adaptive group theoretic algorithm for integer programming problems," *Manage. Sci.*, **17**, 285–306.

Gorry, G. A., J. F. Shapiro, and L. A. Wolsey (1972), "Relaxation methods for pure and mixed integer programming problems," *Manage. Sci.*, **18**, 229–239.

Gould, F. J. and J. W. Tolle (1974), "A unified approach to complementarity in optimization," *Discrete Math.*, **7**, 225–271.

Greenberg, H. (1971), *Integer Programming*, Academic Press.

Grinold, R. C. (1970), "Lagrangean subgradients," *Manage. Sci.*, **17**, 185–188.

Grinold, R. C. (1972), "Steepest ascent for large scale linear programs," *SIAM Rev.*, **14**, 447–464.

Grinold, R. C. (1974), "A generalized discrete dynamic programming model," *Manage. Sci.*, **20**, 1092–1103.

Hadley, G. F. (1961), *Linear Algebra*, Addison-Wesley.

Hadley, G. F. (1962), *Linear Programming*, Addison-Wesley.

Hadley, G. F. (1974), *Nonlinear and Dynamic Programming*, Addison-Wesley.

Hammer, P. L., E. L. Johnson, B. H. Korte, and G. L. Nemhauser (1977), *Studies in Integer Programming*, North-Holland.

Held, M. and R. M. Karp (1962), "A dynamic programming approach to sequencing problems," *J. SIAM,* **10**, 196–210.

Held, M. and R. M. Karp (1970), "The traveling salesman problem and minimum spanning trees," *Oper. Res.*, **18**, 1138–1162.

Held, M. and R. M. Karp (1971), "The traveling salesman problem and minimum spanning trees: part II," *Math. Program.,* **1**, 6–25.

Held, M., P. Wolfe, and H. Crowder (1974), "Validation of subgradient optimization," *Math. Program.,* **6**, 62–88.

Heller, I. and C. B. Tompkins (1956), "An extension of a theorem of Dantzig's, paper 14," in *Linear Inequalities and Related Systems, Annals of Math. Study 38*, H. W. Kuhn and A. W. Tucker, eds., Princeton Univ. Press, pp. 247–254.

Howard, R. A. (1960), *Dynamic Programming and Markov Processes*, MIT Press.

Hu, T. C. (1969), *Integer Programming and Network Flows*, Addison-Wesley.

Hurter, A. P. and R. E. Wendell (1972), "Location and production—A special case," *J. Reg. Sci.*, **12**, 243–247.

Jeroslow, R. G. (1977), "Cutting-plane theory: Disjunctive methods," in *Studies in Integer Programming*, P. L. Hammer, E. L. Johnson, B. H. Korte and G. L. Nemhauser, eds., North-Holland, 293–330.

Jewell, W. S. (1962), "Optimal flows with gains," *Oper. Res.*, **10**, 476–522.

Jewell, W. S. (1963a), "Markov-renewal programming: I. Formulation, finite return models," *Oper. Res.*, **11**, 938–948.

Jewell, W. S. (1963b), "Markov renewal programming: II. Infinite return models, example," *Oper. Res.*, **11**, 949–971.

John, F. (1948), "Extremum problems with inequalities as subsidiary conditions," *Studies and Essays (Courant Anniversary Volume)*, Interscience Publishers, 187–204.

Johnson, E. (1965), "Network flows, graphs and integer programming," *Oper. Res. Cntr. ORC 65-1*, Univ. of California, Berkeley.

Johnson, E. (1973), "Cyclic groups, cutting planes and shortest paths," in *Mathematical Programming*, T. C. Hu and S. Robinson, eds., Academic Press.

Johnson, E. (1975), "Flows in networks," *Operations Research Handbook*, Chap. 4, Elmaghraby, S. E. and J. J. Moder, eds., in press.

Karlin, S. (1959), *Mathematical Methods and Theory in Games, Programming, and Economics*, Vols. 1 and 2, Addison-Wesley.

Karp, R. M. and M. Held (1967), "Finite-state processes and dynamic programming," *SIAM J. Appl. Math.*, **15**, 693–718.

Kiefer, J. (1953), "Sequential minimax search for a maximum," *Proc. Am. Math. Soc.*, **4**, 502–506.

Kruskal, J. B. (1956), "On the shortest spanning subtree of a graph and the traveling salesman problem," *CACM,* **14**, 327–334.

Kuhn, H. W. and A. W. Tucker (1951), "Nonlinear programming," *Econometrica*, **19**, 50–51.

Land, A. H. and A. G. Doig (1960), "An automatic method for solving discrete programming problems," *Econometrica*, **28**, 497–520.

Land, A. H. and S. Powell (1977), "A survey of mixed integer programming codes," Report, London School of Economics.

Lasdon, L. (1968), "Duality and decomposition in mathematical programming," *IEEE Trans. Syst. Sci. Cybernetics*, **4**, 86–100.

Lasdon, L. (1970), *Optimization Theory for Large Systems*, MacMillan.

Lasdon, L. and R. Terjung (1971), "An efficient algorithm for multi-item scheduling," *Oper. Res.*, **19**, 946–969.

Lawler, E. L. (1976), *Combinatorial Optimization: Networks and Matroids*, Holt, Reinhart and Winston.

LeMaréchal, C. (1974), "An algorithm for minimizing convex functions," in *Information Processing*, J. L. Rosenfeld, ed., North-Holland, 552–556.

LeMaréchal, C. (1978), "Nonsmooth optimization and descent methods." RR78-4. International Institute for Applied Systems Analysis, Laxenburg, Austria.

Lemke, C. E. (1965), "Bimatrix equilibrium points and mathematical programming," *Manage. Sci.*, **11**, 681–689.

Lemke, C. E, and J. T. Howson, Jr. (1964), "Equilibrium points of bimatrix games," *SIAM J. Appl. Math.*, **12**, 413–423.

Little, J. D. C., K. G. Murty, D. W. Sweeney, and C. Karel (1963), "An algorithm for the traveling salesman problem," *Oper. Res.*, **11**, 979–989.

Luenberger, D. (1969), *Optimization by Vector Space Methods*, Wiley.

Luenberger, D. (1973), *Introduction to Linear and Nonlinear Programming*, Addison-Wesley.

Magnanti, T. L. (1976), "Optimization for sparse systems," in *Sparse Matrix Computations*, J. R. Bunch and D. J. Rose, eds., Academic Press.

Magnanti, T. L. and B. Golden (1978), "Transportation planning: Network models and their implementation," Tech. Rep. No. 143, Operations Research Center, MIT.

Magnanti, T. L., J. F. Shapiro, and M. H. Wagner (1976), "Generalized linear programming solves the dual," *Manage. Sci.*, **22**, 1195–1203.

Manne, A. S. (1970), "Sufficient conditions for optimality in an infinite horizon development plan," *Econometrica*, **38**, 18–38.

Maurras, J. F. (1973), "Optimization of the flow through networks with gains," *Math. Program.*, **3**, 135–144.

Miliotis, P. (1976), "Integer programming approach to the traveling salesman problem," *Math. Program.*, **10**, 367–378.

Miller, B. L. (1968a), "Finite state continuous time Markovian decision processes with a finite planning horizon," *SIAM J. Control*, **6**, 266–280.

Miller, B. L. (1968b), "Finite state continuous time Markovian decision processes with an infinite planning horizon, *J. Math. Anal. Appl.*, **22**, 552–569.

Miller, B. L. (1971), "On minimizing nonseparate functions defined on the integers with an inventory application," *SIAM J. Appl. Math*, **21**, 166–185.

Mine, H., and S. Osaki (1970), *Markovian Decision Processes*, American Elsevier.

Morin, T. L. and Marsten, R. E. (1976), "Branch and bound strategies for dynamic programming," *Oper. Res.*, **24**, 611–627.

Motzkin, T., and I. J. Schoenberg (1954), "The relaxation method for linear inequalities," *Can. J. Math.*, **6**, 393–404.

Murchland, J. D. (1967), "The once-through method of finding all shortest distances in a graph from a single origin," London Graduate School of Business Studies, Report L135-TNT-56.

Murty, K. G. (1976), *Linear and Combinatorial Programming*, John Wiley.

Nakayama, H., H. Sayama, and Y. Sawargi (1975), "A generalized Lagrangian function and multiplier method," *J. Opt. Theory Appl.*, **17**, 211–227.

Nemhauser, G. L. (1966), *Introduction to Dynamic Programming*, Wiley.

Nemhauser, G. L. and Z. Ullman (1968), "A note on the generalized Lagrange multiplier solution to an integer programming problem," *Oper. Res.*, **16**, 450–452.

Nikaido, H. (1968), *Convex Structures and Economic Theory*, Academic Press.

Norman, R. Z. and M. O. Rabin (1959), "An algorithm for the minimum cover of a graph," *Proc. Amer. Math. Soc.*, **10**, 315–319.

Oettli, W. (1972), "An iterative method, having linear rate of convergence, for solving a pair of dual linear programs," *Math. Program.*, **3**, 302–311.

Orchard-Hays, W. (1968), *Advanced Linear Programming Computing Techniques*, McGraw-Hill.

Polyak, B. T. (1967), "A general method for solving extremal problems," *Sov. Math. Dokl.*, **8**, 593–597.

Polyak, B. T. (1969), "Minimization of unsmooth functionals," *USSR Comput. Math.*, 509–521.

Rockafellar, R. T. (1970), *Convex Analysis*, Princeton University Press.

Ross, S. M. (1970), *Applied Probability Models with Optimization Applications*, Holden-Day.

Rubin, D. S. (1970), "On the unlimited number of faces in integer hulls of linear programs with a single constraint," *Oper. Res.*, **18**, 940–946.

Salkin, H. M. (1975), *Integer Programming*, Addison-Wesley.

Scarf, H. E. (1967), "The approximation of fixed points of a continuous mapping," *SIAM. J. Appl. Math.*, **15**, 1328–1343.

Scarf, H. E. and T. Hansen (1973), *Computation of Econometric Equilibria*, Yale University Press.

Schweitzer, P. J. (1971), "Multiple policy improvements in undiscounted Markov renewal programming," *Oper. Res.*, **19**, 784–793.

Scott, A. J. (1971), *Combinatorial Programming, Spatial Analysis and Planning*, Methuen.

Shapiro, J. F. (1968a), "Dynamic programming algorithms for the integer programming problem I: The integer programming problem viewed as a knapsack-type problem," *Oper. Res.*, **16**, 103–121.

Shapiro, J. F. (1968b), "Turnpike planning horizons for a Markovian decision model," *Manage. Sci.*, **14**, 292–306.

Shapiro, J. F. (1971), "Generalized Lagrange multipliers in integer programming," *Oper. Res.*, **19**, 68–76.

Shapiro, J. F. (1976), "Brouwer's fixed point theorem and finite state space Markovian decision theory," *J. Math. Anal. Appl.*, **49**, 710–712.

Shapiro, J. F. (1977), "A note on the primal-dual and out-of-kilter algorithms for network optimization problems," *Networks*, **7**, 81–88.

Shapiro, J. F. (1978), "Decomposition methods for mathematical programming/economic equilibrium energy planning models," *TIMS Studies in the Management Sciences, 10, Energy Policy*, North-Holland, 63–76.

Shapiro, J. F. and H. M. Wagner (1967), "A finite renewal algorithm for the knapsack and turnpike models," *Oper. Res.*, **15**, 319–341.

Simonnard, M. (1966), *Linear Programming*, Prentice-Hall. Translated from original French edition by W. S. Jewell.

Srinivasan, V. and A. D. Shocker (1973), "Linear programming techniques for multi-dimensional analysis of preferences," *Psychometrika*, **38**, 337–369.

Starr, M. K. and M. Zeleny (eds.) (1977), *Multiple Criteria Decision Making: Volume 6 in Studies in the Management Studies*, North-Holland.

Stoer, J. and C. Witzgall (1970), *Convexity and Optimization in Finite Dimensions I*, Springer-Verlag.

Todd, M. J. (1976), "The computation of fixed points and applications," No. 124, *Lecture Notes* in *Economics and Mathematical Systems*, Springer-Verlag.

Tomlin, J. A. (1966), "Minimum-cost multicommodity network flows," *Oper. Res.*, **14**, 45–51.

Varaiya, P. P. (1972), *Notes on Optimization*, Van Nostrand.

Veinott, A. F., Jr. (1966), "On the finding optimal policies in discrete dynamic programming with no discounting," *Ann. Math. Statist.*, **37**, 1284–1294.

Veinott, A. F., Jr. (1968), "Extreme points of Leontief substitution systems," *Lin. Alg. Appl.*, **1**, 183–194.

Veinott, A. F., Jr. and G. B. Dantzig (1968), "Integral extreme points," *SIAM Rev.*, **10**, 371–372.

Wagner, H. M. (1975), *Principles of Operations Research*, 2nd ed., Prentice-Hall.

Wagner, J. L. and L. M. Falkson (1975), "The optimal nodal location of public facilities with price sensitive demand," *Geograph. Anal.*, **7**, 69–83.

Whittle, P. (1971), *Optimization under Constraints: Theory and Applications of Nonlinear Programming*, Wiley-Interscience.

Wolfe, P. and G. B. Dantzig (1962), "Linear programming in a Markov chain," *Oper. Res.*, **10**, 702–710.

Wolsey, L. A. (1969), "Mixed integer programming: Discretization and the group theoretic approach," Tech. Report No. 42, Operations Research Center, M.I.T.

Zadeh, N. (1973), "More pathological examples for network flow problems," *Math. Program.*, **5**, 217–224.

Zoutendijk, G. (1960), *Methods of Feasible Directions*, Elsevier.

AUTHOR INDEX

SUBJECT INDEX